Statistics:
Probability, Inference, and Decision
Volume I

INTERNATIONAL SERIES IN DECISION PROCESSES

INGRAM OLKIN, Consulting Editor

A Basic Course in Statistics, 2d ed., T. R. Anderson and M. Zelditch, Jr.
Introduction to Statistics, R. A. Hultquist
Applied Probability, W. A. Thompson, Jr.
Elementary Statistical Methods, 3d ed., H. M. Walker and J. Lev
Reliability Handbook, B. A. Koslov and I. A. Ushakov (edited by J. T. Rosenblatt and L. H. Koopmans)
Fundamental Research Statistics for the Behavioral Sciences, J. T. Roscoe
Statistics: Probability, Inference, and Decision, Volumes I and II, W. L. Hays and R. L. Winkler

FORTHCOMING TITLES

Introductory Probability, C. Derman, L. Gleser, and I. Olkin
Probability Theory, Y. S. Chou and H. Teicher
Statistical Inference, 2d ed., H. M. Walker and J. Lev
Statistics for Psychologists, 2d ed., W. L. Hays
Decision Theory for Business, D. Feldman and E. Seiden
Analysis and Design of Experiments, M. Zelen
Times Series Analysis, D. Brillinger
Statistics Handbook, C. Derman, L. Gleser, G. H. Golub, G. J. Lieberman, I. Olkin, A. Madansky, and M. Sobel

Statistics:
Probability, Inference, and Decision

Volume I

William L. Hays
University of Michigan

Robert L. Winkler
Indiana University

HOLT, RINEHART AND WINSTON, INC.

New York · Chicago · San Francisco · Atlanta · Dallas
Montreal · Toronto · London · Sydney

1   2   3   4   5   6   7   8   9

TO PALMA AND DORTH

PREFACE

This book is intended as an introduction to probability theory, statistical inference, and decision theory (no previous knowledge of probability and statistics is assumed). The emphasis is on basic concepts and the theory underlying statistical methods rather than on a detailed exposition of all of the different methods which statisticians find useful, although most of the elementary techniques of statistical inference and decision are discussed. In other words, this book is not meant as a "cookbook"; when a new concept is introduced, every attempt is made to justify and explain the concept, both mathematically and verbally. Our intention is to give the reader a good basic knowledge of the important concepts of probability and statistics. Because of this emphasis on the conceptual framework, the reader should be able to pursue more advanced work in probability and statistics upon the completion of this book, if he so desires.

It has been our experience that the mathematical backgrounds of students approaching a first course in probability and statistics are usually quite varied. Because of the current trend toward requiring some elementary calculus of students in virtually all fields of endeavor, and because the use of calculus makes it possible to discuss continuous probability models and their statistical applications in a more rigorous manner, we have used elementary calculus in this book. This will enable students who have had some calculus to gain a better appreciation of continuous probability models than is possible with college algebra alone. However, the mathematical formulations are accompanied by heuristic explanations, so that the book can also be used profitably by students with no calculus background. In this respect, we have attempted to strike a happy medium between the books using no calculus and the "mathematical statistics" books relying heavily on calculus, incorporating the advantages of both. This has resulted in a longer book, but it should prove quite valuable from a pedagogical standpoint.

The book is divided into three major parts; the first two parts, probability theory and statistical inference and decision, are included in Volume I, and the third part, dealing with specific techniques (regression, analysis of variance, and nonparametric methods), comprises Volume II. A useful feature of the book is the relatively complete development of probability theory in Chapters 1–4. The set-theoretic approach is used, elementary rules of probability are discussed, and both discrete and continuous random variables are covered at some length. Also, because of the later discussion of Bayesian inference and decision theory, quite a bit of space is given to the interpretation of probability. We feel that a sound understanding of basic probability theory is a necessary prerequisite to the study of statistical inference and decision.

The second section of Volume I, statistical inference and decision, is covered in Chapters 5–9. The concept of a sampling distribution is introduced and used in discussions of estimation and hypothesis testing. Next, the Bayesian approach to statistics is discussed; the concept of a prior distribution is introduced and the relationship between Bayesian and "classical," or "sampling theory," techniques is covered. Finally, the concepts of decision theory (payoffs, losses, utilities) are introduced. Decision-making criteria, the value of information, and the distinction between inferential problems and decision-making problems are discussed. We feel that a unique feature of the book is the extensive coverage of Bayesian inference and decision theory and the integration of this material with the previous material on "classical" inferential procedures. Chapters 8 and 9 include many topics that are not generally covered in books of this nature and level.

In Volume II, the third major part of the book is presented. Regression and correlation are discussed extensively, including the theoretical regression curve, simple linear regression, least-squares curve fitting, curvilinear regression, and multiple regression. Next, sampling theory, experimental design, and several analysis of variance models are discussed. Finally, a chapter on nonparametric methods is included.

Numerous problems are included at the end of each chapter, ranging from straightforward applications of the textual material to problems which require considerably more thought. The problems are an integral part of the book, serving to reinforce the reader's grasp of the concepts presented in the text and to point out possible applications and extensions of these concepts. Also, the book is self-contained in the sense that all necessary tables are included at the end of each volume. Additionally, a short appendix with some commonly encountered differentiation and integration formulas is presented in Volume I, and an appendix on matrix algebra, which is used very briefly in the discussion of multiple regression in Chapter 10, is presented in Volume II. Finally, a list of references is provided at the end of each volume for those wishing to pursue any topic further.

As the above summary indicates, there is easily enough material for a two-semester course. The chapters are divided rather closely into sections, however, so that a certain amount of flexibility is afforded. The book could be used for a one-semester or two-quarter course in probability, inference, and decision theory if some sections are deleted. Depending on the background of the students, there are various other possibilities. For instance, if the students have had some probability theory, the book could be used for a one-semester course in statistics, starting with Chapter 5.

We are especially grateful to all of the people who provided help and encouragement in bringing this work to completion. In particular, we are indebted to William A. Ericson and Ingram Olkin for providing numerous valuable comments on an early draft of this book. In addition, there are many people who contribute in an indirect manner to a book such as this, including contributors to the literature in probability and statistics, our colleagues, and our students; we acknowledge their contributions collectively rather than individually. Our greatest thanks, however, go to our wives, Palma and Dorth, for their seemingly endless supply of patience, encouragement, and assistance during the writing of this book.

Ann Arbor, Michigan   William L. Hays

Bloomington, Indiana   Robert L. Winkler

January 1970

CONTENTS

4 Special Probability Distributions 178

8   Bayesian Inference   444

Statistics:
Probability, Inference, and Decision

Volume I

INTRODUCTION

Applications of statistics occur in virtually all fields of endeavor—business, the social sciences, the physical sciences, the biological sciences, education, and so on, almost without end. Although the specific details of the methods differ somewhat in the different fields, the applications all rest on the same general theory of statistics. By examining what the fields have in common in their applications of statistics, we can gain a picture of the general nature of statistics.

To begin, it is convenient to identify three major branches of statistics: descriptive statistics, inferential statistics, and statistical decision theory. Descriptive statistics is a body of techniques for the effective organization, summarization, and communication of data. When the man on the street speaks of "statistics," he usually means data organized by the methods of descriptive statistics. Inferential statistics is a body of methods for arriving at conclusions extending beyond the immediate data. For example, given some information regarding a small subset of a given population, what can be said about the entire population? Finally, statistical decision theory goes one step further; instead of just making inferential statements about the population, the statistician uses the available information to choose among a number of alternative decisions.

The major concern in this book is with inferential statistics and statistical decision theory, both of which are closely related to the concept of uncertainty. In making inferences about some population, the statistician is generally uncertain as to the true characteristics of the population, but he may be able to formulate a good notion about these characteristics. Similarly, a decision maker is usually uncertain about some given conditions that are relevant to his decision-making problem, but by considering the

1

nature of the uncertainty he is able to use some decision criterion to determine his best action. Thus, in one sense, mathematical statistics is a theory about uncertainty. Granted that certain conditions are fulfilled, the theory permits deductions about the likelihood of the various possible outcomes of interest. In this manner, the essential concepts in statistics are based on the theory of probability; the statistician is interested in the probability of particular kinds of outcomes, given that certain initial conditions are met. Therefore, it is necessary for the student to have a basic understanding of probability theory before learning statistics.

It should be emphasized that this book deals with the theory underlying statistical methods rather than with a detailed exposition of all of the different methods that statisticians find useful. In other words, this is not a "cookbook" that will equip the student to meet every possible situation that he might encounter. It is true that many methods will be introduced and we will, in fact, discuss most of the elementary techniques for statistical inference and decision. In the past few years, however, there has been a proliferation of new techniques, particularly in decision theory and in problems involving several variables. This is partly due to the development of computers, which have opened up new avenues of data analysis for the statistician, making it possible for him to answer questions that were formerly unanswerable because of computational complexity. This proliferation of statistical methods can be expected to continue, with some of the current methods being replaced with newer techniques. Thus, we feel that it is better for the student to gain a basic understanding of the theory of statistics than to learn a myriad of specialized methods.

Essentially, the book is divided into three parts. The first part, Chapters 1–4, deals with probability theory, thus laying a foundation for the study of statistics. Chapters 5–9 involve the basic theory of statistical inference and decision, and the third part (Volume II), Chapters 10–12, concerns a number of specialized topics (specialized in comparison with the basic concepts presented in the second part). Throughout the book we have attempted, when introducing a new concept, to present both the relevant mathematics (any formulas, proofs, and so on) *and* a clear explanation in words of the concept. Those mathematical expressions that are of particular importance are denoted by an asterisk following the number of the expression. The numbering system should be self-explanatory; 2.4.1 is the first numbered expression in Section 4 of Chapter 2, for example.

1

SETS
AND
FUNCTIONS

It may seem surprising that a book about statistics starts off with a discussion of sets. Although the study of sets and functions is not usually a part of a course in statistics, these topics actually provide the most fundamental concepts we will use. Set theory will be the basis for our discussion of probability theory, to be introduced in the next chapter. The idea of a function pervades virtually all mathematics and is a key concept in modern scientific work as well. Even at the high-school level most of us are exposed to the notion of a mathematical function, and we know that saying that "Y is a function of X" expresses something about a relation between two things. However, unless the student has a very good background in mathematics, he is usually somewhat vague about the precise meaning of the word "function" used in mathematical or scientific writing. One of the primary purposes of this chapter is to give a very concrete and restricted meaning to this term. It will be shown that the idea of a function is a very simple one, which grows out of the notion of a set.

1.1 SETS

The concept of a set is the starting point for all of modern mathematics, and yet this idea could hardly be more simple:

Any well-defined collection of objects is a set.

The individual objects making up the set are known as the "elements" or "members" of the set. The set itself is the aggregate or totality of its members. If a given object is in the set, then one says that the object is an **element** of, or a **member** of, or **belongs** to, the set.

For example, all students enrolled at a given university in a given year is a set, and any particular student is an element of that set. All living men whose last name is "Jones" is another set. All of the whole numbers between 10 and 10,000 is a set, all animals of the species *Rattus norvegicus* living at this particular moment make up a set, all possible neural pathways in John Doe's brain form a set, and so on, ad infinitum.

It is important to note that in the definition of a set the qualification "well-defined" occurs. This means that *it must be possible, at least in principle, to specify the set so that one can decide whether any given object does or does not belong.* This does not mean that sets can be discussed only if their members actually exist; it is perfectly possible to speak of the set of all women presidents of the United States, for example, even though there are not any such objects at this writing. What *is* required is some procedure or rule for deciding whether an object is or is not in the set; given an object, one can decide if it meets the qualifications of a female president of the United States, and thus the set is well-defined.

The word "object" in the definition can also be interpreted quite liberally. Often, sets are discussed having members that are not objects in the usual sense but rather are "phenomena," or "happenings," or possible outcomes of observation. We will have occasion to use sets of "logical possibilities," all the different ways something might happen, where each distinct possibility is thought of as one member of the set.

1.2 WAYS OF SPECIFYING SETS

In discussing sets, we will follow the practice of letting a capital letter, such as A, symbolize the set itself, with a small letter, such as a, used to indicate a particular member of the set. The symbol "$\in$" is often used to indicate "is a member of"; thus,

$$a \in A$$

is read "a is a member of A."

There are two different ways of specifying a set. The first way is by *listing* all of the members. For example,

$$A = \{1, 2, 3, 4, 5\}$$

is a complete specification of the set A, saying that it consists of the numbers 1, 2, 3, 4, and 5. The braces around the listing are used to indicate that the list makes up a set. Another set specified in a similar way is

$$B = \{\text{orange, grapefruit, lemon, lime, tangerine, kumquat, citron}\},$$

and another is

$$C = \{\text{Roosevelt, Truman, Eisenhower, Kennedy, Johnson, Nixon}\}.$$

In each instance the respective elements of the set are simply listed. These sets are thought of as unordered, since the order of the listing is completely irrelevant so long as each member is included once and only once in the list.

The other way of specifying a set is to give a *rule* that lets one decide whether or not an object is a member of the set in question. Thus, set A may be specified by

$$A = \{a \mid a \text{ is an integer and } 1 \leq a \leq 5\}.$$

(The symbol $\mid$ is read as "such that," so that the expression above is, in words, "the set of all elements a such that a is an integer between 1 and 5 inclusive.") Given this rule, and any potential element of the set A, we can decide immediately whether or not the object actually does belong to the set.

The rule for set B would be

$$B = \{b \mid b \text{ is the name of a kind of cirtrus fruit}\},$$

and for set C,

$$C = \{c \mid c \text{ is a United States President elected between}$$
$$\text{1932 and 1968 inclusive}\}.$$

It is quite possible to specify sets in terms of other sets. For example, we might specify a set D by

$$D = \{d \mid d = a + 15, \text{ for all } a \in A\}.$$

When listed, the elements of D would be

$$D = \{16, 17, 18, 19, 20\}.$$

It is usually far more convenient to specify a set by its rule than by listing the elements. For example, sets such as the following would be awkward or impossible to list:

$$G = \{g \mid g \text{ is a human born in 1955}\},$$
$$X = \{x \mid x \text{ is a positive number}\},$$
$$Y = \{y \mid y \in X \text{ and } y \text{ is an integer}\}.$$

1.3   UNIVERSAL SETS

There are many instances in set theory when it becomes convenient to consider only objects belonging to some "large" set. Then, given that all objects to be discussed belong to this "universal" set, we proceed to talk of particular groupings of elements. The introduction of a universal set acts to set the stage for the kinds of sets that will be introduced. For example, we may wish to deal only with U.S. students enrolled in college in the year 1969, and so we begin by specifying a universal set:

$$W = \{w \mid w \text{ was a U.S. student enrolled in college in 1969}\}.$$

Then particular subsets of W are introduced:

$$A = \{a \mid a \in W \text{ and } a \text{ was a student at Harvard}\},$$
$$B = \{b \mid b \in W \text{ and } b \text{ was classified as a sophomore}\},$$

and so on. Or perhaps the universal set is

$$W = \{w \mid w \text{ is a symbol for a sound}\},$$
$$A = \{a \mid a \in W \text{ and } a \text{ is a letter in the Cyrillic alphabet}\},$$
$$B = \{b \mid b \in W \text{ and } b \text{ is a letter in the Greek alphabet}\},$$

and so on.

Up until this point, the individual member of a set has been denoted by a small letter, a, b, w, and so on, depending on the capital letter used for the set itself. This is not strictly necessary, however, as any symbol that serves as a "place holder" in the rule specifying the set would do as well. In future sections it will sometimes prove convenient to use the same symbol for a member of several sets, so that the neutral symbol x will be used, indicating any member of the universal set under discussion. Furthermore, it is redundant to state that $x \in W$ after the universal set W has been specified, and so this statement will be omitted from the rule for particular subsets of the universal set. Nevertheless, it is always understood that any member of a set is automatically a member of some universe W. In the example just given, the sets are adequately specified by

$$W = \{x \mid x \text{ is a symbol for a sound}\},$$
$$A = \{x \mid x \text{ is a letter in the Cyrillic alphabet}\},$$
and
$$B = \{x \mid x \text{ is a letter in the Greek alphabet}\}.$$

1.4 SUBSETS

Suppose that there is some set A and another set B such that any element which is in B is also in A. Then B *is a subset of* A. This is symbolized by

$$B \subseteq A,$$

read as "B is a subset of A." More formally,

The set B is a subset of the set A if and only if
for each $x \in B$, then $x \in A$.

For example, consider the set A of all male citizens of the United States at this moment. Now let set B be the set of all living U.S. Army generals; the set B is a subset of the set A. Or, let A be the set of all numbers, and let B indicate the set of all whole numbers. Once again, B is a subset of A.

If B is a subset of A *and* there is at least one element of A not in B, then B is a **proper subset** of A, symbolized by

$$B \subset A.$$

In both the examples just given the set B is a proper subset of A, since some U.S. men are not generals, and some numbers are not whole. On the other hand, if every single element in B is in A, *and* every single element in A is in B, so that $B \subseteq A$ *and* $A \subseteq B$, then

$$A = B,$$

the two sets are equivalent or equal. That is, two sets are **equal** if they contain precisely the same elements.

Note the similarity of the symbols "$\subseteq$," for "is a subset of," and "$\subset$," for "is a proper subset of," to the symbols "$\leq$," for "is less than or equal to," and "$<$," for "is less than." The difference in "$\subseteq$" and "$\subset$" reflects the fact that one set can be a subset of another even though the two sets actually are identical or equal element by element; a set can be a proper subset of another only if the two sets are not equal. It should also be noted that the statements "$A \subseteq B$" and "$B \supseteq A$" are equivalent, as are the statements "$A \subset B$" and "$B \supset A$." In each case, the set at the closed end of the "horseshoe" is the subset (or proper subset), and the set at the open end is the larger set. This is similar to the correspondence between "$x \leq y$" and "$y \geq x$," where x and y are numbers.

Given the universal set W for some problem, then every set A to be discussed is a subset of the universal set,

$$A \subseteq W, \text{ for every } A.$$

1.5 FINITE AND INFINITE SETS

In future work it will be necessary to distinguish between finite and infinite sets. For our purposes, a finite set has members equal in number to some specifiable positive integer or to zero. Otherwise, if the number of members is greater than any positive integer you conceivably can name, the set is considered infinite. Thus, the set of all names in the Manhattan telephone book, the set of all books printed in the nineteenth century, the set of all houses in Rhode Island, and the set of all living mammals are all finite sets, even though the number of members each includes is very large. On the other hand, the set of all points lying on a circle, the set of all real numbers (including all rational and irrational, positive and negative numbers), and the set of all intervals into which a straight line may be divided are examples of infinite sets.

In our discussion in the next few sections, all sets will be treated as though they were finite. The general ideas apply to infinite sets as well, but certain qualifications have to be made in some of the definitions, and we will have to overlook these.

1.6 VENN DIAGRAMS

A scheme that is very useful for illustrating sets and for showing relations among them is the Venn diagram (named after the logician J. Venn, 1834–1923). These are sometimes referred to as Euler diagrams (after the mathematician L. Euler, 1707–1783; both men made important contributions to the theory of sets). A Venn diagram pictures a set as all points contained within a circle, square, or other closed geometrical figure.

Since there is an infinite number of possible points within any such figure, Venn diagrams actually represent infinite sets, but they are useful for representing any set. For example, a Venn diagram picturing the universal set W and the subset A would look like Figure 1.6.1. Since A is a subset of W, the area of the circle A is completely included in the area of the rectangle W.

Now consider another set $B \subseteq W$. If $B \subset A$, the Venn diagram would be as shown by Figure 1.6.2. On the other hand, if B were not a subset of A, then Figure 1.6.3 would be the Venn diagram if A and B shared members in common, or perhaps Figure 1.6.4 would be the Venn diagram if the two sets had no members in common.

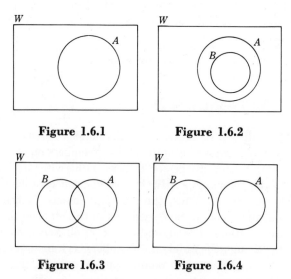

Figure 1.6.1   Figure 1.6.2

Figure 1.6.3   Figure 1.6.4

1.7   THE EMPTY SET

Just as the number "zero" is important in ordinary arithmetic, so the theory of sets requires the notion of the "empty set." **Any set that contains no members is called the empty set.** The empty set is usually identified by the symbol $\varnothing$, a zero with a diagonal slash. We have already had one example of the empty set, the set of all women presidents of the United States. It is not hard to dream up other examples: the set of all months containing 38 days, the set of all people with 8 legs, the set of all numbers not divisible by 1, and so on.

This set has a very special property: **the empty set is a subset of every set.** Notice that the definition of a subset as given in Section 1.4 does not rule out the empty set; since the empty set has no members, the definition of subset is not contradicted, and so $\varnothing \subseteq A$ for every A.

1.8   THE UNION OF SETS

Given some universal set W, and a set of its subsets, A, B, C, and so on, it is possible to "operate" on sets to form new sets, each of which is also a subset of W.

The first of these operations is the **union** of two or more sets. Given W and the two subsets A and B, then the union of A and B is written $A \cup B$ (a useful mnemonic device for the symbol $\cup$ is the u in $union$). By definition,

$A \cup B$ **is the set of all elements that are members of A, or of B, or of both:**

$$A \cup B = \{x \mid x \in A \text{ or } x \in B, \text{ or both}\}.$$

For example, let

$$W = \{x \mid x \text{ is a living American war veteran}\},$$
$$A = \{x \mid x \text{ is a veteran of World War II}\},$$
and
$$B = \{x \mid x \text{ is a veteran of the Korean War}\}.$$

Then

$$A \cup B = \{x \mid x \text{ is a veteran of World War II, or}$$
$$x \text{ is a veteran of the Korean War, or}$$
$$x \text{ is a veteran of both}\}.$$

In the Venn diagram (Figure 1.8.1), the shaded portion shows the union of A and B. Notice that we include the possibility that an element of the union could be a member of *both* A and B. Note further that in this example, $A \cup B \subset W$ (the union is a *proper* subset of the universal set), since there are some living American war veterans who belong neither to A nor B.

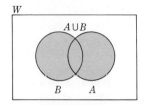

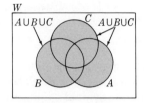

Figure 1.8.1   **Figure 1.8.2**

The idea of the union may be extended to more than two sets. For instance, given W and three sets A, B, and C, then

$$A \cup B \cup C = \{x \mid x \in A, \text{ or } x \in B, \text{ or } x \in C\},$$

as shown in Figure 1.8.2. In general, the union of K sets A_1, A_2, $\cdots$, A_K is defined as follows:

$$\bigcup_{i=1}^{K} A_i = A_1 \cup A_2 \cup \cdots \cup A_K = \{x \mid x \in A_1, \text{ or } x \in A_2, \cdots, \text{ or } x \in A_K\}$$

$$= \{x \mid x \text{ is a member of at least one of the } K \text{ sets } A_1, A_2, \cdots, A_K\}.$$

As an example of the union of three sets, let

$$W = \{x \mid x \text{ is a positive integer, } 1 \leq x \leq 10\},$$
$$A = \{x \mid x \text{ is a perfect square}\} = \{1, 4, 9\},$$
$$B = \{x \mid x \text{ is divisible by 3}\} = \{3, 6, 9\},$$
and $$C = \{x \mid x \text{ is divisible by 7}\} = \{7\}.$$

Then

$$A \cup B \cup C = \{1, 3, 4, 6, 7, 9\}.$$

The union of any set with a subset of itself is simply the larger set: given $B \subseteq A$, then $A \cup B = B \cup A = A$. It follows that

$$A \cup \varnothing = A$$

and $$W \cup A = W.$$

1.9   THE INTERSECTION OF SETS

The verbal rule for the union of two sets always involves the word "or"; the union $A \cup B$ is the set made by finding the elements that are members of A *or* of B or of both. However, what of the set of elements in *both* A and B? This set is included in the union, but we are most often interested in this set by itself as the **intersection** of A and B:

The intersection of sets A and B, written $A \cap B$, is the set of all members belonging to both A and B:

$$A \cap B = \{x \mid x \in A \text{ and } x \in B\}.$$

For example, let

$$W = \{x \mid x \text{ is a chemical compound}\},$$
$$A = \{x \mid x \text{ contains chlorine}\},$$
and
$$B = \{x \mid x \text{ contains oxygen}\}.$$

Then

$$A \cap B = \{x \mid x \text{ contains chlorine } and \text{ oxygen}\}.$$

As another example, consider the sets of numbers

$$W = \{x \mid x \text{ is a positive integer}\},$$
$$A = \{1, 2, 3, 4, 5, 6, 7, 8, 9, 10\},$$
and
$$B = \{8, 9, 10, 11, 12, 13\}.$$

Then

$$A \cap B = \{8, 9, 10\},$$

since only these elements appear in both A and B.

The intersection of two sets A and B is always a subset of their union:

$$A \cap B \subseteq A \cup B.$$

If the intersection of two sets is empty, $A \cap B = \varnothing$, then the sets are said to be **disjoint** or **mutually exclusive.**

The intersection of two sets presented in a Venn diagram appears as Figure 1.9.1.

Whenever B is a subset of A, then the intersection $A \cap B$ or $B \cap A$ is equal to the *smaller* of the two sets, or B. Thus,

$$A \cap \varnothing = \varnothing$$
and
$$A \cap W = A,$$

since $\varnothing$ is a subset of A and A is a subset of W.

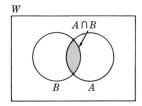

Figure 1.9.1

The intersection may be defined for any number of sets taken together. For example, the intersection of three sets A, B, and C is

$$A \cap B \cap C = \{x \mid x \in A \text{ and } x \in B \text{ and } x \in C\}.$$

Thus, if A were all women, B were all librarians, and C were all persons with blue eyes, then $A \cap B \cap C$ would be the set of all women librarians having blue eyes. In general, the intersection of K sets $A_1, A_2, \cdots, A_K$ is defined as follows:

$$\bigcap_{i=1}^{K} A_i = A_1 \cap A_2 \cap \cdots \cap A_K$$

$$= \{x \mid x \in A_1 \text{ and } x \in A_2 \text{ and } \cdots \text{ and } x \in A_K\}$$
$$= \{x \mid x \text{ is a member of } all \text{ of the } K \text{ sets } A_1, \cdots, A_K\}.$$

The sets $A_1, A_2, \cdots, A_K$ are said to be disjoint, or mutually exclusive, if *all possible pairs* of sets selected from the given K sets are disjoint. Thus, A, B, and C are disjoint if $A \cap B = \varnothing$, $A \cap C = \varnothing$, and $B \cap C = \varnothing$ (see Figure 1.9.2). Note that $A \cap B \cap C = \varnothing$ is *not* a sufficient condition for A, B, and C to be disjoint. In Figure 1.9.3, $A \cap B \cap C = \varnothing$, but the three sets are *not* disjoint, since $A \cap B \neq \varnothing$.

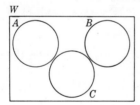

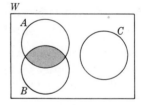

Figure 1.9.2   **Figure 1.9.3**

1.10   THE COMPLEMENT OF A SET

Another operation on sets is called taking the **complement** of a set. Given W and a subset A, then the complement of A is made up of all members of W that are *not* in A. In symbols,

$$\text{the complement of } A = \bar{A} = \{x \mid x \notin A\},$$

where the symbol $\notin$ is read as "not a member of" the set following. For example, if

$$W = \{x \mid x \text{ is a name in the 1969 Detroit, Michigan, telephone directory}\}$$

and

$$A = \{x \mid x \text{ begins with the letter "S"}\},$$

then

$$\bar{A} = \{x \mid x \text{ does not begin with the letter "S"}\}.$$

The intersection of any set and its complement is always empty,

$$A \cap \bar{A} = \varnothing,$$

since no element could be simultaneously a member of both A and $\bar{A}$. The union of A and $\bar{A}$, however, always equals the universal set W:

$$A \cup \bar{A} = W.$$

Notice that the complement of any set is always relative to the universal set; this is why specifying the universal set is so important, since one cannot determine the complement of a set without doing so. For example, suppose that we specified the following set:

$$A = \{x \mid x \text{ has blue eyes}\}.$$

What is $\bar{A}$? That depends on what we assumed W to be. If

$$W = \{x \mid x \text{ is a woman living in the United States}\},$$

then $\bar{A}$ consists of all non-blue-eyed women living in the U.S. However, if

$$W = \{x \mid x \text{ is a person living in the U.S.}\},$$

the $\bar{A}$ is quite a different set, including non-blue-eyed men and children as well. If

$$W = \{x \mid x \text{ is an organism}\},$$

then $\bar{A}$ would include all non-blue-eyed organisms, among them some that have no eyes at all. The complement of a set simply has no meaning without a universal set for reference.

The complement of a set is shown in Figure 1.10.1 as the shaded area in the Venn diagram.

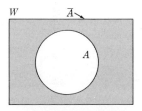

Figure 1.10.1

1.11 THE DIFFERENCE BETWEEN TWO SETS

The difference between two sets is closely allied to the idea of a complement. Given the universal set, the **difference** between sets A and B is

$$A - B = A \cap \bar{B} = \{x \mid x \in A \text{ and } x \in \bar{B}\}.$$

In other words, the difference contains all elements that are members of A *but not* members of B.

Although it makes no difference which set we write first in the symbols for union and intersection, since

$$A \cup B = B \cup A$$

and $\qquad\qquad\qquad\qquad A \cap B = B \cap A,$

the order is very important for the difference between two sets. The difference $A - B$ is *not* the same as the difference $B - A$.

As an example of the two differences, $A - B$ and $B - A$, let

$W = \{x \mid x \text{ was an elected official of the U.S. government in 1969}\}$,
$\quad A = \{x \mid x \text{ was a member of the U.S. Congress}\}$,
and $B = \{x \mid x \text{ was a lawyer}\}$.

Then

$A - B = \{x \mid x \text{ was a member of the U.S. Congress who was not a lawyer}\}$

and

$B - A = \{x \mid x \text{ was a lawyer who was not a member of the U.S. Congress}\}$.

It is easy to see that these two differences are quite different sets; a congressman who is not a lawyer is not the same as a lawyer who is not a congressman.

The Venn diagram for a difference $A - B$ is the shaded area in Figure 1.11.1.

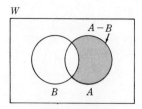

Figure 1.11.1

1.12 THE ALGEBRA OF SETS

In the algebra that everyone studies in high school, there are certain rules which the operations symbolized by "$+$," "$\cdot$," and so on must obey. The study of algebra is largely the study of these basic rules and of the mathematical consequences of the application of these operations to numbers. In the same way, an algebra of sets exists, consisting of a system for forming and manipulating sets by operations of union, intersection, complement, and so on, according to a specific set of rules or postulates. An algebra of sets has a great many similarities to, as well as some important differences from, ordinary algebra. In a course in modern mathematics one learns to use set algebra (often called a Boolean algebra, after the nineteenth-century British mathematician, G. Boole, who first developed it). Although there is little point in dwelling upon set algebra at length in a course in statistics, it may be instructive to list the basic "rules" for an algebra of sets and to give examples of how simple theorems about sets may be proved using those rules. Just as in any formal mathematical system, the basic postulates are used to derive new conclusions by logical argument. Since we are going to be discussing a closely related system in the next chapter, a view of the foundations of this simplest of all mathematical systems may give the reader a somewhat better idea of how postulates are used.

These postulates refer to **abstract sets,** any collections of undefined elements. These statements about abstract sets are assumed true without question; they are simply "postulated." In order for set theory to be a model of sets of *real* objects, it must be true that the conditions specified by the postulates are met by the real sets. For most practical purposes these postulates *are* true of real sets. Every statement that one can make about new sets formed from old by the operations "$\cup$," "$\cap$," and "complement" can be deduced from these postulates.

1.13 POSTULATES AND THEOREMS ABOUT SETS

The postulates of set theory can be stated as follows:

1. **(a) if A and B are sets, then $A \cup B$ is a set;**
    (b) if A and B are sets, then $A \cap B$ is a set.

2. **(a) there is one and only one set $\varnothing$, such that $A \cup \varnothing = A$ for any A;**
    (b) there is one and only one set W, such that $A \cap W = A$ for any set A.

3. **Commutative laws:**
   (a) $A \cup B = B \cup A$;
   (b) $A \cap B = B \cap A$.

4. **Associative laws:**
   (a) $(A \cup B) \cup C = A \cup (B \cup C)$;
   (b) $(A \cap B) \cap C = A \cap (B \cap C)$.

5. **Distributive laws:**
   (a) $A \cup (B \cap C) = (A \cup B) \cap (A \cup C)$;
   (b) $A \cap (B \cup C) = (A \cap B) \cup (A \cap C)$.

6. **For every set A there is one and only one set $\bar{A}$ such that**

$$A \cup \bar{A} = W \text{ and } A \cap \bar{A} = \varnothing.$$

7. (a) $A = B$ and $C = D$ **implies that** $A \cup C = B \cup D$;
   (b) $A = B$ and $C = D$ **implies that** $A \cap C = B \cap D$;
   (c) $A = B$ **implies that** $\bar{A} = \bar{B}$.

8. **There are at least two distinct sets.**

Upon these statements about sets and the operations "∪," "∩," and "complement," an elaborate mathematical structure can be erected. It is very important to note that nowhere in these postulates is there an explicit definition of what is meant by a "set," by "∪," and by "∩." The postulates deal with equivalences among *undefined operations* carried out on *undefined things*. In this sense, the set of postulates is said to be "formal."

It is interesting to note that some of these postulates are true if we happen to be talking about the ordinary algebra of numbers rather than sets, and if "∪" is replaced by "+" and "∩" is replaced by "·." For instance, postulate 1(a) corresponds to a similar postulate about numbers: if x and y are numbers, then $x + y$ is a number. Similarly, postulates 3, 4, and 5(b) have equivalent statements in ordinary algebra:

Commutative laws:

$$x + y = y + x \quad \text{and} \quad x \cdot y = y \cdot x.$$

Associative laws:

$$(x \cdot y) \cdot z = x \cdot (y \cdot z) \quad \text{and} \quad x + (y + z) = (x + y) + z.$$

Distributive law:

$$x \cdot (y + z) = x \cdot y + x \cdot z.$$

Notice, however, that a statement corresponding to postulate 5(a) is *not*, in general, true of numbers:

$$x + (y \cdot z) \text{ is not equal to } (x + y) \cdot (x + z).$$

Thus, the algebra of sets, though similar to ordinary algebra, does differ from it in important respects.

Now an example will be given of how these postulates are used to arrive at new statements, or theorems, not specifically among these original statements. As with any mathematical system, a logical argument is used to arrive at true conclusions *given* that the postulates are true. First of all, a simple but very important theorem will be proved:

Theorem I

For any set A, $A \cap A = A$.

$$
\begin{aligned}
A \cap A &= (A \cap A) \cup \varnothing & &\text{by Postulate 2(a)} \\
&= (A \cap A) \cup (A \cap \bar{A}) & &\text{Postulate 6} \\
&= A \cap (A \cup \bar{A}) & &\text{Postulate 5(b)} \\
&= A \cap W & &\text{Postulate 6} \\
&= A & &\text{Postulate 2(b).}
\end{aligned}
$$

By a series of equivalent statements (the "substitutions" familiar from algebra), each justified by a postulate, we have arrived at a new statement, $A \cap A = A$, which we know must be true whenever the postulates are true.

As a slightly more complicated example of how these postulates are used, together with theorems already proved, consider the following:

Theorem II

If $A \cup B = B$, then $A \cap B = A$.

$$
\begin{aligned}
B &= A \cup B & &\text{Given.} \\
A \cap B &= A \cap (A \cup B) & &\text{Postulate 7(b)} \\
&= (A \cap A) \cup (A \cap B) & &\text{Postulate 5(b)} \\
&= A \cup (A \cap B) & &\text{Theorem I} \\
&= (A \cap W) \cup (A \cap B) & &\text{Postulate 2(b)} \\
&= A \cap (W \cup B) & &\text{Postulate 5(b).}
\end{aligned}
$$

However,

$$
\begin{aligned}
(B \cup W) &= (B \cup W) \cap (B \cup \bar{B}) & &\text{Postulates 2(b) and 6} \\
&= B \cup (W \cap \bar{B}) & &\text{Postulate 5(a)} \\
&= B \cup \bar{B} & &\text{Postulate 2(b)} \\
&= W & &\text{Postulate 6,}
\end{aligned}
$$

so that

$$
\begin{aligned}
A \cap B &= A \cap W & &\text{Substitution} \\
A \cap B &= A & &\text{Postulate 2(b).}
\end{aligned}
$$

Here we have proved a much less obvious statement to be true. Anyone who has studied high-school geometry will appreciate the fact that with an accumulation of such theorems, plus the original postulates, there is virtually no end to the numbers of new theorems that can be proved in this way. The interesting thing is that each of these theorems will be true if one takes sets of real objects and combines and recombines them in the ways represented by ∪ and ∩. Remember that the algebra of sets is not about any particular set of sets, but about "sets" defined abstractly. Propositions about abstract sets will be true of real sets satisfying the postulates exactly as an expression in algebra will be true when the symbols are turned into numbers. Just as algebra is a mathematical model for solving problems about real quantities, and one can find characteristics of real figures using the model of Euclidean geometry, so does the algebra of sets become a useful mathematical model in some real situations. One striking example is the theory underlying the large electronic computers, although very many other instances could be given.

Although this very hurried sketch of set theory cannot possibly give you any real facility with the set language, perhaps it will help when the essential ideas about sets recur in later chapters. In addition, another motive that underlies the discussion of sets is that it permits us to go on to a topic of great importance, both in statistics and in scientific enterprise in general: the study of mathematical relations.

1.14   SET PRODUCTS AND RELATIONS

The business of science or of any other field of knowledge is to discover and describe relationships among things. Everybody knows what is meant by a relationship or connection among objects or phenomena; in order to use language itself we must group our experiences into classes or sets (using the nouns and adjectives) and then state relationships linking one kind of thing with another (using the verbs). However, it is very important that we settle on a formal definition of what makes a **mathematical relation** before turning to other matters. It will be seen that the idea of a mathematical relation is nothing more than an extension of the ideas of set and subset.

Just as a set is defined as a collection of objects, it is also possible to define a special kind of set made up of *pairs of objects*. Let a pair of objects be symbolized by (a, b), where a is a member of some set A, and b is a member of some set B. Each pair of objects is thought of as **ordered,** meaning that the way the objects are listed in a pair is important. Every day we encounter such ordered pairs of objects, where the two members are distinguished from each other by the role they play: husband-wife pairs, right and left hands, first and second movie in a double-feature, and so on. For an ordered pair (a, b), the order of listing is significant for the role played by each element of the pair, so that the pair (a, b) is not necessarily the same as the pair (b, a).

Now suppose that we have two sets, A and B. *All possible* pairs (a, b) are found, each pair associating a member of A with a member of B. This set of all possible pairings of an $a \in A$ with a $b \in B$ is called the **Cartesian product,** or the **set product,** of A and B:

$$A \times B = \{(a, b) \mid a \in A, b \in B\}.$$

The product $A \times B$ is only a set, but this time the elements are the possible pairs, as symbolized by (a, b). The product $B \times A$ would be a different set, with members (b, a).

This idea of a set of ordered pairs originated with the mathematician and philosopher Descartes (1596–1650; Latin: Cartesius), who used it as the foundation of analytic geometry; hence the name *Cartesian* product. The "product" part of the name is attributable to the fact that the total number of possible pairs is always the number of members of A *times* the number of members of B.

As a simple example of a set product, consider these two sets:

$$A = \{\text{Mary, Susan, Tom, Bill}\}$$
and $$B = \{\text{Smith, Jones, Brown, Adams}\}.$$

The product of the two sets, $A \times B$, is nothing more than the set of all possible pairings of one of the first names with one of the last names:

$A \times B = \{$(Mary, Smith)(Mary, Jones)(Mary, Brown)(Mary, Adams)
(Susan, Smith)(Susan, Jones)(Susan, Brown)(Susan, Adams)
(Tom, Smith)(Tom, Jones)(Tom, Brown)(Tom, Adams)
(Bill, Smith)(Bill, Jones)(Bill, Brown)(Bill, Adams)$\}$.

Notice that a pair (b, a) such as (Jones, Mary) is not a member of the product, since the first element must be a member of A (first names) and the second a member of B (last names). This illustrates that the ordering of a pair is important in determining which pairs are included.

The pairs making up $A \times B$ can also be shown graphically, as in Figure 1.14.1. In this graph, each point formed by the intersection of a vertical line with one of the horizontal lines symbolizes one pair in the set $A \times B$.

It is also possible to talk about the product of a set with itself. That is, if we have the set A, then $A \times A$ consists of all possible pairings of a member of A with a member of A. For the example just given,

$A \times A = \{$(Mary, Mary) (Mary, Susan) (Mary, Tom) (Mary, Bill)
(Susan, Mary) (Susan, Susan) (Susan, Tom) (Susan, Bill)
(Tom, Mary) (Tom, Susan) (Tom, Tom) (Tom, Bill)
(Bill, Mary) (Bill, Susan) (Bill, Tom) (Bill, Bill)$\}$.

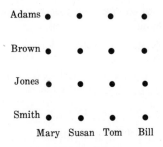

Figure 1.14.1

Again, it *does* make a difference which member comes first in a pair; thus the pair (Mary, Susan) is not considered the same as (Susan, Mary).

One of the most important examples of a Cartesian product is known to anyone who has taken high-school algebra. Suppose that we have the set R, which is the set of all possible **real numbers** (that is, the set of all numbers that can be put into exact one-to-one correspondence with points on a straight line). We take the product $R \times R$, which is the set of all pairings of *two* real numbers. Any given pair might be symbolized (x, y), where x is a number called the value on the X-coordinate, or "abscissa," and y is a number standing for value on the Y-coordinate, or "ordinate." This should be a familiar idea, since the product $R \times R$ is used any time one plots points on a graph or a curve. The first set R is the "X-axis," and the second set R is the "Y-axis" of the "Cartesian coordinates." When a pair such as $(5, 1)$ is used to locate a point on a graph, it means that the number 5 is the element from the first set of numbers, and 1 from the second set. The entire set $R \times R$ consists of all points that one *could* locate in the plane where the two axes of the graph lie. Pictorially, the product $R \times R$ and the pair $(5, 1)$ might be shown as in Figure 1.14.2. Regardless of whether you are dealing with pairs of objects, as in the first example, or pairs of numbers, as in the second, the idea of a Cartesian product is the same: the set of all ordered pairs, the first being a member of one specified set, and the second a member of another.

The sets in a Cartesian product need not contain the same kinds of elements. To illustrate a product of a set of objects with a set of numbers, let

$$A = \{a \mid a \text{ is a man living in the United States}\}$$

and $$G = \{g \mid g \text{ is a weight in pounds}\}.$$

Then

$$A \times G = \{(a, g) \mid a \in A, g \in G\},$$

the set of all possible pairings of a man with a weight. If each man were represented by a point along a horizontal axis, and each weight by a point along a vertical axis, then each intersection of a vertical with a horizontal line stands for a possible man-weight pair, such as (John Jones, 165 pounds).

Given the idea of a product of two sets, then it is finally possible to state what we mean by a **mathematical relation:**

A mathematical relation on two sets A and B is a subset of
    $A \times B$.

In other words, any mathematical relation is a subset of a Cartesian product, *some specific set of pairs out of all possible pairs.* At first glance, this seems to be an extremely trivial idea, but it actually is a remarkably subtle and elegant way to approach a difficult problem.

We often speak loosely of the husband-wife relation, the hand-fits-glove relation, the pitcher-catcher relation, the height-weight relation, the relation of the side of a square to its area, and so on. In each case, the fact of the relation implies that *some* pairs out of all possible pairs make a statement "*a* plays such and such a role relative to *b*" a true one. For any "husband," not all women qualify as "wife"; for any hand, only certain gloves fit; for any side-length of a square, only a certain number can be the area.

On the other hand, a few words of warning are in order before this topic is pushed further. It is important not to confuse the idea of a mathematical relation with the idea of a *true relationship* among objects. All relationships that are true in our experience can be represented as mathematical relations, but it is *not necessarily true* that every mathematical relation we might invent must correspond to a real relationship among things. Furthermore, several different relationships, meaning quite different things in our ordinary experience, may show up as the *same* mathematical relation, in that exactly the same pairs qualify for the relation. For example, it *might* happen that for some set of men and some set of women, the relationship "*a* is married to *b*" would involve exactly the same pairs as "*a* files a joint income tax return with *b*." The two *relations* would be identical, but the *relationship* represented means something quite different in each instance.

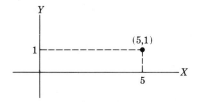

Figure 1.14.2

1.15 SPECIFYING RELATIONS

Being a set, a mathematical relation can be specified in either of the two ways used for any set. In the first place, a relation may be specified by a listing of all pairs that qualify for the relation. In the example of Section 1.14, the set A was four first names, and the set B was four last names. In this circumstance the relation L *might* be

$$L = \{(\text{Mary, Jones})\,(\text{Susan, Brown})\,(\text{Tom, Adams})\,(\text{Bill, Smith})\}.$$

This relation could also be displayed graphically (Figure 1.15.1), the filled dots in the diagram showing the pairs in the relation L; all the remaining pairs $A \times B$ but not in L are shown unfilled.

Quite often, the graph itself serves as a kind of symbolic listing when the pairs are too numerous to list explicitly. For example, for the product $R \times R$, the curve in Figure 1.15.2 stands in place of a listing of (x, y) pairs. In this instance, the number-pair represented by any point falling *on* the curve is *in* the relation. The point's two coordinates denote the particular pair of numbers. This specifies the relation unambiguously, and really serves the purpose of a list. The test of membership for any pair of numbers is whether or not the corresponding point falls exactly on the curve.

Plots of other relations may show up as areas or sectors rather than curves, as for example, in Figure 1.15.3. Here, any pair of numbers (x, y), such as $(7, 5)$, qualifies for the relation if the corresponding point falls into the shaded sector of the graph.

The more usual way of specifying a relation is by a statement of the rule by which a pair qualifies. For example, if A is the set of all women, and B is the set of all men, the relation S might be specified by

$$S = \{(a, b) \in (A \times B) \mid a \text{ is married to } b\}.$$

If A is the set of all men in the United States, and G is the set of all weights in pounds, a relation T might be specified by

$$T = \{(a, g) \in A \times G \mid \text{the weight of } a \text{ is } g\}.$$

If the Cartesian product is $R \times R$, then the most common way to specify a relation is by a mathematical expression giving the qualifications a number-pair must meet: As examples, consider

$$F = \{(x, y) \mid x + y = 10\},$$
$$V = \{(x, y) \mid x^2 - y^2 = 2\},$$
and
$$Z = \{(x, y) \mid 3 \leq x \leq 5,\ -2 \leq y \leq -1\},$$

which are but a few of the countless ways of specifying a relation by a mathematical rule. For these numerical relations, the corresponding graphs are shown in Figures 1.15.4 to 1.15.6.

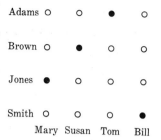

Adams o   o   ●   o

Brown o   ●   o   o

Jones ●   o   o   o

Smith o   o   o   ●

Mary Susan Tom Bill

Figure 1.15.1

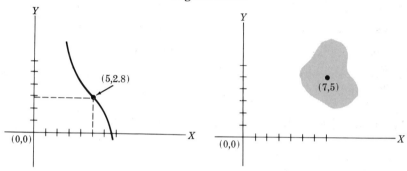

Figure 1.15.2

Figure 1.15.3

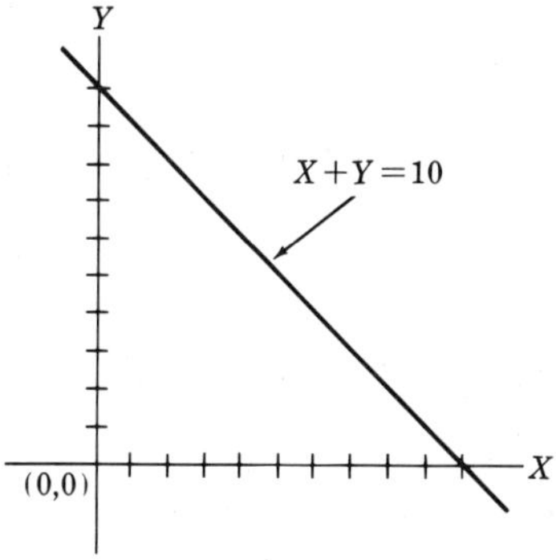

$X+Y=10$

$X^2-Y^2=2$

Figure 1.15.4

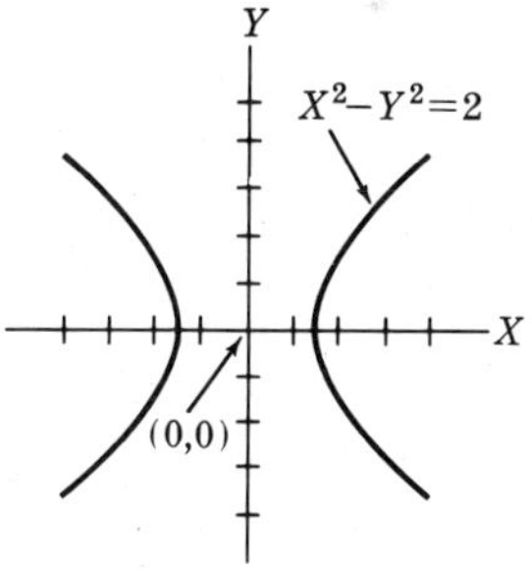

Figure 1.15.5

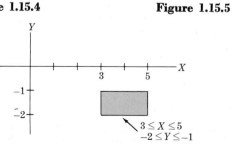

$3 \leq X \leq 5$
$-2 \leq Y \leq -1$

Figure 1.15.6

1.16 THE DOMAIN AND RANGE OF A RELATION

In discussing relations, we need some way to distinguish between the different sets of elements that play different roles. In a relation that is a subset of $A \times B$, it may be true that only *some* elements a enter into one or more (a, b) pairs; what we want is a way to discuss those elements of A that actually *are* paired with b elements in the relation itself. This subset of elements of A actually figuring in the relation is called the **domain,** and it is defined as follows:

Given some relation S, which is a subset of $A \times B$, then

domain of $S = \{a \in A \mid (a, b) \in S,$ for at least one $b \in B\}$.

Notice that **the domain of the relation S is always a subset of A.** The domain includes each element in A that actually plays a role in the relation, vis-à-vis some b.

For example, let A be the set of all men and B the set of all women. Let the relation $S \subseteq A \times B$ be "a is married to b." Then the domain of S is "all married men," since only the subset of married men in A can figure in at least one pair (a, b) in the relation.

As another example, suppose that the product is $R \times R$, and the relation is given by the rule

$$C = \{(x, y) \mid x^2 + y^2 = 4\}.$$

What is the domain of C? The very largest number that x could be is 2 for any (x, y) pair in C, and the very smallest number is -2. Thus, x can be any number between -2 and 2 inclusive, but all other numbers in R are excluded. Hence,

domain of $C = \{x \in R \mid -2 \le x \le 2\}$.

The idea of the **range** of a relation is very similar to that of the domain, except that it applies to members of the set B, the second members of pairs (a, b).

The set of all elements b in B paired with at least one a in A in the relation S is called the range of S:

range of $S = \{b \in B \mid (a, b) \in S,$ for at least one $a \in A\}$.

In the example of the relation "a is married to b," the range is the set of all married women, a subset of B. In the example of the relation C, for y to be a real number the value of y must lie between -2 and 2, and so the range is

range of $C = \{y \in R \mid -2 \le y \le 2\}$.

It is entirely possible for the domain to be equal to A, and for the range to be equal to B, in some examples. In others, the range and the domain will be proper subsets of A and B, respectively.

1.17 FUNCTIONAL RELATIONS

We come now to one of the most important mathematical concepts, from the points of view both of mathematics itself and of its applications. This is the idea of a **functional relation,** or **function.** The definition we will use is:

A relation is said to be a *functional relation* or a *function* if each member of the domain is paired with one and only one member of the range.

That is, in a functional relation each a entering into the relation has *exactly one pair-mate* b. Thus a function is just a special kind of relation. In some mathematical writing this is called a "single-valued function," and a relation is a "multiple-valued function." However, we feel that it is useful to call only the former a "function."

This idea will become clearer if we inspect some relations that are, and some that are not, functions.

Given A as the set of all men and B the set of all women in some society, the relation

$$\{(a, b) \in A \times B \mid \text{``}a \text{ is married to } b\text{''}\}$$

is a function *if the society is monogamous,* so that each man may have only one wife. Here, each member of the domain, a married man, has one and only one wife, a member of the range. If, on the other hand, the society is polygamous so that a man may have two or more wives, then the relation is not necessarily a function.

Consider the relation defined on pairs of real numbers:

$$\{(x, y) \mid x^2 = y\}.$$

This relation is a function; corresponding to each x, there is one and only one y, which is the same as the square of x. Thus, $x = 2$ can be paired only with $y = 4$, $x = 5$ only with $y = 25$, and so on. Contrast this with the relation

$$\{(x, y) \mid x^2 = y^2\}.$$

In this case, the pair $(2, 2)$ qualifies, but so does the pair $(2, -2)$. Each value of x can be associated with *two* values of y by this rule, and so the relation is not a function. Figures 1.17.1–1.17.2 show how these two relations

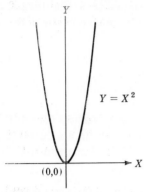

Figure 1.17.1

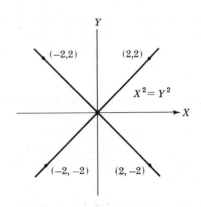

Figure 1.17.2

can be plotted. Notice how in the first example a vertical line drawn from any point on the X-axis will intercept the curve in at most *one* place, while in the second example *two* interceptions are possible. The same criterion may be applied to the plot of any relation between two numbers in order to determine if it is a function.

There is nothing in the definition of a function that limits the number of times an element b *in the range* may appear in a pair. Thus, in the marriage example, the relation is still a function even though the same woman is married to several men, and in the relation with rule $x^2 = y$, the same y value is associated with two x values (for example, $y = 4$ is paired both with $x = 2$ and $x = -2$). It follows that a given rule relating x and y may specify a function for (x, y) pairs but may specify a nonfunctional relation for (y, x) pairs.

Functions are relations, and hence are **sets of pairs,** and the set itself should be distinguished from the rule specifying the set. The distinction is not always clear in mathematical writing, where commonly both the set and the rule are called functions. However, if you remember that the function itself is the set of pairs, you should have little difficulty in deciding which is meant in a discussion using this word.

A very large part of mathematics deals with functions and the properties of their rules as specifications of what the set itself, or certain subsets, must be like. When a mathematician describes the "curve" or plot of a numerical function on the basis of its mathematical rule, he is really saying something about how the set is delimited by the rule. Any curve that can be plotted for a function can be regarded as a symbolic listing of pairs of numbers.

The function idea, of course, applies to products of *any* sets. As we have seen, it is quite possible to describe a function where the elements are pairs of people. Other examples can be given where a variety of other things are paired. However, for mathematicians the most interesting functions involve numbers, either as the range of the function, or as both range and domain.

An important kind of function is one having numbers as the range and a **set of sets** as the domain. This is called a **set function.** For example, suppose that the name of a state in the United States is given to a set of all persons residing in that state on January 1, 1970. Then the whole collection of the fifty state-sets is a set of sets; let us call this set of sets A. For the moment, let the symbol P stand for the set of all *positive* numbers. Then the product $A \times P$ is the set of all pairings of a *set* with a positive number. Consider the relation:

$$\{(a, p) \in A \times P \mid \text{``the total population of } a \text{ was } p \text{ on January 1, 1970''}\}.$$

This relation is a set function, since associated with each and every set of state residents there is one and only one population number p for the date January 1, 1970. One of the most important applications of the idea of a set function is a probability function, to be introduced in the next chapter.

Nevertheless, most of the interesting and useful functions relate numbers to numbers, since their rules can be given quite precise algebraic or symbolic form. The most highly developed parts of mathematics are devoted, essentially, to the study of such functions. Numerical functions are also extremely important in science, where the mathematical function rules form precise statements of the relations between measured quantities. Indeed, the emphasis on functions in mathematics is largely a reflection of their historical importance in physical science.

1.18   VARIABLES AND FUNCTIONAL NOTATION

Because of the importance of functional relations both in mathematics and science a special notation has been developed for discussing functions, and we will find this useful in our study of statistics. Before we deal with this notation, the notion of a "variable" must be clarified:

A variable is a symbol that can be replaced by any one of the elements of some specified set. The particular set is called the range of the variable.

Note carefully that the variable is only a stand-in or placeholder, which can *always* be replaced by a particular element from some set of possibilities. Thus, if x is a variable, wherever x appears in a mathematical expression it can be replaced by *one* element from some specified set. This idea should be a familiar one, since algebra deals largely with variables ranging over the real numbers. Throughout our discussion of set theory, a variable such as X symbolized a set, which could be any member of some *set of sets;* the variable x stood for *some particular member* of the set X.

Variables may range over any well-defined set. For example, the variable C might symbolize any of a set of colors; the variable g might range over the 365 different dates in the year; the variable y might be any one of the infinite set of possible temperatures in Farenheit degrees. When someone says, "You are a no-good blankety-blank!" he is using "blankety-blank" much like a variable, standing for any one of a set of expressions; the range is left to your imagination.

Despite its name, a variable is not something that varies, or wiggles, or scurries around while you work with it. If a variable is used as a symbol in a given mathematical discussion, then replacing the variable by some specific element of its range at one place requires you to replace it with the same element *wherever* the variable appears. The same variable may appear many times in the same discussion, and it retains its identity throughout; if x is replaced by some number in one place, it is replaced by the same number in all other appearances. The name "variable" actually comes from the fact that a symbol represents "various" values. This distinguishes such symbols from mathematical *constants*. Unlike a variable, which can stand for a variety of things, a symbol that can be replaced by *one and only one* number in a given expression is a constant for the expression. For example, consider the well-known mathematical expression

$$c = 2\pi r.$$

The symbols c and r here are variables; each stands for any one of an infinite set of *positive numbers*. However, the symbol π is a constant, which can be replaced only with a particular number. Regardless of the values assumed by r and c, π is the same. Note also, in this instance, that c and r are *functionally related* variables; the value that c can assume is dictated by the value of r (and vice versa).

Most of the variables that will concern us here represent sets of numbers, so that given some range of numbers, x stands for any specific member of the range. For instance, the statement "let X be a variable ranging over the real numbers" means that X is a stand-in for any one of the numbers that can be represented as points on a line. The X-axis in a graph is a representation of the range of this variable. When one number, say 10, is selected to replace X, then we say "X assumes the value 10." Quite often we want to discuss several variables X, Y, Z, and so on, simultaneously. Each of these symbolizes one of a range of numbers, but *not necessarily the same number or even the same range of numbers*. For example, the X-axis and the Y-axis in a graph each symbolize ranges of possible numbers; when X assumes the particular value x and Y the particular value y, we have the pair of numbers (x, y).

Sometimes the range of a numerical variable is specified as other than all real numbers. For example, one might have the variable X, followed by the statement $3 \leq X \leq 5$, meaning that the only values that X symbolizes are those lying between 3 and 5 inclusive (X *ranges* between 3 and 5 inclusive). Or perhaps X is a variable that is integral, meaning that the only numbers that can replace the symbol X are whole numbers.

Variables are the basis for the notation most commonly used for functions. Given the variables x and y, each with specified range, a relation between the variables is a subset of all possible (x, y) pairs. If this relation is a function, then this fact is symbolized by any of the statements such as

$$y = f(x),$$
$$y = G(x),$$
$$y = \varphi(x),$$

and so on. In words, these expressions all say:

> "There is a rule that pairs each possible value
> of x with at most one value of y."

The different Roman and Greek letters that precede the symbol (x) simply indicate different rules, giving different functions or subsets of all (x, y) pairs. *This notation does nothing more than assert that the function rule exists.* It says "Give me the rule, and the value of x, and I'll give you y."

Notice that a symbol $f(x)$ actually stands for a value of y; this is the value of y that is paired with x by the function rule. The symbol $f(x)$ is *not* the function, but an indicator that the function rule exists, turning a value x into some value $f(x)$, or y.

Sometimes the function rule itself is stated:

$$y = f(x) = x^3 - 2x^2 + 4,$$
$$y = G(x) = k \log(x), \quad x > 0,$$

and
$$y = \varphi(x) = \frac{1}{\sqrt{2\pi}\sigma} \exp\left[-\frac{(x - \mu)^2}{2\sigma^2}\right]$$

are but a few of the countless function rules that might be stated for the variables x and y, both ranging over the real numbers. The terms such as k, e, σ, and μ in these rules are simply constants having the same value in a given function rule regardless of the value that x symbolizes. In each example, the function rule stated permits the association of some y value with a given value of x. *Be sure to notice that these expressions are the rules, not the functions.*

Functional notation has been shown here only for the case where both x and y range over numbers, primarily because this is the most common situation in mathematics. However, the same general form can be used for other functions. Suppose that the variable a stands for any one of a set of men, and the variable y stands for any one of the set of positive numbers that are possible scores on an intelligence test. Granted that each man must have one and only one score, then a function exists relating a and y. This could be symbolized by

$$y = f(a)$$

where a designates a particular man. We would be hard put, however, to state the rule for this function, and so we cannot go beyond the simple assertion that a function exists, unless we undertook to list each man paired with his score. In general, such functions are a pain in the neck to specify, and this accounts for some of the elaborate attempts made in science to describe relationships as numerical functions which *can* be given mathematical rules.

Once again, you are warned that the word "function" is often used to refer both to the set of pairs, which is actually the function, *and* the function rule, which specifies the set. In particular, when one says, "Y is a function of X," he is making the assertion symbolized by $Y = f(X)$, that some rule exists pairing no more than one Y with each X. Then he may go ahead to specify the function rule. On the other hand, when he talks about *the* function relating X and Y, he is usually referring to the set of pairings of an X with a Y given by that rule.

1.19   CONTINUOUS VARIABLES AND FUNCTIONS

In statistics it is often necessary to specify that a variable or a function is continuous. Unfortunately, an accurate definition of "continuous" requires more mathematics than some students can command at this point. However, the following way of thinking about continuous variables and functions, although not really adequate as a definition, will serve our purposes.

Consider a variable x. If there are two numbers u and v such that the range of x is $u \le x \le v$, *all real numbers lying between u and v*, then x is said to be continuous in the interval $\langle u, v \rangle$. If the variable x is continuous over all possible intervals defined by pairs of real numbers u and v, then x is said to be continuous over the real numbers.

The idea of a **continuous function** is best conveyed by a picture of a curve without "gaps" or "breaks" in terms of values of y; that is, if a function $y = f(x)$ is continuous in the interval of numbers $u \le x \le v$, then each and every number between $f(u)$ and $f(v)$ also occurs in the range of y values. For each x in the interval, some particular number y can be found by the function rule, if the function is continuous in the interval. A function that is continuous over all possible intervals of real numbers, so that the domain includes all real numbers, is called simply "continuous." Thus, for example, the function given by $y = 3x^2 - 5$ is continuous over all real numbers, since for every possible x there exists a determinable y, and hence all real values of x are in the domain. On the other hand, a function with the rule $y = 1/(x + 1)$ is **discontinuous,** since for $x = -1$ there is no exactly determinable value of y that can be associated with x, and hence the domain does not include $x = -1$.

The study of discontinuous functions is an important part of mathematics, but most of the functions that concern us here will either have a specified number of possible values that x can assume or will be continuous over the real numbers.

1.20   FUNCTIONS WITH SEVERAL VARIABLES

The idea of a numerical function can be extended without difficulty to the case with three or more variables. Here the domain of the function is a set of *pairs* of numbers. Given variables x, y, and z, each ranging over the real numbers, the statement

$$y = f(x, z)$$

means that a rule exists such that given any pair of numbers x and z in the domain, then the value of y is known. For example, it may be that

$$y = f(x, z) = 3x^2 - 2z^2 + 4,$$

so that given this rule, and letting x and z assume some pair of values (x, z), we have completely determined the value of y. Or, it may be that a different rule applies, such as

$$y = g(x, z) = 2x^z,$$

making this second function a very different subset from among all (x, z, y) triples.

In general, if there is some function with three or more variables, and we fix one or more variables at a constant value but leave the other variables free to take on different values, then this also results in a function. For example, consider the function having the following rule:

$$y = xz.$$

Suppose that x were fixed at the value 3, while z can take on any value. Then we have the function

$$y = 3z,$$

which has only two variables. The particular mathematical rule for $y = f(z)$ in this case depends upon the value chosen for x. The same idea applies to functions with any number of independent variables: fixing the value of one or more variables results in a new function, where the fixed values appear as constants in the mathematical rule.

1.21   FUNCTIONS AND PRECISE PREDICTION

In the statistician's use of mathematics, numerical variables symbolize real quantities or magnitudes that can be found (potentially) from measurements of phenomena. Real relationships are summarized as mathematical relations, having some mathematical expression as the specifying rule. When the statistician deals with a true functional relation between quantities, then this is communicated in its most concise and elegant form by a statement of the function rule.

One of the goals of any science is prediction: given some set of known circumstances, what else can we expect to be true? If the scientific relation is functional in form, then *precise* prediction is possible; given the true value of X in the domain, precisely one value of Y will be observed, all other things being constant. However, in this qualification, "all other things being constant," lies one of the central problems of any science.

Each of us, scientist or layman, has learned from infancy that there are relationships in the world about us; things go together, and some things lead to other things. On the other hand, everyone knows that precise prediction of the world about him is virtually impossible. Even given some information about an object or event, we never know *exactly* what its other properties will be. True enough, providing us with information may let us restrict the range of things we *expect* to observe, but the possibility of exact prediction is almost unknown in the everyday world. Nevertheless, the more advanced sciences are very successful in describing relationships *as though* they were mathematical functions. How has this come about? What does the scientist do that makes it possible for him to make precise predictions from known situations, when the world of the man on the street is such a disorderly and unpredictable affair?

It seems safe to say that the world is full of marvelously complicated relationships and that any event we experience must have its character determined by a vast number of influences. Some of these may exert major forces on what we observe, and others may be quite minor in shaping a given event. Most scientists would subscribe to the idea that, ultimately, if we knew the values of *all* of the relevant variables, *all* of the influences that go into the determination of an event, and the rule that relates these variables to the event itself, then precise prediction would always be possible. It may be that given the complete information about the circumstances, all true relations are functional.

In the physical sciences, it may often be possible to obtain complete information about all of the variables relevant to a particular event, and thus to determine functional relationships between variables. By comparison, the social sciences, the behavioral sciences, and the areas comprising the field of business are not so far along in the use of mathematics to specify real relations. The precision and control in experimentation often possible in the physical sciences are not generally attainable in these areas. In situations in which the full range of relevant factors is quite unknown, statistics becomes a most useful tool. The theory of inferential statistics deals with the problem of "error," the failure of an observation to agree with a true situation. Error is the product of "chance," the influence of the innumerable uncontrolled factors determining a particular event. The concept of "probability" is introduced to evaluate the likelihood that a given observation will disagree to a certain extent with a true value. Instead of being certain or nearly certain about the conclusions reached from observations, as the physical scientist often is, the conclusion drawn by the statistician is much like a bet.

The gambler placing a bet is faced with a situation he knows very little about; heaven alone knows the real reasons why a coin comes up heads on a given toss or a particular card comes up on the third draw of a poker game. However, given that some things are true, it can be deduced that some things are more or less *likely* to occur, and still other things *unlikely*, regardless of the reasons why. The gambler knows only that he has a certain probability of being right, and also of being wrong, in a given decision, and the wise gambler places his bets accordingly.

In succeeding sections we will continue to compare the task of the statistician to that of the gambler, and to show how statistical theory aids the statistician in making his own sorts of bets. In order to do this, we need some of the ideas of probability theory, to be introduced in the next chapter.

EXERCISES

1. What is required to determine whether a given object is or is not a member of a given set?

2. What is the role of the universal set in set theory? What is the role of the empty set?

3. Specify the following sets by first listing all their members and then stating a formal rule:
   (a) The set of all positive integers less than or equal to 12.
   (b) The set of all positive even integers less than or equal to 12.
   (c) The set of all odd positive integers less than or equal to 12.
   (d) What is the relation of the sets in parts (b) and (c) to the set in part (a)?

4. Let the set $A = \{0, 1, 2, 3, 4, 5\}$.
   List the elements in each of the following sets that are also in the set A.
   (a) $\{x \mid x + 3 = 5\}$
   (b) $\{x \mid x^2 + 2x + 1 = 0\}$
   (c) $\{x \mid 2x - 3 = 7\}$
   (d) $\{x \mid x^2 > 1\}$
   (e) $\{x \mid x + 2 > 6\}$

5. Let $A = \{-2, -1, 0, 1, 2\}$.
   List the elements of A that are also in each of the following sets:
   (a) $\{x \in A \mid 3x - 6 = 0\}$
   (b) $\{x \in A \mid x^2 - 1 = 0\}$
   (c) $\{x \in A \mid x^2 - 1 > 0\}$
   (d) $\{x \in A \mid 2x^2 - 3x < 0\}$
   (e) $\{x \in A \mid 5x^4 + 3x^3 - 6x^2 + 2x = 0\}$.

6. Let the universal set be the set of real numbers, and let
   $A = \{x \mid x^2 \geq 1\}$, $B = \{x \mid x^2 + 2x + 1 = 0\}$, and $C = \{x \mid 2x - 3 \leq 7\}$.
   Specify the following sets:
   (a) $A \cup B$ (d) $A \cap C$
   (b) $A \cup B \cup C$ (e) $(A \cup \bar{B}) \cap \bar{C}$.
   (c) $C - A$

7. Consider the following sets:

   W = Set of all students at the University of Michigan.
   A = Set of all males.
   B = Set of all graduate students.
   C = Set of all students who take mathematics courses.
   D = Set of all students who take business courses.

   Describe in words the following sets:
   (a) $A \cup (B \cup C)$ (d) $A \cap (\bar{B} \cup C)$
   (b) $A \cup \overline{(B \cup C)}$ (e) $A \cup \overline{(B \cap C)}$.
   (c) $(A \cap \bar{B}) \cup (A \cap B)$

8. A certain journal in statistics makes references to various other journals and also to itself. Suppose that in a sample of 100 articles from this journal the numbers of times the journal referred to journals A, B, and C (itself) is as given below:

A	60
B	40
C	70
A and B	32
A and C	45
B and C	38
A, B, and C	30

   (a) Translate the following verbal statements into the appropriate set

symbols and find the number of times the "event" occurred:
- (i)   The journal refers to itself and B but not to A.
- (ii) The journal refers *only* to itself.
- (iii) The journal refers to A and B but not to itself.

(b)   Find now, in general, set expressions for "the article refers to":
- (i)   Exactly one journal.   (iv) At least two journals.
- (ii) Exactly two journals.   (v)   At most one journal.
- (iii) At least one journal.   (vi) At most two journals.

Also find the number of elements in each of the sets in (i)–(vi).

9. Children and teachers in several kindergarten classes were asked to name their favorite color. The numbers of different responses are given below.

		Red (R)	Blue (B)	Yellow (Y)	Other (O)
(M)	Male student	10	5	3	4
(F)	Female student	6	11	2	1
(T)	Female teacher	3	4	1	1

How many members do each of the following sets have?
O, $R \cup B$, $M \cup Y$, $T \cap (B \cup Y)$, $(\overline{F \cap O})$, $M \cup (F \cap \overline{B})$.

10. Make Venn diagrams to verify that the following theorems on sets hold. Use some systematic shading or coloring to clearly distinguish the relevant sets.
- (a) $A \cap (B \cup C) = (A \cap B) \cup (A \cap C)$
- (b) $(A \cap B) \subseteq A \subseteq (A \cup B)$
- (c) $A - B \subseteq A$
- (d)   if $A \subset B$ and $C \subset D$, then $(A \cup C) \subset (B \cup D)$
- (e)   if $A \supset B$ and $C \supset D$, then $(A \cap C) \supset (B \cap D)$
- (f) $A \cap B = A - (A - B)$
- (g) $A \cup (B - A) = A \cup B$
- (h) $A - (A \cap B) = A - B$
- (i) $A \cap (B - C) = (A \cap B) - C$.

11. If $A = \{0, 1, 2, 3\}$, list all possible subsets of A.

12. What distinguishes a Cartesian product from any other set?

13. When is a relation a function?

14. Let $A = \{2, 3, 4, 8, 9\}$. Graph these *mathematical relations*, each of which are subsets of $A \times A$:
- (a) $\{(2, 9), (4, 9), (8, 9), (9, 9)\}$
- (b) $\{(2, 3), (2, 4), (2, 8), (2, 9)\}$
- (c) $\{(2, 4), (3, 9)\}$
- (d) $\{(2, 3), (3, 4), (4, 8), (8, 9)\}$
- (e) $\{(3, 2), (4, 3), (8, 4), (9, 8)\}$
- (f)   Which of these relations are functions?

15. List all the ordered pairs that can be obtained from the set $X = \{1, 2, 3, 4, 5\}$, corresponding to each of the following relations:
   (a) Is a multiple of.
   (b) Is greater than.
   (c) Is less than.
   (d) Is equal to.
   (e) Is greater by two.
   (f) Is less by two.

16. Suppose four people are sitting at a table. Let U be the set of people sitting at a table, $U = \{A, B, C, D\}$. The people are seated as in the figure.

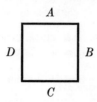

Draw the graphs of the following relations that are subsets of $U \times U$ and specify which of the relations are functions.
   (a) $\{(x, y) \mid x \text{ sits next to } y\}$
   (b) $\{(x, y) \mid x \text{ sits across from } y\}$
   (c) $\{(x, y) \mid x \text{ sits to the right of } y\}$.

17. In each of the following, list the elements of the relation and give its domain and range, where $U = \{1, 2, 3, 4, 5\}$, and (x, y) is a member of $U \times U$.
   (a) $\{(x, y) \mid x = 2\}$
   (b) $\{(x, y) \mid y = 5\}$
   (c) $\{(x, y) \mid y = 2x\}$
   (d) $\{(x, y) \mid x = (1/2)y\}$
   (e) $\{(x, y) \mid y < x\}$.

18. Consider the following sets of letters: $\{a, b, c, d, e\} = A$, $\{u, v, w, x, y, z\} = B$, $C = \{a, e, i, o, u\}$. Let $W = A \cup B \cup C$.
   List the members of:
   (a) $A \cap C$
   (b) $A \cup (B \cap \bar{C})$
   (c) $B \cup \bar{C}$
   (d) $(A \cap B) \cup (\overline{B \cap C})$
   (e) $A - C$
   (f) $A \times C$
   (g) $A \times A$.

19. Given $A = \{x \mid x^2 \le 16\}$, what is the Cartesian product $A \times A$? Draw the graphs of the following relations *on* $A \times A$ and specify which of the relations are functions.
   (a) $\{(x, y) \mid x + y = 3\}$
   (b) $\{(x, y) \mid x^2 + y^2 = 4\}$
   (c) $\{(x, y) \mid x^2 = 16 - y\}$
   (d) $\{(x, y) \mid y^3 = 64 - x\}$
   (e) $\{(x, y) \mid y \ge x + 3\}$.

20. In Exercise 19, find the domains and the ranges of the relations in (a), (b), (c), (d), and (e).

21. For each statement, tell whether it is true (for *any sets* A, B, and C) or false, and if it is false, draw a Venn diagram to demonstrate that it is false.
   (a) $(A - C) \cup (B - C) = (A \cup B) - C$

(b) $A \cap \overline{(A \cap C)} = \varnothing$

(c) $(\bar{A} \cup \bar{B}) \cap C = \overline{(A \cap B)} \cup C$

(d) $A \subseteq (A \cap \bar{B}) \cup B$

(e) $\bar{A} \cup \bar{B} \cup \bar{C} \subset \overline{(A \cup B \cup C)}$.

22. Prove, by set algebra, that

(a) $\overline{(A \cup B)} = (\bar{A} \cap \bar{B})$ (b) $\overline{(A \cap B)} = (\bar{A} \cup \bar{B})$

[*Hint for* (a): Show that $(A \cup B) \cup (\bar{A} \cap \bar{B}) = W$ and that $(A \cup B) \cap (\bar{A} \cap \bar{B}) = \varnothing$, so that the set $(A \cup B)$ must be the complement of $(\bar{A} \cap \bar{B})$.]

23. Given that A, B, and C are *not* mutually exclusive, express their union $A \cup B \cup C$ in terms of the union of three sets which *are* mutually exclusive.

2

ELEMENTARY
PROBABILITY
THEORY

Statistical inference and decision involve statements about probability. Everyone has the words "probable" or "likely" in his vocabulary, and most of us have some notion of the meaning of statements such as "The probability is one-half that the coin will come up heads." However, in order to understand the methods of statistical inference and decision and use them correctly, one must have some grasp of probability theory. This is a mathematical system, closely related to the algebra of sets described in the last chapter. Like many mathematical systems, probability theory becomes a useful model when its elements are identified with particular things, in this case the *outcomes of real or conceptual experiments*. Then the theory lets us deduce propositions about the likelihood of various outcomes, if certain conditions are true.

Originally, of course, the theory of probability was developed to serve as a model of games of chance—in this case the "experiment" in question was rolling a die, or spinning a roulette wheel, or dealing a hand from a deck of playing cards. As this theory developed, it became apparent that it could also serve as a model for many other kinds of things having little obvious connection with games, such as results in the sciences. One feature is common to most applications of this theory, however: the observer is uncertain about what the outcome of some observation will be, and he must eventually infer or guess what will happen. In a sense, then, probability reflects the *uncertainty* about the outcome of an experiment.

In this respect the statistician is like the gambler keeping track of the numbers coming up on a roulette wheel. At any given opportunity he can observe only the tiniest part of the total set of things he would like to know about. Given his human frailty as an observer, he must fall back on logic;

given that certain things are true, he can make deductions about the truth of other things. The statistician's statements about *all* observations of such and such phenomena are on a par with the gambler's; both are deductions about what should be true, if the initial conditions are met. Furthermore, the gambler, the business man, the engineer, the scientist, and, indeed, the man on the street must make decisions based on incomplete evidence. Each does so in the face of uncertainty about how good those decisions will turn out to be. Probability theory alone does not tell any of these people how they should decide, but it does give ways of evaluating the degree of risk one takes for some decisions.

The theory of probability is very closely tied to the theory of sets. Indeed, the main undefined terms in the theory, the "events," are simply sets of possibilities. Before we develop this idea further, we need to talk about the ways that events are made to happen: simple experiments.

2.1 SIMPLE EXPERIMENTS

We shall mean nothing very fancy by the term "simple experiment," and there is no implication that a simple experiment need be anything even remotely resembling a laboratory experiment. **A simple experiment is some well-defined act or process that leads to a single well-defined outcome.** Some simple experiments are tossing a coin to see whether heads or tails comes up, cutting a deck of cards and noting the particular card that appears on the bottom of the cut, opening a book at random and noting the first word on the right-hand page, running a rat through a T maze and noting whether he turns to the right or the left, observing the closing price of a common stock on a particular date, lifting a telephone receiver and recording the time until the dial-tone is heard, asking a person his political preference or brand preference, giving a person a test and computing his score, counting the number of colonies of bacteria seen through a microscope, and so on, literally without end. The simple experiment may be real (actually carried out) or conceptual (completely idealized), but it must be capable of being described. We also require that each performance of the simple experiment have one and only one outcome, that we know when it occurs, and that the set of all possible outcomes can be specified. Any single performance, or trial, of the experiment must eventuate in one of these possibilities.

Obviously, this concept of a simple experiment is a very broad one, and almost any describable act or process can be called a simple experiment. There is no implication that the act or process even be initiated by the experimenter; he need function only in the role of an observer. On the other hand, it is essential that the outcome, whatever it is, be unambiguous and capable of being categorized among all possibilities.

2.2 EVENTS

The basic elements of probability theory are the possible distinct outcomes of some idealized simple experiment. The set of all possible distinct outcomes for a simple experiment will be called the **sample space** for that experiment. Any member of the sample space is called a **sample point,** or an **elementary event.** Every separate and "thinkable" result of an experiment is represented by *one and only one* sample point or elementary event. Any elementary event is *one possible result* of a single trial of the experiment.

For example, we have a standard deck of playing cards. The simple experiment is drawing out one card, haphazardly, from the shuffled deck. The sample space consists of the fifty-two separate and distinct cards that we might draw. If the experiment is stopping people on the street, then the sample space consists of all the different individuals we might possibly stop. If the experiment is reading a thermometer under particular conditions, then the sample space is all the different numerical readings that the thermometer might show.

Seldom, however, is the *particular* elementary event that occurs on a trial of any special interest. Instead, the actual outcome takes on importance only in relation to all possible outcomes. Ordinarily we are interested in the kind or class of outcome an observation represents. The outcome of an experiment is *measured,* at least by allotting it to some qualitative class. For this reason, the main concern of probability theory is with *sets* of elementary events. **Any set of elementary events is known simply as an event, or an event class.**

Imagine, once again, that the experiment is carried out by drawing playing cards from a standard pack. The fifty-two sample points (the distinct cards) can be grouped into sets in a variety of ways. The suits of the cards make up four sets of thirteen cards each. The event "spades" is the set of all card possibilities showing this suit, the event "hearts" is another set, and so on. The event "ace" consists of four different elementary events, as does the event "king," and so on.

If the experiment involves stopping people on the street and noting their eye color, the experimenter may designate seven or eight different eye-color classes. Each such set is an event. The event "blue eyes" is said to "occur" when we run into a person who is a member of the class "blue eyes." If, on the other hand, we find the weight of each person we stop, then there may be a vast number of "weight events," different weight numbers standing for classes into which people may fall. If a person is stopped and weighs exactly 168 pounds, then the event "168 pounds" is said to occur.

In short, **events are sets, or classes, having the elementary events as members.** The elementary events are the raw materials that make up event classes. The occurrence of any member of event class A makes us say that event A has occurred. Since *any* subset of the sample space is an event, then some event *must* occur on each and every trial of the experiment.

2.3 EVENTS AS SETS

The symbol S will be used to stand for the *sample space*, the set of all elementary events that are possible outcomes of some simple experiment. Then capital letters A, B, and so on will represent events, each of which is some subset of S. Notice how the sample space S is used like a universal set; for a given simple experiment, *every event discussed must be some subset of S*. The set S may contain either a finite or an infinite number of elements.

Since any event A is a subset of S, the operations and postulates of set theory carry over directly to operations on events to define other events.

First of all, the set S is an event: S is the "sure event," since it is *bound* to occur (one of its members must turn up on every trial). The event $\emptyset$ is called the "impossible event," since it cannot occur, having no members. It is important to remember that the empty or impossible event $\emptyset$ is, nevertheless, an *event* according to our definition, since the empty set is a subset of every set, so that $\emptyset \subset S$.

By our definition of an event as a subset of S, then if A and B are both events, the union $A \cup B$ is also an event. The event $A \cup B$ occurs if we observe a member of A, or a member of B, or a member of both.

In the same way, $A \cap B$ is an event. The event $A \cap B$ requires an outcome that is *both* an A and a B in order to occur. Notice that any occurrence of $A \cap B$ is automatically an occurrence of $A \cup B$, but that the reverse is not true.

If A is an event, then so is its complement $\bar{A}$, and A and $\bar{A}$ cannot both occur, since $A \cap \bar{A} = \emptyset$. On the other hand, either A or $\bar{A}$ must occur, since $A \cup \bar{A} = S$. The difference between two events, $A - B$, is likewise an event. **In short, each and every subset formed from the elementary events in S is an event.**

Two events A and B, neither equal to $\emptyset$, are said to be **mutually exclusive, or disjoint,** if $A \cap B$ cannot occur—that is, if $A \cap B = \emptyset$. In general,

N events are mutually exclusive, or disjoint, if *all possible pairs* of events selected from the N events are mutually exclusive. Thus, A, B, and C are mutually exclusive if $A \cap B = \varnothing$, $A \cap C = \varnothing$, and $B \cap C = \varnothing$.

Two or more events are said to be **exhaustive** if their union *must* occur; thus, A and $\bar{A}$ are exhaustive events, since $A \cup \bar{A} = S$. If N events A_1, A_2, $\cdots$, A_N are *both mutually exclusive and exhausitive*, we say that they form an *N-fold* **partition** of the sample space S. Note that A and $\bar{A}$ form a two-fold partition of S.

Let us take a concrete example of some events. Suppose that we had a list containing the name of every living person in the United States. We close our eyes, point a finger at some spot on the list, and choose one person to observe. The elementary event is the actual person we see as the result of that choice, and the set S is the total set of possible persons. Suppose that the event A is the set "female," B is the set "red headed," among this total set of persons. If our chosen person turns out to be female, then event A occurs; if not, event $\bar{A}$. If he turns out to be red headed, this is an occurrence of event B; if not, event $\bar{B}$. If the observation shows up as both female and red headed, $A \cap B$ occurs; if either female or red headed, then event $A \cup B$ occurs. If the person is female but not red headed, this is an occurrence of the event $A - B$. Only one thing is sure; we will observe a person living in the United States, and the event S must occur.

Continuing with the above example, suppose that the event C is the set "under 20 years of age," event D is the set "20 to 39 years of age," and event E is the set "at least 40 years of age." The events C, D, and E are mutually exclusive, since no two of them can occur at once. Also, they are exhaustive, since one of them *must* occur. That is, each living person must fall into one and exactly one of the three classes, or events. Thus, the events C, D, and E form a three-fold partition of S.

2.4 FAMILIES OF EVENTS

In the example just given, the choice of the five events A, B, C, D, and E was completely arbitrary, and the example could have been given just as well with any other scheme for arranging elementary events into event classes. Ordinarily, there is some restricted set of events of interest, but any scheme for arranging elementary events into event classes can be used.

In the next section we are going to define probability in terms of events. However, we want to do this in some way so that probability can be discussed regardless of how the sample space is broken into subsets. This is accomplished by considering all possible subsets of sample space S. Given N elementary events in S, there are exactly 2^N event classes that can be formed, including $\varnothing$ and S. *The set $\mathfrak{A}$ consisting of all possible subsets of the sample space S will be called the family of events* in S. This set of subsets $\mathfrak{A}$ will be the basis for the definition of probability to follow.

2.5 PROBABILITY FUNCTIONS

We are now ready for a formal definition of what is meant by probability. This definition will be given in terms of a probability function, a pairing of each event with a positive real number (or zero), its probability. The definition is the basis for a *mathematical* theory of probability. The question of how this probability is to be *interpreted* will be discussed a little later.

Definition:

Given the sample space S, and the family $\mathfrak{A}$ of events in S, a probability function associates with each event A in $\mathfrak{A}$ a real number $P(A)$, the probability of event A, such that the following axioms are true:

1. $P(A) \geq 0$ for every event A.

2. $P(S) = 1$.

3. If there exists some countable set of events, $\{A_1, A_2, \cdots, A_N\}$, and if these events are all mutually exclusive, then

$$P(A_1 \cup A_2 \cup \cdots \cup A_N) = P(A_1) + P(A_2) + \cdots + P(A_N)$$

(the probability of the union of mutually exclusive events is the sum of their separate probabilities).

In essence, this definition states that paired with each event A is a non-negative number, the probability $P(A)$, and that the probability of the sure event S, or $P(S)$, is always 1. Furthermore, if A and B are any two *mutually exclusive* events in the sample space, the probability of their union, $P(A \cup B)$, is simply the sum of their two probabilities, $P(A) + P(B)$.

It is important to remember at this stage that this is a purely formal definition of probability in terms of a function assigning numbers to sets. *Events* have probabilities, and in order to discuss probability we must always discuss the events to which these probabilities belong. When we speak of the probability that a person in the United States has red hair, we are speaking of the probability number assigned to the event "red hair" in the sample space "all persons in the United States." Similarly, when we say that the probability that a coin will come up heads is .50, we are saying that the number .50 is assigned in the probability function to the event "heads" in the sample space "all possible results of tossing a particular coin."

This may seem to be an extremely unmotivated and arbitrary way to discuss probability. We all know that the word "probability" means more than a mere number assigned to a set, and we shall certainly give these probability numbers meaning in the sections to follow. For now, however, let us simply accept this purely formal definition at face value.

Given our definition of a probability function, we can begin to deduce other features that probabilities of events must have. Now we shall proceed to derive some consequences of the formal axioms included in our definition above; not only will this demonstrate that deductions do, in fact, follow directly from the axioms of this formal mathematical system, but also we will find the elementary rules of probability to be derived extremely useful when we begin to calculate probabilities.

First of all, we will give an informal proof of the following statement:

$$P(\bar{A}) = 1 - P(A). \qquad (2.5.1^*)$$

For any event A, the probability of the complementary event "not A" is 1 minus the probability of A. We proceed as follows: we know from the algebra of sets that A and $\bar{A}$ are mutually exclusive $(A \cap \bar{A} = \varnothing)$, and that $A \cup \bar{A} = S$. Then, by Axiom 2 above,

$$P(A \cup \bar{A}) = P(S) = 1.$$

Furthermore, by Axiom 3, since A and $\bar{A}$ are mutually exclusive,

$$P(A \cup \bar{A}) = P(A) + P(\bar{A}).$$

Then it follows that

$$P(A) + P(\bar{A}) = 1,$$

so that $\qquad\qquad P(\bar{A}) = 1 - P(A)$

and $\qquad\qquad P(A) = 1 - P(\bar{A}).$

In this simple way we have proved an elementary theorem in the formal theory of probability.

Next we will show that

$$0 \leq P(A) \leq 1 \qquad (2.5.2^*)$$

for any event A in the family of events S. That is, **the probability of any event must lie between zero and one inclusive.** Suppose that some event A could be found where $P(A) > 1$; would this lead to a contradiction of one or more of our axioms? If so, then under these axioms no event can have a probability greater than 1. From the theorem just proved above,

$$P(A) + P(\bar{A}) = 1.$$

If $P(A)$ should be greater than 1, then it must be true that $P(\bar{A})$ is less than zero; however, Axiom 1 dictates that *any* event must have a probability greater than or equal to zero. Thus, if $P(A)$ is greater than 1 a contradiction is generated, and this means that any event must have a probability lying between zero and 1 inclusive.

The third theorem we shall prove states that

$$P(\varnothing) = 0. \tag{2.5.3*}$$

The probability of the empty or "impossible" event is zero. To show this we recall that

$$\bar{S} = \varnothing,$$

the set of all elementary events not in S is empty. Also, we know that

$$S \cup \bar{S} = S \cup \varnothing = S.$$

Then, by (2.5.1),

$$P(\varnothing) = 1 - P(S)$$
$$= 0,$$

proving the theorem.

Still another important elementary theorem of probability goes as follows: If $A \subseteq B$, where A and B are two events in S, then

$$P(A) \leq P(B). \tag{2.5.4*}$$

In other words, **if the occurrence of an event A implies that an event B occurs, so that the event class A is a subset of the event class B, then the probability of A is less than or equal to the probability of B.** Here, we start by using a set-theoretic result: if $A \subseteq B$, then

$$A \cap B = A$$

and
$$A \cup B = B.$$

Now the set B can be thought of as the union of *two* mutually exclusive sets:

$$B = (A \cap B) \cup (\bar{A} \cap B),$$

or, since A is a subset of B,

$$B = A \cup (\bar{A} \cap B).$$

Then, by Axiom 3,

$$P(B) = P(A) + P(\bar{A} \cap B)$$

and
$$P(B) - P(A) = P(\bar{A} \cap B).$$

Since $P(\bar{A} \cap B)$ cannot be negative, $P(B)$ must be greater than or equal to $P(A)$.

Two immediate and useful consequences of the theorem just proved are the following statements:

For any pair of events, A and B, $P(A \cap B) \leq P(A \cup B)$.   (2.5.5*)

If $A \subseteq B$, then $P(B - A) = P(B) - P(A)$.   (2.5.6*)

The next theorem is actually a most important one for all sorts of probability calculations: given two events A and B, then

$$P(A \cup B) = P(A) + P(B) - P(A \cap B).$$ (2.5.7*)

To prove this, notice that $A \cup B$ can be written in terms of the union of two mutually exclusive sets:

$$A \cup B = A \cup (B - (A \cap B)).$$

By Axiom 3,

$$P(A \cup B) = P(A) + P(B - (A \cap B)).$$

Using (2.5.6),

$$P(B - (A \cap B)) = P(B) - P(A \cap B),$$

and thus

$$P(A \cup B) = P(A) + P(B) - P(A \cap B).$$

In other words, **to find the probability that A or B (or both) occurs, we must know the probability of A, the probability of B, and also the probability that A and B both occur.** Be sure to notice, however, that if A and B are *mutually exclusive* events, so that $P(A \cap B) = 0$, then

$$P(A \cup B) = P(A) + P(B),$$

just as provided by Axiom 3.

The last theorem to be proved here concerns partitions: if $A_1, A_2, \cdots, A_N$ form an N-fold partition of S, then

$$P(A_1) + P(A_2) + \cdots + P(A_N) = 1.$$ (2.5.8*)

In other words, **if N events form a partition of S, then their probabilities must sum to one.** To prove this theorem, recall that if N events form a partition of S, they must be mutually exclusive and exhaustive. Since $A_1, A_2, \cdots, A_N$ are mutually exclusive,

$$P(A_1 \cup A_2 \cup \cdots \cup A_N) = P(A_1) + P(A_2) + \cdots + P(A_N),$$

by Axiom 3. Furthermore, since they are exhaustive,

$$A_1 \cup A_2 \cup \cdots \cup A_N = S.$$

Then, by Axiom 2,

$$P(A_1 \cup A_2 \cup \cdots \cup A_N) = P(S) = 1,$$

and it follows, on substituting $P(A_1) + P(A_2) + \cdots + P(A_N)$ for $P(A_1 \cup A_2 \cdots \cup A_N)$, that

$$P(A_1) + P(A_2) + \cdots + P(A_N) = 1.$$

We have formally deduced several elementary rules of probability from the original three axioms, utilizing set theory. In order to get a good intuitive grasp of these rules, consider a Venn diagram similar to those presented in Chapter 1, but such that the area on the diagram which is covered by any set A corresponds to $P(A)$. Then, by Axiom 2, the area enclosed by the entire Venn diagram representing S is 1. The rules presented above can be seen to be true in terms of the Venn diagram. For example, in Equation (2.5.7), $P(A \cap B)$ is subtracted from $P(A) + P(B)$ in order to avoid "double counting" (that is, the inclusion of the area covered by $A \cap B$ twice—see Figure 2.5.1).

It should be obvious that there is no end to the number of new deductions we could make using the original three axioms and the accumulated theorems such as those just proved. Enormous volumes have been filled with such probability theorems, all deduced in essentially the same way from the same axioms we have been using. Naturally, only some of the most elementary results have been shown here, and these particular results have been chosen because of their simple proofs and because they will be useful for us to have in future sections. The idea to be conveyed, however, is that one can deduce all sorts of consequences which *must* be true of these numbers we call probabilities, if these numbers obey the *formal* definition set forth at the beginning of this section. Not once has it been necessary for us to say what probability "really" means in order to deduce all sorts of rules that probabilities must obey. In short, probability theory can perfectly well be studied strictly as an abstract mathematical system without any interpretation at all.

The simple rules we have deduced and the three basic axioms will all be useful to us in learning to calculate probabilities. In the following sections, we will first discuss the *calculation* of probabilities and then turn to the *interpretation* of these numbers called probabilities.

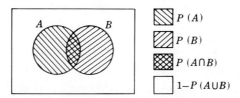

Figure 2.5.1

2.6 A SPECIAL CASE: EQUALLY PROBABLE ELEMENTARY EVENTS

Most simple probability calculations actually rest on the "addition" principle embodied in Axiom 3 of Section 2.5. The elementary events making up the sample space S are conceived as the set of all possible *distinctly different* outcomes of the particular simple experiment. Each such elementary event is an event in its own right, since each constitutes a one-element subset of S. Furthermore, the elementary events are mutually exclusive; if one distinct elementary outcome occurs, no other elementary event may occur. Now under this view, if we somehow knew the probability for each and every elementary event making up the sample space S, then we would also be able to compute the probability of any event in S. Since the elementary events can be regarded as themselves mutually exclusive events, **the probability of any event in S is simply the sum of the probabilities of the elementary events qualifying for that event class;** this is an immediate consequence of Axiom 3. Thus, the probability of any event corresponding to a subset of S can be found provided that the probability associated with each and every elementary event is known.

However, the practical difficulty in actually computing probabilities is in knowing the probabilities of the elementary events themselves. Fortunately, a great many of the simple experiments for which we need to calculate probabilities, particularly games of chance, have a feature that does away with this problem. A great many simple experiments are conducted in such a way that it is reasonable to assume that each and every distinct elementary event has the *same* probability. When tickets are drawn in a lottery, for example, great pains are taken to have the tickets well shaken-up in a tumbler before each one is drawn; this mixing operation makes it reasonable to believe that any particular ticket has the same chance of being drawn as any other. In this case, we say that a ticket is *randomly* selected from the tickets in the tumbler. To select a single item *at random* from a set simply means to select an item in such a manner that all items in the set are equally likely to be selected. (In Chapter 5, we will generalize this definition of random selection to more complex experiments.) Cards are thoroughly shuffled and cut, perfectly balanced dice are thrown in a dice-cage, and, in fact, almost all gambling situations have some feature which makes this equal-chances assumption reasonable. As we shall see presently, even experiments that are not games of chance also may be carried out in such a way that each and every elementary event should have the same probability.

When it can be assumed that each and every one of a finite number of possible elementary events has exactly the same probability, then the probability of any event is particularly easy to compute. That is, imagine a sample space S containing N distinct elementary events. Then, for any event A, let the number of elementary events that are members of A be denoted by $n(A)$. The following principle applies:

If all elementary events in the sample space S have exactly the same probability, then the probability of any event A is given by

$$P(A) = \frac{\text{number of elementary events in } A}{\text{total number of elementary events in } S}$$

$$= \frac{n(A)}{N}. \qquad\qquad (2.6.1^*)$$

It is easy to see that this rule must hold for the probability of any event provided that the various elementary events are equally probable. Suppose that there were N elements in S. Now consider a subset A consisting of exactly two elementary events, $A = \{a_1, a_2\}$. What is the probability of A? Since all such events are equally likely, $P(a_1) = 1/N$ and $P(a_2) = 1/N$. The set A is the union $a_1 \cup a_2$ of two mutually exclusive sets; thus, by Axiom 3, Section 2.5, $P(A) = 1/N + 1/N$, or $2/N$. Proceeding in this way, suppose the set A contained n events. Here $P(A) = 1/N + 1/N + \cdots + 1/N$, or $1/N$ summed n times, so that for any n, $0 \leq n \leq N$,

$$P(A) = \frac{\text{number of elements in } A}{\text{total number of elements}} = \frac{n(A)}{N}.$$

For equally likely elementary events, the probability of an event A is its relative frequency in the sample space.

As an illustration, suppose that a box contains ten marbles. Five of these marbles are white, three are red, and two are black. We perform the simple experiment of drawing a marble out of the box (without looking) in such a way that the marbles are well mixed up, and that there is no reason for any given marble to be favored in our drawing. Now we can identify our experiment with the model of probability we have been discussing. An outcome is the result of our drawing a marble and looking at it. An elementary event is a particular marble, in this instance, and there are exactly ten distinct marbles; hence there are ten elementary events making up the sample space S. The marble events are mutually exclusive, and each has probability $1/10$.

We are concerned with the three events "white," "red," and "black."

Notice that these three events are subsets of S, containing five, three, and two elementary events (marbles), respectively. What is the probability of our drawing a red marble? The answer is given by

$$P \text{ (red)} = \frac{\text{number of red marbles}}{\text{total number of marbles}}$$

$$= \frac{3}{10}$$

so that we may say correctly that the probability of the event "red" in this experiment is .30. In the same way, one can find

$$P \text{ (white)} = .50, \text{ and } P \text{ (black)} = .20.$$

Furthermore, it is easy to see, by applying the rules of Section 2.5, that

P (no color) = .00,
P (red or white) = .30 + .50 = .80,
P (red and white) = .00,
P (red or black) = .30 + .20 = .50,
P (red or white or black) = .30 + .20 + .50 = 1.00,

and so on, for any other event.

Consider another example. A teacher has thirty children in a classroom. In some completely "random" and unsystematic way, the teacher chooses a child to tell his father's occupation. The children are each equally likely to be chosen by the teacher. In this example the elementary events are the different children who might be chosen; there are thirty such elementary events possible, making up the sample space S. Now suppose that there are only four classes of occupations represented in the room: professional, white-collar, skilled labor, and unskilled labor. These four classes will be labeled events A, B, C, and D. If three children represent group A, fifteen, group B, ten, group C, and two, group D, what is the probability that a child chosen will fall into a given group? In other words, what are the probabilities of the four different events? Once again, these probabilities are given by the relative frequencies:

$$P(A) = \frac{\text{number of elementary events in } A}{\text{total number of elementary events}}$$

$$= \frac{3}{30} = .10.$$

In the same way we find:

$$P(B) = \frac{15}{30} = .50,$$

$$P(C) = \frac{10}{30} = .33,$$

and $$P(D) = \frac{2}{30} = .07.$$

Finally, suppose that the experiment in question were that of throwing a fair (that is, unloaded) die, marked with spots from 1 to 6 on its sides. The die is constructed and thrown so that each of the sides is equally likely to appear. There are six distinct outcomes, so that there are six elementary events making up S. The probability of the event "1" is thus 1/6, the event "2" has probability 1/6, and so on. What is the probability of the event "odd number"? An odd number occurs when the die comes up with one, three, or five spots. That is, the desired probability is

P (one or three or five spots) $= P$ (odd number) $= \frac{3}{6} = \frac{1}{2}$.

2.7   COMPUTING PROBABILITIES

Many simple problems in probability can be reduced to problems in counting. For a sample space containing equally probable elementary events, the computation of probability involves two quantities, both of which are *counts of possibilities:* the total number of elementary events, and the number qualifying for a particular event class. The key to solving probability problems is to learn to ask: "How many distinctly different ways can this event happen?" It may be possible simply to list the number of different elementary events that make up the event class in question, but it is often much more convenient to use a rule for finding this number.

There is really no way to become expert in probability computations except by practice in the application of these various counting procedures. Naturally, problems differ widely in the particular principles they involve, but most problems should be approached by these steps:

1. Determine exactly the sample space of elementary events with which this problem deals.

2. Find out how many elementary events make up the sample space. What are all the distinct outcomes that might conceivably occur? If the elementary events are equally probable, then the probability of any single elementary event is one over the total number of elementary events.

3. Decide on the particular events for which probabilities are to be found. How many elementary events qualify for each event class? If no other counting method is available, *list* the elements of these different events and count them.

4. For equally probable elementary events, the probability of any event A is simply the ratio of the number of members of A to the total number of elementary events. Even though elementary events are not equally probable, we can still use the fact that elementary events are mutually exclusive to find the probability of any event A by taking the *sum* of the probabilities of all members of that set.

A game of chance such as roulette shows how probabilities may be computed simply by listing elementary events. A standard roulette wheel has 37 equally spaced slots into which a ball may come to rest after the wheel is spun. These slots are numbered from 0 through 36. Half of the numbers 1 through 36 are red, the others are black, and the zero is generally green.

In playing roulette, you may bet on any single number, on certain groups of numbers, on colors, and so on. One such bet might be on the odd numbers (excluding zero, of course). The only elementary events in the sample space are the numbers 0 through 36 with their respective colors. Which numbers qualify for the event "odd"? It is easy to count these numbers; they form the set

$$\{1, 3, 5, 7, 9, 11, 13, 15, 17, 19, 21, 23, 25, 27, 29, 31, 33, 35\}.$$

Since there are exactly 18 elementary events qualifying as odd numbers, if each slot on the roulette wheel is equally likely to receive the ball, the probability of "odd" is 18/37, slightly less than 1/2.

As a more complicated example that can be solved by listing, let us find the probability that the number contains a "3" as one of the digits. The elementary events qualifying are:

$$\{3, 13, 23, 30, 31, 32, 33, 34, 35, 36\}.$$

Here the probability is 10/37, if the 37 different elementary events are equally likely.

As a final example, consider the probability of a number that is "even and red" on the roulette wheel. These are

$$\{12, 14, 16, 18, 30, 32, 34, 36\}$$

and the probability is 8/37.

There is no end to the examples that might be brought in at this point to show this counting principle in operation, but there is little point in spending more time with it. Just remember that almost all simple problems in probability can be solved in this same way. The important thing to understand is, "When in doubt, make a list." This will usually work, although, as we shall see, other counting methods are often more efficient.

2.8 SEQUENCES OF EVENTS

In many problems, an elementary event may be *the set of outcomes of a series or sequence of observations.* Suppose that any trial of some simple experiment must result in one of K mutually exclusive and exhaustive events, $\{A_1, \cdots, A_K\}$. Now the experiment is *repeated* N times. This leads to a sequence of events: the outcome of the first trial, the outcome of the second, and so on in order through the outcome of the Nth trial. The outcome of the whole series of trials might be the sequence $(A_3, A_1, A_2, \cdots, A_3)$. This denotes that the event A_3 occurred on the first trial, A_1 on the second trial, A_2 on the third, and so on. The place in order gives the trial on which the event occurred, and the symbol occupying that place shows the event occurring for that trial. For example, it might be that the simple experiment is drawing one marble at a time from a box, with replacement after each drawing. The marbles observed may be red, white, or black. For four drawings a sequence of outcomes might be (R, W, B, W), or perhaps (W, W, B, B).

The idea of a sequence also extends to a series of *different* simple experiments. Here the place in the sequence corresponds to the particular experiment being performed. For example, suppose that on the first trial, the experimenter draws a marble from a box, on the second he flips a coin, and on the third he tosses a die. Then the sequence of events observed might be (R, Heads, 6) for example, where the position describes the particular simple experiment, and the symbol in that position tells the observed event.

Remember that each possible sequence is an elementary event if one thinks of the *experiment itself* as the series of trials. Each *experiment* has as its outcome one and only one sequence. Such experiments producing sequences as outcomes are very important to statistics, and so we will begin our study of counting rules by finding how many different sequences a given series of trials could produce. It should be mentioned that the six counting rules to be discussed in the next five sections are merely convenient methods for counting certain possibilities. The rules are not to be interpreted in terms of probabilities, although they will prove useful in the computation of certain probabilities, as we shall see in Section 2.14.

2.9   NUMBER OF POSSIBLE SEQUENCES FOR N TRIALS: COUNTING RULE 1

Suppose that a series of N trials were carried out, and that on each trial any of K events might occur. Then the following rule holds:

Counting Rule 1: If any one of K mutually exclusive and exhaustive events can occur on each of N trials, then there are K^N different sequences that may result from a set of trials.

As an example of this rule, consider a coin's being tossed. Each toss can result only in an H or a T event ($K = 2$). Now the coin is tossed five times, so that $N = 5$. The total number of *possible* results of tossing the coin five times is $K^N = 2^5 = 32$ sequences. Exactly the same number is obtained for the possible outcomes of tossing five coins simultaneously, if the coins are thought of as numbered and a sequence describes what happens to coin 1, to coin 2, and so on.

As another example, the outcome of tossing two dice is a sequence: one number comes up on the first die, and one number comes up on the second. Here $K = 6$ (six different numbers per die), and $N = 2$. There are exactly $6^2 = 36$ different sequences possible as results of this experiment.

2.10   COUNTING RULE 2

Sometimes the number of possible events in the first trial of a series is different from the number possible in the second, the second different from the third, and so on. It is obvious that if there are different numbers K_1, K_2, $\cdots$, K_N of events possible on the respective trials, then the total number of sequences will not be given by Rule 1. Instead, the following rule holds:

Counting Rule 2: If K_1, $\cdots$, K_N are the numbers of distinct events that can occur on trials 1, $\cdots$, N in a series, then the number of different sequences of N events that can occur is $(K_1)(K_2)\cdots(K_N)$.

For example, suppose that for the first trial you toss a coin (two possible outcomes) and for the second you roll a die (six possible outcomes). Then the total number of different sequences would be $(2)(6) = 12$.

Notice that Counting Rule 1 is actually a special case of Rule 2. If the same number K of events can occur on any trial, then the total number of sequences is K multiplied by itself N times, or K^N.

2.11 COUNTING RULE 3: PERMUTATIONS

A rule of extreme importance in probability computations concerns the number of ways that objects may be arranged in order. The rule will be given for arrangements of objects, but it is equally applicable to sequences of events:

Counting Rule 3: The number of different ways that N distinct things may be arranged in order is $N! = (1)(2)(3) \cdots (N-1)(N)$ (where $0! = 1$). An arrangement in order is called a permutation, so that the total number of permutations of N objects is $N!$. The symbol $N!$ is called "N factorial."

As an illustration of this rule, suppose that a classroom contained exactly 10 seats for 10 students. How many ways could the students be assigned to the chairs? Any of the students could be put into the first chair, making 10 possibilities for chair 1. But, given the occupancy of chair 1, there are only 9 students for chair 2; the total number of ways chair 1 and chair 2 could be filled is $(10)(9) = 90$ ways. Now consider chair 3. With chairs 1 and 2 occupied, 8 students remain, so that there are $(10)(9)(8) = 720$ ways to fill chairs 1, 2, and 3. Finally, when 9 chairs have been filled there remains only 1 student to fill the remaining place, so that there are

$$(10)(9)(8)(7)(6)(5)(4)(3)(2)(1) = (10)! = 3628800$$

ways of arranging the 10 students into the 10 chairs.

Now suppose that there are only N different events that can be observed in N trials. Imagine that the occurrence of any given event "uses up" that event for the sequence, so that each event may occur *once and only once* in the sequence. In this case there must be $N!$ different orders in which these events might occur in sequence. Each sequence is a permutation of the N possible events.

To take a homely example, suppose that a man is observed dressing. At each point in our observation there are three articles of clothing he might put on—his shoes, his pants, or his shirt. However, each article can be put on only once. We might observe the following sequences:

> (shoes, pants, shirt)
> (shoes, shirt, pants)
> (shirt, shoes, pants)
> (shirt, pants, shoes)
> (pants, shoes, shirt)
> (pants, shirt, shoes).

The man's putting on any one of these articles of clothing literally uses up that outcome for any sequence of observations, so that there are $N! = (1)(2)(3) = 6$ possible permutation sequences.

For another example, suppose that a teacher has the names of five students in a hat. She draws the names out at random one at a time, *without* replacement. Then there are exactly 5! or 120 different sequences of names that she might observe. If all sequences of names are equally likely to be drawn, then the probability for any one sequence is 1/120.

What is the probability that the name of any given child will be drawn first? Given that a certain child is drawn first, the order of the remainder of the sequence is still unspecified. There are $(N - 1)! = 4! = 24$ different orders in which the other children can appear. Thus the probability of any given child being first in sequence is $24/120 = .20$ or 1/5.

Suppose that exactly *two* of the names in the hat are girls'. What is the probability that the first two names drawn belong to the girls? Given the first two names are girls', there are still 3! = 6 ways three boys may be arranged. The girls' names may themselves be ordered in two ways. Thus, the probability of the girls' names being drawn first and second is $(2)(6)/120 = 1/10$.

2.12   COUNTING RULE 4: ORDERED COMBINATIONS

Sometimes it is necessary to count the number of ways that r objects might be selected from among some N objects in all $(r \leq N)$. Furthermore, each different *arrangement* of the r objects is considered separately. Then the following rule applies:

Counting Rule 4: The number of ways of selecting and arranging r objects from among N distinct objects is

$$\frac{N!}{(N - r)!}.$$

The reasoning underlying this rule becomes clear if a simple example is taken. Consider a classroom teacher, once again, who has 10 students to be assigned to seats. This time, however, imagine that there are only 5 seats. How many different ways could the teacher select 5 students and arrange them into the available seats? Notice that there are 10 ways that the first seat might be filled, 9 ways for the second, and so on, until seat 5 could be filled in 6 ways. Thus there are

$$(10)(9)(8)(7)(6)$$

ways to select students to fill the 5 seats. This number is equivalent to

$$\frac{10!}{5!} = \frac{N!}{(N - r)!},$$

the number of ways that 10 students out of 10 may be selected and arranged, but divided by the number of arrangements of the 5 *unselected* students in the 5 *missing* seats.

As an example of the use of this principle in probability calculations, consider the following: in lotteries it is usual for the first person whose name is drawn to receive a large amount, the second some smaller amount, and so on, until some r prizes are awarded. This means that some r names are drawn in all, and the *order* in which those names are drawn determines the size of the prizes awarded to individuals. Suppose that in a rather small lottery 40 tickets had been sold, each to a different person, and only 3 were to be drawn for first, second, and third prizes. Here $N = 40$ and $r = 3$. How many different assignments of prizes to persons could there be? The answer, by Counting Rule 4, is

$$\frac{40!}{(40-3)!} = \frac{40!}{37!} = (38)(39)(40) = 59{,}280.$$

On how many of these possible sequences of winners would a given person, John Doe, appear as first, second, *or* third prize winner? If he were drawn first, the number of possible selections for second and third prize would be $39!/37! = 1482$. Similarly, there would be 1482 sequences in which he could appear second, and a like number where he could appear third. Thus, the probability that he appears in a sequence of three drawings, winning first, second, or third prize is (by Axiom 3, Section 2.5)

$$P\text{ (first, second, or third)} = \frac{3(1482)}{59{,}280} = \frac{3}{40}.$$

2.13   COUNTING RULES 5 AND 6: COMBINATIONS

In a very large class of probability problems, we are not interested in the *order* of events, but only in the number of ways that r things could be selected from among N things, *irrespective of order*. We have just seen that the total number of ways of selecting r things from N and ordering them is $N!/(N-r)!$ by Rule 4. Each set of r objects has $r!$ possible orderings, by Rule 3. A combination of these two facts gives us

Counting Rule 5: The total number of ways of selecting r distinct combinations of N objects, irrespective of order, is

$$\frac{N!}{r!(N-r)!} = \binom{N}{r}.$$

The symbol $\binom{N}{r}$ is *not* a symbol for a fraction, but instead denotes the number of combinations of N things, taken r at a time. Sometimes the number of combinations is known as a "binomial coefficient," and occasionally $\binom{N}{r}$ is replaced by the symbols $^N C_r$ or $_N C_r$. However, the name and symbol introduced in Rule 5 will be used here.

It is helpful to note that

$$\binom{N}{r} = \binom{N}{N-r}.$$

Thus, $\binom{10}{3} = \binom{10}{7}$, $\binom{50}{49} = \binom{50}{1}$, and so on.

As an example of the use of this rule, suppose that a total of 33 men were candidates for the board of supervisors in some community. Three supervisors are to be elected at large; how many ways could 3 men be selected from among these 33 candidates? Here, $r = 3$, $N = 33$, so that

$$\binom{N}{r} = \frac{(33)!}{3!(30)!} = \frac{(1\cdot 2\cdots\cdot 32\cdot 33)}{(1\cdot 2\cdot 3)(1\cdot 2\cdots\cdot 29\cdot 30)}.$$

Canceling in numerator and denominator, we get

$$\binom{33}{3} = \frac{(31)(32)(33)}{6} = (31)(16)(11) = 5456.$$

If all sets of three men are equally likely to be chosen, the probability for any given set is $1/5456$.

Because of their utility in probability calculations, a table of $\binom{N}{r}$ values for various values of N and r is included in the Appendix. Although this table shows values of N only up to 20, and values of r up to 10, other values can be found by the relation given above, and also by the relation known as Pascal's rule:

$$\binom{N}{r} = \binom{N-1}{r-1} + \binom{N-1}{r}.$$

Still other values may be worked out from the table of factorials also included in the Appendix.

Rule 5 can be interpreted as the total number of ways of dividing a set of N objects into two subsets of r and $(N - r)$ objects, since the determination of the first subset of r objects uniquely determines the second subset of $(N - r)$ objects, which consists of the objects remaining after the first subset has been selected. This can be generalized to the situation in which the set of N objects is to be divided into K subsets, where $K \geq 2$:

Counting Rule 6: The total number of ways of dividing a set of N objects into K mutually exclusive and exhaustive subsets

of n_1, n_2, $\cdots$, n_{K-1}, and n_K objects, respectively, where

$$n_1 + n_2 + \cdots + n_K = N,$$

is

$$\frac{N!}{n_1!n_2!\cdots n_K!}.$$

Here the denominator is equal to the product of the $n_i!$'s. It can readily be seen that Rule 5 is a special case of Rule 6, with $K = 2$, $n_1 = r$, and $n_2 = N - r$.

The proof of Rule 6 requires successive applications of Rule 5. Suppose we select the first subset of n_1 objects from the N objects, then select the second subset of n_2 objects from the remaining $(N - n_1)$ objects, then select the third subset of n_3 objects from the remaining $[N - (n_1 + n_2)]$ objects, and so on. But, from Rules 5 and 2, the number of ways in which we can do this is

$$\binom{N}{n_1}\binom{N - n_1}{n_2}\binom{N - (n_1 + n_2)}{n_3}$$

$$\cdots\binom{N - (n_1 + n_2 + \cdots + n_{K-2})}{n_{K-1}}\binom{N - (n_1 + n_2 + \cdots + n_{K-1})}{n_K}.$$

Rewriting this in terms of factorials, we get

$$\frac{N!}{n_1!(N - n_1)!}\cdot\frac{(N - n_1)!}{n_2!(N - (n_1 + n_2))!}\cdot\frac{(N - (n_1 + n_2))!}{n_3!(N - (n_1 + n_2 + n_3))!}$$

$$\cdots\frac{(N - (n_1 + n_2 + \cdots + n_{K-2}))!}{n_{K-1}!(N - (n_1 + \cdots + n_{K-1}))!}\cdot\frac{(N - (n_1 + \cdots + n_{K-1}))!}{n_K!(N - (n_1 + \cdots + n_K))!}.$$

Canceling like terms, we arrive at the following result:

$$\frac{N!}{n_1!n_2!\cdots n_K!(N - (n_1 + n_2 + \cdots + n_K))!}.$$

But since $n_1 + n_2 + \cdots + n_K = N$, the last term in the denominator is equal to $0!$, which is defined to be 1, so that the required number of ways is

$$\frac{N!}{n_1!n_2!\cdots n_K!}.$$

As an example of the use of Rule 6, consider an extension of the previous example. Suppose that a total of 33 men were candidates for the board of supervisors. Three supervisors are to be elected at large, with the fourth and fifth highest vote-getters being designated as alternates. How many ways can the 33 candidates be divided into 3 commissioners, 2 alternates, and 28 remaining candidates? Here $N = 33$, $K = 3$, $n_1 = 3$, $n_2 = 2$, and $n_3 = 28$, so that the number of combinations is

$$\frac{N!}{n_1!n_2!n_3!} = \frac{33!}{3!2!28!} = \frac{(1 \cdot 2 \cdot \cdots \cdot 32 \cdot 33)}{(1 \cdot 2 \cdot 3)(1 \cdot 2)(1 \cdot 2 \cdot \cdots \cdot 27 \cdot 28)}$$

$$= \frac{(29 \cdot 30 \cdot 31 \cdot 32 \cdot 33)}{6 \cdot 2} = (29 \cdot 5 \cdot 31 \cdot 16 \cdot 33) = 2,373,360.$$

2.14   SOME EXAMPLES: POKER HANDS

These six rules provide the basis for many calculations of probability when the number of elementary events is finite, especially when sampling is without replacement. One of the easiest examples of their application is the calculation of probabilities for poker hands.

The game of poker as discussed here will be highly simplified and not very exciting to play; the player simply deals five cards to himself from a well-shuffled standard deck of fifty-two cards. Nevertheless, the probabilities of the various hands are interesting and can be computed quite easily.

The particular hands we will examine are the following:

(a)   one pair, with three different remaining cards
(b)   full house (three of a kind, and one pair)
(c)   flush (all cards of the same suit).

First of all, we need to know how many different *hands* may be drawn. A given hand of five cards will be thought of as an elementary event; notice that the order in which the cards appear in a hand is immaterial. By Rule 5, since there are 52 different cards in all and only 5 are selected, then there are

$$\binom{52}{5} = \frac{52!}{5!47!}$$

different hands that might be drawn. If all hands are equally likely, then the probability of any given one is $1/\binom{52}{5}$.

There are 13 numbers that a pair of cards may show (counting the picture cards), and each member of a pair must have a suit. Thus, by Rule 5 there are $13\binom{4}{2}$ different pairs that might be observed. The remaining cards must show three of the twelve remaining numbers and each of the three cards may be of any suit. By Rules 1 and 5 there are $\binom{12}{3}$ (4) (4) (4) ways of filling out the hand in this way. Multiplying, we find that there are

$$13\binom{4}{2}\binom{12}{3}4^3$$

different ways for this event to occur. Thus,

$$P \text{ (one pair)} = \frac{13\binom{4}{2}\binom{12}{3}4^3}{\binom{52}{5}}.$$

This number can be worked out by writing out the factorials, canceling in numerator and denominator, and dividing. It is approximately equal to .42. The chances are roughly four in ten of drawing a single pair in five cards, if all possible hands are equally likely to be drawn.

This same scheme can be followed to find the probability of a full house. There are thirteen numbers that the three of a kind may have, and then twelve numbers possible for the pair. The three of a kind must represent three of four suits, and the pair two of four suits. This gives

$$13\binom{4}{3}12\binom{4}{2}$$

different ways to get a full house. The probability is

$$P \text{ (full house)} = \frac{13\binom{4}{3}12\binom{4}{2}}{\binom{52}{5}},$$

which is about .0014.

Finally, a flush is a hand of five cards all of which are in the same suit. There are exactly four suit possibilities, and a selection of five out of thirteen numbers that the cards may show. Hence, the number of different flushes is $4\binom{13}{5}$, and

$$P \text{ (flush of five cards)} = \frac{4\binom{13}{5}}{\binom{52}{5}},$$

or about .002.

The probabilities of the various other hands can be worked out in a similar way. The point of this illustration is that it typifies the use of these counting rules for actually figuring probabilities of complicated events, such as particular poker hands. Naturally, a great deal of practice is usually necessary before one can visualize and carry out probability calculations from "scratch" with any facility. Nevertheless, many probability calculations depend upon counting how many ways events of a certain kind can occur.

2.15   THE FREQUENCY INTERPRETATION OF PROBABILITY

Each of the examples in Sections 2.6–2.14 illustrates how the outcomes of an experiment are identified with the elementary events in a sample space, and how probabilities may be computed. However, we have still not really answered the question of what these probabilities mean. What does it mean to say that the probability of drawing a red marble from the box described in Section 2.6 is 3/10? What does it mean to say that the probability of the event "odd number" in one throw of a die is 1/2? What does it mean to say that the probability of obtaining a full house in a five-card poker hand is .0014? In this section, we will consider one possible interpretation of probability—the **long-run frequency interpretation.** Later in the chapter, we will discuss a second possible interpretation—the **subjective,** or **personalistic interpretation.**

In the first of the examples mentioned in the preceding paragraph, if one were to keep drawing marbles out of the box, and after each draw were to replace the marble in the box before drawing again, then in the *long run,* after very many observations, what *proportion of the marbles drawn should be red*? It seems reasonable that one should expect about 3 in 10 of such observations to be red. In the same way, in the second example, if one threw the die repeatedly, then it seems reasonable that in the long run, over a large number of such observations, about 1/2, or one in two of the throws should result in an odd number.

Finally, in the poker example, one should expect a full house about 14 times in 10,000 in the long run. In these examples, we have interpreted the **probability of an event** as the **relative frequency of occurrence of that event to be expected in the long run.**

Simple examples of this sort all involve equally likely elementary events, and in many applications of probability theory, especially to games of chance, this assumption is made. On the other hand, as we demonstrate in the next section, the same idea applies even when elementary events are *not* equally likely: **in the long run, if the experiment of interest is repeated many times under identical conditions, the relative frequency of occurrence approaches the probability for any event.**

2.16 BERNOULLI'S THEOREM

The connection between probability and long-run relative frequency is a simple and appealing one, and it does form a tie between that basically undefined notion, "the probability of an event," and something we can actually observe, a relative frequency of occurrence. A statement of probability tells us what to *expect* about the relative frequency of occurrence of an event given that enough sample observations are made at random. In the long run, the relative frequency of occurrence of event X should approach the probability of this event, if independent trials are made at random over an indefinitely long sequence. As it stands, this idea is a familiar and reasonable one, but it will be useful to have a more exact statement of this principle for use in future work. The principle was first formulated and proved by James Bernoulli in the early eighteenth century, and it often goes by the name "Bernoulli's theorem." A more or less precise statement of Bernoulli's theorem goes as follows:

If the probability of occurrence of the event X is p, and if N trials are made, independently and under exactly the same conditions, then the probability that the relative frequency of occurrence of X differs from p by any amount, however small, approaches zero as the number of trials grows indefinitely large.

The theorem may also be represented by the following mathematical expression:

$$P\left(\left|\frac{f}{N} - p\right| > \epsilon\right) \to 0, \text{ as } N \to \infty, \qquad (2.16.1)$$

where f stands for the frequency of occurrence of event X among the N trials, p is the probability of X, ϵ is some arbitrarily small positive number, and $N \to \infty$ indicates that the number of trials increases until it approaches an infinite number. The mathematical statistician says that the proportion f/N approaches the value p "in probability" as N becomes indefinitely large.

In effect, the theorem says this: imagine some N trials made at random and in such a way that the outcome on any one trial cannot possibly influence the probability of outcomes on any other trial. That is, N independent trials are made. This act of taking N independent trials of the same simple experiment may itself be regarded as another experiment in which the possible outcomes are the numbers of times event X occurs out of N trials. Each possible such outcome has a probability. Then the most probable result of carrying out N trials is a proportion of occurrences of X that is very nearly the value of p for one trial (equal to p when that proportion can occur; see Section 4.8). Although this is the most probable outcome of N trials of the same experiment, this does not mean, however, that the proportion of X occurrences among any N trials *must* be p; the proportion actually observed might be any number between 0 and 1. Nevertheless, given more and more trials, the relative frequency of X occurrences may be expected to come closer and closer to p.

In one way, this principle is what people have in mind when they refer to "the law of averages." Most of us have the intuition that the relative number of times an event occurs over repeated opportunities reflects its probability. Given enough experimental trials or sample observations from the sample space, we expect the relative frequency of occurrence to balance out to equal the probability.

However, *do not fall into the error of thinking that an event is ever due to occur on any given trial.* If you toss a coin 100 times, you expect about 50 percent of the tosses to show the event "heads," since the theoretical probability of that event for a fair coin is .50. This does not mean, however, that for any given one hundred trials (or one thousand, or million, or billion trials) the coin *must* show 50 percent heads. This need not be true at all. Every one of your 100 tosses *could* result in the event heads, or none of them might result in this event—the coin does not ever have to come up heads in any finite number of tosses. Only in an infinite number of tosses must the relative frequency equal .50. The only thing which we can say with assurance is that .50 is the relative frequency of heads we should *expect* to observe in any given number N of tosses, and that it is increasingly probable that we observe close to 50 percent heads as the N grows larger. But on any finite number of tosses of the coin, the relative frequency of heads observed can be anything. This same is true for the occurrence of any event in any simple experiment in which observations are made at random and in which the probability p of the event X is other than 1 or 0.

The above argument can be restated in slightly different terms; the key point is the independence of trials, which essentially means that the outcome on any single trial has no bearing on the outcomes on any other trials. In terms of tossing a fair coin, we may say that the fact that it comes up heads on one toss bears no relationship to the outcome of the previous toss, the next toss, or any other toss. Thus, an event cannot be thought of as "due" to occur. But if this is the case, how does Bernoulli's theorem allow for unusual occurrences such as 100 heads in a row? It does so not by compensating for the 100 heads in a row with 100 tails in a row, but rather by a "swamping" procedure, which works as follows. If the first 100 tosses of a sequence are heads, this is an unusual event (provided that the coin is fair). Suppose, however, that in the next 1000 trials, we observe 500 heads and 500 tails. For all 1100 tosses, we have 600 heads and 500 tails, which does not seem as unusual as 100 heads and 0 tails. If we tossed the coin a million more times, with 500,000 heads and 500,000 tails, the total number of heads and tails would be 500,600 and 500,500. This result does not seem too unusual to us at all. As the sample size has increased, the original "run" of 100 straight heads has been "swamped" by the subsequent trials, even though the subsequent trials resulted in exactly 50 percent heads and 50 percent tails.

It is interesting to note at this point that in some situations we might have cause to believe the *opposite* of the "event is due to occur" argument. If we met a stranger in a bar and he produced a coin and proceeded to toss 100 straight heads, we would *not* think that a tail was "due". We would be much more likely to suspect that the coin was not fair and that the probability of a head on a single toss was greater than 1/2 (if the coin were two-headed, the probability of a head would be 1). This amounts to questioning the assumption of a fair coin, and demonstrates an important use of Bernoulli's theorem: if we have no idea what the probability of a particular event is, and we observe a large number of independent trials, we can with some confidence claim that the probability of the event is close to the relative frequency of the event in the sequence of trials.

Although it is true that the relative frequency of occurrence of any event must exactly equal the probability only for an infinite number of independent trials, this point must not be overstressed. Even with relatively small numbers of trials we have very good reason to expect the observed relative frequency to be quite close to the probability. The rate of convergence of the relative frequency to the probability is very rapid, even at the lower levels of sample size, although the probability of a small discrepancy between relative frequency and probability is much smaller for extremely large than for extremely small samples. A probability is not a curiosity requiring unattainable conditions, but rather a value that can be estimated with considerable accuracy from a sample. A good bet about the probability of an event is the actual relative frequency we have observed from some N trials, and the larger N is, the better the bet.

In a later section, we will actually prove the main features of Bernoulli's theorem, but this must wait until Chapter 5. As yet, we lack the vocabulary to give a precise meaning to terms, such as "at random" and "independent events," that figure in this theorem (recall that in Section 2.6 we defined "at random" for the case in which there was only a single trial, not for the case with N trials). Nevertheless, accepting this principle at face value for the moment, we can show a simple example of its use.

2.17 AN EXAMPLE OF SIMPLE STATISTICAL INFERENCE

Imagine that we were faced with the following problem: we are given a coin and asked to decide if the coin is fair (that is, if the probability of heads is .50). This is to be decided through the repeated simple experiment of tossing the coin. If the coin actually is fair, then in an infinite number of tosses the relative frequency of the event heads *must* occur with probability of exactly .50. However, on any given number of tosses, we should only *expect* a relative frequency of .50 to occur. The larger the number of tosses, the more nearly should we expect the relative frequency to approach .50.

Now we toss the coin four times; suppose that four heads occur. This would cast only slight doubt upon the fairness of the coin, since it is relatively likely that the relative frequency of heads will depart widely from .50 on such a small number of trials. However, if after 500 tosses, the number of heads were 500, we would have very serious doubts about the fairness of the coin; although this *could* happen, it is far from what one *expects*. Nevertheless, we still could not be sure. If we tossed the coin 500 million times, still with 100 percent heads, we could be practically certain that the probability of a head for this coin is not .50—*practically* certain, that is, but still not *completely* certain. Only if we tossed the coin an infinite number of times would we be absolutely certain that the coin was not fair, given that the long-run relative frequency differed from .50 in any way.

This hypothetical example exhibits one of the basic features of statistical inference as applied to many real problems. One starts out with a theoretical state of affairs that can be represented by a statement of the probability for one or more events. He wishes to use empirical evidence to check the truth of that theoretical situation. Therefore he conducts an experiment in which the event or events in question are possible outcomes, and he takes some N observations at random. The theoretical probabilities tell him what to expect with regard to the relative frequencies of the events in question, and he compares the obtained relative frequencies against those expected. To the extent that the obtained relative frequencies of events depart widely from the expected, he has some evidence that the theory is not true. Moreover, given any degree of departure of the observed from the expected relative frequencies, he is more certain that the theory is not true the larger the sample N. However, *he can never be completely sure in his judgment that the theory is false unless he has made an infinite number of observations.*

By the same token, he can never be completely sure that the theory is true. An extremely biased coin (short of a coin which must come up heads with probability 1 or tails with probability 1) *can* give exactly 50 percent heads on any set of N trials. Even though the evidence appears to agree well with the theory, this in no way implies that the theory must be true. Only after an infinite number of trials, when the relative frequencies of occurrence of events matches the theoretical probability exactly, can one assert with complete confidence that the theory is true. Increased numbers of trials lend confidence to our judgments about the true state of affairs, but we can always be wrong, short of an infinite number of observations. A point of interest is that although the "theory" in this example was given as "the probability of heads is .50," it is doubtful that anyone would believe that the probability is *exactly* .50. We are actually interested in whether the probability is *near* .50, where by "near" we may mean, for example, that the probability is between .49 and .51, or between .495 and .505. This point is of great importance in the theory of statistical inference and decision, and we will discuss it at length in Chapters 7–9.

The principle embodied in Bernoulli's theorem can be used not only to compare empirical results with theoretical probabilities, but also to let us *estimate* true probabilities of events by observing their relative frequencies over some limited number of trials. For example, suppose that once again we had a box containing marbles of different solid colors. We do not know how many marbles there are in the box, and we do not know how many different colors, but our task is to find out the probability of each color. We conduct the simple experiment of mixing up the marbles well, drawing one out, observing its color, and putting it back (random sampling with replacement). Now suppose that the first marable is white. The observed relative frequencies of colors are white 1.00, other .00. We wish to decide about the probabilities in such a way that our estimate would, if correct, have made the occurrence of this particular sample result have the greatest prior probability. What probability of the event "white" would make drawing a white marble most likely? The answer is $P(\text{white}) = 1.00$ and $P(\text{other}) = .00$, so that this is our first estimate of the probabilities of the colors of the marbles.

However, we now draw a second marble at random with replacement. This marble turns out to be red, and our best guess about the probabilities is now white .50, red .50, and other .00, as these probabilities would make the occurrence of the actual sample of two marbles most likely. If we kept on drawing, observing, and replacing marbles for 100 trials, shaking the box so that the marbles (elementary events) are equally likely to be chosen on a trial, we might begin to get a reasonably clear picture of the probabilities of the colors:

Color	Relative frequency
White	.24
Red	.50
Blue	.26

By the time we had made 10,000 observations, we could be even more confident that the probabilities are close to the relative frequencies:

Color	Relative frequency
White	.24
Red	.50
Blue	.24
Green	.02

and so on. Given an infinite number of trials, we could specify the relative frequencies (that is, the probabilities) of the marbles in the box precisely. The larger the number of observations, the less do we expect to "miss" in our estimates of what the box contains. However, for fewer than an infinite number of observations, the observed relative frequencies *need not* reflect the probabilities of the colors exactly, though they may.

It is important to note that since we are drawing samples *with replacement* in this situation, the fact that there may be only some finite numbers of distinct elementary events (marbles in this case) in no way prevents us from taking a very large or even an infinite number of sample observations with replacement after each trial. Thus Bernoulli's theorem applies perfectly well to situations in which there may be a very small number of distinct elementary events possible, so long as one can draw an unlimited number of samples or make an unlimited number of trials of the experiment (as in coin tossing or die rolling).

The statistician making observations is doing something quite similar to drawing marbles from a box. He cannot see "into the box" and observe all such phenomena about which he wishes to generalize, but he can observe samples, drawn from among all such elementary events, and generalize from what he actually observes. His generalization is much like a bet, and how good the bet is depends to a great extent on how many sample observations he makes. He can never be completely sure that his generalization is the correct one, but he can make the chance of being wrong very small by making a sufficient number of observations.

2.18 THE SUBJECTIVE INTERPRETATION OF PROBABILITY

Long-run relative frequency is but one interpretation that can be given to the formal notion of probability. It is important to remember that this is an *interpretation* of the abstract model. The model per se is a system of relations among and rules for calculating with numbers that happen to be called probabilities. There is no "true meaning" of probability, any more than there is a true meaning of the symbol x as used in algebra. Probability is an abstract mathematical concept that takes on meaning in the ordinary sense only when it is identified with something in our experience, such as the relative frequency of real events.

The probability concept acquired its interpretation as relative frequency because it was originally developed to describe certain games of chance where plays (such as spinning a roulette wheel or tossing dice or dealing cards) are indeed repeated for very many trials. Similarly, there are situations in which the statistician makes many observations under the same conditions, and so the mathematical theory of probability can be given a relative frequency interpretation in this case as well.

On the other hand, there are many events which can be thought about in a probabilistic sense but which cannot have a probability in the relative frequency interpretation. Indeed, an everyday use of the probability concept does not have this relative frequency connotation at all. We say "It will probably rain tomorrow," or "The Packers will probably win the championship," or "I am unlikely to pass this test." Extending this notion, we might say that "The chances are two in three that it will rain tomorrow," or "The odds are even that the Packers will win the championship," or "I only have about one chance in ten of passing this test." These statements appear to be probability statements, and their meanings should be clear to most listeners, but it is very difficult to see how they describe long-run relative frequencies of outcomes of simple experiments repeated over and over again. Each statement describes the speaker's **degree of belief** about an event that will occur once and once only. In each case it will not be possible to observe repetitive trials of the experiment, so the probability statements cannot be explained in terms of the long-run frequency interpretation of probability.

An interpretation of probability which enables one to explain the probability statements in the preceding paragraph is the **subjective, or personalistic interpretation. In this approach, a probability is interpreted as a measure of degree of belief, or as the quantified judgment of a particular individual.** The probability statements above thus represent the judgments of the individual making the statements. Since a probability, in this interpretation, is a measure of degree of belief rather than a long-run frequency, it is perfectly reasonable to assign a probability to an event which is nonrepetitive. As a result, we can think of a probability as representing the individual's judgment concerning what will happen in a single trial of the event in question in any given situation rather than a statement about what will happen in the long run.

It is not necessary, of course, for an event to be nonrepetitive for the subjective interpretation of probability to be applicable. Consider once more the three examples which we discussed from a long-run frequency viewpoint in Section 2.15: drawing marbles from a box, throwing a die, and dealing a poker hand. In each of these examples, the probability of any particular event could be interpreted as a degree of belief, although the experiments are all capable of being repeated. For instance, an individual might have a certain degree of belief that a full house will occur *on a particular deal* in a game of poker. In the die example, an individual who is willing to accept certain judgmental assumptions (the die is "fair", the method of tossing the die does not favor any side or sides of the die, and so on) would feel that the probability of the die coming up odd *on a particular throw* is one-half. Notice that this is a probability statement about a particular throw, not about a long sequence of throws. In understanding this, it might be helpful to recall that the actual *computation* of the probabilities in the three examples (marbles, die, and poker) was done without reference to any interpretation of probability.

Because they represent degrees of belief, subjective probabilities may vary from person to person for the same event. Different individuals may have different degrees of belief, or judgments. Such differences may be due to differences in background, knowledge, and/or information about the event in question. For example, you might state that the probability of rain tomorrow is 1/10, and the meteorologist at the Weather Bureau might state that the probability of rain tomorrow is 1/2. According to the subjective interpretation of probability, both of these values properly can be thought of as probabilities and subjected to the usual mathematical treatment. Such a difference in probabilities is allowable, provided that you and the meteorologist both feel that your respective probabilities accurately reflect your respective judgments concerning the chance of rain tomorrow. In addition, the quantification of these judgments is not as arbitrary as it might first appear. The consideration of betting odds can be extremely helpful in an attempt to accurately represent such judgments, as we shall see in the next section.

In some respects, subjective probability can be thought of as an *extension* of the frequency interpretation of probability. As we have seen, the use of long-run frequencies is based on certain *assumptions* which are stated in Bernoulli's theorem. One important assumption is that the trials comprising the "long run" are independent. Essentially, this means that the outcome on any one trial will in no way affect the outcome on any other trial. Another assumption is that the trials are conducted under identical conditions. Although in later chapters we will discuss techniques for statistically investigating assumptions such as these, it should be pointed out that the decision as to whether the assumptions seem reasonable in any given situation is ultimately a subjective decision. Thus there is even an element of subjectivity in the long-run frequency interpretation of probability. If a person feels that the assumptions are reasonable, it is perfectly acceptable for him to make his subjective probability for a given event equal to the probability arrived at by the frequency approach. If he does not feel that the assumptions are reasonable or if he has some other information (other than frequencies) about the event in question, his subjective probability may differ from the frequency probability. In this respect (and in the respect that the subjective interpretation allows one to make probability statements about nonrepetitive events), subjective probability can be thought of as an extension of long-run frequency probability.

For example, suppose that you are interested in the probability that a given person will be involved in an automobile accident (as the driver of a car, not as a passenger) during the next year. You know from recent data that among drivers in the United States of the same age as the person of interest, the frequency of drivers involved in one or more accidents in a year is 12/100. If you feel that the given person is representative of drivers in his age group (and this is a subjective judgment on your part), you might be willing to assign the probability .12 to the event that the person will be involved in an accident in the next year. On the other hand, you might feel that he is somewhat reckless, so that the probability should be, say, .20; or you might want to look up the historical frequencies for drivers in the same state (as opposed to all drivers in the United States), or for drivers of a particular type of car, and so on. The point is that relative frequencies may be quite useful in the assessment of probabilities, but the ultimate assessment is subjective, and thus the subjective interpretation of probability can be thought of as an extension of the long-run frequency interpretation.

2.19 PROBABILITY AND BETTING ODDS

The subjective interpretation of probability is sometimes criticized on the grounds that since subjective judgment is involved, the resulting probabilities are not "objective," but are in some respects arbitrary. However, there are "objective" procedures which can take away much of the force of this argument. **If we think of the probabilities as a representation of odds at which a person would be willing to bet, the element of arbitrariness is eliminated.** For example, if you say that "The chances are 2 in 3 that Green Bay will win the championship game of the National Football League," this implies that

$$P \text{ (Green Bay wins championship)} = 2/3$$

and P (Green Bay does not win championship) $= 1/3$.

Thus, Green Bay is twice as likely to win the championship as it is to lose the championship, in your opinion. This means that you think that 2-to-1 odds would be a fair bet on this event:

$$\begin{cases} A \text{ bets \$2 that Green Bay will win.} \\ B \text{ bets \$1 that Green Bay will not win.} \end{cases}$$

Your probabilities indicate that you think that this bet is "fair"; that is, you are indifferent between A's side of the bet and B's side of the bet. Although we cannot at this point show formally why this is true, we can give a brief intuitive explanation. If you take A's side of the bet, you feel that you have a probability of 2/3 of winning \$1 (the amount your opponent, B, is wagering), since that is your probability that Green Bay will win. Similarly, you have a probability of 1/3 of losing \$2 (the amount you, as A, are wagering). But $(2/3)(\$1)$ is equal to $(1/3)(\$2)$, so the amount you expect to gain from the bet is \$0. It turns out that if you take B's side of the bet, the amount you expect to gain is still \$0. As a result, you should be indifferent between the two sides of the bet, and you consider the bet to be "fair". Conversely, only if you are prepared to act as though this is a fair bet can you say that the probabilities are 2/3 and 1/3.

If we think of subjective probabilities in terms of betting odds, we see that they are not really arbitrary, even though they may differ from person to person. The formal relationship between probabilities and odds is as follows: **If the probability of an event is equal to p, then the odds in favor of that event are p to $(1-p)$.** If the chances are two in three that it will rain tomorrow, the odds in favor of rain are 2/3 to 1/3, or 2 to 1. If you only have about one chance in ten of passing a particular test, the odds in favor of your passing the test are 1/10 to 9/10, or 1 to 9.

It is also possible to convert odds into probabilities: **If the odds in favor of an event are a to b, then the probability of that event is equal to $a/(a+b)$.** If the odds are even that the Dow-Jones Averages will rise tomorrow, the odds are a to a, so the probability that the Dow-Jones Averages will rise tomorrow is $a/(a+a)$, or 1/2. If we know the probability of an event, we can calculate the odds in favor of the event, and vice versa.

Other devices, similar in nature to betting odds, have been suggested for use in the quantification of judgments. Examples are hypothetical lotteries and standard gambles. For instance, suppose that you are offered a choice between Lottery A and Lottery B:

Lottery A: You win \$100 with probability 1/2.
                     You win \$0 with probability 1/2.

Lottery B: You win \$100 if it rains tomorrow.
                     You win \$0 if it does not rain tomorrow.

If you choose Lottery B, then you must feel that the probability of rain tomorrow is greater than $1/2$; if you choose Lottery A, then you must feel that this probability is less than $1/2$; if you are indifferent between the two lotteries, then you must feel that the probability of rain tomorrow is equal to $1/2$. These conclusions are based on the assumption that you prefer a greater chance at winning $100 to a lesser chance. Suppose that you choose Lottery A. Now consider the same lotteries, except that the probabilities in Lottery A are changed to $1/4$ and $3/4$. If you still prefer Lottery A, then your subjective probability of rain is less than $1/4$. Presumably we could keep changing the probabilities in Lottery A until you are just indifferent between Lotteries A and B; if this happens when the probabilities are .1 and .9, then your subjective probability of rain is .1. In a similar manner, you can assess your probability for any event.

It should be noted that the use of betting situations and lotteries to assess probabilities depends on certain reasonable "axioms of coherence," or "axioms of consistency." It can be shown that if a person behaves in accordance with these axioms, then his assessed probabilities must satisfy the mathematical definition of probability given in Section 2.5. For example, if a person obeys the axioms of coherence, it is impossible for him to assess the probability of rain tomorrow as $1/2$ and the probability of no rain tomorrow as $3/4$, because these two probabilities must sum to one. Thus, the axioms of coherence justify the application of the usual mathematical manipulations of probabilities to subjective probabilities. We will not present all of the axioms of coherence, but we will give two examples of such axioms. The axiom of *transistivity* states that preferences are transitive. That is, if you prefer X to Y and Y to Z, then you must prefer X to Z. This seems intuitively reasonable. The axiom of *substitutability* states that if you are indifferent between X and Y, then X can be substituted for Y as a prize in a lottery or as a stake in a bet without changing your preferences with regard to the bets or lotteries in question. For example, if you are indifferent between receiving $2 and receiving a ticket to a concert, then the concert ticket can be substituted for the $2 as a stake in the bet (concerning Green Bay) presented earlier in this section without affecting your choice between A's side of the bet and B's side of the bet.

Devices such as betting situations and lotteries have the common property of resulting in statements that can be converted into probabilities. They represent the "behavioral" approach to the quantification of judgment; that is, they force the person who is attempting to determine his subjective probabilities to state what *action* he would take in a particular decision-making situation rather than just to state directly what the probability is. His probabilities can then be inferred from his behavior (that is, his decisions). In this regard, the subjective interpretation of probability should have great applicability in decision-making situations.

The subjective interpretation is also valuable in making decisions because it allows probability statements for nonrepetitive events. Important decisions often depend on the outcomes of nonrepetitive experiments. For example, an investment decision may depend on the probability that Congress will raise taxes or the probability that the stock market will go up in the next year. A decision concerning a price increase for a certain product may depend on the probability that other firms producing the product will refuse to go along with the increase. These examples refer to nonrepetitive events. In the next section, we present an example of the use of probabilities in decision-making, an example in which the subjective interpretation of probability is most useful. A more complete discussion of this area will be presented in Chapters 8 and 9.

2.20   A SIMPLE DECISION-MAKING EXAMPLE

Consider the following problem: you are preparing to leave the house in the morning, and you must decide whether or not to carry an umbrella. You are wearing your new suit, which would be ruined if it got wet. On the other hand, it is a nuisance to have to carry an umbrella with you all day. Should you take the umbrella or not?

This problem is a very simple one—most of us make numerous decisions of this nature every day, and we do so on an intuitive, informal basis. It should be instructive, however, to consider the problem in a more formal manner in order to illustrate the statistical theory of decision-making. Many important decisions are similar in nature to the problem of whether or not to carry an umbrella, and the same formal theory can be used for important decisions and for less important decisions. When the decision is a minor one, such as the umbrella example, we usually do not apply the formal theory because of the time and effort required to do so. For important decisions, the time and effort required to implement the theory may be minor considerations in comparison with the potential gains or losses to be realized from the decision.

In order to apply the formal theory to the umbrella example, it is necessary to specify a few more details of the problem. Suppose that the suit is worth $60, and that it would be completely worthless if it were rained upon. Furthermore, suppose that you feel that the cost of carrying the umbrella, in terms of the inconvenience and the possibility of leaving the umbrella somewhere and thus losing it, is equivalent to $1. This implies that you would be willing to pay $1 to avoid carrying the umbrella, and it demonstrates the consideration of nonmonetary factors in the analysis. You have attempted to express the nonmonetary factors in terms of an equivalent amount of money. The consequences of your decision, which depend on whether or not you carry an umbrella and on whether or not it rains, can be expressed in a *payoff table*, as follows:

Event

	Rain	No rain
Carry umbrella	−$1	−$1
Do not carry umbrella	−$60	$0

Your decision:

The cost of carrying an umbrella is $1, whether it rains or not (a cost of $1 is the same as a payoff of −$1). If you do not carry an umbrella, the cost is $60 if it rains and your suit is ruined, and $0 if it does not rain.

The payoffs in the above table are important factors in your decision. Another important factor is the chance of rain. If you are sure that it will rain, you would certainly carry your umbrella. On the other hand, if you are positive that it will not rain, you would not carry the umbrella. These cases are of little interest because the decision is obvious, so let us consider the case in which you are uncertain as to whether or not it will rain. Your uncertainty can be expressed in terms of probabilities P (Rain) and P (No rain). If you do not have access to a weather report or to information regarding past frequencies of rain for the particular location and time of year, the probabilities will have to be based on your subjective judgments. Suppose that you look out the window and see a few scattered dark clouds. On the basis of this, you assess the probability of rain to be 1/10. This is a subjective, or degree-of-belief probability, and it implies that you feel that the event "No rain" is nine times as likely (9/10 to 1/10) as the event "Rain." It also implies that you would be indifferent between Lotteries A and B in the preceding section if the probabilities in Lottery A were 9/10 and 1/10.

Formal decision theory prescribes certain criteria for making decisions based on inputs regarding the possible consequences, or payoffs, and inputs regarding the probabilities for the various possible events. We will discuss these criteria in a later chapter. At this point, however, we can offer an informal, intuitive discussion of how you might make your decision in the umbrella example. If you carry the umbrella, you are certain to incur a cost equivalent to $1. If you do not carry the umbrella, you will incur a cost of $60 with probability 1/10 and a cost of $0 with probability 9/10. If you do not carry the umbrella, then your payoff is uncertain, depending on whether it rains or not. To evaluate this payoff, we use the concept of mathematical expectation, which will be discussed at length in Chapter 3, to compute the "expected value" of the second action, "do not carry the umbrella." Obtaining a payoff of −$60 with probability 1/10 and a payoff of $0 with probability 9/10 corresponds to an "expected" payoff of −$60(1/10) + $0(9/10), or −$6. Comparing this with the payoff which you will obtain if you carry the umbrella, −$1, we see that your decision should be to carry the umbrella.

The umbrella example demonstrates the use of probabilities in decision-making. In actual decision-making situations, the events of interest are often nonrepetitive. For example, consider decisions regarding a bet on a football game, the purchase of some common stock, the drilling of an oil well, or the medical treatment of someone who is ill. In each of these cases, it is not possible to observe repeated trials of the situation at hand, although it may be possible to observe trials involving similar situations, such as previous football games, past performance of stocks, the outcomes of other oil-drilling ventures, or the reaction of other patients with similar symptoms to various medical treatments. Because of the nonrepetitive nature of the events, the probabilities used as inputs to the decision-making procedure must be subjective rather than frequency probabilities.

Of course, not all decision-making situations involve nonrepetitive events, so the frequency interpretation of probability may be applicable in some cases. Often we have both information regarding long-run frequencies *and* subjective information. In Chapter 8 we will discuss the possibility of *combining* sample information (frequencies) and subjective information. In this way, it will be possible to utilize all available information which is relevant to the decision-making situation.

In the remainder of the book we will often work with probabilities without specifying whether they are based on long-run frequencies or on subjective judgments. In many cases we will simply use the probabilities as inputs to statistical procedures without worrying about their interpretations. This is permissible because, as we have explained, probability is an abstract mathematical concept which can be studied entirely apart from any interpretations. As we pointed out in Section 2.18, the subjective interpretation of probability can be thought of as an extension of the long-run frequency interpretation in the sense that both information involving observed frequencies *and* any other available information can be taken into consideration in determining subjective probabilities. Because of this and because of the value of subjective probabilities in decision-making situations, we favor the subjective interpretation of probability. Nevertheless, with the exception of the material relating to Bayesian inference and decision theory (Chapters 8 and 9), the interpretation of probabilities will not be explicitly discussed unless it seems useful to do so for a particular discussion, example, or exercise.

2.21 JOINT EVENTS

The idea of independent events has already figured in our discussion, and we have dealt only with events sampled independently. Now this concept of independence will be elaborated. However, the concepts of joint events and conditional probability must first be introduced.

As we have seen, the elementary events making up the sample space may be grouped into events in a variety of ways. For example, if the sample space is a set of outcomes of drawing one card from a deck of playing cards, we might find the probability of the event "red suit," or the event "picture card," or the event "ten." However, notice that there is nothing to keep any given card from qualifying for two or more events, provided those events are not mutually exclusive. For example, an observed card might be both a red card and a picture card. When we draw such a card, the joint event "red *and* picture card" occurs.

The same idea applies to any sample space: **Any event that is the intersection of two or more events is a joint event.** Thus, given the event A and the event B, the joint event is $A \cap B$. The event $A \cap B$ occurs when any elementary event belonging to this set is actually observed.

Since a member of $A \cap B$ must be a member of set A, and also of set B, both A and B events occur when $A \cap B$ occurs. Furthermore, a member of $A \cap B$ is a member of $A \cup B$, and so whenever the joint event occurs, $A \cup B$ occurs as well.

Provided that the elements of S are all equally likely to occur, the probability of the joint event is found in the usual way:

$$P(A \cap B) = \frac{\text{number of elementary events in } A \cap B}{\text{total number of elementary events}}. \quad (2.21.1)$$

With playing cards, the joint event "red and three" is the set of all cards having a red suit and three spots. There are exactly two cards, the three of diamonds and the three of hearts, meeting this qualification, and so

$$P(\text{red and three}) = \frac{2}{52} = \frac{1}{26}.$$

As another example, imagine a simple experiment consisting of rolling a die *and* tossing a penny. Each elementary event consists of the number that comes up on the die and the side that comes up on the penny. The sample space is shown below:

Each point in the diagram is one elementary event, a possible outcome of this experiment, and there are exactly $(2)(6) = 12$ such outcomes.

Consider the joint event "coin comes up heads, and die comes up with an odd number." There are exactly three elementary events that qualify: (heads, 1) (heads, 3), and (heads, 5). Since there are twelve elementary events in all,

$$P \text{ (H and odd) } = \frac{3}{12} = \frac{1}{4}.$$

This simple experiment illustrates that an elementary event can also be a joint event: the event (head, 1) is the intersection of the event "heads" and the event "1," and so on. This is very often the case in experiments where all events of interest are not mutually exclusive; each elementary event is the intersection of two or more events. Referring to the example of cards once again, we see that each distinct card represents the intersection of some suit-event with some number or picture-event.

Many situations give simple examples of joint events; imagine that someone samples students on a college campus at random and asks each his age, his sex (if necessary), and his field of concentration. Let the event A be that the student is 21 years or older, the event B that he is male, and C that he is an English major. The set of all students on the campus who are 21 or over and male English majors is $A \cap B \cap C$. The probability that a student sampled at random represents this joint event is

$$P \text{ } (A \cap B \cap C) = \frac{\text{number of students in } A \cap B \cap C}{\text{total number of students}}. \qquad (2.21.2)$$

2.22 COMBINING PROBABILITIES OF JOINT EVENTS

For any event A there is also an event $\bar{A}$, consisting of all of the elementary events not in the set A. In Section 2.5 it was pointed out that

$$P(A) + P(\bar{A}) = 1.$$

Furthermore, along with $A \cap B$ we can discuss the other joint events:

$A \cap \bar{B}$, consisting of all elementary events in A and not in B;
$\bar{A} \cap B$, consisting of all elementary events not in A but in B;
$\bar{A} \cap \bar{B}$, consisting of all elementary events neither in A nor in B.

These four events are all mutually exclusive, so that

$$P(A \cap B) + P(A \cap \bar{B}) = P(A),$$
$$P(A \cap B) + P(\bar{A} \cap B) = P(B), \qquad (2.22.1)$$
$$P(\bar{A} \cap B) + P(\bar{A} \cap \bar{B}) = P(\bar{A}),$$
and $\qquad P(A \cap \bar{B}) + P(\bar{A} \cap \bar{B}) = P(\bar{B}).$

Another principle for combining the probabilities of joint events also comes from rule 2.5.7 given in Section 2.5. Suppose that there are two events A and B, and that we wish to determine the probability of the occurrence of the event A or B or both. That is, we wish to determine the probability of the event class $A \cup B$. As we saw in Section 2.5, this probability is given by

$$P(A \cup B) = P(A) + P(B) - P(A \cap B). \qquad (2.22.2)$$

In other words, the probability of the occurrence of an A event or a B event or an elementary event falling into *both* classes is given by the probability of A plus the probability of B *minus* the probability of the compound event A and B. Recall that this is true because

$$P(A) + P(B) = P(A \cap \bar{B}) + P(\bar{A} \cap B) + 2P(A \cap B),$$

and $$P(A \cup B) = P(A \cap \bar{B}) + P(\bar{A} \cap B) + P(A \cap B).$$

Thus, $P(A \cap B)$ must be subtracted from $P(A) + P(B)$ to find the probability of the union $A \cup B$. This rule for finding the probability of the union of two event classes is sometimes called the "or" rule, because $P(A \cup B)$ can be read as the probability of A *or* B *or* both.

The most important case of the *or* rule is that in which the two events A and B are *mutually exclusive*, so that the probability of the joint event $A \cap B$ is zero. Here, Axiom 3 of Section 2.5 applies directly. In this case, for mutually exclusive events,

$$P(A \cup B) = P(A) + P(B). \qquad (2.22.3)$$

As an illustration of the use of the *or* rule for combining probabilities, think of a city school system, where we are going to sample one pupil at random. In this school system, we know that 35 percent of the pupils are left-handed, so that we know also that the probability is .35 for the event "left-handed." In the same way we know that the probability is .51 of our observing a girl, and that the probability is .10 of observing a girl who is left-handed. What is the probability of observing *either* a left-handed student *or* a girl (or both)? The answer according to the *or* rule is

$$P \text{ (left-handed or girl)} = .35 + .51 - .10 = .76.$$

Now suppose that .25 is the probability of observing a left-handed boy, and .10 is the probability of observing a left-handed girl. What is the probability of observing either a left-handed boy or a left-handed girl? It is not possible for an elementary event to fall both into the set of left-handed boys and the set of left-handed girls: these two events are mutually exclusive. Thus the desired probability is

$$P \text{ (left-handed boy or left-handed girl)} = .25 + .10 = .35,$$

which is just the probability of "left-handed" alone.

The *or* rule for events may be generalized to more than two events. Consider three events A, B, and C. Then

$$P(A \cup B \cup C) = P(A) + P(B) + P(C) - P(A \cap B) - P(A \cap C)$$
$$- P(B \cap C) + P(A \cap B \cap C). \quad (2.22.4)$$

Similar expressions exist for the union of more than three events, but we will not have occasion to use them in our elementary work. If three or more events are mutually exclusive, then the union is just the sum of the separate probabilities:

$$P(A \cup B \cup C) = P(A) + P(B) + P(C), \quad (2.22.5)$$

as given in Axiom 3, Section 2.5.

2.23   CONDITIONAL PROBABILITY

Suppose that in sampling children in a school system, we restrict our observations only to girls. Here, there is a new sample space including only part of the elementary events in the original set. In this new sample space, consisting only of girls, what is the probability of observing a left-handed person? If all girls in the school system are equally likely to be observed, it is obvious that

$$P \text{ (left-handed, among the girls)} = \frac{\text{number left-handed girls}}{\text{total number of girls}}.$$

This probability can be found from the *original* probabilities for the total sample space; first of all we know that

$$P \text{ (left-handed and girl)} = \frac{\text{number left-handed girls}}{\text{total number of pupils}} = .10$$

and that

$$P \text{ (girl)} = \frac{\text{total number of girls}}{\text{total number of pupils}} = .51.$$

It follows that

$$P \text{ (left-handed, among the girls)} = \frac{P \text{ (left-handed and girl)}}{P \text{ (girl)}} = \frac{.10}{.51} = .196,$$

since the two probabilities put into ratio each have the same denominator.

The probability just found, based on only a part of the total sample space, is called a conditional probability. A more formal definition follows:

Let A and B be events in a sample space made up of a finite number of elementary events. Then the conditional proba-

bility of B given A, denoted by $P(B \mid A)$, is

$$P(B \mid A) = \frac{P(A \cap B)}{P(A)}, \qquad (2.23.1^*)$$

provided that $P(A)$ is not zero.

Notice that the conditional probability symbol, $P(B \mid A)$, is read as "the probability of B *given* A." In the example it was given that the observation would be a girl; the value desired was the probability of "left-handed," given that a girl were observed. It is necessary that $P(A) > 0$; it makes no sense to speak of the probability of B given A if A is an impossible event.

For any two events A and B, there are two conditional probabilities that may be calculated:

$$P(B \mid A) = \frac{P(A \cap B)}{P(A)}$$

and

$$P(A \mid B) = \frac{P(A \cap B)}{P(B)}.$$

We might find the probability of left-handed, given girl ($.10/.51 = .19$) or the conditional probability of girl, given left-handed ($.10/.35 = .29$). In general, these two conditional probabilities will not be equal, since they represent quite different sets of elementary events.

As another example, take the probabilities found in the example for a coin and a die in Section 2.21. Here, the probability of a tail coming up on the coin is .50, and the probability of an odd number on the die *and* a tail on the coin is .25. The probability that the die comes up with an odd number, *given* that the coin comes up tails is

$$P \text{ (odd} \mid \text{tails)} = \frac{.25}{.50} = .50.$$

The probability that the coin comes up tails and the die comes up with a 1 or a 2 can be found from the example to be 1/6 or .167. If we wish to know the probability that the die comes up with a 1 or a 2 given that the coin comes up tails, the probability is

$$P \text{ (1 or 2} \mid \text{tails)} = \frac{.167}{.50} = .33.$$

In this section we have discussed the concept of the conditional probability of one event given a second event. This notion can be extended to allow more "given" events. For example, we can speak of the probability of event C given events A and B. From the definition of conditional probability, then, since the occurrence of A and B is equivalent to the occurrence of $A \cap B$, we have

$$P(C \mid A, B) = P(C \mid A \cap B) = \frac{P(A \cap B \cap C)}{P(A \cap B)}. \qquad (2.23.2)$$

Here the new sample space is $A \cap B$, since we are given A and B. In general, the new (or reduced) sample space in conditional probability consists of the intersection of the events that are "given."

It is possible to use conditional probabilities in the computation of probabilities of joint events. Multiplying both sides of Equation (2.23.1) by $P(A)$, we get

$$P(A \cap B) = P(B \mid A)P(A). \qquad (2.23.3^*)$$

Similarly,

$$P(A \cap B) = P(A \mid B)P(B). \qquad (2.23.4^*)$$

Extending this to the intersection of three events, we will prove that

$$P(A \cap B \cap C) = P(C \mid A, B)P(B \mid A)P(A). \qquad (2.23.5)$$

To prove this, note from Equation (2.23.2) that:

$$P(C \mid A, B) = \frac{P(A \cap B \cap C)}{P(A \cap B)}.$$

Thus,

$$P(C \mid A, B)P(B \mid A)P(A) = \left(\frac{P(A \cap B \cap C)}{P(A \cap B)}\right)\left(\frac{P(A \cap B)}{P(A)}\right)P(A).$$

Canceling terms, we get

$$P(C \mid A, B)P(B \mid A)P(A) = P(A \cap B \cap C),$$

which is what we set out to prove. In a similar manner, we can determine the probability of the intersection of any number of events.

In previous sections we computed probabilities with no mention of conditional probabilities. In a way, however, all probabilities are conditional. That is, they may be conditional upon some assumptions, upon the definition of the sample space, or upon numerous similar factors. For example, when speaking of the probability of a head coming up on a single flip of a coin as 1/2, we really should write:

$$P \text{ (head} \mid \text{fair coin)} = 1/2.$$

Similarly, when speaking of the probability of a full house in a five card poker hand, we should write:

$$P \text{ (full house} \mid \text{standard deck of 52 cards which is well shuffled)} = 0.0014.$$

Because it is bothersome to write out such long expressions, we usually do not do so, assuming that the "given" conditions are understood by the reader. Since the "given" conditions are not other events in the same sample space, it is not necessary to use Equation (2.23.1) to compute probabilities which are conditional only in the sense indicated in this paragraph. This is not because the equation is not applicable, but because it is trivial in such a case. If the "given" condition A in Equation (2.23.1) is an assumption which we are accepting as being true, then the effect of A is reflected in the construction of the sample space. As a result,

$$P(A) = 1 \quad \text{and} \quad P(A \cap B) = P(B),$$

so we usually will not write the probability expression in conditional form, even though it is clearly a conditional probability. In some situations, particularly when there are competing "hypotheses," or sets of assumptions, we *will* write the probability in conditional form. At any rate, you should keep in mind that all probabilities are conditional upon some factors, even if these factors are not explicitly noted in the probability statement for the sake of convenience.

2.24 RELATIONS AMONG CONDITIONAL PROBABILITIES

The conditional probability $P(B \mid A)$ can either be larger or smaller than $P(B)$; there is no necessary relation between the size of $P(B)$ and the size of $P(B \mid A)$, other than that they both must lie between zero and one. Since conditional probabilities are, as the name implies, probabilities, they must obey the basic rules of probability.

For any event and its complement,

$$P(B \mid A) + P(\bar{B} \mid A) = 1. \tag{2.24.1*}$$

This is easily seen to be true if one remembers that

$$P(A \cap B) + P(A \cap \bar{B}) = P(A),$$

so that
$$\frac{P(A \cap B)}{P(A)} + \frac{P(A \cap \bar{B})}{P(A)} = 1.$$

If there are three events, A, B, and C, then

$$P(B \mid A) + P(C \mid A) - P(B \cap C \mid A) = P(B \cup C \mid A), \tag{2.24.2}$$

since *within the set A* it is also true that

$$P(A \cap B) + P(A \cap C) - P(A \cap B \cap C) = P[(A \cap B) \cup (A \cap C)],$$

so that

$$\frac{P(A \cap B)}{P(A)} + \frac{P(A \cap C)}{P(A)} - \frac{P(A \cap B \cap C)}{P(A)} = \frac{P[A \cap (B \cup C)]}{P(A)}.$$

If we know the conditional probability $P(B \mid A)$ and also $P(A)$, then we can also find the joint probability $P(A \cap B)$:

$$P(A)P(B \mid A) = P(A \cap B) \qquad (2.24.3^*)$$

and

$$P(B)P(A \mid B) = P(A \cap B),$$

since, by definition,

$$P(B \mid A) = \frac{P(A \cap B)}{P(A)}$$

and

$$P(A \mid B) = \frac{P(A \cap B)}{P(B)}.$$

It follows that

$$P(A)P(B \mid A) = P(B)P(A \mid B) \qquad (2.24.4)$$

and that

$$\frac{P(B \mid A)}{P(A \mid B)} = \frac{P(B)}{P(A)}.$$

2.25 BAYES' THEOREM

The relation among various conditional probabilities are embodied in the theorem named for Bayes, an English clergyman who did early work in probability theory. **The simplest version of this theorem follows: for two events A and B, the following relation must hold:**

$$P(A \mid B) = \frac{P(A \cap B)}{P(B)}$$

$$= \frac{P(B \mid A)P(A)}{P(B \mid A)P(A) + P(B \mid \bar{A})P(\bar{A})}. \qquad (2.25.1^*)$$

To prove this, recall that

$$P(B) = P(A \cap B) + P(\bar{A} \cap B).$$

Substituting this into the expression for $P(A \mid B)$, we get

$$P(A \mid B) = \frac{P(A \cap B)}{P(A \cap B) + P(\bar{A} \cap B)}.$$

But, from Section 2.24,

$$P(A \cap B) = P(B \mid A)P(A)$$

and
$$P(\bar{A} \cap B) = P(B \mid \bar{A})P(\bar{A}),$$

so that

$$P(A \mid B) = \frac{P(B \mid A)P(A)}{P(B \mid A)P(A) + P(B \mid \bar{A})P(\bar{A})}.$$

To illustrate Bayes' theorem, suppose you awake in the middle of the night with a headache and stumble into the bathroom without turning on the light. In the dark you grab one of three bottles containing pills and take one. An hour later you become very ill, and the thought strikes you that one of the three bottles contains a poison, the other two, aspirin. With a few assumptions this can be made into a problem in conditional probability. (Before you jump to any unflattering conclusions about a professor's thoughts as he lies dying, be assured that a doctor most certainly would be called. If you will, this is an exercise to fill the time until the doctor comes.)

Let us assume that you have a medical text handy, and that it shows that 80 percent of normal individuals show your symptoms after taking the poison, and that only 5 percent have these symptoms from taking aspirin. Let B stand for the event "symptoms," and let A stand for the event "took poison," with $\bar{A}$ for the event "took aspirin." In effect, the sample space is restricted to those individuals who took *either* aspirin or poison. According to your information, you assume that

$$P(B \mid A) = .80$$
and
$$P(B \mid \bar{A}) = .05.$$

If each bottle had equal probability of being chosen in the dark, then

$$P(A) = 1/3$$
and
$$P(\bar{A}) = 2/3.$$

Given these figures, what is the probability that someone in this situation has taken poison, given that he has the symptoms? This is

$$P(A \mid B) = \frac{(.80)(.33)}{(.80)(.33) + (.05)(.67)} = .89.$$

It is a good thing that the doctor *was* called! The conditional probability. of "took poison, given the symptoms" is high.

Although this example is dramatic, it hardly illuminates the real uses to which this idea can be put, provided that one knows the necessary initial probabilities. Consider this example: at a university it is decided to try out a new placement test for admitting students to a special mathematics class. Experience has shown that in general only 60 percent of students applying for admission actually can pass this course. Heretofore, each student applying for the course was admitted. Of the students who passed the course, some 80 percent passed the placement test beforehand, whereas only 40 percent of those who failed the course could pass the placement test initially. If this test is to be used for placement, and only students who pass the test are to be admitted to the course, what is the probability that such a student will pass the course?

First of all, event A is defined to be "passes the course," and it is assumed that $P(A) = .60$. The event B is "passes the test." The conditional probabilities are taken to be

$$P(B \mid A) = .80 \qquad P(\bar{B} \mid A) = .20$$
$$P(B \mid \bar{A}) = .40 \qquad P(\bar{B} \mid \bar{A}) = .60.$$

However, we want to know the conditional probability, $P(A \mid B)$, that a student will pass the course, *given* that he has passed the test. Bayes' theorem shows that

$$P(A \mid B) = \frac{(.80)(.60)}{(.80)(.60) + (.40)(.40)}$$

$$= \frac{.48}{.48 + .16} = .75.$$

Thus, the probability is .75 for a student's passing the course *given* that he passed the test. If we wanted to find other conditional probabilities, we could also apply the theorem. For example,

$$P(\bar{A} \mid \bar{B}) = \frac{(.60)(.40)}{(.60)(.40) + (.20)(.60)}$$

$$= \frac{.24}{.24 + .12} = .67.$$

Notice that the probability of a student's passing the course given that he passes the test is greater than the probability in general of passing the course. Any student is a better bet to pass the course if he has passed the placement test.

What is the probability of being *right* if students are admitted or refused the course strictly on the basis of this test? The probability of a correct decision can be found by

$$P(A \cap B) + P(\bar{A} \cap \bar{B}) = P \text{ (correct)}.$$

By Equation (2.24.3) this is

$$P \text{ (correct)} = P(B \mid A)P(A) + P(\bar{B} \mid \bar{A})P(\bar{A})$$
$$= (.80)(.60) + (.60)(.40) = .72.$$

In other words, the administrator will have a probability of .72 of being right about a student's proper placement if he uses the test. If he does not use the test and simply admits all students to the course, the probability of being right (of the student passing the course) is only .60. Thus, the test does something for the administrator; its use allows him to increase the probability of being right about a given student selected at random.

Bayes' theorem can be put into much more general form: if $\{A_1, A_2, \cdots, A_J\}$ **represents a set of** J **mutually exclusive events, and if**

$$B \subset A_1 \cup A_2 \cup \cdots \cup A_J,$$

so that

$$P(B) < P(A_1) + P(A_2) + \cdots + P(A_J),$$

then

$$P(A_j \mid B)$$

$$= \frac{P(B \mid A_j)P(A_j)}{P(B \mid A_1)P(A_1) + \cdots + P(B \mid A_j)P(A_j) + \cdots + P(B \mid A_J)P(A_J)}$$

$$(2.25.2^*)$$

for any event A_j **in the set. (Notice that** A_j **and** A_J **do not necessarily symbolize the same event.)**

Bayes' theorem will be utilized in Chapters 8 and 9 to revise prior probabilities on the basis of sample information. The following terminology will be used: $P(A_i)$ represents a prior probability; $P(B \mid A_i)$ represents the likelihood function, which concerns the sample information; and $P(A_i \mid B)$ represents a posterior probability. The terms "prior" and "posterior" are relative to the observed sample. In terms of the above example, the prior probabilities of the administrator were $P(\text{pass}) = .60$ and $P(\text{fail}) = .40$. After seeing some sample information, the posterior probabilities given that a student passed the exam were .75 and .25, and the posterior probabilities given that a student failed the exam were .33 and .67.

Bayes' theorem is a key feature of a modern, decision-theoretic approach to statistics. We will discuss this approach and compare it with the more traditional, or "classical" approach in Chapters 8 and 9.

2.26 INDEPENDENCE

Now we are ready to take up the topic of independence. The general idea used heretofore is that independent events are those having nothing to do with each other: the occurrence of one event in no way affects the probability of the other event. But how does one know if two events A and B are independent? If the occurrence of the event A has nothing whatever to do with the occurrence of event B, then we should expect the conditional probability of B given A to be exactly the same as the probability of B, $P(B \mid A) = P(B)$. Likewise, the conditional probability of A given B should be equal to the probability of A, or $P(A \mid B) = P(A)$. The information that one event has occurred does not affect the probability of the other event, when the events are independent.

The condition of independence of two events may also be stated in another form: if $P(B \mid A) = P(B)$, then

$$\frac{P(A \cap B)}{P(A)} = P(B), \qquad (2.26.1^*)$$

so that
$$P(A \cap B) = P(A)P(B).$$

In the same way,

$$P(A \mid B) = P(A) \qquad (2.26.2^*)$$

leads to the statement that

$$P(A \cap B) = P(A)P(B).$$

This fact leads to the usual definition of independence:

Events A and B are independent if, and only if, the joint probability $P(A \cap B)$ is equal to the probability of A times the probability of B:

$$P(A \cap B) = P(A)P(B). \qquad \mathbf{(2.26.3^*)}$$

For example, suppose that you go into a library and at random select one book (each book having an equal likelihood of being drawn). The sample space consists of all distinct books that a person might select. Suppose that the proportion of books then on the shelves and classified as "fiction" is exactly .15, so that the probability of selecting such a book is also .15. Furthermore, suppose that the proportion of books having red covers is exactly .30. If the event "fiction" is independent of the event "red cover," the probability of the joint event is found very easily:

P (fiction and red cover) $= P$ (fiction)P (red cover) $= (.15)(.30) = .045.$

Your chances of coming up with a red-covered piece of fiction are forty-five in one thousand.

The example in which a coin is tossed and a die rolled also illustrates independent events. In this example, the probability that the coin comes up heads is .50, and the probability that the die comes up with an odd number is also .50. If we assume that the two events are independent, then the probability of the event "heads and odd number" is given by

P (heads and odd number) $= P$ (heads)P (odd number) $= (.50)(.50) = .25.$

If you look at the sample space for this experiment, pictured in Section 2.21, you will find that this number .25 agrees with the probability for the joint event "heads and odd number" calculated directly. When all the elementary events in this sample space are equally likely, these two events actually are independent.

It is certainly not true that all events must be independent. It is very easy to give examples where the joint probability of two events is not equal to the product of their separate probabilities. For example, suppose that children were selected at random. A child is observed and his hair color noted, and also whether or not he is freckled. For our purposes, the event A consists of the set of all children having red hair, and the event B is the set of all children who are freckled. Is it reasonable that the event A will be independent of the event B? The answer is no; everyone knows that among red heads freckles are much more common than among children in general. One would expect $P(B \mid A)$ to be *greater* than $P(B)$ in this case, so that it should *not* be true that $P(A \cap B) = P(A)P(B)$. The events "red hair" and "freckled" *do* tend to occur together and are not ordinarily regarded as independent.

This example suggests one of the uses of the concept of independent events. The definition of independence permits us to decide whether or not events *are associated* or *dependent* in some way:

If, for two events A and B, $P(A \cap B)$ is not equal to $P(A)P(B)$, then A and B are said to be associated or dependent.

For example, consider a sample space with elementary events consisting of all the adult male persons in the United States. We wish to answer the question, "Is making over twenty thousand dollars a year independent of having a college education?" Let us call event A "makes over twenty thousand dollars a year," and B "has a college education." Now suppose that the proportion of adult males in the United States who make more than twenty thousand a year is .06; if each adult male is equally likely to be selected, the probability of event A is thus .06. Suppose also that the proportion of adult males with a college education is .30, so that $P(B) = .30$. Finally, the probability of an event $A \cap B$, our observing an adult male making over twenty thousand a year and who has a college education is .04. If the events were independent, this probability $P(A \cap B)$ *should* be $(.06)(.30) = .018$; the actual probability is .04. This leads immediately to the conclusion that these two events are *not* independent (or *are* associated). The probability that event B will occur given event A is much greater than it should be if the events were independent: $P(B \mid A) = .04/.06 = .67$, whereas if the events were independent, it should be true that

$$P(B \mid A) = P(B) = .30.$$

This idea of independence is very important to many of the statistical techniques to be discussed in later chapters. The question of association or dependence of events is central to the scientific question of the relationships among the phenomena we observe (literally, the question of "what goes with what"). The techniques for studying the presence and degree of relatedness among observations will depend directly upon this idea of comparing probabilities for joint events with the probabilities when events are independent.

However, the concept of association of events must not be confused with the idea of causation. Association, as used in statistics, means simply that the events are not independent, and that a certain correspondence between their joint and separate probabilities does not hold. When events A and B are associated, it need not mean that A causes B or that B causes A, but only that the events occur together with probability different from the product of their separate probabilities. This warning carries force whenever we talk of the association of events; association may be a consequence of causation, but this need not be true.

Sometimes we are interested in the independence of more than two events, as, for example, when the assumption is made that the trials of a sample are independent. We say that K events $A_1, A_2, \ldots, A_k$ **are mutually independent if for all** i, j, k, **and so on, the following conditions hold (where** i, j, k, **and so on, must all be different):**

$$P(A_i \cap A_j) = P(A_i)P(A_j),$$
$$P(A_i \cap A_j \cap A_k) = P(A_i)P(A_j)P(A_k),$$
$$P(A_i \cap A_j \cap A_k \cap A_l) = P(A_i)P(A_j)P(A_k)P(A_l),$$

$$\cdot \qquad\qquad \cdot$$
$$\cdot \qquad\qquad \cdot \qquad\qquad (2.26.4^*)$$
$$\cdot \qquad\qquad \cdot$$

and $\qquad P(A_1 \cap A_2 \cap \cdots \cap A_k) = P(A_1)P(A_2) \cdots P(A_k).$

In other words, the probability of the intersection of any two or more (up to all K**) of the** K **events must equal the product of the respective probabilities of the individual events.** If only the first condition $[P(A_i \cap A_j) = P(A_i)P(A_j)]$ holds, we say that the K events are **pairwise independent.** It is possible for K events to be pairwise independent but *not* mutually independent.

It should be pointed out that it is possible to speak of *conditional* independence or nonindependence. If H is some event such that

$$P(A \cap B \mid H) = P(A \mid H)P(B \mid H), \qquad (2.26.5^*)$$

then A and B are said to be **conditionally independent,** where their independence is conditional on the event H. Similarly, if (2.26.5) is *not* satisfied, then A and B are said to be conditionally nonindependent, conditional upon H. Alternatively, A and B are conditionally independent given

H if

$$P(A \mid B, H) = P(A \mid H), \qquad (2.26.6)$$

and conditionally nonindependent given *H* if this equation is not satisfied.

For example, let *A* represent the occurrence of an act of sabotage, and let *B* represent the assassination of a government official. Suppose that two alternative hypotheses are considered: H_W, corresponding to the event that some country is preparing to start a war with our country; and H_P, corresponding to the event that peace will prevail. We may consider the events *A* and *B* to be independent conditional upon H_P being true but nonindependent conditional upon H_W being true. That is, given the peace hypothesis, the probability of an act of sabotage given that a government official has been assassinated may be considered to be the same as the probability of an act of sabotage given no knowledge of any assassination. On the other hand, under the war hypothesis, the two events might be believed to be related.

Just as all probabilities are, in a sense, conditional (Section 2.23), all independence or nonindependence can be thought of as conditional upon certain assumptions or hypotheses. Even in a simple experiment such as tossing a coin, the claim that the trials are independent rests upon assumptions such as the assumption that the way the person flips the coin does not depend on the previous outcome. Just as with conditional probabilities, it will be convenient for us not always to write out the given conditions under which events are independent or nonindependent. If it is important in any specific case, we will do so.

2.27 REPRESENTING JOINT EVENTS IN TABLES

Sometimes it is convenient to list joint events by means of a table, with each cell representing one joint possibility. For example, consider a sample space consisting of all individual students at a particular college campus. Each student is either male or female, of course, and when asked his opinion on some question, a student will respond either "yes" or "no." The possible joint events are as follows:

Male	(male and yes)	(male and no)
Female	(female and yes)	(female and no)
	Yes	No

It might be that on this particular campus, the probability of our observing a male student is .55 and of observing a female .45. Furthermore, suppose that the probability of obtaining a "yes" answer is .40, and .60 for obtaining a "no." Then the table of probabilities follows this general pattern:

P (Male) = .55	P (Male and Yes)	P (Male and No)
P (Female) = .45	P (Female and Yes)	P (Female and No)

$$P \text{ (Yes)} = .40 \qquad P \text{ (No)} = .60$$

The values of P(Male), P(Female), P(Yes), and P(No) are called the **marginal probabilities** (they get the name from the obvious circumstance of appearing in the *margin* of the table). Each marginal probability is a sum of all the joint probabilities in some particular row or column of the table:

$$P(\text{Male}) = P(\text{Male and Yes}) + P(\text{Male and No}),$$
$$P(\text{Female}) = P(\text{Female and Yes}) + P(\text{Female and No}),$$
$$P(\text{Yes}) = P(\text{Male and Yes}) + P(\text{Female and Yes}),$$
and $$P(\text{No}) = P(\text{Male and No}) + P(\text{Female and No}).$$

Ordinarily the event classes that appear together along any margin of the table are *mutually exclusive and exhaustive.* Thus, the events "Yes" and "No" are mutually exclusive and exhaustive and so are the events "Male" and "Female." Any set of mutually exclusive and exhaustive event classes that make up one margin of the table can be called an **attribute** or a **dimension.** This table embodies two attributes, "the sex of the student" and "the response of the student," respectively.

Suppose that the two attributes of the table were independent; this means that any event along one margin must be independent of every event along the other margin. Here, this is tantamount to saying that the sex of the student has absolutely nothing to do with how he answers the question. **If independence exists, the probability of each joint event** (as given in a cell of the table) **must be equal to the product of the probabilities of the corresponding marginal events.**

For example, if the two attributes are independent, then the probability P(Male and Yes) must be equal to the product of the two marginal probabilities:

$$P(\text{Male and Yes}) = P(\text{Male})P(\text{Yes}) = (.55)(.40) = .22.$$

In the same way the joint probabilities for each of the other cells may be found from products of marginal probabilities, and the following table should be correct:

$P(\text{Male})$ $= .55$	$P(\text{Male})P(\text{Yes})$ $= (.55)(.40) = .22$	$P(\text{Male})P(\text{No})$ $= (.55)(.60) = .33$
$P(\text{Female})$ $= (.45)$	$P(\text{Female})P(\text{Yes})$ $= (.45)(.40) = .18$	$P(\text{Female})P(\text{No})$ $= (.45)(.60) = .27$

$$P(\text{Yes}) = .40 \qquad P(\text{No}) = .60$$

It is important to remember, however, that these will be the correct joint probabilities only if the attributes *are* independent. It might be that the following table is the true one:

$P(\text{Male})$ $= .55$	$P(\text{Male and Yes})$ $= .10$	$P(\text{Male and No})$ $= .45$
$P(\text{Female})$ $= .45$	$P(\text{Female and Yes})$ $= .30$	$P(\text{Female and No})$ $= .15$

$$P(\text{Yes}) = .40 \qquad P(\text{No}) = .60$$

When this is true it is safe to assert that for this sample space the sex of the student *is* associated with how he will answer the question. Look at the conditional probabilities: here, the probability of the student's answering "yes" if he is a male is

$$P(\text{Yes} \mid \text{Male}) = \frac{.10}{.55} = .18,$$

and the probability of a "yes" answer if the student is a female is

$$P(\text{Yes} \mid \text{Female}) = \frac{.30}{.45} = .67.$$

Thus a female is much more likely to answer "yes" to this question than is a male. On the other hand, had the two attributes been independent, these two conditional probabilities *should* have been the same:

$$P(\text{Yes} \mid \text{Male}) = \frac{.22}{.55} = .40$$

and $$P(\text{Yes} \mid \text{Female}) = \frac{.18}{.45} = .40,$$

indicating that sex gives no information about how a person tends to respond to the question.

This idea of representing probabilities in tabular form can be extended to any number of attributes, each consisting of any number of event classes. For example, consider four mutually exclusive and exhaustive classes labeled $\{A_1, A_2, A_3, A_4\}$, which we may call the attribute A, and another set of classes $\{B_1, B_2, B_3\}$, which is the attribute B. Then the joint events consisting of the intersection of an event class from A with an event class from B may be represented by the table

B_1	$A_1 \cap B_1$	$A_2 \cap B_1$	$A_3 \cap B_1$	$A_4 \cap B_1$
B_2	$A_1 \cap B_2$	$A_2 \cap B_2$	$A_3 \cap B_2$	$A_4 \cap B_2$
B_3	$A_1 \cap B_3$	$A_2 \cap B_3$	$A_3 \cap B_3$	$A_4 \cap B_3$
	A_1	A_2	A_3	A_4

(Notice the similarity of this formulation of joint events to the idea of a Cartesian product introduced in Chapter 1; the set of all joint-event pairs is nothing more than the Cartesian product $A \times B$.)

Associated with any joint-event class there is a probability, such as $P(A_1 \cap B_1)$ or $P(A_1 \cap B_2)$. Each of the marginal probabilities is a sum of the probabilities in a row or a column of the table: thus,

$$P(B_1) = P(A_1 \cap B_1) + P(A_2 \cap B_1) + P(A_3 \cap B_1) + P(A_4 \cap B_1),$$

and so on for the other marginal probabilities.

Given that the attributes are independent, then the joint probabilities are the products of the marginal probabilities, and the table becomes:

$P(B_1)$	$P(A_1)P(B_1)$	$P(A_2)P(B_1)$	$P(A_3)P(B_1)$	$P(A_4)P(B_1)$
$P(B_2)$	$P(A_1)P(B_2)$	$P(A_2)P(B_2)$	$P(A_3)P(B_2)$	$P(A_4)P(B_2)$
$P(B_3)$	$P(A_1)P(B_3)$	$P(A_2)P(B_3)$	$P(A_3)P(B_3)$	$P(A_4)P(B_3)$
	$P(A_1)$	$P(A_2)$	$P(A_3)$	$P(A_4)$

Again it must be emphasized, however, that the joint probabilities are equal to the products of the marginals *only* when the attributes represented by the margins are independent.

EXERCISES

1. Explain, briefly, the connection between the theory of sets and probability theory.

2. A black die (B) and a white die (W) are each tossed. Let b indicate the value of the B die, and let w indicate the value of the W die. (That is, let the number of spots coming up on a die be its numerical value for that toss). Specify the sample space completely, and then find the probabilities for the events indicated.
   (a)   the event $b = w$
   (b)   the event $b \leq w$
   (c)   the event $b = 6$ and $w < 6$
   (d)   the event $[(b = 3) \cup (w = 3)]$
   (e)   the event $[(b + w = 7) \cap (w - b = 3)]$
   (f)   the event $[(b + w = 7) \cup (w - b = 3)]$
   (g)   the event $[(b = 6) \cap (w < 6)] \cup [(b \neq 1) \cap (w = 1)]$.

3. A letter is selected at random from the English alphabet. What is the probability that:
   (a)   the letter is a vowel
   (b)   the letter is a consonant
   (c)   the letter occurs in the last ten positions of the alphabet, in the conventional ordering

(d) the letter is a consonant falling between the two vowels "a" and "i" in the conventional ordering

(e) two letters are drawn at random from the alphabet, and the first is replaced before the second is drawn. What is the sample space for this experiment? Find the probability that the first letter drawn precedes the second letter (the two letters not being identical) in the conventional order of the alphabet.

4. Two distinguishable coins are tossed two times each. Find the probability for each of the following events, given that the coins are fair. Start by specifying the elementary events making up the sample space.
(a) at least two heads occur
(b) at most two heads occur
(c) the number of heads for coin 1 is less than the number of heads for coin 2.

5. Why is it that almost all elementary discussions of, and computations involving, probability rely on the assumption of equally probable elementary events? Is there anything about the theory of probability itself that makes this assumption necessary? What if elementary events cannot be assumed equally probable; does the axiomatic theory of probability still apply?

6. A fair coin is tossed N times, and each time either a head or a tail occurs. If the elementary events for this experiment are viewed as the possible results of N tosses, how many different elementary events are there?

7. A young man has a little black book that contains the names and phone numbers of girls who are potential dates. Suppose that he has the names of *3 redheads, 18 brunettes,* and *6 blondes* in his book, and that the book has a total of 30 pages. He decides to select one page at random from the book in order to determine whom he will call for a date on a particular night, and each name appears on a separate page.
(a) What is the probability that he will call a blonde? a redhead? a brunette? (Here, count a blank page as "no call.")
(b) What is the probability that he will sample a page containing the name either of a *redhead or a brunette?* What is the probability that the page will *not* contain the name of a blonde?

8. Show by a Venn diagram that
(a) $P(A \cap \bar{B}) = P(A) - P(A \cap B)$
(b) if $B \subseteq A$, then $P(A - B) = P(A) - P(B)$
(c) $P(A \cap B) + P(A \cap C) = P[A \cap (B \cup C)] + P(A \cap B \cap C)$.

9. Let A_1, A_2, and A_3 be events defined on some specific sample space S. Find expressions involving only *union, intersection,* and *complementation* for the probability that:
(a) *exactly two* of A_1, A_2, A_3 occur simultaneously
(b) *at least two* of A_1, A_2, A_3 occur simultaneously
(c) *at most two* of the events A_1, A_2, A_3 occur simultaneously.

10. Compute the probabilities for (a)–(c) in the preceding exercise if:

(a) $P(A_1) = 1/2, \quad P(A_2) = P(A_3) = 1/4$

$P(A_1 \cap A_2) = P(A_1 \cap A_3) = P(A_2 \cap A_3) = 1/8$

$P(A_1 \cap A_2 \cap A_3) = 1/32;$

(b) $P(A_1) = 1/3, P(A_2) = 2/3$

$P(A_1 \cap A_2) = P(A_1 \cap A_3) = P(A_2 \cap A_3) = P(A_1 \cap A_2 \cap A_3) = 0.$

11. A card is drawn from a well-shuffled standard playing deck. What is the probability that it is:

(a) either a spade or the queen of hearts

(b) an even numbered card (not counting face cards) but not a spade or a heart

(c) a spade which is not the ace or king, or a heart which is not the queen or jack

(d) a numbered card that is not a perfect square (ignoring jack, queen, king, and ace as non-numbered cards).

12. A famous problem in probability theory is the following: if an "ace" is the occurrence of the number "one" on the toss of a die, find the probabilities of

(a) obtaining *at least* one ace in four tosses of a fair die

(b) obtaining *at least* one "double ace" in twenty-four tosses of a *pair* of fair dice.

If you were given a chance to gamble on event (a) or event (b), which would you prefer to gamble on? (Assume that you *must* choose one of the two gambles.)

13. Find a general formula, similar to Equation (2.5.7), for the probability of the union of three events, $P(A \cup B \cup C)$. [*Hint*: See Exercise 23, Chapter 1.]

14. Suppose that a fair die is thrown until a six appears. Find the probability that this will take more than three throws.

15. Prove that $P(A \cap B) \leq P(A) \leq P(A \cup B)$ for any events A and B.

16. In one year, the Parliament of a European country included 45 members of the Liberal party, 38 members of the Conservative Party, and 15 members of the Labour Party. The Prime Minister wished to appoint a commission of 10 persons from Parliament, consisting of 5 Liberals, 3 Conservatives, and 2 Laborites. How many different such committees *might* he appoint? (Leave the answer in symbolic form, and do not bother to work out the exact figure.) If the Prime Minister decided simply to choose a commission of 10 without regard to party, and he did that at random, what is the probability of getting 5 Liberals, 3 Conservatives, and 2 Laborites on the commission? What is the probability of getting *no* Conservatives? At least 2 Conservatives? (Again, leave answers in symbolic form.)

17. If there are 5 people in a room, what is the probability that no two of them have the same birthday? In general, if there are r people in a room, what is the probability that no two of them have the same birthday? What assumptions did you make in computing these probabilities?

18. If we draw five cards from a well-shuffled standard deck of 52 cards, what is the probability of obtaining
    (a) four of a kind
    (b) a royal flush (Ace, King, Queen, Jack, and Ten from the same suit).

19. In how many *distinguishable* ways can the letters in the word "sassafras" be ordered?

20. Prove Pascal's rule:

$$\binom{N}{r} = \binom{N-1}{r-1} + \binom{N-1}{r}.$$

21. Prove that

$$N\binom{N}{r-1} = r\binom{N}{r} + (r-1)\binom{N}{r-1}.$$

22. How many ways can a starting team of 11 players be chosen from a squad of 30 football players, ignoring the position to be played by each player? If the squad consists of 10 backs and 20 linemen, in how many ways can a starting team of 4 backs and 7 linemen be chosen?

23. State how many ways can 8 persons be seated in a row of 8 seats if
    (a) there are no restrictions on the seating arrangement
    (b) person A must sit next to person B, but there are no restrictions on the remaining six persons
    (c) there are four men and four women, and no two men (or two women) can sit next to each other
    (d) there are four married couples, and each couple must sit together.

24. How many ways can 8 persons be seated around a circular table, taking into account only the *relative* location of the 8 persons (that is, if all 8 persons get up and move one chair to the left, the new arrangement is considered to be indistinguishable from the old arrangement)?

25. Suppose that you want to arrange 3 statistics books, 2 mathematics books, and 4 novels on a bookshelf. Calculate how many arrangements are possible if
    (a) the books can be arranged in any manner
    (b) the statistics books must be together, the mathematics books must be together, and the novels must be together
    (c) the novels must be together, but the other books can be arranged in any manner.

26. The so-called "gambler's fallacy" goes something like this: in a dice game, for example, a seven has not turned up in quite a few rolls of a pair of honest dice; now a seven is said to be "due" to come up. Why is this a fallacy?

27. Bernoulli's theorem has occasionally been called "the link between the mathematical concept of probability and the real world about us." Comment on this proposition.

28. Take a fair die and throw it 300 times, recording the number which appears at each throw. Find the frequency of each of the six possible events after 10, 20, 50, 100, and 300 throws, and comment on these results in light of Bernoulli's theorem.

29. Find the probability of event A if the odds in favor of A are:
    (a) 2 to 1   (b) 1 to 2   (c) 3 to 7.

30. Find the odds in favor of event A if the probability of event A is
    (a) .50   (b) .20   (c) .875.

31. Explain why the subjective interpretation of probability can be thought of as an extension of the long-run frequency interpretation. Are there any restrictions on the types of events for which subjective probabilities can be used?

32. (a) What is your subjective probability that it will rain tomorrow?
    (b) What in your opinion are the odds in favor of rain tomorrow?
    (c) Are your answers to (a) and (b) consistent? If not, why not?
    (d) When "precipitation probabilities" are given out to the public, how do you interpret them? How do you think the average person interprets them?

33. Three football teams are fighting for the league championship. A sportswriter claims that the odds are even, or 1 to 1, that Team A will win the championship, the odds are even that Team B will win the championship, and the odds are even that Team C will win the championship. Is this reasonable? Explain your answer.

34. In the preceding problem, the sportswriter states that he is willing to bet for or against Team A at even odds, and likewise for Teams B and C. Even without knowing anything about the teams, would you like to bet with him? Explain.

35. In what sense is it true that a joint event emerging from a compound experiment can always be regarded as an elementary event?

36. As we saw in Chapter 1 of the text, a relation is a subset of a Cartesian product, such as $A \times B$. In a *statistical relation*, what would be the elements of the sets A and B making up the Cartesian product?

37. Consider an experiment involving a Red and a Black die. One of the two dice is selected at random, and then the selected die is rolled.
    (a) Let $C = \{R, B\}$ and $G = \{1, 2, 3, 4, 5, 6\}$. Find $C \times G = S$, the sample space for the above experiment.
    (b) Display this sample space in graphic form.
    (c) Find the elementary events (which are also joint events) making up the following events, and also the probability of these events:
        (i)   the die selected is R and comes up with an even number
        (ii) the die selected is black and comes up either with an odd number or a 6

      (iii)   the die selected comes up either with a 2 or a 3

      (iv)   the die selected is either red or black and comes up with a number divisible by 3.

38. Suppose that you have three dice, Red, Black, and Yellow. Now one of the dice is selected and rolled.

    (a)   Find the sample space S for this experiment.

    (b)   Find the elementary events associated with each of the following events, as well as the probabilities of these events:

        (i)   the die selected is Yellow and comes up with a number less than 5

      (ii)   the die selected is Red and comes up with an odd number, or is Black and comes up with an even number.

    (c)   Show that the color of the die and the number appearing on a roll are independent events. [*Hint*: Show, by counting elementary events, that the probability of any number and color combination is the product of the number probability and the color probability in this experiment.]

39. Let A be the event "a person is a male," and B the event "a person is blue-eyed" when individuals are chosen at random from some well-defined group of persons. Also let

$$P(A) = .60$$
$$P(B) = .20$$
$$P(A \mid B) = .10.$$

Find: (a) $P(A \cup B)$ (c) $P(\bar{A} \cup \bar{B}), P(\overline{A \cap B})$

        (b) $P(A \cup \bar{B}), P(\bar{A} \cup B)$ (d) $P\{(A \cap \bar{B}) \cup (\bar{A} \cap B)\}.$

40. If $P(A) = .6$, $P(B) = .15$, and $P(B \mid \bar{A}) = .25$, see if you can complete the following table:

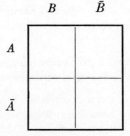

Also, find the following probabilities:

    $P(B \mid A)$ $P(A \cup B)$

    $P(A \mid B)$ $P\{(A \cap \bar{B}) \cup (\bar{A} \cap B)\}.$

41. Consider some sample space, in which events A_1, A_2, and A_3 are defined. You are given the following probabilities:

    (a) $P(A_1) = 1/2, P(A_2) = P(A_3) = 1/4$

       $P(A_1 \cap A_2) = P(A_1 \cap A_3) = P(A_2 \cap A_3) = 1/8$

       $P(A_1 \cap A_2 \cap A_3) = 1/32.$

       Find $P(A_1 \mid A_2)$, $P(A_1 \cap A_2 \mid A_3)$, $P(A_1 \mid A_2 \cap A_3)$,

       $P(A_1 \cup A_2 \mid A_3).$

(b)   Would you say that A_1, A_2, and A_3 are independent events?

(c)   Are $\bar{A}_1$, $\bar{A}_2$, and $\bar{A}_3$ independent events?

42.   In Exercise 41, imagine that $P(A_1) = 1/3$, $P(A_2) = 2/3$, and that $P(A_3) = 0$. Also that $P(A_1 \cap A_2) = P(A_1 \cap A_3) = P(A_2 \cap A_3) = 0$. Are A_1, A_2, and A_3 independent events here?

43.   Explain the difference between mutually exclusive events and independent events. Is it possible for two events A and B to be both mutually exclusive *and* independent? Explain.

44.   If $P(A) = 0$, does it make any sense to consider the conditional probability $P(B \mid A)$? Explain.

45.   Show that if $A \cap B = \varnothing$, then $P(A \mid A \cup B) = \dfrac{P(A)}{P(A) + P(B)}$.

46.   Suppose that a survey of 200 people at a certain college town has provided the following information about a proposal to allow more liberalized sale of beer:

	Male		Female		
	Student	Nonstudent	Student	Nonstudent	
Favor liberalized sale of beer	70	10	40	0	120
Do not favor liberalized sale	5	30	10	20	65
No opinion	5	0	10	0	15
Totals	80	40	60	20	200

If a person were selected at random from this group, what is the probability that the person

(a)   is a male

(b)   is a female who favors the liberalized sale of beer

(c)   is a female nonstudent

(d)   is either a female or a nonstudent

(e)   is a male who has no opinion, or a female who does not favor the liberalized sale of beer?

47.   Give an example of three events which are pairwise independent but not mutually independent. [*Hint:* Consider a simple experiment, such as the throwing of two dice.]

48.   Suppose that the probability that a student passes a quiz is .80 if he studies for the quiz and .50 if he does not study. If 60 percent of the students in a particular class studied for a given quiz, and a student who is chosen randomly from the class passes the quiz, what is the probability that this student studied for the quiz?

49. The probability that 1 percent of the items produced by a certain process are defective is .80, the probability that 5 percent of the items are defective is .10, and the probability that 10 percent of the items are defective is .10. An item is randomly chosen, and it is defective. Now what is the probability that 1 percent of the items are defective? That 5 percent are defective? That 10 percent are defective? Suppose that a *second* item is randomly chosen from the output of the process, and it too is defective. Following this second observation, what are the probabilities that 1, 5, and 10 percent, respectively, of the items produced by the process are defective?

50. In a football game, suppose that a team has time for one more play, and that they need to score a touchdown on this play to win the game. As a fan, you feel that there are only three possible plays that can be used, and that the three plays are equally likely to be used. The plays are a long pass, a screen pass, and an end run. Based on past experience in similar situations, you feel that the probabilities of getting a touchdown with these plays are as follows:

Play	*P (touchdown | play)*
Long pass	.50
Screen pass	.30
End run	.10

The team scores a touchdown on the play. What is the probability that they used the long pass? The screen pass? The end run?

51. Suppose that the probability that a person with tuberculosis will have a positive reading on a tuberculin skin test is .98. Furthermore, suppose that the probability that a person without tuberculosis will have a positive reading is .10. If .1 percent of the population of a given city has tuberculosis, what is the probability that a person selected from that population will have tuberculosis, given that he has a positive reading on the tuberculin skin test? Intuitively, this probability may seem quite low—explain why it is so low.

52. Suppose that Urn A is filled with 700 red balls and 300 green balls, and that Urn B is filled with 700 green balls and 300 red balls. One of the two urns is selected at random (that is, the two urns are equally likely to be selected). The experiment consists of selecting a ball at random (that is, all balls are equally likely) from the chosen urn, recording its color, and then replacing it and thoroughly mixing the balls again. This experiment is conducted three times, and each time the ball chosen is red. What is the probability that the urn which was chosen was Urn A?

3

PROBABILITY DISTRIBUTIONS

Given the definitions of "event," "sample space," and "probability" in the last chapter, we are now ready to take up the first major set of concepts in probability and statistics. First of all, the concept of a random variable will be introduced, and then we will discuss probability distributions of random variables. Finally, measures for summarizing probability distributions will be discussed.

3.1 RANDOM VARIABLES

We defined the **sample space** as **the set of all elementary events that are possible outcomes of some simple experiment.** These outcomes may be expressed in terms of numbers (for instance, the number coming up on one throw of a die; the temperature in degrees centigrade at a given place and time; the price of a certain stock on a given day) or they may be expressed in nonnumerical terms (for instance, the color of a ball drawn from an urn; the sex of a respondent to a questionnaire; the occurrence or nonoccurrence of rain on a particular day). Imagine that a single number could be assigned to each and every possible elementary event in a sample space S. Even if the elementary events themselves are already expressed in terms of numbers, we may think of a reassignment of a single number to each elementary event. Thus, each elementary event in S is associated with one and only one number. If we throw a die, the elementary events may correspond to the six possible faces that can come up. If we wished, we could assign to each elementary event the number of spots that appears on the face that comes up; or we could assign the *square* of the number that comes up; or we might even assign the number 1 if a six comes up, 2 if a five comes up, and so on, up to 6 if a one comes up. In all of these cases, each elementary event is associated with exactly one number. The most natural choice in the die example, of course, is to assign the number of spots on the face which comes up. The other examples point out the fact that this is not the only choice.

If the elementary events are already expressed conveniently in terms of numbers, as in the die example, these numbers themselves may be assigned to the events. On the other hand, numbers may be assigned to events as the result of some measurement process (for example, the use of a scale to measure weights), or they might even be assigned arbitrarily. It is important to note that more than one elementary event can be associated with a given number. In the die example, we could assign the number one to the events "1 comes up," "3 comes up," and "5 comes up," and assign the number zero to the remaining events. Thus the number one corresponds to the event "odd number comes up" and the number zero corresponds to the event "even number comes up." This is an example of how more than one elementary event can be associated with a given number. It is also an example of the assignment of numbers to events which are not originally expressed in terms of numbers. The events "odd number comes up" and "even number comes up" are expressed in nonnumerical terms. In a similar manner, in the rain example, we could assign the number one to the event "rain" and the number zero to the event "no rain."

The above examples show how a number can be assigned to each elementary event in a sample space. Suppose that the symbol X is used to stand for any particular number assigned to any given elementary event. Thus X is a variable, and any value of X is also an event, since there will be some set of elementary events assigned this value. Furthermore, since a value of X is an event, this event will have a probability $P(X)$ for the sample space S. We will use the following terminology: **If the symbol X represents a function which associates a real number with each and every elementary event in a sample space S, then X is called a random variable which is defined on the sample space S.** In other words, a random variable X is a real-valued function defined on a sample space. We will use *upper case* letters, such as X, Y, and Z, to denote random variables, and *lower case* letters, such as x, y, z, a, b, c, and so on, to denote particular *values* of random variables. The expression $P(X = x)$ symbolizes the probability that the random variable X takes on the particular value x. Where there is no chance for confusion, $P(X = x)$ will be abbreviated $P(x)$.

Although the term "random variable" is rather awkward, it will be used here because of its popularity in statistical writing. The terms "uncertain quantity," "chance variable," or "stochastic variable" are sometimes used. These all mean precisely the same thing, however: a symbol for numerical events each having a probability.

Given the random variable X, other events may be defined. Thus, given two numbers a and b,

$$(a \leq X \leq b)$$

is the event of the random variable X taking on a value between the numbers a and b (inclusive). This event has some probability $P(a \leq X \leq b)$. Furthermore, there is the event

$$(X \leq a),$$

which states that the random variable takes on a value less than or equal to the number a; this has probability $P(X \leq a)$.

This leads us to an alternate definition of a random variable:

A quantity symbolized by X is said to be a random variable if for every real number a there exists a probability $P(X \leq a)$ that X takes on a value that is less than or equal to a.

As we shall see, the definition stated above has its advantages when the variable X is continuous, ranging over all the real numbers, although there is nothing in this definition that says that the random variable *must* be continuous.

As an example of a random variable, suppose that X symbolizes the height of an American man, measured to the nearest inch. Here, there is some probability that $X \leq 60$, or $P(X \leq 60)$, since there is presumably some set of American men with heights less than or equal to 60 inches. Furthermore, there is some probability $P(70 \leq X \leq 72)$, since there is a set of American men having heights between five feet ten inches and six feet, inclusive. There are also some probabilities $P(X \leq 1)$ and $P(X \leq 120)$. These probabilities are almost surely zero and one, respectively, but the point is that they *do* exist. Here, X symbolizes the numerical value (height in inches) assigned to an American man (an elementary event). This symbol X represents any one of many different such values, and for any arbitrary number a there is some probability that the value of X is less than or equal to the value of a.

3.2 PROBABILITY DISTRIBUTIONS

Consider the set of events $\{A_1, A_2, \ldots, A_N\}$, and suppose that they form a partition of the sample space S. That is, they are mutually exclusive and exhaustive. The corresponding set of probabilities, $\{P(A_1), P(A_2), \ldots, P(A_N)\}$, is a probability distribution. In general, **any statement of a probability function having a set of mutually exclusive and exhaustive events for its domain is a probability distribution.**

As a simple example of a probability distribution, imagine a sample space of all boys of high-school age. A boy is selected at random and classified as "right-handed," "left-handed," or "ambidextrous." The probability distribution might be:

Class	P
Right	.60
Left	.30
Ambidextrous	.10
	1.00

Or, perhaps, the height of a boy drawn at random is measured. Some seven class intervals are used to record the height of any boy, and the probability distribution might be:

Height in inches	P
78–82	.002
73–77	.021
68–72	.136
63–67	.682
58–62	.136
53–57	.021
48–52	.002
	1.000

Here each class interval is an event.

The set of all possible elementary events form a partition of S. As a result, the set consisting of the probabilities of all of the elementary events in a sample space is a probability distribution. If we throw a fair die, the distribution would be:

Face coming up on die	P
1	1/6
2	1/6
3	1/6
4	1/6
5	1/6
6	1/6
	——
	1.0

However, suppose that we are interested in the *square* of the number of dots showing on the die. If we define the random variable X to be the square of the number of dots showing, then we can represent the probability distribution of X as follows:

x	$P(X = x)$
1	1/6
4	1/6
9	1/6
16	1/6
25	1/6
36	1/6
	——
	1.0

All values of X other than the values listed in the table have probabilities of zero. Recall that a random variable defined on a sample space associates a real number with each and every elementary event in the sample space. In the example, the probability distribution on the elementary events determines the probability distribution of the random variable X. This exemplifies the fact that, in general, **a set of probabilities, or a probability distribution, for the elementary events in S determines a set of probabilities, or a probability distribution, for any random variable defined on S.** This introduces the important concept of a probability distribution, or simply a distribution, of a random variable.

In the above example, each value of the random variable has the same probability. This need not always be true. Consider the random variable Y, which is defined as the number of heads occurring in two flips of a fair coin:

y	$P(Y = y)$
0	1/4
1	1/2
2	1/4

In general, the various values of a random variable are not necessarily equally probable.

In these examples we have specified probability distributions of random variables by simply listing the possible values of the random variables and the corresponding probabilities. If the possible values are few in number, this is an easy way to specify a probability distribution. On the other hand, if there is a large number of possible values, a listing may become cumbersome. In the extreme case, where we have an infinite number of possible values (for example, all real numbers between zero and one), it is clearly impossible to make a listing. Fortunately, there are other ways to specify a probability distribution; in some instances we may be able to specify a mathematical function from which the probability of any value or interval of values can be computed, or we may be able to express the distribution in graphical form. These methods will be discussed in the following sections. First, it is necessary to distinguish between two types of random variables, discrete and continuous.

3.3 DISCRETE RANDOM VARIABLES

There are very many situations where the random variable X can assume only a particular *finite* or *countably infinite* set of values; that is, the possible values of X are finite in number or they are infinite in number but can be put in a one-to-one correspondence with the positive integers [informally, they can be counted, such as the whole numbers $(1, 2, 3, \ldots)$]. For example, suppose that the simple experiment consists of selecting one United States family for observation and noting the number of children in that family; an elementary event is one particular United States family, and associated with each family is a number X. Here, X can assume only certain values between, say, 0 and 30, and these values must be integers. The numbers 3.68, -17, or π simply cannot be associated appropriately with any elementary event in the sample space. The random variable X can assume only a *finite* set of values. As another example, suppose that the simple experiment consists of tossing a fair coin repeatedly until a head appears. Let X be the number of tosses it takes to get the first head. The first head could appear on the first toss, in which case X would be 1; it might not appear until the one-hundredth toss, in which case X would be 100; in general, it might not appear until the Kth toss, where K is any whole number. Here we have a *countably infinite* set of values for X. **If the random variable X can assume only a particular finite or countably infinite set of values, it is said to be a discrete random variable.**

Not all random variables are discrete. Consider the simple experiment which consists of recording the temperature at a given place and time. Suppose that we can measure the temperature exactly; that is, our measuring device allows us to record the temperature to any number of decimal points. If X is the temperature reading, it is not possible for us to specify a finite or countably infinite set of values as the only possible values for X. If we specify a finite set of values, it is a simple matter to find more values that can be assumed by X, just by taking one of the finite set of values and adding more decimal places. For example, if one of the finite set of values is 75.965, we can determine values 75.9651, 75.9652, and so on, which are also possible values of X. What is being pointed out here is that the possible values of X consist of the set of real numbers, a set which contains an infinite (and uncountable) number of values. In this case, it is said that X is a **continuous random variable.** We will give a more formal definition of a continuous random variable after we finish discussing discrete random variables.

Given a discrete random variable taking on only the K different values $a_1, a_2, \cdots, a_K$, the following two conditions must be satisfied:

1. $P(X = a_i) \geq 0, \qquad i = 1, 2, \cdots, K.$ \hfill (3.3.1*)

2. $\sum_{i=1}^{K} P(X = a_i) = 1.$ \hfill (3.3.2*)

That is, the probability of each value that X can assume must be non-negative and the sum of the probabilities over all of the different values must be 1.00. These two conditions follow directly from the axioms of probability presented in Section 2.5.

Quite often one is interested in finding the probability that the obtained value of X will fall *between* two particular values, or in some interval. For example, the number of spots coming up on a pair of dice is a discrete random variable taking on only the whole number values 2 through 12 (the probabilities of 0 and of 1 are both zero, of course). What is the probability that X, the number of spots, is between 3 and 5 inclusive?

In order to find such probabilities, we rely once again on Rule 3 of Section 2.5. The various possible values of X are mutually exclusive events, and so

$$P(3 \leq X \leq 5) = P(3 \cup 4 \cup 5) = P(3) + P(4) + P(5).$$

If X is the number of spots on a pair of dice, then the random variable can be shown to have the distribution of Table 3.3.1.

Table 3.3.1

x	$P(X = x)$
12	1/36
11	2/36
10	3/36
9	4/36
8	5/36
7	6/36
6	5/36
5	4/36
4	3/36
3	2/36
2	1/36
	36/36

Using these probabilities, we find

$$P(3 \leq X \leq 5) = P(X = 3) + P(X = 4) + P(X = 5)$$
$$= \frac{2 + 3 + 4}{36} = \frac{1}{4}.$$

Similarly,

$$P(9 \leq X \leq 10) = P(X = 9) + P(X = 10) = \frac{7}{36}.$$

In general, the probability that X falls in the interval between any two numbers a and b, inclusive, is found by the sum of probabilities for X over *all possible* values between a and b, inclusive:

$$P(a \leq X \leq b) = \text{sum of } P(X = c) \text{ for all } c \text{ such that } a \leq c \leq b$$
$$= \sum_{a \leq c \leq b} P(X = c). \qquad (3.3.3^*)$$

By the same argument,

$$P(a \leq X) = \text{sum of } P(X = c) \text{ for all } c \text{ such that } a \leq c.$$

Furthermore,

$$P(X < a) = 1 - P(a \leq X) = \sum_{c<a} P(X = c). \qquad (3.3.4)$$

For this example,

$$P(5 \leq X) = P(X = 5) + \cdots + P(X = 12) = \frac{4}{36} + \cdots + \frac{1}{36}$$

$$= \frac{30}{36} = \frac{5}{6},$$

and

$$P(X < 5) = P(X = 4) + P(X = 3) + P(X = 2)$$

$$= \frac{3}{36} + \frac{2}{36} + \frac{1}{36} = \frac{1}{6}.$$

3.4 GRAPHS OF PROBABILITY DISTRIBUTIONS OF DISCRETE RANDOM VARIABLES

The probability distribution of a discrete random variable can be specified by a listing of the possible values of the random variable together with the corresponding probabilities. Such a listing is often represented by a graph. Suppose that the set of possible values of a random variable X can be denoted by $\{a_1, a_2, \cdots, a_K\}$, and that

$$\boldsymbol{p_i} = P(X = a_i), \qquad \text{for } i = 1, 2, \cdots, K.$$

Then the probability distribution can be represented by the set of pairs of values (a_i, p_i), where $i = 1, 2, \ldots, K$. We can graph such pairs in the Cartesian plane, as shown in Figure 3.4.1. The graph can be thought of as placing a mass of probability p_i at the point a_i on the X-axis. As a result, such a graph is often referred to as the graph of a **probability mass**

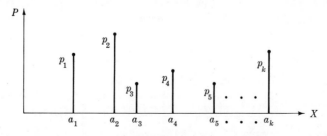

Figure 3.4.1

function (PMF), or simply a **probability function.** The advantage of such a graph over a listing is that it is generally easier to read and gives the reader a better notion of the nature of the probability distribution. In constructing graphs of PMF's, we will not always draw in the vertical axis, which is understood to represent probability.

Consider the PMF of the random variable Y from Section 3.2: the number of heads in two flips of a fair coin (Figure 3.4.2). As another example, consider the PMF of the random variable X specified in Table 3.3.1: the number of spots on a pair of dice (Figure 3.4.3).

The probabilities, or masses, in any PMF must be nonnegative, of course, and their sum over all values of the random variable must be one.

In some cases it may be possible to express the PMF in the form of a mathematical function. The simplest example concerns the case in which there are exactly K possible values of the random variable X, the integers 1 through K, and each value has probability $1/K$. If we denote the value of the PMF at the point $X = a$ by $P(X = a)$, we can write:

$$P(X = a) = \begin{cases} 1/K & \text{if } a = 1, 2, \cdots, K, \\ \\ 0 & \text{elsewhere.} \end{cases}$$

It is important to specify that the PMF is equal to zero at all values other than 1, 2, $\cdots$, K. Since the PMF is a function on the real line, it must be defined at all points on the real line.

Consider once again the random variable Y, the number of heads in two flips of a fair coin. The PMF of Y can be represented by

$$P(Y = b) = \begin{cases} \dfrac{\dbinom{2}{b}}{4} & \text{if } b = 0, 1, \text{ or } 2, \\ \\ 0 & \text{elsewhere.} \end{cases}$$

There are four possible outcomes (hence the 4 in the denominator), and the number of ways of getting b heads in 2 flips is $\binom{2}{b}$. By calculating the specific values of the PMF from the above formula, you can verify for

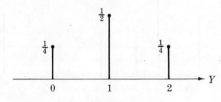

Figure 3.4.2

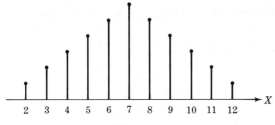

Figure 3.4.3

yourself that the numbers are equal to those in the listing of the distribution of Y given in Section 3.2.

In this section, we have seen how a discrete random variable can be represented in graphical or functional form. Whenever it is possible to use one or both of these two forms, they are usually preferable to a simple listing. In most cases they require less space than a listing and they are easier to understand. When we discuss the most important classes of discrete random variables in Chapter 4, we will rely almost exclusively on functional and graphical representations of probability distributions.

3.5   CONTINUOUS RANDOM VARIABLES

For our purposes, a continuous random variable may be defined as follows:

A random variable X is said to be continuous if for every pair of values u and v such that $P(X \le u) < P(X \le v)$, it is true that $P(u < X < v) > 0$.

Notice that for a random variable to be continuous in an interval limited by the values u and v, it is necessary that every nonzero interval in this range have a nonzero probability. For this to be true, there must be an infinite number of possible values that the random variable can assume.

It is very difficult to find real examples of continuous random variables; this concept is really an idealization that is never quite true, but has enormous mathematical utility. The closest one might come to a simple example is to imagine an infinitely large set of men, each of whom is weighed on a fantastically accurate spring balance, giving his weight to any number of decimal places. Obviously, no man will weigh as little as zero pounds, and it is hardly possible to find a man weighing many tons. However, between *some* limits any class interval of nonzero size should have nonzero probability. The weights would then be a continuous random variable between the specified limits. This is similar to the temperature example in Section 3.3.

In Section 3.3 it was possible for us to discuss discrete random variables in terms of probabilities $P(X = x)$, the probability that the random variable takes on *exactly* some value. However, consider the example of an infinite set of men weighed with any degree of accuracy. What is the probability that a man will be observed at random weighing *exactly* 160 pounds? This event is an interval of values smaller than 159.5–160.5, or 159.95–160.05, or any other interval $160 \pm .5i$, where i is not exactly zero. Since the smaller the interval, the smaller the probability, the probability of exactly 160 pounds is, in effect, zero. **In general, for continuous random variables, the occurrence of any *exact* value of X may be regarded as having zero probability.**

For this reason, one does not usually discuss the probability per se for a value of a continuous random variable; probabilities are discussed only for *intervals* of X values in a continuous distribution. Instead of the probability that X takes on some value a, we deal with the so-called **probability density** of X at a, symbolized by

$$f(a) = \text{probability density of } X \text{ at } a.$$

A fairly "rough" definition of a probability density can be given as follows: for any distribution it is proper to speak of the probability associated with an interval. Imagine an interval with limits a and b. Then the probability of that interval is

$$P(a \leq X \leq b) = P(X \leq b) - P(X < a).$$

Now suppose that we let the size of the interval be denoted by

$$\Delta X = b - a,$$

so that

$$b = a + \Delta X.$$

Then the probability of the interval *relative* to ΔX is just

$$\frac{P(a \leq X \leq b)}{\Delta X} = \frac{P(X \leq a + \Delta X) - P(X < a)}{\Delta X}.$$

Now suppose that we fix a, but allow ΔX to vary; in fact, let ΔX approach zero in size. What happens to the probability of this interval relative to the interval size? Both numbers will change, and the ratio will change, but this ratio will approach some **limiting value** as ΔX comes close to zero. That is, we can speak of the limit of $P(a \leq X \leq b)/\Delta X$ as ΔX approaches zero. *This limit gives the probability density of the variable X at value a.* Loosely speaking, one can say that the probability density at a is the *rate of change* in the probability of an interval with lower limit a, for minute changes in the size of the interval. This rate of change will depend on two things: the function rule assigning probabilities to intervals such as $(X \leq a)$ and the particular "region" of X values we happen to be talking about. Rather than talk about probabilities of X values per se, for continuous random variables it is mathematically far more convenient to discuss the probability density and reserve the idea of probability only for the discussion of *intervals* of X values. This need not trouble us especially, as we will generally be interested only in intervals of values in the first place. We will even continue to speak loosely of the probabilities for different X values. However, distribution functions plotted as smooth curves are really plots of probability *densities*, and the only probabilities represented are the areas cut off by nonzero intervals under these curves. Furthermore, when we come to look at a function rule for a continuous random variable, we will actually be looking at the rule for a density function.

If the random variable is discrete, then we can say that

$$P(X = a) = f(a) = \text{the probability density of } X \text{ at value } a.$$

Intervals can always be given probabilities, regardless of whether the random variable is discrete or continuous. For continuous random variables, the probability of any interval depends on the probability density associated with each value in the interval. The probability of any continuous interval is given by

$$P(a \leq X \leq b) = \int_a^b f(x) \, dx. \qquad (3.5.1^*)$$

Since the definite integral of a function over some interval corresponds to the area under the curve of the function in that interval, **we can say that the probability of an interval is the same as the area cut off by that interval under the curve for the probability densities, when the random variable is continuous, and the total area is equal to 1.00.** The expression $f(x) \, dx$ can be thought of as the area of a minute interval with midpoint $X = x$, somewhere between a and b. When the number of such intervals approaches an infinite number and their size approaches zero, the sum of all these areas is the *entire* area cut off by the limits a and b. Since there is an infinite number of such intervals, this sum is expressed by the definite integral sign

$$\int_a^b .$$

Note that this agrees with the definition of the probability of an interval for a discrete variable. In Section 3.3 it was stated that the probability that X lies between two numbers, a and b inclusive, is a sum of probabilities. Since a definite integral is analogous to a sum, the probability of an interval for a continuous random variable is analogous to the probability of an interval for a discrete random variable.

As noted above, when X is continuous, $P(X = a) = 0$ and $P(X = b) = 0$. As a result, the probability of an interval for a continuous random variable is the same whether or not the end points of the interval are included. Obviously, inclusion or exclusion of the end points will not affect the area given by

$$\int_a^b f(x)\ dx.$$

In the discrete case, on the other hand, the end points *do* make a difference. If X is the number coming up on one throw of a fair die, for instance,

$$P(2 < X < 4) = P(X = 3) = 1/6,$$

while

$$P(2 \leq X \leq 4) = P(X = 2) + P(X = 3) + P(X = 4) = 3/6 = 1/2.$$

A probability density function must satisfy two requirements:

1. $f(x) \geq 0$ for all x.   (3.5.2*)

2. $\int_{-\infty}^{\infty} f(x)\ dx = 1.$ (3.5.3*)

These conditions are analogous to the requirements imposed on a discrete PMF: $P(X = a) \geq 0$ and $\sum_a P(X = a) = 1$. In order to clarify the concept of a continuous random variable and to emphasize the analogy between discrete and continuous random variables, a graphical analysis is quite valuable.

3.6 GRAPHS OF PROBABILITY DISTRIBUTIONS OF CONTINUOUS RANDOM VARIABLES

It is impossible to specify the probability distribution of a continuous random variable by a listing of the possible values of the random variable and the corresponding probabilities. This is because the number of possible values is infinite (and not countably infinite), and as we pointed out in the last section, the probability for any exact value of a continuous random variable is zero. However, the probability distribution of a continuous random variable can be represented by a **probability density function** (PDF). Before giving examples of PDF's, we will present a graphical version of the development of a density function. This will parallel the development of the previous section.

Suppose that an experiment consists of selecting a boy at random and measuring his height, which is represented by the random variable X. Suppose that we can measure a boy's height to any number of decimal

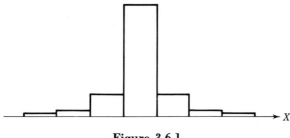

Figure 3.6.1

places, so that X is a continuous random variable. Now suppose that we divide the set of all possible heights into seven mutually exclusive and exhaustive intervals, as we did in Section 3.2. We can graphically represent the probability distribution given in Section 3.2 by means of a histogram (a bar graph). This is similar to a PMF, with the exception that the mass is thought of as being spread over a number of intervals instead of being concentrated at a number of points. The vertical axis is still in terms of probability, and the horizontal axis is still in terms of X (see Figure 3.6.1). If probability is represented by the mass, or area under the histogram, then the probability of any one of the seven specified intervals is equal to the area of the respective bar of the histogram.

Recall that it was assumed that X was continuous. Therefore, the choice of the intervals used in the histogram is arbitrary. In particular, we could choose any interval size i, where i is greater than zero, and still have a nonzero probability for all intervals of size i that lie within some outer limits, in this case 48 and 82 (from the height data in Section 3.2). This follows from the definition of a continuous random variable. For an interval size smaller than the $i = 5$ used above, say $i = 1$, our histogram might look like that shown in Figure 3.6.2. Here, the number of intervals is larger and the histogram area for each is smaller. Suppose that the interval size were made *extremely* small, $i = \Delta X$. Again, we can do this because X is continuous. In this circumstance each class interval will have an extremely small corresponding area, and the area for the histogram as a whole would be very nearly the same as enclosed in a smooth curve (Figure

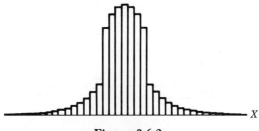

Figure 3.6.2

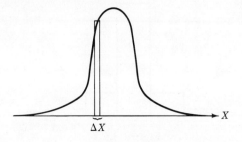

Figure 3.6.3

3.6.3). The width of the bar shown on the graph is a representative class interval. Notice that the area cut off by the interval with width ΔX under the curve is almost the same as the area of the histogram bar itself; that is, the area under the smooth curve for that interval is almost the same as the probability.

Finally, imagine that the interval size is made *very nearly zero*. As the class-interval size ΔX approaches zero, the probability associated with any class interval and the area cut off by the interval under the curve should agree increasingly well. Notice that as the interval size ΔX approaches zero, the probability associated with any interval must also approach zero, since the corresponding area under the curve is steadily reduced. As ΔX approaches zero, the smooth curve approaches the probability density function, or PDF, of X.

A continuous probability distribution is always represented as a smooth curve erected over the horizontal axis, which itself represents the possible values of the random variable X. A curve for some hypothetical distribution is shown in Figure 3.6.4. The two points a and b marked off on the horizontal axis represent limits to some interval. The shaded portion between a and b shows

$$P(a \leq X \leq b) = \int_{a}^{b} f(x) \, dx \qquad (3.6.1^*)$$

the probability that X takes on a value between the limits a and b. Re-

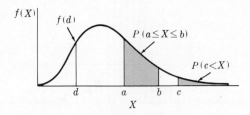

Figure 3.6.4

member that this probability corresponds to an *area* under the curve: in any continuous distribution, the probability of an interval can be represented by an area under the density curve. The total area under the curve represents 1.00, the probability that X takes on *some* value, and the curve can never be below the horizontal axis, since negative probabilities are not admissible. In other words, Equations (3.5.2) and (3.5.3) must be satisfied.

For an example of the graph of a specific PDF, consider a random variable that is uniformly distributed over the interval from zero to one. That is, if we divide the interval from zero to one into subintervals of equal width i, then these subintervals have the same probabilities, and we can do this for any value of i. The probability distribution of X can be represented in functional form:

$$f(x) = \begin{cases} 1 & \text{if } 0 \le x \le 1, \\ 0 & \text{elsewhere.} \end{cases}$$

Notice that $f(x)$ satisfies the two requirements of a PDF, since $f(x)$ is nonnegative and

$$\int_{-\infty}^{\infty} f(x)\, dx = \int_{-\infty}^{0} 0\, dx + \int_{0}^{1} 1\, dx + \int_{1}^{\infty} 0\, dx$$

$$= \int_{0}^{1} dx = 1 - 0 = 1.$$

The graph of $f(x)$ is as shown in Figure 3.6.5. Note that the total area under $f(x)$ is one, since it is simply the area of the unit square. In this example, it is easy to calculate the probability of any interval. Suppose that $0 \le a \le b \le 1$. Then

$$P(a \le X \le b) = \int_{a}^{b} f(x)\, dx = b - a.$$

It should be pointed out that it is possible for a random variable to be discrete over part of its range and continuous over another part of its range.

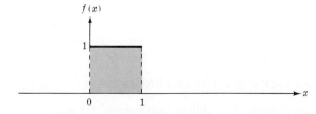

Figure 3.6.5

Suppose that $P(X = 2) = \frac{1}{2}$, and that the remaining probability $(1 - \frac{1}{2})$ is uniformly distributed over the unit interval from zero to one. The graph of the distribution of X is a combination of a PMF and a PDF:

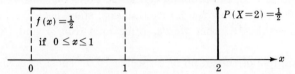

Although distributions of this nature do arise occasionally in practice, we will confine our discussion to distributions of random variables which are either discrete over the entire range or continuous over the entire range.

3.7 CONTINUOUS RANDOM VARIABLES AS AN IDEALIZATION

A continuous distribution of a random variable is a theoretical state of affairs. Continuous distributions can never be observed, but are only idealizations. In the first place, no set of potential observations is actually infinite in number; the statistician can deal only with finite numbers of real things, and so obtained data distributions are always discrete. Regardless of the size of his sample, it will never be large enough to permit each of the possible real numbers in any interval to be observed as values of the random variable. In the second place, measurements are imprecise. Not even in the most precise work known can accuracy be obtained to *any* number of decimals. This puts a limit to the actual possibility of encountering a continuous distribution in practice.

Why, then, do statisticians so often deal with these idealizations? The answer is that, mathematically, continuous distributions are far more tractable than discrete distributions. The function rules for continuous distributions are relatively easy to state and to study using the full power of mathematical analysis. This is not usually true for discrete distributions. On the other hand, continuous distributions are very good approximations to many truly discrete distributions. This fact makes it possible to organize statistical theory about a few such idealized distributions and find methods that are good approximations to results for the more complicated discrete situations. Nevertheless, the student should realize that these continuous distributions are mathematical abstractions that happen to be quite useful; they do not necessarily describe "real" situations exactly.

3.8 CUMULATIVE DISTRIBUTION FUNCTIONS

We have seen that the probability distribution of a random variable X can be represented by a probability mass function (PMF) in the discrete case and by a probability density function (PDF) in the continuous case.

There is a useful alternative method for specifying a probability distribution for both discrete and continuous random variables. This involves the concept of cumulative probabilities.

The probability that a random variable X takes on a value at or below a given number a is often written as

$$F(a) = P(X \leq a). \tag{3.8.1*}$$

The symbol $F(a)$ denotes the particular probability for the interval $X \leq a$; the general symbol $F(x)$ is sometimes used to represent the function relating the various values of X to the corresponding *cumulative* probabilities. This function is called the **cumulative distribution function** **(CDF)**.

A CDF must satisfy certain mathematical properties, the most important of which are:

1. $0 \leq F(x) \leq 1.$ $\qquad\qquad$ (3.8.2*)

2. If $a < b$, $F(a) \leq F(b)$. $\qquad$ (3.8.3*)

3. $F(\infty) = 1$ and $F(-\infty) = 0.$ $\qquad$ (3.8.4*)

The first property reflects the fact that $F(x)$ is a probability, $P(X \leq x)$, and properties 2 and 3 follow directly from the definition of a CDF.

Given the PMF of a discrete random variable or the PDF of a continuous random variable, the cumulative distribution function is uniquely determined. If X is discrete,

$$F(a) = P(X \leq a) = \sum_{X \leq a} P(X = a). \tag{3.8.5*}$$

If X is continuous,

$$F(a) = P(X \leq a) = \int_{-\infty}^{a} f(x)\, dx. \tag{3.8.6*}$$

Consider the example in which X represented the number of heads occurring in two flips of a fair coin. The PMF was

$$P(X = a) = \begin{cases} \frac{1}{4} & \text{for } a = 0 \text{ or } a = 2, \\ \frac{1}{2} & \text{for } a = 1, \\ 0 & \text{elsewhere.} \end{cases}$$

To compute the CDF, note that $F(a)$ will be one for all values of a greater than or equal to 2, since X is certain to be less than or equal to all of these values. Also, $F(a)$ will be zero for all negative values of X, since X cannot possibly be smaller than a negative number (it only takes on the values 0, 1, and 2). Similarly, we can determine the values of the CDF between 0 and 2, with the result shown in Figure 3.8.1.

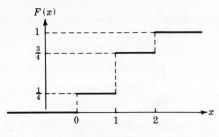

Figure 3.8.1

In the discrete case, the CDF is a step function: a step, or discontinuity in the CDF, occurs at any point a such that $P(X = a) > 0$. As a result, there will be K such "steps" in the CDF of X if X can take on any one of K values.

The CDF for the coin example can be expressed in functional form:

$$F(a) = \begin{cases} 1 & \text{for } a \geq 2, \\[2mm] \frac{3}{4} & \text{for } 1 \leq a < 2, \\[2mm] \frac{1}{4} & \text{for } 0 \leq a < 1, \\[2mm] 0 & \text{for } a < 0. \end{cases}$$

The CDF, like the PMF, is a function defined on the real line, and in the example it is assigned exactly one value for each real number. Note that the four intervals in the expression for $F(a)$ form a partition of the real line. Each step in the CDF of a discrete random variable occurs at a point a such that $P(X = a) > 0$; furthermore, the amount of the step is equal to $P(X = a)$. To see this graphically, compare the graphs of the PMF and the CDF in the coin example (Figures 3.4.2 and 3.8.1). It can be seen that $F(a)$ has a step of $1 - (\frac{3}{4}) = \frac{1}{4}$ at $a = 2$, a step of $(\frac{3}{4}) - (\frac{1}{4}) = \frac{1}{2}$ at $a = 1$, and a step of $(\frac{1}{4}) - 0 = \frac{1}{4}$ at $a = 0$. Thus, if we know the CDF of a discrete random variable, we can determine the corresponding PMF, and vice versa.

For an example of the CDF of a continuous random variable, recall the situation in which X was uniformly distributed over the interval from zero to one:

$$f(x) = \begin{cases} 1 & \text{if } 0 \leq x \leq 1, \\[2mm] 0 & \text{elsewhere.} \end{cases}$$

We see that $F(x)$ is equal to 1 for values of X greater than 1 and equal to

0 for values of X less than zero. For values of X between 0 and 1,

$$F(a) = \int_0^a f(x)\,dx = \int_0^a 1\,dx = a.$$

Thus the graph of the CDF of X is as shown in Figure 3.8.2.

Observe that the CDF in this example is a continuous function rather than a step function. This is true for all continuous random variables. In fact, the definition of a continuous random variable can be given in terms of the CDF: **A random variable X is said to be continuous if its cumulative distribution function $F(x)$ is a continuous function.** On the other hand, **the CDF of a discrete random variable is always a step function rather than a continuous function.**

If we know the CDF of a continuous random variable, it is possible to find the density function, $f(x)$. The relationship is given by

$$f(x) = \frac{d}{dx} F(x) = F'(x). \tag{3.8.7*}$$

In general, the derivative of the CDF is equal to the PDF. This follows from a fundamental theorem of calculus which requires that $F(x)$ be a continuous function. The reader not familiar with this theorem can safely take the above statement for granted and not worry about its derivation.

In the previous example, it follows that

$$F'(x) = \begin{cases} \dfrac{d}{dx}\, x = 1 & \text{for } 0 \le x \le 1, \\[3em] \dfrac{d}{dx} \cdot 1 = 0 & \text{for } x > 1, \\[3em] \dfrac{d}{dx} \cdot 0 = 0 & \text{for } x < 0. \end{cases}$$

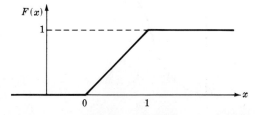

Figure 3.8.2

This is identical to $f(x)$ as given in Section 3.6. In geometric terms, the value of the density function at any point x is equal to the slope of the line tangent to the CDF at the point x. In the example, $F(x)$ is a straight line for $0 \leq x \leq 1$. At any point x, the tangent line is identical to the line representing $F(x)$. Then, since this line has a given slope which is obviously constant for all x in the unit interval, the density function must simply be constant for all x in the unit interval. Compare the graphs of the PDF and CDF (Figures 3.6.5 and 3.8.2).

In general (that is, for distributions other than uniform distributions), density functions are not constant (graphically, they are not horizontal lines). This means that, in general, cumulative distribution functions are not straight lines, as in the above example. Cumulative distribution functions for continuous random variables often have more or less the characteristic S-shape shown in Figure 3.8.3.

The probability that a continuous random variable takes on any value between limits a and b can be found from

$$P(a \leq X \leq b) = F(b) - F(a).$$

This is seen easily if it is recalled that $F(b)$ is the probability that X takes on value b or below, $F(a)$ is the probability that X takes on value a or below; their difference must be the probability of a value between a and b.

In the inequalities such as $a \leq X \leq b$, the "less than or equal to" sign is used. These could just as well be written with "less than" signs alone, however, and the statements would still be true for a *continuous* distribution, as we have pointed out. The reason is that for a continuous random variable the probability that X equals any exact number is regarded as zero, and thus the probabilities remain the same whether or not a or b or both are considered inside or outside of the interval. However, for discrete variables, $<$ and $\leq$ signs may lead to different probabilities.

The symbol $F(x)$ can be used to represent the cumulative probability that X is less than or equal to x either for a continuous or a discrete random variable. All random variables have cumulative distribution functions.

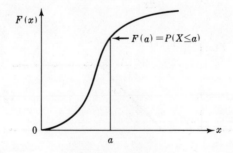

Figure 3.8.3

However, remember that symbols such as $f(x)$ are read differently for continuous and discrete variables. For a continuous variable, $f(x)$ means the *probability density* for X at the exact value x; this is not a probability. On the other hand, if the symbol $f(x)$ is applied to a discrete random variable, this *can* be read as the probability that $X = x$. To avoid possible confusion the density notation $f(x)$ will be reserved for *continuous* variables in all of the following, and $P(x)$ will be used both for discrete variables and for *intervals* of continuous variables, as in $P(a \leq X \leq b)$. Occasionally in mathematical statistics the terms "distribution function" or "probability function" are used to refer only to the *cumulative* distribution of a random variable, and "density function" is used where we have used "probability distribution." However, we believe that the terms used here are simpler for the beginning student to learn and use.

Tables of theoretical probability distributions are most often given in terms of the cumulative distribution, or $F(x)$, for a random variable. Given these cumulative probabilities for various values of X it is easy to find the probability for any interval by subtraction. In particular, when we come to use the table of the so-called "normal" distribution, we will find that cumulative probabilities are shown.

3.9 JOINT PROBABILITY DISTRIBUTIONS

We are often interested not just in *one* random variable, but in the relationship between two or more random variables. In a sample of American men, we might be interested in the relationship between X, the height of a man, and Y, the weight of a man. A medical researcher might be interested in the relationship between X, the number of cigarettes smoked by a person, and Y, the occurrence or nonoccurrence of cancer in that person. A stock market analyst might be interested in the relationship between X, the volume of sales for a given stock on a particular day, and Y, the closing price of that stock on that day. Examples of this nature are numerous. In Chapter 2, *joint events* were discussed; now we will introduce the concept of **joint probability distributions of two or more random variables.**

Suppose that we have two discrete random variables:

X, ranging over the values $a_1, a_2, \cdots, a_J$

and Y, ranging over the values $b_1, b_2, \cdots, b_K$.

We can discuss the joint event that X takes on some value a *and* that Y takes on some value b: this is the event $(X = a, Y = b)$, with probability $P(X = a, Y = b)$. The set of all such events, together with their probabilities, comprise the joint probability distribution of X and Y. Note that in this case there will be JK such events; of course, some may have zero probability, since there may be some combinations $(X = a, Y = b)$ which cannot possibly occur. For example, suppose that an experiment consists of tossing two fair dice. Let X represent the number coming up on the first die, let Y represent the number coming up on the second die, and let Z represent the sum of the numbers on the two dice. There are six possible values for X: 1, 2, 3, 4, 5, 6; there are six possible values for Y: 1, 2, 3, 4, 5, 6; and there are eleven possible values for Z: 2, 3, 4, 5, 6, 7, 8, 9, 10, 11, 12. There are $6 \cdot 6 = 36$ possible combinations (X, Y), each having probability 1/36. However, there are *not* $11 \cdot 6 = 66$ possible combinations (X, Z). The combination $(X = 6, Z = 3)$ is impossible, for example; if a six shows on the first die, the sum of the numbers on the two dice obviously must be greater than six, and hence cannot be three.

As in the case of a single random variable, a joint probability distribution of two discrete random variables can be represented by a listing, in functional form, or by a graph. The listing is similar to the listing of probabilities of joint events, presented in Section 2.27. For example, suppose that two coins are flipped, with X representing the outcome of the first toss ($X = 1$ if heads; $X = 0$ if tails) and Y representing the outcome of the second toss ($Y = 1$ if heads; $Y = 0$ if tails). The joint probability distribution of X and Y can be represented as follows:

$$
\begin{array}{cc|c|c|c}
 & & \multicolumn{2}{c}{Y} & \\
 & & 0 & 1 & \\
\hline
 & 0 & \frac{1}{4} & \frac{1}{4} & \frac{1}{2} \\
\hline
X & 1 & \frac{1}{4} & \frac{1}{4} & \frac{1}{2} \\
\hline
 & & \frac{1}{2} & \frac{1}{2} &
\end{array}
$$

The probabilities within the table comprise the joint distribution over the four possible events. This can be represented graphically, although the graph here must represent three, rather than two dimensions (Figure 3.9.1). This is merely an extension of the idea of the graph of a PMF of a single random variable, except that here probabilities are associated with *pairs* of values. Since there are two random variables, the total mass, or probability, must be distributed over the JK possible joint events. Here there are four possible joint events, each with probability $\frac{1}{4}$.

Even though our main interest is in the *relationship between two random variables*, we still might want to know something about each of the two random variables taken individually. If X is the height of a man and Y is the weight of the same man, we might be interested both in the height-weight relationship and in height and weight as separate random variables. Fortunately, if we know the joint distribution of X and Y, it is possible

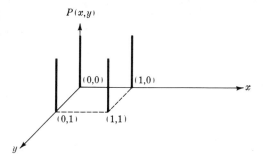

Figure 3.9.1

to find the individual distributions. For example, in the coin example represented by the above table, the entries in the margins of the table of the joint distribution of X and Y represent the probabilities for X and Y taken separately. For obvious reasons, these are known as *marginal probabilities*. It is a simple matter to derive marginal probabilities from joint probabilities:

$$P(X = a_j) = \sum_{k=1}^{K} P(X = a_j, Y = b_k). \qquad (3.9.1)$$

$$P(Y = b_k) = \sum_{j=1}^{J} P(X = a_j, Y = b_k). \qquad (3.9.2)$$

In other words, the marginal probabilities simply represent the sum of the appropriate joint probabilities (that is, the sum of all joint probabilities involving the value of X (or Y) for which the marginal probability is desired).

The set of all values of X together with their marginal probabilities is called the **marginal distribution** of X; the marginal distribution of Y is defined similarly. Notice that this distribution is determined from the joint distribution of X and Y. However, since the effect of the values of Y upon the values of X is summed out by Equation (3.9.1), the marginal distribution is simply the probability distribution of a single random variable, X. If we considered X alone and determined its distribution without looking at any other variables, as in Section 3.3, the result would be identical to the marginal distribution derived from $P(x, y)$.

Using joint probabilities and marginal probabilities, we can determine **conditional probabilities,** just as in Chapter 2:

$$P(X = a \mid Y = b) = \frac{P(X = a, Y = b)}{P(Y = b)}. \qquad (3.9.3)$$

$$P(Y = b \mid X = a) = \frac{P(X = a, Y = b)}{P(X = a)}. \qquad (3.9.4)$$

For example, consider a sample space of American men. Let the random variable X be the hat size of a man, and let Y be his shoe size. Imagine that we know that

$$P(X = 7) = .30$$

and

$$P(Y = 9) = .20.$$

Furthermore,

$$P(X = 7, Y = 9) = .10.$$

Then the probability that a man wears hat size 7 given that his shoe size is 9 is found to be

$$(X = 7 \mid Y = 9) = \frac{.10}{.20} = .50.$$

The set of values of $P(X = a \mid Y = b)$ for all values of a and a fixed value of b is called the **conditional distribution of X given that $Y = b$.** In general, this distribution is different from the marginal distribution of X. Notice that the value b must be specified before the conditional distribution of X given that $Y = b$ can be determined.

In the discrete case, joint probability distributions are analogous to joint events. The same ideas also apply to continuous random variables as well, except that the definitions are stated in terms of probability density functions. Instead of a series of probabilities of the form $P(X = a, Y = b)$, one considers a joint density function $f(x, y)$, such that

$$f(x, y) \geq 0 \qquad \text{for all real } x, y \qquad (3.9.5)$$

and

$$\int_{-\infty}^{\infty} \int_{-\infty}^{\infty} f(x, y) \, dx \, dy = 1. \qquad (3.9.6)$$

Of particular interest are joint events such as $(a \leq X \leq b, c \leq Y \leq d)$, which amount to joint intervals of values for two random variables. By use of the joint density function (3.9.6), we find that

$$P(a \leq X \leq b, c \leq Y \leq d) = \int_{c}^{d} \int_{a}^{b} f(x, y) \, dx \, dy. \qquad (3.9.7)$$

Given the joint density function $f(x, y)$ of the continuous random variables X and Y, it is possible to determine the density functions (or **marginal density functions**) of both X and Y, as follows:

$$f(x) = \int_{-\infty}^{\infty} f(x, y) \, dy \qquad (3.9.8)$$

and

$$f(y) = \int_{-\infty}^{\infty} f(x, y) \, dx. \qquad (3.9.9)$$

Notice the similarity between this operation and the derivation of marginal distributions in the discrete case [Equations (3.9.1) and (3.9.2)]. To get the marginal probabilities for the various values of X in the discrete case, we sum the joint probabilities over the possible values of Y. To get the marginal density for X in the continuous case, we integrate the joint density over the possible values of Y.

Once we have the joint density and the marginal densities, it is possible to determine the **conditional densities** in a manner parallel to the determination of conditional probabilities:

$$f(x \mid Y = y) = \frac{f(x, y)}{f(y)} \qquad (3.9.10)$$

and

$$f(y \mid X = x) = \frac{f(x, y)}{f(x)} . \qquad (3.9.11)$$

Here $f(x \mid Y = y)$ symbolizes the density for the random variable X, given that the random variable Y assumes the value y. In general, the conditional density of one random variable given a particular value of a second random variable is simply the joint density of the two random variables divided by the marginal density of the second random variable, where both densities are evaluated at the given value of the second random variable.

For example, suppose that the joint density of X and Y is given by

$$f(x, y) = \begin{cases} 4xy & \text{if } 0 \leq x \leq 1, 0 \leq y \leq 1, \\ 0 & \text{elsewhere.} \end{cases}$$

The marginal densities of X and Y are

$$f(x) = \int_{-\infty}^{\infty} f(x, y) \, dy = \int_{0}^{1} 4xy \, dy = \left[\frac{4xy^2}{2} \right]_{y=0}^{y=1} = 2x$$

and

$$f(y) = \int_{-\infty}^{\infty} f(x, y) \, dx = \int_{0}^{1} 4xy \, dx = \left[\frac{4x^2y}{2} \right]_{x=0}^{x=1} = 2y.$$

The conditional densities are

$$f(x \mid y) = \frac{f(x, y)}{f(y)} = \frac{4xy}{2y} = 2x$$

and

$$f(y \mid x) = \frac{f(x, y)}{f(x)} = \frac{4xy}{2x} = 2y.$$

Note that in this example, $f(x, y) = f(x)f(y)$; the joint density of X and

Y is equal to the product of the marginal densities of X and Y. This will not be true in general; when it *is* true, we say that the random variables X and Y are independent, as we shall see in the next section. Because of this independence, it is also true that the marginal density of X is equal to the conditional density of X given *any* value of Y, and similarly for the marginal and conditional densities of Y.

The relationship among *several* random variables is very often of interest. Extending the examples at the beginning of the section, we might be interested in the relationship among three random variables: X, the height of a man; Y, the weight of a man; and Z, the age of a man. A medical researcher might be interested in the relationship among four variables: X, the number of cigarettes smoked by a person; Y, the occurrence or nonoccurrence of cancer in that person; Z, the age of that person; and W, the occurrence or nonoccurrence of heart disease in that person. A stock market analyst might be interested in the relationship among X, the volume of sales for a given stock on a particular day; Y, the closing price of that stock on that day; Z, the volume of sales for the entire stock market on that day; and W, the closing price of some other particular stock on that day. Clearly there are many situations in which the relationship among several random variables is of interest.

The representation of joint probability distributions of three or more random variables is a straightforward extension of the analysis presented for two random variables. In the discrete case, suppose that we have K random variables $X_1, X_2, \cdots, X_K$. The joint distribution of the K random variables can then be represented by

$$P(X_1 = a_1, X_2 = a_2, \cdots, X_K = a_K),$$

where the probability function is defined for all possible combinations of values $(a_1, a_2, \cdots, a_K)$. In accordance with the general axioms of probability, this function must satisfy two requirements:

1. $P(X_1 = a_1, X_2 = a_2, \cdots, X_K = a_K) \geq 0$ for all $(a_1, a_2, \cdots, a_K)$

(3.9.12)

and

2. $\displaystyle\sum_{a_1} \sum_{a_2} \cdots \sum_{a_K} P(X_1 = a_1, X_2 = a_2, \cdots, X_K = a_K) = 1.$ (3.9.13)

To determine the marginal probability function for any single one of the K random variables, we simply sum $P(X_1 = a_1, X_2 = a_2, \cdots, X_K = a_K)$ over the remaining $K - 1$ random variables. For instance, if we wanted to find $P(X_1 = a_1)$,

$$P(X_1 = a_1) = \sum_{a_2} \sum_{a_3} \cdots \sum_{a_K} P(X_1 = a_1, X_2 = a_2, \cdots, X_K = a_K).$$

(3.9.14)

Similarly, to determine the joint probability function for any particular pair of random variables chosen from $X_1, X_2, \cdots, X_K$, just sum $P(X_1 = a_1, X_2 = a_2, \cdots, X_K = a_K)$ over the remaining $K - 2$ random variables.

The continuous case parallels the discrete case, with definite integrals replacing sums. Suppose that $X_1, X_2, \cdots, X_K$ are continuous random variables. Their joint density function can be represented by

$$f(x_1, x_2, \cdots, x_K),$$

where $\qquad f(x_1, x_2, \cdots, x_K) \geq 0$ for all real $x_1, x_2, \cdots, x_K \qquad$ (3.9.15)

and $\qquad \displaystyle\int_{-\infty}^{\infty} \int_{-\infty}^{\infty} \cdots \int_{-\infty}^{\infty} f(x_1, x_2, \cdots, x_K) \, dx_1 \, dx_2 \cdots dx_K = 1.$ (3.9.16)

[The reader not familiar with the technique of multiple integration need not be overly concerned about the actual process of evaluating definite multiple integrals. It suffices to say that multiple integration is an extension of the integration of functions of one variable to allow us to integrate functions of several variables. Think of this process as being analogous to a multiple summation such as that presented in Equation (3.9.13).]

To find the probability of any particular K-dimensional interval, it is necessary to integrate the joint density over that interval:

$$P(a_1 \leq X_1 \leq b_1, a_2 \leq X_2 \leq b_2, \cdots, a_K \leq X_K \leq b_K)$$

$$= \int_{a_k}^{b_k} \int_{a_{k-1}}^{b_{k-1}} \cdots \int_{a_2}^{b_2} \int_{a_1}^{b_1} f(x_1, x_2, \cdots, x_K) dx_1 \, dx_2 \cdots dx_K \qquad (3.9.17)$$

To determine the marginal density function for any single one of the K random variables, we integrate the joint density over the remaining $K - 1$ random variables. For instance, if we wanted to find the marginal density of X_K, $f_K(x_K)$,

$$f_K(x_K) = \int_{-\infty}^{\infty} \int_{-\infty}^{\infty} \cdots \int_{-\infty}^{\infty} f(x_1, x_2, \cdots, x_K) \, dx_1 \, dx_2 \cdots dx_{K-1}.$$

In this section we have introduced the concepts of joint, marginal, and conditional probability distributions. These distributions are important whenever we are interested in the relationship among two or more variables. For instance, suppose that we wanted to predict X, the sales of a particular product during a given time period. The relevant probability distribution would be the distribution of X. However, suppose that we could obtain information regarding the disposable income of consumers during this time period (call this Y). Since the knowledge of the value of Y during this period should help us to predict X, we might now be interested in the joint distribution of X and Y. From this joint distribution, we could compute the conditional distribution of X given that $Y = y$. But we *know* that

$Y = y$, so this conditional distribution is the relevant distribution for the prediction of X. Similarly, if we could find out the value of a third related variable, say Z, then the distribution of interest would be the conditional distribution of X given that $Y = y$ and $Z = z$.

This example illustrates the potential use of distributions involving two or more random variables. Problems such as this will be discussed in Chapter 10, and we will defer until then other specific examples of joint, marginal, and conditional distributions.

3.10 INDEPENDENCE OF RANDOM VARIABLES

We defined the independence of events in Section 2.26, and now we will define independence of random variables in both the discrete and continuous cases. As we shall see, the concept of independence is critical in any attempt to discuss the relationship among random variables.

The definition of independence of discrete random variables is very similar to the definition of independence of events:

Two discrete random variables are independent if, and only if,

$$P(X = a, Y = b) = P(X = a)P(Y = b) \qquad (3.10.1)$$

for all pairs of values a and b.

The same idea applies to continuous random variables as well, except that the definition is stated in terms of probability densities:

A random variable X with density $f(x)$ at value x and a random variable Y with density $f(y)$ at value y are independent if, and only if, for all (x, y),

$$f(x, y) = f(x)f(y), \qquad (3.10.2^*)$$

where $f(x, y)$ is the joint density for the event (x, y).

Alternatively, X and Y are independent if and only if for all (x, y) such that $X = x$ and $Y = y$,

$$f(x \mid y) = f(x), \qquad (3.10.3)$$

or, equivalently, $\qquad\qquad f(y \mid x) = f(y). \qquad (3.10.4)$

That is, the random variables are independent if and only if the conditional densities are everywhere equal to the marginal densities. All three definitions [(3.10.2), (3.10.3), and (3.10.4)] can be shown to be equivalent by using the definition of a conditional density function.

For any random variables, one often has occasion to consider joint intervals such as

$$[(a \leq X \leq b) \text{ and } (c \leq Y \leq d)],$$

some value of X lying between a and b and some value of Y between c and d. For independent random variables,

$$P(a \leq X \leq b \text{ and } c \leq Y \leq d) = P(a \leq X \leq b)P(c \leq Y \leq d),$$

$$(3.10.5)$$

just as for any independent events.

The concept of independence can be extended to more than two random variables. The K random variables X_1, X_2, $\cdots$, X_K are independent if, and only if, their joint mass (density) function is equal to the product of their mass (density) functions. Formally, if X_1, X_2, $\cdots$, X_K are discrete random variables, they are independent if, and only if,

$$P(X_1 = a_1, X_2 = a_2, \cdots, X_K = a_K)$$

$$= P(X_1 = a_1)P(X_2 = a_2)\cdots P(X_K = a_K) \quad (3.10.6)$$

for all sets of values $(a_1, a_2, \cdots, a_K)$. Similarly, if X_1, X_2, $\cdots$, X_K are continuous random variables with probability density functions $f_1(x_1)$, $f_2(x_2)$, $\cdots$, $f_K(x_K)$, they are independent if, and only if,

$$f(x_1, x_2, \cdots, x_K) = f_1(x_1)f_2(x_2)\cdots f_K(x_K) \quad (3.10.7)$$

for all sets of values $(x_1, x_2, \cdots, x_K)$, where $f(x_1, x_2, \cdots, x_K)$ is the joint density for the event $(x_1, x_2, \cdots, x_K)$. It should also be noted that the concepts of pairwise independence and conditional independence, defined in Section 2.26 for events, also apply to random variables by a straightforward extension of the definitions in Section 2.26.

The idea of independent random variables permeates all of statistical theory. In some instances, X and Y are different kinds of measurements, and one is interested in the association between the attributes they represent. However, equally important is the situation where a series of observations are made, each observation producing one value of the *same* random variable X. The series of outcomes can be regarded as a set of random variables X_1, X_2, X_3, and so on, each having exactly the same distribution. The subscripts refer only to the order of sampling in the sequence, the value observed first, second, and so on. For example, we might be measuring the same individual over time, or drawing several cases from the same group of people measured in the same way. If these observations are independent, then for, say, X_1 and X_2,

$$f(x_2 \mid x_1) = f(x_2)$$

and

$$f(x_1 \mid x_2) = f(x_1);$$

the distribution of X is the same regardless of which observation in the sequence or in the total sample we are considering. This amounts to saying that the chances for different intervals of values of X to occur stay the same regardless of where we are in the sampling process. This essentially is the concept of independence which was invoked in the discussion of Bernoulli's theorem in Section 2.16.

3.11 INDEPENDENCE OF FUNCTIONS OF RANDOM VARIABLES

The idea of independence of random variables can be extended to functions of random variables as well. That is, suppose we have a random variable X with a density $f(X)$, and an *independent* random variable Y with density $f(Y)$, as before.

Now imagine some function of X: associated with each possible X value is a new number $V = t(X)$. For example, it might be that

$$V = aX + b,$$

so that V is some linear function of X. Or, perhaps

$$V = 3X^2 + 2X,$$

and so on for any other function rule. Furthermore, let there be some function of Y,

$$W = s(Y),$$

so that a new number W is associated with each possible value of Y. *Then V and W are independent random variables, provided that X and Y are independent.* This principle applies both to continuous and to discrete random variables, and to all the ordinary functions studied in elementary mathematical analysis. We will make extensive use of this principle in later sections on sampling random variables.

It is especially interesting to apply this principle to measurements as functions of underlying magnitudes of properties. Imagine that the true magnitudes on two properties for some sample space of objects are represented by the random variables X and Y, and that the numerical measurements assigned to these objects are represented by V and W. Furthermore, suppose that V is some function of X, and W is some function of Y. Then if the underlying variables X and Y are independent, the measurements V and W must be independent as well. Conversely, if V and W are not independent, X and Y are not independent.

3.12 SUMMARY MEASURES OF PROBABILITY DISTRIBUTIONS

As we have seen, the probability distribution of a random variable, whether discrete or continuous, can be represented in several alternative ways. If a graphical analysis is desired, a distribution can be represented by a probability mass function (PMF) in the discrete case and by a probability density function (PDF) in the continuous case; or it can be represented by a cumulative distribution function (CDF) in either case. In many cases it is also possible to represent a probability distribution in functional notation, provided there is some relatively simple mathematical function that can describe the PMF, PDF, or CDF. In the discrete case, a probability distribution can be represented by a simple listing of all possible values of the random variable and the probability corresponding to each value (provided that the random variable takes on only a finite number of values). This last form of representation could also be used as an approximation in the continuous case if the possible values of the random variable are grouped into a finite set of mutually exclusive and exhaustive intervals on the real line.

Thus, there are several ways to represent the probability distribution of a random variable. Each of these methods has some disadvantages: if there are numerous possible values for a discrete random variable, a listing or a graph might be inconvenient; if the mathematical function corresponding to a distribution is quite complicated or if no function can be found to represent the distribution, the use of functional notation might be infeasible. Even if these disadvantages do not apply in a particular case, it may be quite time-consuming to represent the entire probability distribution of interest. Consequently, it is often easier and more efficient to look only at certain characteristics of distributions than to attempt to specify the distribution as a whole. Such characteristics can be summarized into one or more numerical values, conveying some, though not all, of the information in the entire distribution.

Two such general characteristics of any distribution are its measures of central tendency (or location) and dispersion (or variability). Indices of central tendency are ways of describing the "typical" or the "average" value of the random variable. Indices of dispersion, on the other hand, describe the "spread" or the extent of the difference among the possible values of the random variable.

In the following sections we will discuss several measures of central tendency and dispersion. In many cases, it is possible to adequately describe a probability distribution with a few such measures. It should be remembered, however, that these measures serve only to summarize some important features of the probability distribution; in general, they do not completely describe the entire distribution. In some situations in statistical analysis the entire distribution is of genuine interest, and the summary measures are not sufficient for the problem being attacked. On the other hand, as we will see in Chapter 4, if we know that the distribution is one of a number of frequently encountered types of distributions, it may be possible to get a complete picture of the distribution from a few summary measures.

One of the most common and most useful summary measures of a probability distribution is a measure known as the *expectation* of a random variable. The concept of expectation plays an important role not only as a useful summary measure, but also as a central concept in the theory of probability and statistics.

3.13   THE EXPECTATION OF A RANDOM VARIABLE

A very prominent place in theoretical statistics is occupied by the concept of the mathematical expectation of a random variable X. **If the distribution of X is discrete, then the expectation (or expected value) of X is defined to be**

$$E(X) = \sum_x xP(X = x) = \sum_x xP(x), \qquad (3.13.1^*)$$

where the sum is taken over all of the different values that the variable X can assume, and

$$\sum_x P(x) = 1.00.$$

For a continuous random variable X ranging over all the real numbers, the expectation is defined to be

$$E(X) = \int_{-\infty}^{\infty} xf(x) \, dx, \qquad (3.13.2^*)$$

where $\qquad \int_{-\infty}^{\infty} f(x) \, dx = 1.00.$

Note the similarity between the discrete and continuous cases. In the discrete case, $E(X)$ is a sum of products of values of X and their probabilities; in the continuous case, $E(X)$ is an integral (which is, after all, a sum in the limiting case) of products of values of X and the corresponding values of the probability density function of X. A mathematical note of interest is that the sum in Equation (3.13.1) or the integral in Equation (3.13.2) may not converge to a finite value, in which case we say that the expectation does not exist. Since the expectations which we will deal with in this book exist in all but a few relatively unimportant cases, we need not concern ourselves with this mathematical problem.

As the discussion in Sections 2.19 and 2.20 suggests, the idea of expectation of a random variable is closely connected with the origin of statistics in games of chance. Gamblers were interested in how much they could "expect" to win in the long run in a game, and in how much they should wager in certain games if the game was to be "fair." Thus, expected value originally meant the expected long-run winnings (or losings) over repeated play; this term has been retained in mathematical statistics to mean the long-run average value for any random variable over an indefinite number of samplings. This holds whether a large number of samplings will actually be conducted or whether the situation is a one-trial affair and we consider hypothetical repetitions of the situation. Over a long series of trials, we can "expect" to observe the expected value. At any *single* trial, we in general cannot "expect" the expected value; usually the expected value is not even a possible value of the random variable for any single trial, as the following examples illustrate.

The use of expectation in a game of chance is easy to illustrate. For example, suppose that someone were setting up a lottery, selling 1000 tickets at \$1 per ticket. He is going to give a prize of \$750 to the winner of the first draw. Suppose now that you buy a ticket. How *good* is this ticket in the sense of *how much you should expect to gain?* Should you have bought it in the first place? You can think of your chances of winning and losing as represented in a probability distribution where the outcome of any drawing falls into one of two event categories:

Class	Prob.
Win	1/1000
Do not win	999/1000

Translated into the amount of money gained (the random variable X), and with a loss regarded as a negative gain, this distribution becomes

x	$P(x)$
\$749	1/1000
−\$ 1	999/1000

Applying Equation (3.13.1), we find that

$$E(X) = \sum xP(x) = 749(1/1000) + (-1)(999/1000)$$
$$= .749 - .999 = -.25.$$

This amount, a *minus* 25 cents, is the amount which you can expect to gain by buying the ticket, meaning that if you played the game over and over indefinitely, in the long run you would be poorer by a quarter per trial. Should you buy the lottery ticket? Probably not, if you are going to be strictly rational about it; the mean winnings (the expected value) is certainly not in your favor.

On the other hand, suppose that the prize offered were $1000, so that the gain in winning is $999. Now the expected value is

$$E(X) = 999(1/1000) + (-1)(999/1000) = 0.$$

Here the game is more worth your while, as there at least is no amount of money to be lost *or* gained in the long run. Such a game is often called "fair." Obviously, truly fair lotteries and other games of chance are hard to find, since their purpose is to make money for the proprietors, not to break even or lose money. If the prize were $2000, you would likely jump at the opportunity, as in this case the expected value woud be exactly the *gain* of one dollar. This statement requires the assumption, made in Chapter 2, that the expected value of a gamble is a good criterion for deciding whether or not to accept the gamble. We will discuss this point at some length in Chapter 9.

In figuring odds in gambling situations, one uses the expected value to find what constitutes a fair bet; that is, a bet where $E(X)$ of the probability distribution of gains and losses is zero. For instance, it is known that in a particular game the odds are 4 to 1 *against* winning. This means that the probability of winning is $1/5$ and that of losing is $4/5$. It costs the player exactly $1 to play the game once. How much should the amount he gains by winning be in order to make this a fair game with expectation of zero? The expectation is

$$E(X) = \text{(gain value)}P(\text{win}) + \text{(loss value)}P(\text{lose})$$

$$= \text{(gain value)}(1/5) - \$1(4/5).$$

Setting $E(X)$ equal to zero and solving gives

$$\text{(gain value)}(1/5) - \$1(4/5) = 0,$$

or   gain value $= \$4.$

In short, one should stand to gain $4 for $1 put up if the game is to be fair. In general, if odds are A to B against winning, the game or bet is fair when B dollars put up gains A dollars.

In betting situations, the random variable is, of course, gains or losses of amounts of money, or of other things having utility value for the person. Nevertheless, the same general idea applies to any random variable; the expectation is the long-run average value that one should observe. The expectation can be thought of in this way even if the experiment (whether a betting situation or not) is strictly a one-trial affair and will not be repeated. No assumption has been made regarding the *interpretation* of the probabilities used to determine $E(X)$.

3.14   THE ALGEBRA OF EXPECTATIONS

In essence, the expectation of a random variable, whether the random variable is discrete or continuous, is a kind of weighted sum of values, and thus the rules for the algebraic treatment of expectations are basically extensions and applications of the rules of summation. These rules apply either to discrete or to continuous random variables if particular boundary conditions exist; for our purpose these rules can be used without our going further into these special qualifications. Proofs will be presented for the discrete case only, since the proofs for the continuous case are virtually identical, with integrals replacing summation signs. The student is advised to become familiar with these rules. If he does so, he should have little trouble in following simple derivations in mathematical statistics such as this book contains.

RULE 1

If $g(X)$ is some function of X, then

$$E(g(X)) = \sum_x g(x)P(x) \qquad \textbf{in the discrete case} \qquad (3.14.1^*)$$

and

$$E(g(X)) = \int_{-\infty}^{\infty} g(x)f(x)\, dx \quad \textbf{in the continuous case.} \quad (3.14.2^*)$$

From the original definition of expectation in the discrete case,

$$E(g(X)) = \sum_x g(x)P(g(x)).$$

But the probability that the function of X, $g(X)$, will take on a particular value, say $g(a)$, is the same as the probability that X will take on the value a, and therefore $P(g(x)) = P(x)$, which gives us Rule 1. Technically, this last statement is not absolutely correct, since it is possible that two possible values of X might result in the same value of $g(X)$, say $g(a) = g(b) = c$. Then $P(g(X) = c) = P(X = a) + P(X = b)$. Since both of these possibilities are accounted for in the determination of $E(g(X))$, Rule 1 is correct as stated.

RULE 2

If a is some constant number, then

$$E(a) = a. \qquad (3.14.3^*)$$

That is, if the same constant value a were associated with each and every elementary event in some sample space, the expectation or mean of the values would most certainly be a. For a discrete random variable X,

$$E(a) = \sum_x aP(x) = a \sum_x P(x) = a.$$

Rule 2 says that if you will gain \$$a$ no matter what event occurs, your expected gain is obviously \$$a$.

RULE 3

If a is some constant real number and X is a random variable with expectation $E(X)$, then

$$E(aX) = aE(X). \qquad (3.14.4^*)$$

Suppose that a new random variable is formed by multiplying each value of X by the constant number a. Then the expectation of the new random variable is just a times the expectation of X. This is very simple to show for a discrete random variable X:

$$E(aX) = \sum_x axP(x) = a \sum_x xP(x) = aE(x).$$

RULE 4

If a is a constant real number and X is a random variable, then

$$E(X + a) = E(X) + a. \qquad (3.14.5^*)$$

This can be shown very simply for a discrete variable. Here,

$$E(X + a) = \sum_x (x + a)P(x)$$

$$= \sum_x xP(x) + a \sum_x P(x)$$

$$= E(X) + a \sum_x P(x)$$

$$= E(X) + a.$$

The expectations of functions of random variables, such as

$$E[(X + 2)^2],$$
$$E(\sqrt{X + b}),$$

and $\qquad\qquad\qquad E(b^X),$

to give only a few examples, are subject to the same algebraic rules as summations. That is, the operation indicated within the punctuation is to be carried out *before* the expectation is taken. It is most important that this be kept in mind during any algebraic argument involving summations. In general,

$$E[(X + 2)^2] \neq [E(X) + E(2)]^2,$$
$$E(\sqrt{X}) \neq (\sqrt{E(X)}),$$
$$E(b^X) \neq b^{E(X)},$$

and so forth.

The next few rules concern two (or more) random variables, symbolized by X and Y.

RULE 5

If X and Y are random variables with joint probability function $P(x, y)$ in the discrete case or joint density function $f(x, y)$ in the continuous case, and if $g(X, Y)$ is some function of X and Y, then

$$E(g(X, Y)) = \sum_x \sum_y g(x, y)P(x, y) \quad \textbf{in the discrete case} \quad (3.14.6^*)$$

and

$$E(g(X, Y)) = \int_{-\infty}^{\infty} \int_{-\infty}^{\infty} g(x, y)f(x, y) \, dx \, dy \quad \textbf{in the continuous case.}$$

$$(3.14.7^*)$$

Rule 5 is simply an extension of Rule 1 to two random variables. Similar extensions can be made for more than two random variables.

RULE 6

If X is a random variable with expectation $E(X)$, and Y is a random variable with expectation $E(Y)$, then

$$E(X + Y) = E(X) + E(Y). \quad (3.14.8^*)$$

Verbally, this rule says that the expectation of a sum of two random variables is the sum of their expectations. Once again, the proof is simple for two discrete variables X and Y. Consider the new random variable $(X + Y)$. The probability of a value of $(X + Y)$ involving a *particular* X value and a *particular* Y value is the joint probability $P(x, y)$. Thus,

$$E(X + Y) = \sum_x \sum_y (x + y)P(x, y).$$

Notice that here the expectation involves the sum over all possible *joint* events (X, Y). This could be written as

$$E(X + Y) = \sum_x \sum_y (x + y)P(x, y)$$
$$= \sum_x \sum_y xP(x, y) + \sum_x \sum_y yP(x, y).$$

However, for any fixed x,

$$\sum_y P(x, y) = P(x),$$

and for any fixed y, $$\sum_x P(x, y) = P(y).$$

Thus,

$$E(X + Y) = \sum_x xP(x) + \sum_y yP(y) = E(X) + E(Y).$$

In particular, one of the random variables may be in a functional relation to the other. For example, let $Y = 3X^2$. Then

$$E(X + Y) = E(X + 3X^2)$$
$$= E(X) + E(3X^2)$$
$$= E(X) + 3E(X^2).$$

Rule 6 says that if you expect to gain \$$a$ from one bet and \$$b$ from another bet, then your expected gain from both bets is \$$(a + b)$.

This principle lets one *distribute* the expectation over an expression which itself has the form of a sum. We will make a great deal of use of this principle.

This rule may also be extended to any finite number of random variables:

RULE 7

Given some finite number of random variables, the expectation of the sum of those variables is the sum of their individual expectations. Thus,

$$E(X_1 + X_2 + \cdots + X_n) = E(X_1) + E(X_2) + \cdots + E(X_n). \quad (3.14.9^*)$$

In particular, some of the random variables may be in functional relations to others. Let $Y = 6X^4$, and let $Z = \sqrt{2X}$. Then

$$E(X + 6X^4 + \sqrt{2X}) = E(X) + 6E(X^4) + E(\sqrt{2X}).$$

The next rule is a most important one that applies only to *independent* random variables:

RULE 8

Given the random variable X with expectation $E(X)$ and the random variable Y with expectation $E(Y)$, then if X and Y are independent,

$$E(XY) = E(X)E(Y). \quad (3.14.10^*)$$

This rule states that if random variables are *statistically independent*, the expectation of the product of these variables is the product of their separate expectations. An important corollary to this principle is:

If $E(XY) \neq E(X)E(Y)$, the variables X and Y are not independent.

The basis for Rule 8 can also be shown fairly simply for discrete variables. Since X and Y are independent, $P(x, y) = P(x)P(y)$. Then

$$E(XY) = \sum_x \sum_y xyP(x)P(y) = \sum_x \sum_y xP(x)yP(y).$$

However, for any fixed x, $yP(y)$ is perfectly free to be any value, so that

$$E(XY) = \sum_x xP(x) \sum_y yP(y) = E(X)E(Y).$$

By extension of Rule 8 to any finite number of random variables we have:

RULE 9

Given any finite number of random variables, if all the variables are independent of each other, the expectation of their product is the product of their separate expectations; thus,

$$E(X_1X_2\cdots X_n) = E(X_1)E(X_2)\cdots E(X_n). \qquad (3.14.11^*)$$

3.15 MEASURES OF CENTRAL TENDENCY

Imagine a probability distribution of a random variable, X. If you were asked to state *one value* that would best communicate the central tendency, or location, of the distribution as a whole, which value should you choose? Could you determine a rule that would tell you what value to pick for *any* given distribution? You probably could not, since there are several alternatives available, each of which has certain advantages and certain disadvantages. The three most commonly used alternatives are: (1) the expectation of X, $E(X)$, which is also called the **mean** of the distribution; (2) the value midway (in terms of probability) between the smallest possible value and the largest possible value, which is called the **median;** and (3) the value of X at which the PMF or PDF reaches its highest point, which is called the **mode.** We will briefly mention other measures of central tendency, but our primary concern will be with the mean, the median, and the mode.

3.16 THE MEAN AS THE "CENTER OF GRAVITY" OF A DISTRIBUTION

In many important instances of probability distributions, the mean or expectation is a parameter. That is to say, the mean enters into the function rule assigning a probability or probability density to each possible value of X. It will sometimes be convenient to use still another symbol when we are dealing with the mean of a random variable, especially in its role as a parameter. The small Greek letter mu, μ, will stand for the mean of a probability distribution of a random variable X. That is,

$$E(X) = \mu = \text{mean of the distribution of } X. \qquad (3.16.1^*)$$

In much of what follows, small Greek letters will be used to indicate parameters of probability distributions, while Roman letters will stand for sample values. The word "parameter" will always indicate a characteristic of a probability distribution and the word "statistic" will denote a summary value calculated from a sample. Thus, given some sample space and the random variable X, the mean of the distribution of X is a parameter, μ, and the mean of any given sample, M, is a statistic.

The mean of a distribution parallels the physical idea of a center of gravity, or balance point, of ideal objects arranged in a straight line. For example, imagine an ideal board having zero weight. Along this board are arranged stacks of objects at various positions. The objects have uniform weight and differ from each other only in their positions on the board. The board is marked off in equal units of some kind, and each object is assigned a number according to its position. A drawing of the board and the objects looks like a PMF (or a histogram, if you will). This is shown in Figure 3.16.1. Now given this idealized situation, at what point would a fulcrum placed under the board create a state of balance? That is, what is the point at which the "push" of objects on one side of the board is exactly equal to the push exerted by objects on the other side? This is found from the mean of the positions of the various objects:

$$\mu = 2\left(\frac{2}{10}\right) + 8\left(\frac{1}{10}\right) + 10\left(\frac{3}{10}\right) + 15\left(\frac{2}{10}\right) + 18\left(\frac{1}{10}\right) + 20\left(\frac{1}{10}\right)$$

$$= 11.$$

Here the board would exactly balance if a fulcrum were placed at the position marked 11. Note that since there were piles of uniform objects at various positions on the board, this center of gravity was found in exactly

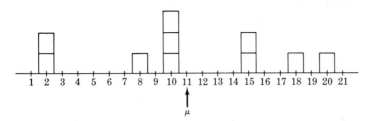

Figure 3.16.1

the same way as for the mean of a distribution, since each position (value of x) was, in effect, multiplied by the proportion of objects at that position (the probability of that value), and then these products were summed.

In short, the position of any object on the board is analogous to a possible value of X. The mean is then like the center of gravity, or balance point. The mean is that score about which deviations in one direction exactly equal deviations in the other. The tendency for values in a distribution to differ from the mean in one way is exactly balanced by the tendency to differ in the opposite way.

This property of the mean is bound up in the statement that **the expectation of the deviation about the mean is zero in any distribution.** A deviation from the mean is simply the signed difference between a possible value of X and the mean score:

$$d = X - \mu. \tag{3.16.2}$$

In order to show that the expectation of this deviation must be zero, suppose that we had a distribution of a random variable X. Then

$$E(X - \mu) = E(X) - E(\mu)$$
$$= E(X) - \mu,$$

since μ is a constant. But by definition, $\mu = E(X)$, so

$$E(X - \mu) = 0. \tag{3.16.3*}$$

Thus, the expectation of the deviation about the mean is zero.

Suppose once again that you were asked to state *one value* to communicate the central tendency of a distribution. If you guessed the mean, your expected error, $E(X - \mu)$, would be zero, though the mean itself might not even be a possible value of X (as in the board example).

For example, let X represent the income (in dollars) of a person chosen at random from the employees of a particular company. Suppose that the distribution is as follows:

x	$P(x)$
6,000	.10
8,000	.40
10,000	.30
15,000	.10
25,000	.05
50,000	.05

The mean of the distribution is

$$E(X) = \sum_x xP(x)$$

$$= 6{,}000(.10) + 8{,}000(.40) + 10{,}000(.30) + 15{,}000(.10)$$

$$+ 25{,}000(.05) + 50{,}000(.05) = 12{,}050.$$

The PMF in Figure 3.16.2 illustrates the idea of the mean as the center of gravity in this example.

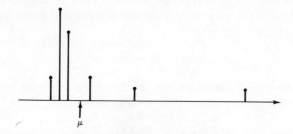

Figure 3.16.2

For an example of the calculation of the mean for a continuous random variable, consider the uniform distribution on the unit interval from zero to one:

$$f(x) = \begin{cases} 1 & \text{if } 0 \leq x \leq 1, \\ 0 & \text{elsewhere.} \end{cases}$$

From the definition of the expectation of a continuous random variable,

$$E(X) = \int_{-\infty}^{\infty} xf(x)\,dx = \int_0^1 x\,dx = \left[\frac{x^2}{2}\right]_0^1 = \frac{1}{2} - 0 = \frac{1}{2}.$$

3.17   THE MEDIAN

We defined the median informally as the value of X which is midway (in terms of probability) between the smallest possible value and the largest possible value. In terms of a continuous distribution, this is easy to visualize: the median is the point at which the total area of one under the PMF is divided equally into two segments, each of which has an area of one-half. In other words, the probability that X is less than the median is $\frac{1}{2}$, and the probability that X is greater than the median is also $\frac{1}{2}$. Recall that in the case of a continuous random variable the probability that X is exactly equal to any single value is zero. This is not true for a discrete random variable, so a more formal definition is required for a median. For our purposes, the following definition will suffice:

If

$$P(X \le a) \ge .50 \quad \textbf{and} \quad P(X \ge a) \ge .50, \qquad (3.17.1^*)$$

then a is called a median of the distribution of X.

In the continuous case, this amounts to requiring that

$$\int_{-\infty}^{a} f(x)\, dx = \int_{a}^{\infty} f(x)\, dx = .50, \qquad (3.17.2)$$

and there is a unique median since $f(x)$ is a continuous function. For example, consider once again the uniform distribution over the unit interval:

$$f(x) = \begin{cases} 1 & \text{if } 0 \le x \le 1, \\ 0 & \text{elsewhere.} \end{cases}$$

If a is the median, then

$$\int_{-\infty}^{a} f(x)\, dx = \int_{0}^{a} dx = .50.$$

But

$$\int_{0}^{a} dx = a - 0 = a,$$

so $a = .50$. This is reasonable, since at $X = .50$ the unit square which represents the PDF of X is divided into two rectangles, each with area one-half. The reader can verify for himself that

$$\int_{a}^{\infty} f(x)\, dx = .50.$$

Although we have seen that a continuous distribution has one and only one median, note that in the definition of a median for a discrete random variable, the term "a median" was used instead of "the median." This is because it is possible for a discrete distribution to have more than one point a which satisfies the definition. For example, consider the distribution of incomes presented in Section 3.16. Notice that

$$P(X \leq 8000) = .50, \qquad P(X \geq 8000) = .90,$$

$$P(X \leq 8500) = .50, \qquad P(X \geq 8500) = .50,$$

and $\qquad P(X \leq 9153) = .50, \qquad P(X \geq 9153) = .50.$

In fact, $P(X \leq a) \geq .50$ and $P(X \geq a) \geq .50$ are satisfied for any value of a such that

$$8000 \leq a \leq 10{,}000.$$

It is *not* satisfied at the value 10,001, since

$$P(X \geq 10{,}001) = .20.$$

If you were asked to state *one value* to communicate the central tendency of the distribution of X, and you wanted to use the median, which value would you choose? Any value between 8000 and 10,000 can be thought of as a median; if a single number is desired, a reasonable choice is to take the midpoint of the interval from 8000 to 10,000, which is

$$\frac{8000 + 10{,}000}{2} = 9000.$$

In general, if the median can assume any value in an interval, we will adopt the convention of taking the midpoint of the interval and calling it the median, even though it is technically only one of numerous possible values which satisfy the definition of a median.

On the other hand, it should be pointed out that in many cases there *is* a unique median in the discrete case as well as in the continuous case. Suppose that the distribution of incomes was slightly different, so that $P(X = 8000) = .39$ instead of .40, and $P(X = 10{,}000) = .31$ instead of .30. Then the only value satisfying the definition of a median is 10,000.

The median is but one of a class of summary measures known as **fractiles:**

If

$$F(a) = P(X \leq a) \geq f \quad \textbf{and} \quad P(X \geq a) \geq 1 - f, \quad (3.17.3^*)$$

then a is called an f fractile of the distribution of X.

For the distribution of incomes, 7000 is a .10 fractile, since

$$P(X \leq 7000) = .10 \quad \text{and} \quad P(X \geq 7000) = .90.$$

A median, of course, is simply a .50 fractile. Other fractiles which are encountered frequently are the quartiles (.25 and .75 fractiles), the deciles (.10, .20, $\cdots$, .90 fractiles), and the percentiles (.01, .02, .03, $\cdots$, .99 fractiles). The reader can easily verify that for the distribution of incomes, the quartiles are 8000 and 10,000 and the first and ninety-ninth percentiles are 6000 and 50,000. Since fractiles are defined in terms of the cumulative distribution function, $F(x)$, they can readily be determined from a graph of the CDF. In the discrete case, of course, there may be more than one f fractile for any particular value of f, as we have seen for $f = .50$. The median is the most reasonable fractile to use as a measure of central tendency; uses for other fractiles will be discussed in later sections of the book.

3.18   THE MODE

The **mode,** which is defined as **the value of X at which the PMF or PDF of X reaches its highest point,** is generally the easiest of the three measures of central tendency to compute and interpret. If a graph of the PMF or PDF or a listing of the possible values of X along with their probabilities is available, the determination of the mode is quite simple. If the distribution is expressed in functional form, the determination of the mode may not be quite so easy; in the continuous case, for example, it may require the use of calculus to determine the point at which the PDF attains its maximum value.

From the listing of the distribution of incomes presented in Section 3.16, the mode of the distribution occurs at $X = 8000$. Suppose that the probabilities were changed slightly so that $P(X = 10,000) = .00$ and $P(X = 15,000) = .40$. That is, all employees earning 10,000 are given a raise in pay to 15,000. This change results in two modes, 8000 and 15,000, in the new distribution. By generalizing from this example, you can see that it is possible for a distribution to have several modes, in which case the mode is probably of doubtful value as a measure of central tendency. An extreme example of this is the uniform distribution on the unit interval from zero to one. The PDF is a horizontal line over the interval from zero to one. Technically, we could say that every value of X between zero and one is a mode of the distribution since at each point the definition of mode is satisfied. In this circumstance, however, it is more common to say that there is *no mode*, since the PDF is no higher at any given point in the unit interval than it is at any other point.

3.19 "BEST GUESS" INTERPRETATIONS
OF THE MEAN, MODE, AND MEDIAN

We have seen that for any distribution, the mean provides a best guess about any randomly selected value of the random variable, provided that one is interested in making the signed error zero on the average. This follows from Equation (3.16.3). The mean is also a best guess if one wishes to make the average *squared error* as small as possible, as we will see in Section 3.23. However, both the mode and the median can also be given the same kind of interpretation as best guesses if, instead of wishing to make the expected error or the expected squared error as small as possible, we want something else to be true about the errors.

First of all, suppose that there were some distribution for which you had to guess the value of X picked at random, and you wanted to be *absolutely right* with the highest possible probability. Then you should guess the mode rather than the mean; since it is the most probable value, guessing the mode guarantees the greatest likelihood of hitting the value "on the nose" for a case drawn at random (if X is continuous, this is not quite accurate due to a technicality which we will ignore here).

On the other hand, it might be that in guessing the value drawn at random you are not interested in being exactly right most often, nor in making the signed error or the squared error smallest on the average, but instead you want to make the expected *absolute error* as small as possible. Here the *sign* of the error is unimportant; the *size* of the error in absolute terms is what matters. Then the best guess is the median, which makes the expected absolute error as small as possible. The median is the typical value in this sense; it is closest on the average to all of the scores in the distribution.

There is really no way to say which is the best measure of central tendency in general terms. This depends very much on what one is trying to do and on what one wants to communicate in summary form about the distribution. Each of the measures of central tendency is, in its way, a best guess, but the sense of "best" differs with the way error is regarded. If both the size of the errors and their signs are considered, and we want to minimize the average error or the average squared error, then the mean serves as a best guess. If a miss is as good as a mile, and one wants to be right as often as possible, then the mode is indicated. If one wants to come as close as possible on the average, irrespective of the sign of the error, then the median is a best guess. We will discuss the idea of a summary measure as a "best bet" at greater length in Chapters 6 and 9.

From the point of view of purely descriptive statistics, as apart from inferential work, the median is a most servicable measure. Its property of representing the typical (most nearly like) value makes it fit the requirements of simple and effective communication better than the mean in many contexts. In the income example, the median (9000) appears to be a "better" measure of central tendency than the mean (12,050). This is primarily because the mean is strongly affected by the large incomes, even though they have small probabilities. We will discuss this in greater detail in the next section.

On the other hand, the median is usually inferior to the mean when our purpose is to make inferences beyond the sample. The median has mathematical properties making it difficult to work with, whereas the mean is mathematically tractable. For this and other reasons, mathematical statistics has taken the mean as the focus of most of its inferential methods, and the median is relatively unimportant in inferential statistics. Nevertheless, as a description of a given distribution, the median is extremely useful in communicating the "typical" value of the random variable.

The mode is generally inferior to the median and the mean as a measure of central tendency. This is because (1) a distribution may be multimodal or (2) if it is unimodal, it is possible for the mode to be at an "extreme" value rather than at a "typical" value, as demonstrated in Figure 3.19.1.

The three measures of central tendency considered here are by no means the only possible measures. Another possible measure is the mid-range, which is the mid-point of the possible range of values. If the smallest possible value of X is s and the largest possible value is l, then the mid-range is $(s + l)/2$. A variation of this is the mid-interquartile range, which is the mid-point of the interval from the .25 fractile of X to the .75 fractile of X: (.25 fractile + .75 fractile)/2. The advantage of measures such as the mid-range is the ease with which they can be computed. A disadvantage is the fact that the mid-range depends solely on the extreme values and not on any probabilities. Another class of measures is the class of trimmed means, which essentially are means computed after some extreme values of the random variable have been eliminated or adjusted toward the center of the distribution. Trimmed means are less sensitive to extreme values than is the untrimmed mean, μ.

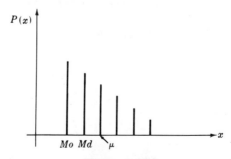

Figure 3.19.1

There are many potential measures of central tendency, then, each having its advantages and disadvantages. It is impossible to state that one measure is "best" in all situations. *The choice of a measure must depend on the particular distribution and on what the statistician is trying to get across about the distribution.* The discussion in this section and the following section should give the reader some idea of the considerations involved in the choice of a measure.

3.20 RELATIONS BETWEEN CENTRAL TENDENCY MEASURES AND THE "SHAPES" OF DISTRIBUTIONS

In discussions of probability distributions, it is often expedient to describe the general "shape" of the distribution curve. Although the terms used here can be applied to any distribution, it will be convenient to illustrate them by referring to graphs of continuous distributions.

First of all, a distribution may be described by the number of relative maximum points it exhibits, its "modality." This usually refers to the number of "humps" apparent in the graph of the distribution. Strictly speaking, if the density (or probability or frequency) is greatest at one point, then that value is *the* mode, regardless of whether other relative maxima occur in the distribution or not. Nevertheless, it is common to find a distribution described as bimodal or multimodal whenever there are two or more pronounced humps in the curve, even though there is only one distinct mode. Thus, a distribution may have no modes (Figure 3.20.1), may be unimodal (Figure 3.20.2), or may be multimodal (Figure 3.20.3). Once again, notice that the possibility of multimodal distributions lowers the effectiveness of the mode as a description of central tendency.

Another characteristic of a distribution is its symmetry, or conversely, its skewness. A distribution is **symmetric** only if it is possible to divide its graph into two "mirror-image halves," as illustrated in Figure 3.20.4. Note that in the graph of Figure 3.20.2 there is no point at which the distribution may be divided into two similar parts, as in the first example. When a distribution is symmetric, it will be true that the mean and the median are equal in value. It is not necessarily true, however, that the

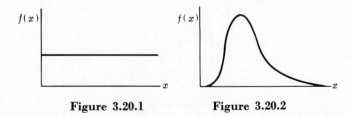

Figure 3.20.1   Figure 3.20.2

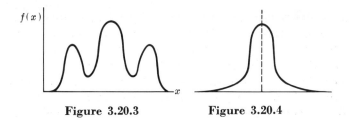

Figure 3.20.3   Figure 3.20.4

mode(s) will equal either the mean or the median; witness the example of Figure 3.20.5. On the other hand, a nonsymmetric distribution is sometimes described as **skewed,** which means that the length of one of the **tails** of the distribution, relative to the central section, is disproportionate to the other. For example, the distribution in Figure 3.20.6 is skewed to the right, or **skewed positively.** In a positively skewed distribution, the bulk of the cases fall into the lower part of the range of scores, and relatively few show

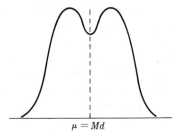

$$\mu = Md$$

Figure 3.20.5

extremely high values. This is reflected by the relation

$$\text{Mean} > \text{Median}$$

usually found in a positively skewed distribution.

On the other hand, it is possible to find distributions skewed to the left, or **negatively skewed.** In such a distribution, the long tail of the distribution occurs among the low values of the variable. That is, the bulk of the

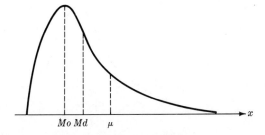

$Mo\ Md$ μ

Figure 3.20.6

distribution shows relatively high scores, although there are a few quite low scores (Figure 3.20.7). Generally in a negatively skewed distribution, Median > Mean. Thus, a "rough and ready" way to describe the skewness of a distribution is to find the mean and median; if Mean > Median, then you can conclude that the distribution is skewed to the right (positively). If Median > Mean, then you should conclude that negative skewness exists. If a more accurate determination is needed, other indices reflecting skewness are available; however, we will not discuss such indices. A word of warning: if a distribution is symmetric, then Mean = Median, but the fact that Mean = Median does not necessarily imply that the distribution is symmetric.

Describing the skewness of a distribution in terms of measures of central tendency again points up the contrast between the mean and the median as measures of central tendency. The mean is much more affected by the extreme cases in the distribution than is the median. Any alteration of the scores of cases at the extreme ends of a distribution will have no effect at all on the median so long as the rank order of the scores is roughly preserved; only when scores near the center of the distribution are altered is there likelihood of altering the median. This is not true of the mean, which is very sensitive to changes at the extremes of the distribution. The alteration of the probability for a single extreme case in a distribution may have a profound effect on the mean. It is evident that the mean follows the skewed tail in the distribution, while the median does so to less extent. *The occurrence of even a few very high or very low possible values can seriously distort the impression of the distribution given by the mean, provided that one mistakenly interprets the mean as the typical value.* If you are dealing with a non-symmetric distribution, and you want to communicate the typical value, the median would probably be a better measure than the mean. On the other hand, in spite of the distribution's shape, the mean always communicates the same thing: the point about which the sum of deviations is zero.

The distribution of incomes presented in Section 3.16 demonstrates the above point. The graph of the PMF of this distribution is presented in Figure 3.20.8. The distribution is positively skewed, and Mean > Median.

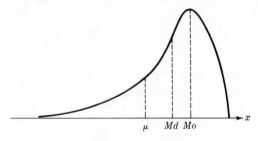

Figure 3.20.7

Figure 3.20.8

Differences in shape are only rough, qualitative ways of distinguishing among distributions. The only adequate description of a theoretical distribution is its function rule. Distributions that look similar in their graphic form may be very different functions. Conversely, distributions that appear quite different when graphed actually may belong to the same family, as we will see in Chapter 4. A description in terms of modality and skewness is sometimes useful for giving a general impression what a distribution is like, but it does not communicate its essential character.

3.21 MEASURES OF DISPERSION: THE VARIANCE

An index of central tendency summarizes only one special aspect of a distribution, be it mode, median, mean, or any other measure. It is easy to find two distributions which have the same mean but which are not at all similar in any other respect. Any distribution has at least one more important feature that must be summarized in some way. Distributions exhibit *spread* or *dispersion*, the tendency for observations to depart from central tendency. If central tendency measures are thought of as good bets about observations in a distribution, then measures of spread represent the other side of the question; dispersion reflects the "poorness" of central tendency as a description of a randomly selected value, the tendency of values *not* to be like the "average."

The mean is a good bet about the location of a random variable, but *no single value need be exactly like the mean*. A *deviation* from the mean expresses how "off" the mean is as a bet about a particular value, or how much *in error* is the mean as a description of this value:

$$d = X - \mu.$$

In the same way, we could talk about a deviation from the median, if we wished,

$$d' = X \cdot - Md,$$

or even perhaps from the mode. It is quite obvious that the larger such deviations are, on the whole, from a measure of central tendency, the more cases differ from each other and the more spread the distribution shows. What we need is an index (or set of indices) to reflect this spread or variability.

First of all, why not simply take the expectation of the deviation about the mean as our measure of variability:

$$E(X - \mu) = \begin{cases} \displaystyle\sum_{x} (x - \mu) P(x) & \text{if } X \text{ is discrete,} \\[2em] \displaystyle\int_{-\infty}^{\infty} (x - \mu) f(x) \, dx & \text{if } X \text{ is continuous.} \end{cases}$$

This will not work, however, because in Section 3.16 it was shown that in any particular probability distribution, the expected deviation from the mean must be zero.

The device used to get around this difficulty is to take the *square* of each deviation from the mean, and then to find the expectation of the squared deviation:

$$\sigma^2 = E(X - \mu)^2. \tag{3.21.1*}$$

For any distribution, the index σ^2, equal to the expectation of the squared deviation from the mean, is called the variance of the distribution. *The variance reflects the degree of spread, since σ^2 will be zero if and only if there is only one possible value of X: the mean. The more that the values tend to differ from each other and the mean, the larger will the variance be.*

Using the algebra of expectations presented in Section 3.14, it is possible to derive useful rules for dealing with variances. The rules apply to both discrete and continuous random variables, provided that certain boundary conditions are satisfied. As in Section 3.14, we need not concern ourselves with the boundary conditions in a discussion such as this.

RULE 1

The variance of a random variable is equal to the expectation of the square of the random variable minus the square of the expectation of the random variable:

$$\text{var } (X) = \sigma^2 = E(X^2) - [E(X)]^2. \tag{3.21.2*}$$

By definition,

$$\begin{aligned} \text{var } (X) = \sigma^2 &= E(X - \mu)^2 \\ &= E(X^2 - 2\mu X + \mu^2) \\ &= E(X^2) - E(2\mu X) + E(\mu^2). \end{aligned}$$

But μ is a constant, so

$$\text{var}(X) = E(X^2) - 2\mu E(X) + \mu^2.$$

Then, since $\mu = E(X)$,

$$\text{var}(X) = E(X^2) - 2[E(X)]^2 + [E(X)]^2$$
$$= E(X^2) - [E(X)]^2.$$

RULE 2

If a is some constant real number, and if X is a random variable with expectation $E(X)$ and variance σ_X^2, then the random variable $(X + a)$ has variance σ_X^2:

$$\text{var}(X + a) = \text{var}(X) = \sigma_X^2. \tag{3.21.3*}$$

This can be shown as follows:

$$\text{var}(X + a) = E[(X + a) - E(X + a)]^2.$$

But

$$E(X + a) = E(X) + a,$$

so that

$$\text{var}(X + a) = E[(X + a) - (E(X) + a)]^2$$
$$= E[X - E(X)]^2$$
$$= \sigma_X^2.$$

RULE 3

If a is some constant real number, and if X is a random variable with variance σ_X^2, the variance of the random variable aX is

$$\text{var}(aX) = a^2\sigma_X^2. \tag{3.21.4*}$$

In order to show this, we take

$$\text{var}(aX) = E[(aX)^2] - [E(aX)]^2$$
$$= a^2 E(X^2) - a^2[E(X)]^2$$
$$= a^2(E(X^2) - [E(X)]^2)$$
$$= a^2\sigma_X^2.$$

In short, adding a constant value to each value of a random variable leaves the variance unchanged, but multiplying each value by a constant multiplies the variance by the square of the constant. This is reasonable, since adding a constant merely shifts the distribution along the X-axis. This is quite obvious in a graphical analysis; the dispersion of the distribution remains the same, although the location is shifted. On the other hand, multiplying by a constant changes the scale as well as the location. A value two units from the mean becomes $2a$ units from the mean if each value of the random variable is multiplied by a. The change in scale results in a change in dispersion.

3.22 THE STANDARD DEVIATION

Although the variance is an adequate way of describing the degree of variability in a distribution, it does have one important drawback. The variance is a quantity in squared units of measurement. For instance, if a random variable is expressed in terms of inches, then the mean is some number of inches, and a deviation from the mean is a difference in inches. However, the square of a deviation is in *square-inch units,* and thus the variance, being a mean squared deviation, must also be in square inches. Naturally, this is not an insurmountable problem; taking the positive square root of the variance gives an index of variability in the original units. **The square root of the variance for a distribution is called the standard deviation and is an index of variability in the original measurement units.** The standard deviation of a random variable will be denoted by σ:

$$\sigma = \sqrt{\sigma^2} = \sqrt{E(X - \mu)^2}. \qquad (3.22.1^*)$$

For examples of the computation of the variance and standard deviation in the discrete and continuous cases, consider the discrete distribution of incomes presented in Section 3.16 and the continuous uniform distribution on the unit interval. For the distribution of incomes, we found $E(X)$ to be 12,050. Furthermore,

$$
\begin{aligned}
E(X^2) &= \sum_x x^2 P(x) \\
&= (6000)^2(.10) + (8000)^2(.40) + (10{,}000)^2(.30) \\
&\quad + (15{,}000)^2(.10) + (25{,}000)^2(.05) + (50{,}000)^2(.05) \\
&= 237{,}950{,}000.
\end{aligned}
$$

Then

$$
\begin{aligned}
\text{var}\,(X) &= E(X^2) - E(X)^2 \\
&= 237{,}950{,}000 - (12{,}050)^2 = 92{,}747{,}500,
\end{aligned}
$$

and $\sigma = \sqrt{92{,}747{,}500} = 9631.$

This example illustrates a common computational problem. The numbers involved in these calculations are quite large and thus are cumbersome to work with. However, by use of Rule 3 from Section 3.21, the calculations can be expressed in terms of smaller and more convenient numbers. Suppose that $Y = X/1,000$. Then $E(Y) = E(X)/1000 = 12.05$, and

$$E(Y^2) = \sum_y y^2 P(y)$$

$$= 6^2(.10) + 8^2(.40) + 10^2(.30) + 15^2(.10) + 25^2(.05)$$

$$+ 50^2(.05) = 237.95.$$

Then

$$\text{var } (Y) = E(Y^2) - E(Y)^2$$

$$= 237.95 - (12.05)^2 = 92.7475.$$

But $Y = X/1000$, so $X = 1000Y$, and by Rule 3,

$$\text{var } (X) = \text{var } (1000Y) = (1000)^2 \text{ var } (Y) = 92,747,500.$$

For the continuous uniform distribution on the unit interval, we have $E(X) = .50$, and

$$E(X^2) = \int_{-\infty}^{\infty} x^2 f(x) \, dx = \int_0^1 x^2 \, dx = \frac{x^3}{3} \Big|_0^1 = \frac{1}{3}.$$

Hence

$$\text{var } (X) = E(X^2) - E(X)^2 = \left(\frac{1}{3}\right) - \left(\frac{1}{2}\right)^2 = \left(\frac{1}{3}\right) - \left(\frac{1}{4}\right) = \frac{1}{12},$$

and $\sigma = \sqrt{\dfrac{1}{12}}.$

The reader acquainted with elementary physics may recognize not only that the mean is the center of gravity of a physical distribution of objects, but also that the variance is the **moment of inertia** of a distribution of mass. Furthermore, the standard deviation corresponds to the radius of gyration of a mass distribution; the standard deviation is analogous to a resultant force away from the mean. These physical concepts and their associated mathematical formulations have influenced the course of theoretical statistics very strongly, and have helped to shape the form of statistical inference as we will encounter it.

3.23 THE MEAN AS THE "ORIGIN" FOR THE VARIANCE

This question may already have occurred to the reader: "Why is the variance, and hence the standard deviation, always calculated in terms of deviations from the *mean*? Why couldn't one of the other measures of central tendency be used as well?" The answer lies in the fact that **the expected squared deviation (that is, the variance) is smallest when calculated from the mean.** That is, if the disagreement of any value with the mean is indicated by the square of its difference from the mean, then on the average the mean agrees better with the scores than any other single value one might choose.

This may be demonstrated as follows: suppose that we choose some arbitrary value C, and calculate a "pseudo-variance"

$$\sigma_c^2 = E(X - C)^2. \qquad (3.23.1)$$

Adding and subtracting μ within the parentheses would not change the value of σ_c^2 at all. However, if we did so, we could expand each squared deviation as follows:

$$(X - C)^2 = (X - \mu + \mu - C)^2 = [(X - \mu) + (\mu - C)]^2$$
$$= (X - \mu)^2 + 2(X - \mu)(\mu - C) + (\mu - C)^2.$$

Substituting into the expression for σ_c^2, we have

$$\sigma_c^2 = E(X - \mu)^2 + 2E(X - \mu)(\mu - C) + E(\mu - C)^2.$$

But $(\mu - C)$ is a constant with regard to the expectations, so that

$$\sigma_c^2 = E(X - \mu)^2 + 2(\mu - C)E(X - \mu) + (\mu - C)^2.$$

The first term on the right above is simply σ^2, the variance about the mean, and the second term is zero, since the expected deviation from μ is zero. On making these substitutions, we find

$$\sigma_c^2 = \sigma^2 + (\mu - C)^2. \qquad (3.23.2)$$

Since $(\mu - C)^2$ is a squared number, it can only be positive or zero, and so σ_c^2 must be greater than or equal to σ^2. The value of σ_c^2 can be equal to σ^2 only when μ and C are the same. In short, we have shown that the variance calculated about the mean will always be smaller than about any other point.

If we are going to use the mean to express the central tendency of a distribution, and if we let the standard deviation indicate the extent of error we stand to make in guessing the mean, then this error-quantity is at its minimum when we guess the mean in place of any other single value. However, it is a curious fact that if we appraise error by taking the absolute difference (disregarding sign) between a value and a measure of central tendency, then **the expected absolute deviation is smallest when the median is used.** This is one of the reasons that squared deviations rather than absolute deviations figure in the indices of variability when the mean is used to express central tendency. When the median is used to express central tendency it is often accompanied by the expected absolute deviation from the median to indicate dispersion, rather than by the standard deviation, which is more appropriate to the use of the mean. The expected absolute deviation is simply

$$\text{A.D.} = E \mid X - Md \mid, \tag{3.23.3}$$

where the vertical bars indicate a disregarding of sign. Analogically speaking this measure is to the median as the standard deviation is to the mean.

In addition to the variance, the standard deviation, and the absolute deviation, there are other measures of dispersion which for the most part are not as "good" as the above measures but are often easier to calculate. The range is defined as the difference between the largest possible value of the random variable and the smallest possible value. This is extremely simple to determine. Unfortunately, the range may not be particularly informative, since it considers only the two extreme values and not the probabilities of the various possible values. Furthermore, without any change in the probabilities, the range can be severely affected by shifting one of the extreme values even further away from the other values. Finally, for many continuous random variables, the range is infinite.

These problems are reduced somewhat by using the interquartile range instead of the range. The interquartile range is the difference between the .75 and .25 fractiles of the distribution, and is not affected by the extreme values as is the range. Another possibility is the interdecile range, the difference between the .90 and .10 fractiles of the distribution. All of these ranges are reasonably easy to calculate in most cases, and hence could be useful as "rough and ready" measures of dispersion. In general, however, we will use the variance and standard deviation as measures of dispersion because they are mathematically tractable, whereas the other measures have mathematical properties which make them more difficult to work with.

3.24 MOMENTS OF A DISTRIBUTION

A truly mathematical treatment of distributions would introduce not only the mean and the variance, but also a number of other summary characteristics. These are the so-called "**moments**" of a distribution, which are simply **the expectations of different powers of the random variable.** Thus, the first moment about the origin of a random variable X is

$$E(X) = \text{the mean.}$$

The second moment about the origin is

$$E(X^2),$$

the third $$E(X^3),$$

and so on. When the mean is subtracted from X before the power is taken, then the moment is said to be **about the mean;** the variance

$$E[X - E(X)]^2$$

is the second moment about the mean;

$$E[X - E(X)]^3$$

is the third moment about the mean, and so on.

Just as the mean describes the "location" of the distribution on the X-axis, and the variance describes its dispersion, so do the higher moments reflect other features of the distribution. For example, the third moment about the mean is used in certain measures of degree of **skewness;** the third moment will be zero for a symmetric distribution, negative for skewness to the left, positive for skewness to the right. The fourth moment indicates the degree of "peakedness" or **kurtosis** of the distribution, and so on. These higher moments have relatively little use in elementary applications of statistics, but they are important for mathematical statisticians in the study of the properties of distributions and in arriving at theoretical distributions fitting observed data. The entire set of moments for a distribution will ordinarily determine the distribution exactly, and distributions are sometimes specified in this way when their general function rules are unknown or difficult to state.

Moments can be defined not only for simple distributions of a single variable, such as $P(x)$ or $f(x)$, but also for conditional distributions such as $P(x \mid y)$ or $f(x \mid y)$. Moments of conditional distributions are based on the concept of **conditional expectation,** which is identical to expectation as we have defined it except for the fact that the expectation is taken with regard to a conditional distribution:

$$E(X \mid y) = \sum_x xP(x \mid y) \qquad \text{in the discrete case,} \qquad (3.24.1^*)$$

and $\qquad E(X \mid y) = \int_{-\infty}^{\infty} xf(x \mid y) \, dx \quad \text{in the continuous case.} \qquad (3.24.2^*)$

Recall that if X and Y are independent random variables, then

$$P(x \mid y) = P(x) \qquad \text{in the discrete case}$$

and $\qquad f(x \mid y) = f(x) \qquad \text{in the continuous case.}$

But if this is true, then from our definition of conditional expectation,

$$E(X \mid Y) = E(X). \qquad (3.24.3)$$

Thus, if X and Y are independent, the conditional expectation of X given Y is equal to the unconditional expectation of X. By interchanging X

and Y, it is easy to see that this also implies that $E(Y \mid X) = E(Y)$ if X and Y are independent. Conditional expectations are extremely useful when we know the value of one variable and would like to predict the value of another variable, as we shall see in Chapter 10.

3.25   MOMENTS OF JOINT DISTRIBUTIONS

The concept of moments of a distribution can be extended to joint distributions of two or more random variables. Of particular importance are the expectations of products of powers of the random variables. That is, for a joint distribution of two random variables X and Y, the moments are of the form $E(X^r Y^s)$, where r and s are nonnegative integers. The moment $E(XY)$, where $r = 1$ and $s = 1$, is one moment of special interest in probability theory and statistics.

Recall from Section 3.14 that if X and Y are independent random variables, then

$$E(XY) = E(X)E(Y).$$

This fact has particular use in measures of the relationship between two random variables. Such measures are designed to reflect the direction and the strength of the relationship, and the extent to which one variable may be used to predict another. A moment of the joint distribution of X and Y which reflects the direction of their relationship is the **covariance,** defined as follows:

$$\text{cov } (X, Y) = E[(X - \mu_X)(Y - \mu_Y)]. \tag{3.25.1*}$$

This can be put into somewhat more convenient form. Expanding and taking expectations, we find that

$$\text{cov } (X, Y) = E(XY - \mu_Y X - \mu_X Y + \mu_X \mu_Y)$$
$$= E(XY) - \mu_Y E(X) - \mu_X E(Y) + \mu_X \mu_Y$$
$$= E(XY) - \mu_X \mu_Y - \mu_X \mu_Y + \mu_X \mu_Y,$$

or $$\text{cov } (X, Y) = E(XY) - E(X)E(Y). \tag{3.25.2*}$$

If X and Y are independent, $E(XY) = E(X)E(Y)$, so cov $(X, Y) = 0$. That is, the direction of the relationship between X and Y is neither positive nor negative. However, suppose that X and Y are positively related, meaning that a higher value of X *tends* to correspond with a higher value of Y, and a lower value of X *tends* to correspond with a lower value of Y. For example, let X be the income of a given person and let Y represent the number of years of schooling of that person. Then the covariance of X and Y is positive. If X and Y are negatively related (that is, a higher value of X *tends* to correspond with a lower value of Y, and vice versa), then the covariance is negative. For example, let X be the price of steak and let Y be the amount of steak sold by a particular store. The sign of the covariance, then, gives us information as to the direction of the relationship between X and Y.

Unfortunately, the magnitude of the covariance depends very much on the units of measurement for X and Y. Thus it is very hard to interpret in terms of the *strength* of the relationship between X and Y. For example, X might be in units of inches and Y in units of dollars; the covariance is then in units of "inch-dollars." The problem is solved by dividing this covariance by a product involving the same units of measurement as X and Y, which results in an index which is not "contaminated" by the units of measurement. This measure which corrects for the scaling of X and Y and thus reflects the strength as well as the direction of the relationship is the **correlation coefficient,** which is usually denoted by the Greek letter rho:

$$\rho = \frac{\operatorname{cov}(X, Y)}{\sigma_X \sigma_Y}. \tag{3.25.3*}$$

It can be shown that

$$-1 \le \rho \le +1. \tag{3.25.4*}$$

If the correlation coefficient is equal to $+1$, X and Y are said to be perfectly correlated; if it is -1, they are said to be perfectly negatively correlated; and if it is 0, they are said to be uncorrelated. For values of ρ other than $+1$, -1, and 0, a ρ closer to $+1$ or -1 implies a stronger relationship, and a ρ closer to zero ordinarily implies a weaker relationship. A point of great importance is that ρ measures the strength of the *linear* relationship between two variables. Thus, X and $Y = 2X + 1$ are perfectly correlated because they have a perfect *linear* relationship; on the other hand, X and $Z = X^2$ would *not* be perfectly correlated because their relationship is not linear (see Figure 3.25.1). We will pursue the concept of correlation

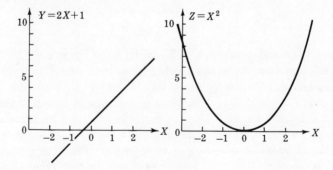

Figure 3.25.1

in greater detail in Chapter 10. The purpose of introducing it here was to show that important summary measures exist for joint distributions as well as for distributions of a single random variable.

The covariance of two random variables is useful in the determination of the variance of their sum or difference as well as in the determination of their correlation coefficient. In Section 3.14 we proved that $E(X + Y) = E(X) + E(Y)$. Can we determine a general formula like this for the variance of the sum of two random variables? It turns out that we can, and that in general, the variance of a sum is not equal to the sum of the variances. From the definition of variance,

$$\text{var } (X + Y) = E[(X + Y) - E(X + Y)]^2.$$

But $E(X + Y) = E(X) + E(Y) = \mu_X + \mu_Y$, so

$$\begin{aligned}
\text{var } (X + Y) &= E[(X + Y) - (\mu_X + \mu_Y)]^2 \\
&= E[(X - \mu_X) + (Y - \mu_Y)]^2 \\
&= E[(X - \mu_X)^2 + 2(X - \mu_X)(Y - \mu_Y) + (Y - \mu_Y)^2] \\
&= E(X - \mu_X)^2 + 2E(X - \mu_X)(Y - \mu_Y) + E(Y - \mu_Y)^2.
\end{aligned}$$

But the first and last terms are simply σ_X^2 and σ_Y^2, and the middle term is twice the covariance, cov (X, Y). Therefore,

$$\text{var } (X + Y) = \text{var } (X) + \text{var } (Y) + 2 \text{cov } (X, Y). \quad (3.25.5^*)$$

In a similar manner, it can be shown that

$$\text{var } (X - Y) = \text{var } (X) + \text{var } (Y) - 2 \text{cov } (X, Y). \quad (3.25.6^*)$$

If X and Y are independent, their covariance is equal to zero, and we have

$$\text{var } (X + Y) = \text{var } (X) + \text{var } (Y) \quad (3.25.7^*)$$

and

$$\text{var } (X - Y) = \text{var } (X) + \text{var } (Y). \quad (3.25.8^*)$$

It is interesting to note that unless X and Y are independent, their covariance has an influence on the variance of their sum and the variance of their difference. If X and Y are positively related (covariance greater than zero), the variance of their sum is *greater* and the variance of their difference is *less* than if X and Y were independent. If they are negatively related, the variance of their sum is less and the variance of their difference is greater than if they were independent.

For example, suppose that

$$\text{var } (X) = 10, \text{var } (Y) = 20, \text{ and cov } (X, Y) = 5.$$

Then

$$\text{var } (X + Y) = 10 + 20 + 2(5) = 40$$

and

$$\text{var } (X - Y) = 10 + 20 - 2(5) = 20.$$

Since cov (X, Y) is positive, large values of X tend to be associated with large values of Y, and small values of X with small values of Y. Thus, when X and Y are added, some of the resulting values are quite large (large value of X plus large value of Y) and some are quite small (small value of X plus small value of Y), implying a large variance. The variance of $X - Y$ is smaller because the values resulting from (large value of X minus large value of Y) are somewhat similar to the values resulting from (small value of X minus small value of Y). Of course, the situation is reversed when the covariance is negative.

3.26 THE RELATIVE LOCATION OF A VALUE IN A PROBABILITY DISTRIBUTION: STANDARDIZED RANDOM VARIABLES

A major use of the mean and standard deviation is in transforming random variables into standardized random variables, showing the *relative status* of possible values of a random variable. If you are given the information, "John Doe has a score of 60," you really know very little about what this score means. Is this score high, low, middling, or what? However, if you know something about the distribution of scores, you can judge the location of the score in the distribution. The score of 60 gives quite a different picture when the mean is 30 and the standard deviation 10 than when the mean is 65 and the standard deviation is 20.

If X is a random variable, then the corresponding **standardized random variable** is

$$z = \frac{X - E(X)}{\sigma} = \frac{X - \mu}{\sigma}. \qquad (3.26.1^*)$$

The standardized random variable is a deviation from expectation, relative to the standard deviation. For any value of X, the corresponding value of z tells you how many standard deviations the value of X is away from the mean, $E(X)$. In general, we will denote a *standardized* random variable *and* its values by a lower case z. It should be clear from the context whether we are referring to the random variable or to a possible value of the random variable.

The two distributions mentioned above give quite different z values to the score of 60:

$$z_1 = \frac{60 - 30}{10} = 3,$$

and

$$z_2 = \frac{60 - 65}{20} = -.25.$$

The conversion of raw values to z values is handy when one wishes to emphasize the *location* or *status* of a value in the distribution, and in future sections we will deal with standardized random variables when this aspect of any random variable is to be discussed. The value of z corresponding to any value of X is found in precisely the same way whether the distribution is discrete or continuous.

Changing each of the values in a distribution to a standardized value creates a distribution having a "standard" mean and "standard" standard deviation: **the mean of a distribution of a standardized random variable is always 0, and the standard deviation is always 1.** This is easily shown as follows:

$$E(z) = E\left(\frac{X - \mu}{\sigma}\right) = \frac{1}{\sigma} E(X - \mu) = 0, \qquad (3.26.2^*)$$

since μ and σ are constants over the various possible values of X and the expected deviation from the mean is always zero. The standard deviation of a standardized random variable is always the square root of the variance:

$$\mathrm{var}\ (z) = E[z - E(z)]^2 = E(z^2),$$

by Equation (3.25.2). But then

$$\mathrm{var}\ (z) = E\left(\frac{X - \mu}{\sigma}\right)^2 = E\left[\frac{(X - \mu)^2}{\sigma^2}\right] = \frac{1}{\sigma^2} E(X - \mu)^2.$$

By definition, $E(X - \mu)^2 = \sigma^2$, so

$$\mathrm{var}\ (z) = \frac{\sigma^2}{\sigma^2} = 1,$$

and

$$\sigma_z = \sqrt{\mathrm{var}\ (z)} = 1. \qquad (3.26.3^*)$$

The general shape of the probability distribution is not changed by the transformation to standardized values. That is, the standardized distribution corresponding to a positively skewed random variable is also positively skewed, and so on. Often in statistics it is easier to work with a standardized than with a nonstandardized random variable, as we will see in later chapters.

3.27 TCHEBYCHEFF'S INEQUALITY

There is a very close connection between the size of deviations from the mean and probability, holding for distributions having finite expectation and variance. The following relation is called **the Tchebycheff inequality,** after the Russian mathematician who first proved this very general principle:

$$P(|X - \mu| \geq b) \leq \frac{\sigma^2}{b^2}, \qquad (3.27.1^*)$$

the probability that a random variable X will differ absolutely from its expectation by b or more units ($b > 0$) is *always* less than or equal to the ratio of σ^2 to b^2. Any deviation from expectation of b or more units can be *no more probable* than σ^2/b^2.

This relation can be clarified somewhat by dealing with the deviation in σ units, making the random variable standardized. If we let $b = k\sigma$, then the following version of the Tchebycheff inequality is true:

$$P\left(\frac{|X - \mu|}{\sigma} \geq k\right) \leq \frac{1}{k^2}, \qquad (3.27.2^*)$$

the probability that a standardized random variable has *absolute* magnitude greater than or equal to some positive number k is *always* less than or equal to $1/k^2$. Thus, given a distribution with some mean and variance, the probability of drawing a case having a standardized value of 2 or more (disregarding sign) must be *at most* 1/4. The probability of a standardized value of 3 or more must be no more than 1/9, the probability of 10 or more can be no more than 1/100, and so on.

This last form of the Tchebycheff inequality (3.27.2) is quite simple to prove for a discrete random variable. Let

$$z = \frac{X - \mu}{\sigma}.$$

Then $P(z) = P(X)$, $E(z) = 0$, and $\sigma_z^2 = 1$. Now the probability of a z value equaling or exceeding in absolute value any arbitrary positive number is, for a discrete variable,

$$P(|z| \geq k) = \sum_{(z \geq k)} P(z) + \sum_{(z \leq -k)} P(z).$$

Multiplying both sides of this equation by k^2 we have

$$k^2 P(|z| \geq k) = \sum_{(z \geq k)} k^2 P(z) + \sum_{(z \leq -k)} k^2 P(z).$$

However, each and every value of z represented in the sum on the right of the equation above is greater than or equal to k in absolute value. Thus

$$\sum_{(z \geq k)} k^2 P(z) + \sum_{(z \leq -k)} k^2 P(z) \leq \sum_{(z \geq k)} z^2 P(z) + \sum_{(z \leq -k)} z^2 P(z).$$

Furthermore,

$$\sum_{(z \geq k)} z^2 P(z) + \sum_{(z \leq -k)} z^2 P(z) \leq \sigma_z^2,$$

since

$$\sigma_z^2 = \sum_z z^2 P(z),$$

the sum being taken over *all* values of z, not just the values such that $z \geq k$ or $k \leq -k$. Then

$$k^2 P(|z| \geq k) \leq \sigma_z^2,$$

or

$$P(|z| \geq k) \leq \frac{1}{k^2}.$$

When the value of k^2 is less than or equal to 1.00, the Tchebycheff inequality itself is quite trivial. Here, the value of 1 divided by k^2 must be a number no less than 1.00, and we have a statement that we know *must* be true for any probability:

$$P(|z| \geq k) \leq v \quad \text{for any } v \geq 1.00.$$

On the other hand, for any value of k greater than 1.00, the Tchebycheff inequality is not trivial, and it does set broad limits to the probability associated with extreme intervals of z values, or of deviations of X values from the mean in terms of standard deviation units.

As an example of how this principle might be applied, consider a random variable X with a mean of 100 and a standard deviation of 25. Suppose that we know nothing about the distribution except the mean and variance. It is impossible to make exact probability statements about X, since we do not know the distribution exactly. By using the Tchebycheff inequality, however, we can make approximate probability statements in the form of probability inequalities. Consider the value 175—what is the probability that X will be at least this far from the mean (in either direction)? The standardized value relative to 175 for this distribution is

$$z = \frac{X - \mu}{\sigma} = \frac{175 - 100}{25} = 3.$$

Now by setting $k = 3$ in the Tchebycheff inequality, we find

$$P(|z| \geq 3) \leq \frac{1}{9}.$$

The probability of observing a value 3 or more standard deviations from the mean is *no more* than 1/9, regardless of the distribution of scores.

Within how many standard deviations from the mean must *at least one half* of the probability fall? That is, we want the absolute value k of z for which

$$P(|z| \geq k) \leq \frac{1}{2}.$$

By Equation (3.27.2) the value is $\sqrt{2}$, or about 1.4. Regardless of how X is actually distributed, we can state that one half or more of the probability must lie within the approximate limits $100 - (1.4)(25)$ and $100 + (1.4)(25)$, or 65 to 135.

Although this principle is very important theoretically, it is not extremely powerful as a tool in applied problems. The Tchebycheff inequality can be strengthened somewhat if we are willing to make assumptions about the general form of the distribution, however. For example, if we assume that the distribution of the random variable is both *symmetric* and *unimodal*, then the relation becomes

$$P(|z| \geq k) \leq \frac{4}{9}\left(\frac{1}{k^2}\right). \tag{3.27.3*}$$

For such distributions, we can make somewhat "tighter" statements about how large the probability of a given amount of deviation may be. For the example given above, if the distribution were unimodal-symmetric a value differing by three or more standard deviations from the mean should be observed with probability *no greater than* $(4/9)(1/9)$, or about .05. Similarly, we could find that at least $1 - 4/9$ or $5/9$ of the probability falls within one standard deviation to either side of the mean. Furthermore, $1 - (4/9)(1/4)$ or $8/9$ must fall within two standard deviations, and so on.

Thus, by adding assumptions about the form of the distribution to the general principle relating standardized values to probabilities we are able to make stronger and stronger probability statements about a sample result's departures from the mean. In order to make very precise statements one has to make even stronger assumptions about the distribution of the random variable, unless he is dealing with very large samples of cases, as we shall see. This chapter has concluded with the introduction of the Tchebycheff inequality to suggest that the mean and the standard deviation play a key role in the theory of statistical inference. The mean and standard deviation are, of course, useful devices for summarizing distributions if that is our purpose, although there are situations where other measures of central tendency and variability may do equally well or better. The really paramount importance of mean and standard deviation does not emerge until one is interested in making *inferences*, involving the estimation of parameters, or assigning probabilities to sample results. Here, we shall find that most "classical" statistical theory is erected around these two indices, together with their combination in standardized scores.

EXERCISES

1. The random variable X has the following probability distribution:

x	$P(X = x)$
-1	.2
0	.3
3	.2
4	.2
6	.1

(a)   graph the PMF and the CDF of X
(b)   explain the relationship between the PMF and the CDF
(c)   find $P(X \geq 2.5)$
(d)   find $P(-1 < X < 4)$
(e)   find $P(-1 \leq X \leq 4)$
(f)   find $P(X < -3)$
(g)   find $P(X = 1)$.

2. The random variable X has the probability distribution given by the rule

$$P(X = a) = \begin{cases} 1/k & \text{for } a = 1, 2, 3, 4, 5, \\ 0 & \text{elsewhere.} \end{cases}$$

(a)   is X discrete or continuous?
(b)   find k
(c)   graph the distribution of X in two ways
(d)   find the distribution of $Y = (X - 3)^2$ and graph it.

3. The cumulative distribution function of X is given by the rule

$$F(x) = \begin{cases} 1 & \text{if } x \geq 2, \\ \frac{1}{4} & \text{if } 1 \leq x < 2, \\ 0 & \text{if } x < 1. \end{cases}$$

(a)   find the corresponding PMF
(b)   find $P(1 < X < 2)$.

4. The density function of X is given by

$$f(x) = \begin{cases} kx(1 - x) & \text{for } 0 < x < 1, \\ 0 & \text{elsewhere.} \end{cases}$$

(a)   find k and graph the density function
(b)   find $P(\frac{1}{4} < X < \frac{1}{2})$
(c)   find $P(-\frac{1}{2} \leq X \leq \frac{1}{4})$
(d)   find the CDF and graph it.

5. The CDF of X is given by

$$F(x) = \begin{cases} 1 & \text{for } x \geq 2, \\ x^2/4 & \text{for } 0 \leq x < 2, \\ 0 & \text{for } x < 0. \end{cases}$$

(a) find $f(x)$, the density function, and show that it satisfies the two require-
    ments for a density function
(b) graph $f(x)$ and $F(x)$.

6. Suppose that the face value of a playing card is regarded as a random variable,
    with an ace counting as 1 and any face card (Jack, Queen, King) counting
    as 10. You draw one card at random from a well-shuffled deck. Construct
    a PMF showing the probability distribution for this random variable and
    find the probabilities of the following events:
    (a) $P(X \leq 6)$ (d) $P(X \text{ is an even number})$
    (b) $P(4 < X)$ (e) $P(X \neq 2 \cap X \neq 8)$.
    (c) $P[(2 \leq X \leq 7) \cup (X = 10)]$

7. The density function of X is given by

$$f(x) = \frac{2(3 - x)}{9} \quad \text{for } 0 \leq x \leq 3,$$

$$= 0 \qquad\qquad \text{elsewhere.}$$

Without using integration, show that the area under the curve is equal to
one and find $P(1 < X < 1.5)$ and $P(X > 2)$. [*Hint:* Use the fact that the
area within a triangle is equal to $\frac{1}{2}$ of the product of the base and the height.]

8. Suppose that the random variable X represents the number of heads occurring
    in three independent tosses of a fair coin. Represent the distribution of X
    (a) by a listing
    (b) by a graph of the PMF
    (c) by a graph of the CDF.
    Do the same for Y, the number of heads occurring in *four* independent tosses
    of a fair coin.

9. Discuss the proposition: "All observed numerical events represent values of
    discrete variables, and continuous variables are only an idealization."

10. Why is it necessary to deal with probability densities rather than probabilities
    such as $P(X = a)$ when the variable under consideration is continuous?

11. The density function of X is given by the rule

$$f(x) = \begin{cases} x & \text{for } 0 < x \leq 1, \\ 2 - x & \text{for } 1 < x < 2, \\ 0 & \text{elsewhere.} \end{cases}$$

(a)   find the CDF
(b)   prove that $f(x)$ satisfies the two requirements for a density function
(c)   find $P(\frac{1}{2} < X < \frac{3}{2})$.

12. Does the function

$$f(x) = \begin{cases} 2x/3 & \text{for } -1 \le x \le 2, \\ 0 & \text{elsewhere,} \end{cases}$$

satisfy the two requirements for a density function?

13. Suppose that a person agrees to pay you \$10 if you throw at least one six in four tosses of a fair die. How much would you have to pay him for this opportunity in order to make it a "fair" gamble?

14. Explain what is meant by the statement, "on any single trial, we do not expect the expectation."

15. For the distribution of Exercise 1, find the mean, the mode, and the median. Is the distribution skewed positively or negatively?

16. For the distribution of Exercise 2, find the mean, the mode, the median, and the mid-range and compare these measures of location, or central tendency. Is this distribution symmetric or skewed?

17. Find the following fractiles of the distribution in Exercise 1:
(a)   .25   (b)   .75   (c)   .01
(d)   .33   (e)   .90.

18. Explain why it is possible for a discrete random variable to have more than one median.

19. For the distribution of Exercise 5, find the mean, the mode, and the median, and comment on the relative value of these measures in this example as measures of the "typical value" of the random variable.

20. Can you determine the mode of a distribution by examining the CDF? Explain for both discrete and continuous distributions.

21. Suppose that X represents the daily sales of a particular product, and that the probability distribution of X is as follows:

x	$P(X = x)$
7,000	.05
7,500	.20
8,000	.35
8,500	.19
9,000	.12
9,500	.08
10,000	.01

Find the expectation and the variance of daily sales. [*Hint*: To find the variance, it is easiest to first find the variance of a different variable, such as $Y = (X - 7000)/1000$.] If the net profit resulting from sales of X items can be given by

$$Z = 5X - 38,000,$$

find the expectation and the variance of net profit, Z.

22. For the distribution of Exercise 1, find the variance, the standard deviation, the expected absolute deviation, and the range. Also, find these same values under the assumption that the largest value of X is 12 rather than 6. In light of your results, comment on the relative merit of the different measures of dispersion.

23. For the distribution of Exercise 3, find
    (a) $E(X)$ (b) $E(X^2)$ (c) $E(4X + 12)$
    (d) var (X) (e) var $(4X + 12)$.

24. Find the mean and variance of the random variables X and Y in Exercise 8.

25. What do we mean when we say that the expectation of a random variable can be thought of as a center of gravity?

26. If a random variable has a mean of ten and a variance of zero, graph its distribution.

27. Prove that in general, $E(X^2)$ is not equal to $[E(X)]^2$. [*Hint*: If they are equal, what can be said about the variance, σ^2?]

28. Criticize the following statement: If the mean of a distribution is 50, and the standard deviation is 10, then the "best bet" about any case drawn at random is 50, and, on the average, one can expect to be in error by 10 points.

29. If the density function of X is given by

$$f(x) = \begin{cases} a + bx^2 & \text{for } 0 \le x \le 1, \\ 0 & \text{elsewhere,} \end{cases}$$

and $E(X) = \frac{2}{3}$, find a and b.

30. Find the following fractiles of the distribution in Exercise 5:
    (a) .01   (b) .05   (c) .25
    (d) .40   (e) .70   (f) .85.

31. From your own main area of interest, think of two variables that would be expected to have reasonably symmetric distributions, two variables that would be expected to have positively skewed distributions, and two variables that would be expected to have negatively skewed distributions.

32. The range is the easiest to compute of all of the measures of dispersion which we discussed. In view of this, why is the range not preferred to the standard deviation, which is much more difficult to compute?

33. What advantage does the standard deviation have over the variance as a measure of dispersion?

34. Suppose that the joint distribution of X and Y is represented by the following table:

Y

	1	2	3	4
1	.12	.18	.24	.05
X **2**	.06	.09	.12	.03
3	.02	.03	.04	.01

(a) graph the joint PMF of X and Y (you will need a three-dimensional graph)

(b) are X and Y independent? Explain your answer.

(c) determine the marginal distributions of X and Y

(d) determine the conditional distribution of X given that $Y = 2$. What does this tell you about the conditional distribution of X given *any* particular value of Y?

(e) find $E(X)$, $E(Y)$, $E(X + Y)$, $E(XY)$, and $E(4X - 2Y)$

(f) find cov (X, Y), var $(X + Y)$, and var $(X - Y)$

(g) using the distribution obtained in (d), find $E(X \mid Y = 2)$.

35. Complete the following table, given that X and Y are independent.

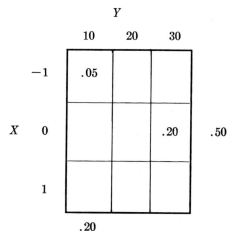

Y

	10	20	30	
−1	.05			
X **0**			.20	.50
1				
	.20			

36. Complete the following table, given that $P(X = 1 \mid Y = 2) = \frac{1}{3}$ and $P(Y = 3 \mid X = 2) = \frac{1}{2}$.

	Y				
X	1	2	3	4	
1				.15	.60
2				.05	
	.10		.50	.20	

37. From your major field of interest, make up two examples of variables that you might reasonably expect to be independent. Also, make up four examples of variables that should be associated to some degree, two examples involving variables with a positive relationship and two examples involving variables with a negative relationship.

38. There is a real danger in confusing the idea of causation with that of statistical association. Why do you think these two concepts are so often confused? For example, for centuries it was thought that swampy air when breathed "caused" malaria (hence the name, literally "bad air"). Comment on what this example shows about statistical association and causation.

39. For the distribution of Exercise 2, find the correlation coefficient of X and $Y = (X - 3)^2$. In this example, if we know X, we can easily determine Y with certainty, and vice versa. If this is true, why is the correlation coefficient not equal to $+1$ or -1? What does this example illustrate about correlation and association?

40. If cov $(X, Y) = 1.0$, what can we say about the relationship between X and Y? If cov $(Z, W) = 2.0$, does this mean that the relationship between Z and W is *stronger* than the relationship between X and Y? Explain.

41. Prove that if X and Y are independent, then their covariance and their correlation coefficient are equal to zero [*Note:* The reverse implication is not true in general.]

42. Prove that cov $(aX, bY) = ab$ cov (X, Y).

43. If var $(X) = 50$, var $(X + Y) = 80$, and var $(X - Y) = 40$, find var (Y) and cov (X, Y).

44. Suppose that X and Y are continuous random variables with joint density function given by the rule

$$f(x, y) = \begin{cases} k(x + y) & \text{for } 0 \le x \le 2, 0 \le y \le 2, \\ 0 & \text{elsewhere.} \end{cases}$$

(a) find k
(b) find the marginal density functions of X and Y
(c) find the conditional density function of X, given that $Y = 1$
(d) find the conditional density function of X, given that $Y = \frac{1}{2}$
(e) are X and Y independent?
(f) find $E(X)$, $E(Y)$, $E(XY)$, cov (X, Y), and ρ.

45. For the distribution in Exercise 1, find the first three moments about the origin and the first three moments about the mean.

46. Suppose that on a final exam in statistics the mean was 50 and the standard deviation was 10. Find the following:
   (i) the standardized (z) scores of students receiving the following grades: 50, 25, 0, 100, 64.
   (ii) the raw grades corresponding to standardized scores of

$$-2, \quad 2, \quad 1.95, \quad -2.58, \quad 1.65, \quad .33.$$

47. For the distribution in Exercise 2, find the distribution of the corresponding *standardized* random variable, z. In what ways are the distributions of X and z similar and in what ways are they dissimilar?

48. The covariance is not a good measure of the strength of a relationship between two random variables because it depends so much on the units of measurement of the two variables. However, if X and Y are standardized random variables, this problem should be eliminated. Is it? [*Hint*: Compare cov (X, Y) and ρ.]

49. For the distribution of Exercise 1, calculate $P(|X - \mu| \geq 2)$ and compare this with the upper bound for this probability obtained from Tchebycheff's inequality.

50. For the distribution of Exercise 2, calculate $P(|X - \mu| \geq 1.5)$ and compare this with a) the upper bound obtained from Tchebycheff's inequality and with b) the upper bound obtained from the stronger form of Tchebycheff's inequality (3.27.3).

51. A rough computational check on the accuracy of a standard deviation is that around six times the standard deviation should, in general, include almost the entire range of values for the distribution on which it is based. Do you see any reason why this rule should work? *Must* it be true?

4

SPECIAL
PROBABILITY
DISTRIBUTIONS

In probability and statistics there are certain classes, or families, of probability distributions that are frequently encountered. These classes of distributions have a wide range of applicability, and as a result it is very useful to study them in their most general form. First, some specific types of discrete distributions will be developed and discussed, and then some continuous distributions will be covered. The first discrete distribution to be introduced is the binomial, and we will dwell somewhat longer on the binomial than on some of the other distributions, both because it is an important distribution and because it is valuable to show in some detail how a theoretical distribution of a discrete random variable can be constructed.

4.1 BERNOULLI TRIALS

The simplest probability distribution is one with only two event classes. For example, a coin is tossed and one of two events, heads or tails, must occur, each with some probability. Or a normal human being is selected at random and his sex noted; the outcome can be only male or female. Such experiments consisting of a series of independent trials, each of which can eventuate in only one of two outcomes, are usually called Bernoulli trials (after J. Bernoulli), and we will call the two event classes and their associated probabilities a **Bernoulli process.**

In general, one of the two events is called a "success" and the other a "failure," or "nonsuccess." These names serve only to tell the events apart, and are not meant to bear any connotation of "goodness" of the event. In the discussion to follow, the symbol p will stand for the probability of a success, and $q = 1 - p$ for the probability of a failure. Thus, in tossing a fair coin, let a head be a success. Then $p = \frac{1}{2}$, $q = 1 - p = \frac{1}{2}$. If the coin is biased, so that heads is twice as likely to come up as tails, then $p = \frac{2}{3}$, $q = \frac{1}{3}$. In the following, take care to distinguish between p, standing for the probability of a success, and $P(x)$, standing for the probability of some value of a random variable.

4.2 SAMPLING FROM A BERNOULLI PROCESS

Suppose that some sample space exists fitting a Bernoulli process. Furthermore, suppose either that we sample independently with replacement, or that an infinite number of elementary events exist, so that for each sample observation out of N trials, p and q are unchanged.

Now we proceed to make N independent observations. Let N be, say, 5. How many different sequences of five outcomes could be observed? The answer, by Rule 1, is $2^5 = 32$. *However, here it is not necessarily true that all sequences will be equally probable.* The probability of a given sequence depends upon p and q, the probabilities of the two events. Fortunately, since trials are independent, one can compute the probability of any sequence by the application of Equation (2.26.3).

We want to find the probability of the particular sequence of events

$$(S, S, F, F, S),$$

where S stands for a success and F for failure. The probability of first observing an S is p. If the second observation is independent of the first, then by Equation (2.26.3),

$$\text{Probability of } (S, S) = p \cdot p = p^2.$$

The probability of an F on the third trial is q, so that the probability of (S, S) followed by F is $p^2 q$. In the same way, the probability of $(S, S, F, F) = p^2 q^2$ and that of the entire sequence is $p^2 q^2 p = p^3 q^2$.

The same argument shows that the probability of the sequence (S, F, F, F, F) is pq^4, that of (S, S, S, S, S) is p^5, of (F, S, S, S, F) is $p^3 q^2$, and so on.

Now if we write out all of the possible sequences and their probabilities, an interesting fact emerges: **the probability of any given sequence of N independent Bernoulli trials depends only on the number of successes and p, the probability of a success.** That is, regardless of the *order* in which successes and failures occur in a sequence, the probability is

$$p^r q^{N-r}, \tag{4.2.1}$$

where r is the number of successes, and $N - r$ is the number of failures.

Suppose that in a sequence of 10 trials, exactly 4 successes occur. Then the probability of that particular sequence is

$$p^4 q^6.$$

If $p = \frac{2}{3}$, then the probability can be worked out from

$$\left(\frac{2}{3}\right)^4 \left(\frac{1}{3}\right)^6.$$

The same procedure would be followed for any r successes out of N trials for any p.

For example, if we toss a fair coin ($p = \frac{1}{2}$) six times, what is the probability of observing three heads followed in order by three tails? The answer is

$$p^3 q^3 = \left(\frac{1}{2}\right)^3 \left(\frac{1}{2}\right)^3 = \frac{1}{64}.$$

This is also the probability of the sequence (H, T, H, T, H, T), of the sequence (H, T, T, T, H, H), and of any other sequence containing exactly three successes or heads.

The probabilities just found are for *particular sequences*, arrangements of r successes and $N - r$ failures in a certain order. What we have found is that if we want to know the probability of a particular sequence of outcomes of independent Bernoulli trials, that sequence will have the same probability as any other sequence with exactly the same number of successes, given N and p.

In most instances, however, one is not especially interested in particular sequences *in order*. We would like to know probabilities of given numbers of successes *regardless* of order in which they occur. For example, when a coin is tossed five times, there are several sequences of outcomes where exactly two heads occur:

$$(H, H, T, T, T)$$
$$(H, T, H, T, T)$$
$$(H, T, T, H, T)$$
$$(H, T, T, T, H)$$
$$(T, H, T, T, H)$$
$$(T, H, H, T, T)$$
$$(T, H, T, H, T)$$
$$(T, T, H, H, T)$$
$$(T, T, H, T, H)$$
$$(T, T, T, H, H).$$

Each and every one of these different sequences must have the same probability, p^2q^3, since each shows exactly two successes and three failures. Notice that there are

$$\binom{N}{r} = \binom{5}{2} = 10$$

different such sequences, exactly as Counting Rule 5 from Chapter 2 gives for the number of ways 5 things can be taken 2 at a time.

What we want now is the probability that $r = 2$ successes will occur *regardless* of order. This could be paraphrased as "the probability of the sequence (H, H, T, T, T) *or* the sequence (H, T, H, T, T) *or* any other sequence showing exactly 2 successes in 5 trials." Such "or" statements about mutually exclusive events recall Axiom 3 in Section 2.5: if A and B are mutually exclusive events, then $P(A \cup B) = P(A) + P(B)$. Thus, the probability of 2 successes in any sequence of 5 trials is $P(2$ successes in 5 trials$) = p^2q^3 + p^2q^3 + \cdots + p^2q^3 = \binom{5}{2}p^2q^3$ since each of these sequences has the same probability and there are $\binom{5}{2}$ of them.

Generalizing this idea for any r, N, and p, we have the following principle: **in sampling from a Bernoulli process with the probability of a success equal to p, the probability of observing exactly r successes in N independent trials is**

$$P(r \text{ successes} \mid N, p) = \binom{N}{r}p^r q^{N-r}. \qquad (4.2.2^*)$$

For example, imagine that in some very large population of animals, 80 percent of the individuals have normal coloration and only 20 percent are albino (no skin and hair pigmentation). This may be regarded as a Bernoulli process with "albino" being a success and "normal" a failure. Suppose that the probabilities are $.20 = p$ and $.80 = q$. A biologist manages to sample this population at random, catching three animals. What is the probability that he catches one albino? Here, $N = 3, r = 1$, so that

$$P(1 \text{ albino in 3 animals}) = \binom{3}{1}(.20)^1(.80)^2$$

$$= .384.$$

If the sampling is random, and if the population is so large that sampling without replacement still permits one to regard the results of the successive trials as independent, then the biologist has about 38 chances in 100 of observing exactly 1 albino in his sample of 3 animals.

4.3   NUMBER OF SUCCESSES AS A RANDOM VARIABLE: THE BINOMIAL DISTRIBUTION

When samples of N trials are taken from a Bernoulli process, the number of successes is a discrete random variable. Since the various values are counts of successes out of N observations, the random variable can take on only the whole values from 0 through N. We have just seen how the probability for any given number of successes can be found. Now we will discuss the distribution pairing each possible number of successes with its probability. This distribution of number of successes in N trials is called the **binomial distribution.** This is the first distribution we have studied that can easily be described not only by listing or graphic methods but also by its mathematical rule.

A binomial distribution can be illustrated most simply as follows: consider a simple experiment repeated independently five times. Each trial must result in only one of two outcomes and the result of five trials is a sequence of outcomes like those in the preceding sections. However, we are not at all interested in the order of the outcomes, but only in the number of successes in the set of trials, which we call X.

In order to find the probability for each value of the discrete random variable, given N and p, let us begin with the largest value, $X = 5$. By Counting Rule 5, there must be $\binom{5}{5} = 1$ possible sequence where all of the outcomes are successes. The probability of this sequence is $p^5 q^0 = p^5$, so that we have

$$P(X = 5 \mid N = 5, p) = \binom{5}{5} p^5 = p^5.$$

For four successes, we find that, by Counting Rule 5,

$$\binom{5}{4} = \frac{5!}{4!1!} = \frac{(1)\,(2)\,(3)\,(4)\,(5)}{(1)\,(2)\,(3)\,(4)\,(1)} = 5,$$

so that four successes can appear in five different sequences. Each sequence has probability $p^4 q^1$. Thus,

$$P(X = 4 \mid N = 5, p) = \binom{5}{4} p^4 q^1 = 5p^4 q^1.$$

Going on in this way, we find

$$P(X = 3 \mid N = 5, p) = \binom{5}{3} p^3 q^2 = 10p^3 q^2,$$

$$P(X = 2 \mid N = 5, p) = \binom{5}{2}p^2q^3 = 10p^2q^3,$$

$$P(X = 1 \mid N = 5, p) = \binom{5}{1}pq^4 = 5pq^4,$$

and
$$P(X = 0 \mid N = 5, p) = \binom{5}{0}q^5 = q^5.$$

(Note that $\binom{5}{0} = \dfrac{5!}{0!5!} = 1$, since 0! is 1 by definition.)

To take a concrete instance of this binomial distribution, let the experiment be that of tossing a fair coin five times. Then $p = \frac{1}{2}$, and the binomial distribution for the number of heads is

r	$P(r)$	
5	$(1/2)^5$	$= 1/32$
4	$5(1/2)^4(1/2)$	$= 5/32$
3	$10(1/2)^3(1/2)^2$	$= 10/32$
2	$10(1/2)^2(1/2)^3$	$= 10/32$
1	$5(1/2)(1/2)^4$	$= 5/32$
0	$(1/2)^5$	$= 1/32$
		$32/32$

Notice that the probabilities over all values of X sum to 1.00, just as they must for any probability distribution.

Now consider another example that is formally identical to this last one but provides different probability values. Suppose that among American male college students who are undergraduates, only one in ten is married. A sample of five male students is drawn at random. Let X be the number of married students observed. (We will assume the total set of students to be large enough that sampling can be without replacement without affecting the probabilities and that observations are independent.) Here, $p = .10$

and the distribution is

r	$P(r)$		
5	$(1/10)^5$	$=$	1/100,000
4	$5(1/10)^4(9/10)$	$=$	45/100,000
3	$10(1/10)^3(9/10)^2$	$=$	810/100,000
2	$10(1/10)^2(9/10)^3$	$=$	7,290/100,000
1	$5(1/10)(9/10)^4$	$=$	32,805/100,000
0	$(9/10)^5$	$=$	59,049/100,000
			100,000/100,000

Contrast this distribution with the preceding one; when p was $\frac{1}{2}$ the distribution showed the greatest probability for $X = 2$ and $X = 3$, with the probabilities diminishing gradually both toward $X = 0$ and toward $X = 5$. On the other hand, in the second distribution, the most probable value of X is 0, with a steady decrease in probability for the values 1 through 5. The distribution over such values of X is very different in these two situations, even though the probabilities are found by exactly the same *formal* rule. This illustrates that **the binomial is actually a family of theoretical distributions, each following the same mathematical rule for associating probabilities with values of the random variable, but differing in particular probabilities depending upon N and p.**

The general definition of the binomial distribution can be stated as follows:

Any random variable X with probability function given by

$$P(X = r) = \binom{N}{r} p^r q^{N-r}, \qquad 0 \le r \le N, \qquad (4.3.1^*)$$

is said to have a binomial distribution with parameters N and p.

The unspecified mathematical constants such as N and p that enter into the rules for probability or density functions indicate **parameters.** Families of distributions share the same mathematical rule for assigning probabilities or probability densities to values of X; in these rules the parameters are simply symbolized as constants. Actually assigning values to the parameters, such as we did for p and N in the distributions above, gives some particular distribution belonging to the family. Thus, *the* binomial distribution usually refers to the family of distributions having the same rule, and a binomial distribution is a particular one of this family found by fixing N and p. Table V in the Appendix gives values of the binomial probability function for various values of N and p.

Almost all theoretical distributions of interest in statistics can be specified by stating the function rule. The way this simplifies the discussion of distributions will be obvious as we go along; indeed, continuous distributions cannot really be discussed at all except in terms of their function rule.

4.4   THE BINOMIAL DISTRIBUTION AND THE BINOMIAL EXPANSION

In algebra you were very likely taught how to expand an expression such as $(a + b)^n$ by the following rule:

$$(a + b)^n = a^n + \frac{n!}{(n - 1)!1!} a^{n-1}b + \frac{n!}{(n - 2)!2!} a^{n-2}b^2 + \cdots$$

$$+ \frac{n!}{1!(n - 1)!} ab^{n-1} + b^n.$$

For example, $(a + b)^3 = a^3 + 3a^2b + 3ab^2 + b^3$ according to this rule. This is the familiar "binomial theorem" for expanding a sum of two terms raised to a power.

Notice that the various probabilities in the binomial distribution are simply terms in such a binomial expansion. Thus, if we take $a = p$, $b = q$, and $n = N$,

$$(p + q)^N = p^N + \binom{N}{N - 1}p^{N-1}q + \binom{N}{N - 2}p^{N-2}q^2 + \cdots + q^N.$$

Since $p + q$ must equal 1.00, then $(p + q)^N = 1.00$, and the sum of all of the probabilities in a binomial distribution is 1.00.

4.5   PROBABILITIES OF INTERVALS IN THE BINOMIAL DISTRIBUTION

In Chapter 3 we saw how to find a probability that a value of a random variable lies in an interval, such as $P(2 \leq X \leq 8)$, the probability that the random variable X takes on some value between 2 and 8 inclusive. This idea is easy to extend to a binomial variable.

Consider the binomial distribution shown in Table 4.5.1, where $p = .3$ and $N = 10$.

Table 4.5.1

r	$\binom{N}{r} p^r q^{N-r} = P(X = r)$
10	.00001
9	.00014
8	.00145
7	.00900
6	.03676
5	.10292
4	.20012
3	.26683
2	.23347
1	.12106
0	.02824
	1.00000

First of all we will find the probability, $P(1 \leq X \leq 7)$, that X lies between the values 1 and 7 inclusive. This is given by the sum

$$
\begin{array}{ll}
P(X = 1) & .12106 \\
+P(X = 2) & .23347 \\
+P(X = 3) & .26683 \\
+P(X = 4) & .20012 \\
+P(X = 5) & .10292 \\
+P(X = 6) & .03676 \\
+P(X = 7) & .00900 \\
\hline
 & .97016 = P(1 \leq X \leq 7).
\end{array}
$$

The probability is about .97 that an observed value of X will lie between 1 and 7 inclusive. Thus, if we are drawing random samples of 10 observations with replacement from a Bernoulli distribution where the probability of a success is .3, we should be very likely to observe a number of successes between 1 and 7 inclusive.

By the same token, we can find

$$ P(8 \leq X) = P(X = 8) + P(X = 9) + P(X = 10) = .00160. $$

The chances are less than two in a thousand of observing eight or more successes, *if* $p = .30$.

Notice that

$$P(X = 0, \text{ or } 8 \leq X) = .02824 + .00160 = .02984,$$

which is the same as

$$1 - P(1 \leq X \leq 7) = 1 - .97016 = .02984,$$

so that this is the probability that X falls *outside* the interval bounded by 1 and 7.

Binomial distributions can also be put into cumulative form. Thus, the probability that X falls at or below a certain value a is the probability of the interval $X \leq a$. For this particular distribution, the corresponding cumulative distribution is

r	$P(X \leq r)$
10	1.00000
9	.99999
8	.99985
7	.99840
6	.98940
5	.95264
4	.84972
3	.64960
2	.38277
1	.14930
0	.02824

In this distribution, we see that about 65 percent of samples of 10 should show 3 or fewer successes, about 85 percent should have 4 or fewer, and 99.8 percent should have 7 or fewer. Every sample (100 percent) must have ten or fewer successes, of course, since $N = 10$.

It must be reemphasized that the binomial distribution is *theoretical.* It shows the probabilities for various numbers of successes out of N trials *if* independent random samplings are carried out from a Bernoulli distribution and *if* p is the probability of a success. Given a different value of p or of N (or of both) the probabilities will be different. Nevertheless, regardless of N or p, the probabilities are found by the same binomial rule.

4.6 THE MEAN AND VARIANCE OF THE BINOMIAL DISTRIBUTION

To find the mean and variance of the binomial distribution, first consider the situation in which there is only one Bernoulli trial. Then $P(X = 1) = p$, $P(X = 0) = 1 - p$, and we have

$$E(X) = 1 \cdot p + 0 \cdot (1 - p) = p,$$

$$E(X^2) = 1^2 \cdot p + 0^2 \cdot (1 - p) = p,$$

and $\text{var}(X) = E(X^2) - E(X)^2 = p - p^2 = p(1 - p).$

The binomial distribution is simply the distribution of the number of successes in N independent Bernoulli trials. By the rules of expectation, the mean of the binomial distribution is the sum of the means of the individual trials, or p summed N times:

$$E(X) = Np. \tag{4.6.1*}$$

Similarly, because of the independence of the trials, the variance of the binomial distribution is the sum of the variances of the individual trials, or $p(1 - p)$ summed N times:

$$\text{var}(X) = Np(1 - p) = Npq. \tag{4.6.2*}$$

For example, if $N = 10$ and $p = \frac{1}{2}$, then

$$E(X) = (10)(\tfrac{1}{2}) = 5$$

and $\text{var}(X) = (10)(\tfrac{1}{2})(\tfrac{1}{2}) = 2.5.$

If $N = 25$ and $p = .3$, the mean is $(25)(.3) = 7.5$ and the variance is $(25)(.3)(.7) = 5.25$. Notice that the mean *can* be some value that X cannot take on; the number of successes actually observed may be greater than or less than Np, but Np serves as a good summary measure of the location, or central tendency, of the random variable X.

Alternatively, the mean and variance of the binomial distribution can be determined directly from the probability functions,

$$P(X = r) = \binom{N}{r} p^r q^{N-r} = \frac{N!}{r!(N - r)!} p^r q^{N-r}, \quad \text{where } r = 0, 1, \cdots, N.$$

From the definition of expectation,

$$E(X) = \sum_r rP(r) = \sum_{r=0}^{N} r \left[\frac{N!}{r!(N - r)!} p^r q^{N-r} \right]. \tag{4.6.1}$$

Now, notice that this expression could be factored somewhat, canceling

MEAN AND VARIANCE OF THE BINOMIAL DISTRIBUTION

r in the numerator and denominator and bringing an N and a p outside the brackets:

$$E(X) = \sum_{r=0}^{N} Np \left[\frac{(N-1)!}{(r-1)!(N-r)!} p^{r-1} q^{N-r} \right].$$

For $r = 0$, the expression being summed in Equation (4.6.1) is equal to zero, and thus it is possible to change the index of summation, r, so that it only goes from 1 to N rather than from 0 to N:

$$E(X) = \sum_{r=1}^{N} Np \left[\frac{(N-1)!}{(r-1)!(N-r)!} p^{r-1} q^{N-r} \right]$$

$$= Np \sum_{r=1}^{N} \frac{(N-1)!}{(r-1)!(N-r)!} p^{r-1} q^{N-r}.$$

But the summation consists of the summation of all of the terms of a binomial distribution with parameters $N - 1$ and p, and thus the sum must be equal to one (if you do not see this, rewrite the expression in terms of $N' = N - 1$ and $r' = r - 1$). Since the sum is one,

$$E(X) = Np.$$

We can find the variance of the binomial distribution in a similar manner. First, we will determine $E(X^2 - X)$:

$$E(X^2 - X) = E[X(X-1)] = \sum_{r} r(r-1)P(r).$$

For the binomial distribution,

$$E(X^2 - X) = \sum_{r=0}^{N} r(r-1) \left[\frac{N!}{r!(N-r)!} p^r q^{N-r} \right].$$

The expression being summed is equal to zero if $r = 0$ or $r = 1$, so we can sum from $r = 2$ to $r = N$ instead of from $r = 0$ to $r = N$. Canceling the $r(r-1)$ and factoring $N(N-1)p^2$ from the expression within brackets, we get:

$$E(X^2 - X) = N(N-1)p^2 \sum_{r=2}^{N} \frac{(N-2)!}{(r-2)!(N-r)!} p^{r-2} q^{N-r}.$$

Once again the summation consists of the summation of all of the terms of a binomial distribution, this time a binomial distribution with parameters $N - 2$ and p, so the sum is equal to one. As a result,

$$E(X^2 - X) = N(N-1)p^2.$$

But

$$E(X^2 - X) = E(X^2) - E(X),$$

so $E(X^2) = E(X^2 - X) + E(X) = N(N - 1)p^2 + Np.$

Therefore,

$$\begin{aligned}
\text{var } (X) &= E(X^2) - [E(X)]^2 = N(N - 1)p^2 + Np - (Np)^2 \\
&= N(N - 1)p^2 + Np - N^2p^2 \\
&= Np^2(N - 1 - N) + Np.
\end{aligned}$$

$$\text{var } (X) = -Np^2 + Np = Np(1 - p) = Npq. \qquad (4.6.3*)$$

Thus the variance of the binomial distribution is equal to Npq, and the standard deviation is equal to $\sqrt{Npq}$.

4.7 THE BINOMIAL DISTRIBUTION OF PROPORTIONS

Quite often researchers are interested not in the number of successes that occur for some N trial observations, but rather in the *proportion* of successes, X/N. The proportion of successes is also a random variable, taking on fractional (or decimal) values between 0 and 1.00. Such sample proportions will be designated by the capital letter P to distinguish them from p, the probability of a success.

The probability of any given proportion P of successes among N cases sampled from a given Bernoulli distribution is exactly the same as the probability of the number of successes; that is,

$$P\left(P = \frac{r}{N}\right) = P(X = r). \qquad (4.7.1)$$

For instance, if $N = 6$ and $r = 4$,

$$P(X = 4) = P\left(P = \frac{4}{6}\right) = \binom{6}{4}p^4q^2.$$

The distribution of sample proportions is therefore given by the binomial distribution, the only difference being that each possible value of X becomes a value of $P = X/N$, so that prob.$(P = a)$ = prob.$(X = Na)$ for any particular value a. The mean and variance of P are

$$E(P) = E\left(\frac{X}{N}\right) = \frac{1}{N}E(X) = \frac{Np}{N} = p \qquad (4.7.2)$$

and $\text{var } (P) = \text{var }\left(\dfrac{X}{N}\right) = \dfrac{1}{N^2}\text{var } (X) = \dfrac{Npq}{N^2} = \dfrac{pq}{N}. \qquad (4.7.3)$

In the further discussion of the binomial distribution in later chapters, care will be taken to specify whether the random variable is regarded as X or P, since the arithmetic is slightly different in the two situations. Nevertheless, the probability of any given P is the same as for the corresponding X, given the sample size N.

4.8 THE FORM OF A BINOMIAL DISTRIBUTION

Although the mathematical rule for a binomial distribution is the same regardless of the particular values of N and p entering into the expression, the "shape" of a histogram or other representation of a binomial distribution will depend upon N and p. In general, the probabilities increase for increasing values of X until some maximum point is reached, and then for X values beyond this point the probabilities decrease once again. This makes the picture given by the distribution show a single "hump" somewhere between $X = 0$ and $X = N$, with probabilities gradually decreasing on either side of this maximum point. The exact location of this highest probability relative to X depends, of course, on N and p.

One device for locating where this maximum point must occur in any binomial distribution is to find the ratio of probabilities for successive values of X. That is, we take $X = r'$ and $X = r$, where $r' = r - 1$, and put these probabilities in ratio:

$$\frac{P(X = r \mid N, p)}{P(X = r' \mid N, p)} = \frac{\binom{N}{r} p^r q^{N-r}}{\binom{N}{r'} p^{r'} q^{N-r'}} = 1 + \frac{(N + 1)p - r}{qr}. \quad (4.8.1)$$

When this ratio is greater than 1.00, the probabilities are successively increasing; when the ratio is less than 1.00, the probabilities are decreasing; when the ratio is exactly equal to 1.00, the successive probabilities are exactly the same.

Now consider the situation where X can assume the exact value Np (that is, Np is an integer). If we take $r = Np$, we find that the ratio (4.8.1) is a number greater than 1.00, indicating that the probability for $X = Np$ is greater than the probability for $X = Np - 1$. On the other hand, if we take $r = Np + 1$ we find this ratio is less than 1, meaning that the probability for $Np + 1$ is less than that for Np. In this way it can be shown that when Np is an integer, the probability associated with $X = Np$ is greater than that for any other value of X; in this situation, the maximum point in a graph of the binomial distribution occurs at $X = Np$. Recall that this is also the mean of the distribution. Therefore, one finds that in this special situation the value $P = p$ is the most probable relative frequency of occurrence of an event with probability p.

On the other hand, it may be that X cannot attain the exact value Np, since Np need not be an integer. However, by exploring the changes in the ratio of probabilities, we would find that in this situation the value $X = r$ with the highest probability is *no more* than p or q away from the value of Np; that is

$$-q \leq r - Np \leq p,$$

where r is the value of X having highest probability in the binomial distribution. Actually, the statement interpreting Bernoulli's theorem in Section 2.16 should be read as meaning "the most probable proportion of the event's occurrences to the total number of trials is also p, whenever that value of X/N can actually occur." Nevertheless, the most probable relative frequency is always a number *close* to the true value of p, even though this exact value may not occur for some values of N.

The PMF of the binomial distribution is symmetric if $p = \frac{1}{2}$, positively skewed if $p < \frac{1}{2}$, and negatively skewed if $p > \frac{1}{2}$. For any value of p other than $\frac{1}{2}$, the skewness is reduced as N becomes larger. For a given value of p, the mean and variance become larger as N increases. For a given value of N, the mean becomes larger as p increases and the variance becomes larger as p gets closer to $\frac{1}{2}$ (in the extreme cases, when $p = 0$ or $p = 1$, the variance is equal to zero, as can be seen by using the formula var $(X) = Npq$). Examples of binomial PMF's are presented in Figure 4.8.1.

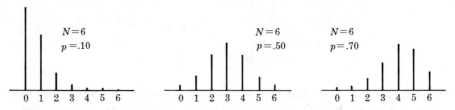

Figure 4.8.1

4.9 THE BINOMIAL AS A SAMPLING DISTRIBUTION

A binomial distribution is one example of a very important kind of theoretical distribution encountered again and again in statistics. These are sampling distributions, where a random variable X denotes a possible value of some measure *summarizing* the outcomes of N distinct observations. In the binomial case, this random variable is either the number of successes out of N trials or perhaps P, the proportion of successes out of the N trials. In other contexts the random variable may be some other number summarizing a sample result, such as an average, or some index of variability. There are many different kinds of sampling distributions, depending on the basic sample space and the way that samples are summarized. Nevertheless, the essential idea is the same: a sample of N observations is drawn at random, so that each and every distinct possible sample has an equal probability of occurrence. A number is attached to the sample, in some way summarizing the N outcomes observed. In principle, different values will be found for different samples, and each sample value may be regarded as a value of a random variable. The sampling distribution is a theoretical statement of the probability of observing various intervals of values of such a random variable over all possible samples of the same size N drawn at random.

For example, imagine that all United States adults could be classified into two categories:

Success = prefers foreign movies

Failure = prefers U.S. movies.

We will assume that each person has one of these two preferences, and that there is an almost infinite number of such persons. This population could be arranged in a Bernoulli distribution, pairing each of the two classes with its proportion, p or q.

We sample United States adults at random, asking each his preference. Suppose that our random sample consists of exactly 10 adults. The number of successes observed could be any number from 0 through 10, and there is some probability that can be associated each possible number, given the true p. The pairing of different sample values one might get and the probability of each is the sampling distribution. Note well, however, that the sampling distribution depends upon how the sampling was done, how we attached the value to a sample, and most of all, what the *true* situation is. If only .30 of the population of adults prefer foreign movies, then the sampling distribution for number of successes is given by Table 4.5.1. This tells us that if we observe 10 successes out of 10 cases, then we have encountered an event that should occur only once in 100,000 samples of 10 people, *if p is .30*. On the other hand, for $p = .30$, we should expect 3 out of 10 to occur relatively often; about 26 in 100 samples should give us this exact value. About 70 in 100 samples should give us 3, 4, or 5 people preferring foreign movies, and so on. In short, a sampling distribution allows one to judge the probability of a particular kind of sample result, given that something is true in general about the population being sampled. Before one can find the sampling distribution, however, he must postulate something true about the general situation; in this case, we postulated that exactly 30 percent of United States adults prefer foreign movies. Given that this is true, the sampling distribution specifies the probabilities of various numbers of adults in the sample who show this preference.

In the next section a little preview of statistical inference and decision will be given in which the binomial rule provides a sampling distribution. This example cannot be complete at this time simply because all the necessary ingredients have not been introduced. On the other hand, we feel that it is important that the student get some feel for the use of theoretical distributions as early as possible, to provide a basis for the more technical development to follow.

4.10   A PREVIEW OF A USE OF THE BINOMIAL DISTRIBUTION

It is now possible to point ahead to an important use of the binomial distribution. This example will deal with a quality control situation in which we must decide, on the basis of sample information, whether or not a process is "out of control" and should be adjusted.

Suppose that the manager of a plant is concerned with a particular manufacturing process within his plant. He knows that from time to time the process goes haywire, or out of control, and produces too many defective items. If it is out of control, it can be corrected with an adjustment. However, the manager would have to call in a mechanic to make this adjustment, and this involves an expenditure of money. On the other hand, if the process is in fact out of control, it is expensive to repair the defective items. If the process is not out of control, then the manager *does not* want to pay to have it adjusted; if it is out of control, he *does* want to do so.

In order to attack this problem, the manager must be more precise about what he means by "out of control." Let us assume that the process can be thought of as a Bernoulli process with parameter p, the probability of any single item from the process being defective. Furthermore, assume that the process is in control if p is no greater than .10; otherwise it is out of control. In other words, the manager has a hypothesis: the process is in control, or in terms of p, p is no greater than .10. If he could be sure that the hypothesis was true, he would take one action (not adjust the process), and if he could be sure that the hypothesis was false, he would take a different action (call the mechanic to adjust the process). Under the manager's hypothesis, the sample space for each item produced by the process is:

Outcome	P (Outcome)
Defective	.10
Nondefective	.90

Note that we have taken p to be equal to .10, whereas the manager's hypothesis included all values of p less than or equal to .10. This simplifies the problem somewhat and considers the worst possible situation which is still consistent with the hypothesis.

Assume that the manager has absolutely no idea as to whether or not the process is out of control, but that he *is* able to observe a sample of ten items from the process. Assuming the above distribution *and* assuming that the process is independent and stationary (that is, the trials are independent and p remains the same from trial to trial), the manager can calculate the probabilities of the various possible sample results for a sample of $N = 10$, using the binomial distribution. We say that he is using the binomial distribution as a sampling distribution; he is using it to calculate a probability distribution for X, the number of defectives in the sample. The resulting distribution is presented in Table 4.10.1. It should

Table 4.10.1

r	$P(X = r) = \binom{10}{r} (.10)^r (.90)^{10-r}$
0 defectives	.3487
1	.3874
2	.1937
3	.0574
4	.0112
5	.0015
6	.0001
7 or more	.0000

be pointed out that the probability of seven or more defective items is not *exactly* zero; to four decimal places, it is zero, however.

Given this theoretical distribution of possible outcomes, we turn to the actual results of the sample. The manager finds that there are five defectives in the sample of ten items. What is the probability that exactly this result should have come up by chance, given that the manager's hypothesis is true? It is the binomial probability for five successes in ten trials, given that p is .10. From Table 4.10.1, this probability is .0015. This sample result is not a likely thing to occur if p is in fact equal to .10.

However, we should be interested in the probability not only of obtaining exactly 5 defective items, but rather in the probability of obtaining *this many or more defective items*. The manager should ask himself, "What is the probability of a sample result which is *at least* this "bad," that is, a sample result with at least five defective items?" The answer to this question involves the probability of an interval:

$$P(X \geq 5) = P(X = 5) + P(X = 6) + P(X = 7)$$

$$+ \cdots + P(X = 10) = .0016.$$

Given that $p = .10$, five or more defectives should occur only about 16 times in 10,000 independent replications of the experiment. Does this unlikely result cast any doubt on the manager's hypothesis that the process is in control? The answer is yes. For a hypothesis to seem reasonable, it should forecast results that agree fairly well with actual outcomes. If the manager had observed no defectives or one defective in the sample, then there would be little cause to doubt the hypothesis that the process is in control. Even two defectives does not seem to be too unlikely a result, although it may be enough to cause the manager to call the mechanic, particularly if it is very costly to repair defective items. Five defectives, however, is so unlikely that the manager is almost sure to call the mechanic, unless the mechanic's fee is many times greater than the cost of repairing defective items.

Note that the sample results may cast doubt on the hypothesis; they do *not* disprove it. The results obtained, unlikely as they seem, could occur by chance. Thus, the manager could be making the wrong decision if he calls the mechanic. The chances are good that it is the right decision, but he cannot be sure of this. If he calls the mechanic, a measure of the amount of risk he runs of being wrong by abandoning his hypothesis on the basis of the sample evidence is given by the probability of sample results as extreme or more extreme than those actually obtained, if the hypothesis is true. In this case, his chance of being wrong in calling the mechanic is .0016. If this probability had been higher, he would be less likely to call the mechanic. If he had observed one defective item, for example, the chance of being wrong if he calls the mechanic is .6513. In this case the results are not so incompatible with his hypothesis, and he is less likely to decide to call the mechanic.

The ultimate decision to call or not to call the mechanic depends not only on the sample results, but also on the relevant costs. That is, it depends on the cost of calling the mechanic and the cost of repairing defective items. If the former is much greater than the latter, the manager might decide to stand pat even if the sample result is quite unlikely under the hypothesis. If the latter is much greater than the former, the manager might decide to call the mechanic even if the sample result is not too unlikely under the given hypothesis. We shall discuss this type of question in Chapter 9.

Although this little example should not be taken as a model of sophisticated statistical (and decision-theoretic) practice, it does suggest the use of a theoretical distribution of sample results as an aid in making inferences and decisions on the basis of sample information. The binomial distribution is only one of a number of such sampling distributions we shall employ in this general way.

4.11 NUMBER OF TRIALS AS A RANDOM VARIABLE: THE PASCAL AND GEOMETRIC DISTRIBUTIONS

In the development of the binomial distribution, it was assumed that a fixed number N of Bernoulli trials would be observed, and the random variable of interest was X, the number of successes in N trials. In some situations, the sampling procedure may be slightly different; a sequence of independent Bernoulli trials may be observed, where the observations continue until a fixed number R of successes is observed. Here N is not fixed, but is a random variable. The distribution of N can be determined by the same line of reasoning used to find the theoretical binomial distribution. For any particular sequence with R successes in N trials, the probability has been found to be

$$p^R q^{N-R}.$$

To determine the binomial distribution, it was necessary to "count" the number of possible sequences having R successes in N trials: $\binom{N}{R}$. If the sampling is such that the number of successes, R, is predetermined instead of the number of trials, N, then it must be true that the final trial resulted in a success. Otherwise, the trial would not have been observed, since the Rth success would have occurred on a previous trial. If the last trial must be a success, then exactly $R - 1$ successes must have occurred in the previous $N - 1$ trials. But the number of possible sequences with $R - 1$ successes in $N - 1$ trials is $\binom{N-1}{R-1}$, so we have the following principle: **in sampling from a Bernoulli process with the probability of a success equal to p, the probability that it will take exactly N trials to observe R successes is**

$$P(N \text{ trials} \mid R, p) = \binom{N-1}{R-1} p^R q^{N-R}, \qquad 0 < R \le N. \quad (4.11.1^*)$$

This distribution, the distribution of the number of trials until the Rth success, is called the **Pascal distribution,** after the French mathematician Pascal. The mean and variance of the Pascal distribution can be shown to be equal to

$$E(N) = \frac{R}{p} \quad \text{and} \quad \text{var}(N) = \left(\frac{R}{p}\right)\left(\frac{1}{p} - 1\right) = \frac{R(1-p)}{p^2}. \quad (4.11.2^*)$$

The distribution is positively skewed, as can be seen from Figure 4.11.1.

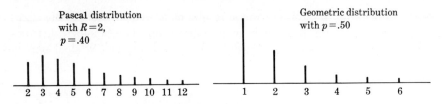

Figure 4.11.1

Consider the example in Section 4.2 of the albino and normal animals, where p, the probability of an albino, is equal to .20. If the biologist takes a sample of size 3, the probability of observing exactly one albino is a binomial probability:

$$P \text{ (1 albino in 3 animals)} = \binom{3}{1} (.20)^1 (.80)^2 = .384.$$

If the biologist decides to sample until he observes a single albino, however, the probability that this will take exactly 3 trials is a Pascal probability:

$$P \text{ (3 animals to get 1 albino)} = \binom{2}{0} (.20)^1 (.80)^2 = .128.$$

Even though they describe the same sample, the binomial and Pascal probabilities differ because the sampling *plan* is not the same; in one case, the number of trials is fixed before observing the sample, and in the other case, the number of successes is fixed in advance. We say that the *stopping rules* are different: in binomial sampling, we stop sampling after a fixed number of trials (N); in Pascal sampling, we stop sampling after a fixed number of successes (R).

The last example demonstrates a special case of the Pascal distribution: the case in which $R = 1$, and sampling continues until the first success. This special case is called the **geometric distribution,** and its probability function is given by

$$P \text{ (N trials until the first success} \mid p) = pq^{N-1}. \qquad (4.11.3^*)$$

This is reasonable, since the only sequence satisfying the given condition is the sequence consisting of $N - 1$ failures followed by a success, and the probability of this sequence must be pq^{N-1} because of the independence of the trials. The geometric distribution is applicable in gambling situations in which a gambler is interested in the probability of going, say 10 trials (a trial might be a spin of a roulette wheel, a "deal" in a card game, a throw of a pair of dice, and so on) until the first successful gamble. If the gambler knows p, the probability of winning on any given trial, he can compute the above probability by using the geometric distribution. For instance, if $p = \frac{1}{2}$,

$$P \text{ (10 trials until the first win} \mid p = \tfrac{1}{2}) = (\tfrac{1}{2})(\tfrac{1}{2})^9 = 1/1024.$$

By noting that the mean and variance of the geometric distribution are equivalent to the mean and variance of the Pascal distribution with $R = 1$, since the geometric distribution is a special case of the Pascal distribution, the gambler can compute the expected number of trials until the first win and the variance of the number of trials until the first win:

$$E(N) = \frac{1}{p} = \frac{1}{\frac{1}{2}} = 2 \quad \text{and} \quad \text{var } (N) = \left(\frac{1}{p}\right)\left(\frac{1}{p} - 1\right) = \left(\frac{1}{\frac{1}{2}}\right)\left(\frac{1}{\frac{1}{2}} - 1\right) = 2.$$

The binomial, Pascal, and geometric distributions all involve Bernoulli trials. More specifically, they all require independent trials with p, the probability of success, remaining constant from trial to trial. If this last requirement is met, it is said that the process is *stationary*. In applications, p often varies from trial to trial, and thus the process is not stationary. An example of a nonstationary process might be a tennis match, in which the two players alternate serving. When the first player serves, his probability of winning that particular game may be .75, but when his opponent serves, the first player's probability of winning the game may be only .40. Even if we accept the assumption of independence (which is quite questionable given the psychological and physiological factors involved in "hot streaks" or "cold streaks" in sporting events), the distributions we have considered are not *directly* applicable to the tennis match. (They could be made applicable by considering the match as a combination of two separate Bernoulli processes, depending on the server.)

It is important to remember that the theoretical distributions we have discussed thus far in this chapter require the assumption that sampling is done from a stationary and independent Bernoulli process. Only if this assumption is satisfied are the three theoretical distributions applicable in the form in which they have been presented. Of course, a mathematical model is seldom a perfect representation of a real-world situation, and yet such models *are* applied to real-world situations. If the assumptions of a model are *approximately* satisfied, then the model still may be useful. Because of the inherent complexity of most real-world problems, it is usually necessary to settle for approximate results. Throughout our study of probability and statistics, we will rely on approximations and on assumptions which are seldom perfectly satisfied. The student should always remember, however, that statistical techniques are based on certain assumptions and that in any application of the techniques the results are only as good as the assumptions.

4.12   THE MULTINOMIAL DISTRIBUTION

The basic rationale underlying the binomial distribution can be generalized to situations with more than two event classes. This generalization is known as the "multinomial distribution," having the following rule:

Consider K classes, mutually exclusive and exhaustive, and with probabilities p_1, p_2, $\cdots$, p_K. If N observations are made independently and at random, then the probability that exactly n_1 will be of kind 1, n_2 of kind 2, $\cdots$, and n_K of kind K, where $n_1 + n_2 + \cdots + n_K = N$, is given by

$$\frac{N!}{n_1!n_2!\cdots n_K!}\,(p_1)^{n_1}(p_2)^{n_2}\cdots(p_K)^{n_K}. \qquad (4.12.1^*)$$

Note that the first factor of Equation (4.12.1) corresponds to Counting Rule 6 of Section 2.13. Also observe that the binomial distribution corresponds to the multinomial distribution with $K = 2$.

Although the multinomial rule is relatively easy to state, a tabulation or graph of this distribution is very complicated; here a sample result is not a single number, as for the binomial, but rather a set of $(K - 1)$ numbers (once n_1, n_2, $\cdots$, n_{K-2}, and n_{K-1} are known, n_K is fixed, since the sum of the n_i's must equal N).

For example, think of colored marbles mixed together in a box, where the following probability distribution holds:

Color	p
Black	.40
Red	.30
White	.20
Blue	.10
	1.00

Now suppose that 10 balls were drawn at random and with replacement. The sample shows 2 black, 3 red, 5 white, and 0 blue. What is the probability of a sample distribution such as this? On substituting into the multinomial rule, we have

$$\frac{10!}{(2!)(3!)(5!)(0!)}\,(.4)^2(.3)^3(.2)^5(.1)^0$$

$$= \frac{1\cdot2\cdot3\cdot4\cdot5\cdot6\cdot7\cdot8\cdot9\cdot10}{(1\cdot2)(1\cdot2\cdot3)(1\cdot2\cdot3\cdot4\cdot5)(1)}\,(.4)^2(.3)^3(.2)^5,$$

since 0! and $(.1)^0$ are both equal to 1. Working out this number, we find that .087 is the probability of the *sample* distribution

Black	2
Red	3
White	5
Blue	0
	—
	10

if the probability distribution given above is the true one. Using this multinomial rule, one could work out a probability for *each possible sample distribution*.

This raises an interesting possibility. Given any discrete probability distribution, one might work out the probability of all possible sample distributions for N observations. Then, in terms of these probabilities, the disagreement of sample and theoretical distributions could be evaluated. Although this is possible in principle, the drawbacks should be obvious. For any but small numbers of possible events and small N, the sheer number of different sample distributions is fantastically large. Ordinarily we are not interested in the probability of our sample alone, but rather in that of samples *as deviant or more so* than ours. This makes for too much computation to be practical in most instances. Furthermore, we run into trouble in trying to formulate a similar scheme for *continuous* theoretical distributions. For these reasons, the multinomial distribution plays a rather small role in theoretical statistics. The probability of obtaining an entire sample distribution is not usually of interest; instead, we will be finding probabilities for various indices summarizing a sample distribution.

4.13   THE HYPERGEOMETRIC DISTRIBUTION

Another theoretical probability distribution deserves passing mention for the same reason. The multinomial rule (and the binomial, of course) assumes either that the sampling is done with replacement or that the sample space is infinite, so that the basic probabilities do not change over the trials made. However, suppose that one were sampling from a finite space *without* replacement; then the probabilities would change for each observation made. By a series of arguments very similar to those used for finding probabilities of poker hands in Section 2.14, we could arrive at a new rule for finding the probabilities of sample results.

This rule describes the hypergeometric distribution, and can be stated as follows:

Given a sample space containing a finite number T of elements, suppose that the elements are divided into K mutually

exclusive and exhaustive classes, with T_1 in class 1, T_2 in class 2, $\cdots$, T_K in class K. A sample of N observations is drawn at random without replacement, and is found to contain n_1 of class 1, n_2 of class 2, $\cdots$, n_K of class K. Then the probability of occurrence of such a sample is given by

$$\frac{\binom{T_1}{n_1}\binom{T_2}{n_2}\cdots\binom{T_K}{n_K}}{\binom{T}{N}},\qquad (4.13.1^*)$$

where $n_1 + n_2 + \cdots + n_K = N$ and $T_1 + T_2 + \cdots + T_K = T$.

For an illustration of the use of the hypergeometric rule, let us return to the problem of drawing marbles at random from a box, but this time *without* replacement. Suppose there had been 30 balls in the box originally, with the following frequencies of colors:

Color	f
Black	12
Red	9
White	6
Blue	3
	30

Notice that the relative frequencies are the same as the probabilities in the previous example. Now ten balls are drawn at random without replacement. We want the probability that the *sample* distribution is:

Color	f
Black	2
Red	3
White	5
Blue	0
	10

Using the hypergeometric rule, we get

$$\frac{\binom{12}{2}\binom{9}{3}\binom{6}{5}\binom{3}{0}}{\binom{30}{10}},$$

which works out to be about .0011. This is not, of course, the same probability as we found for this sample result using the multinomial rule, since the entire sampling scheme is assumed different in this second example. This illustrates that *the sampling scheme adopted makes a real difference in the probability of a given result.*

However, this is a practical consideration only when the basic sample space contains a finite and small number of elementary events. When there is a very large number of elementary events in the sample space, the selection and nonreplacement of a particular unit for observation has negligible effect on the probabilities of events for successive samplings. For this reason, the hypergeometric probabilities are very closely approximated by binomial or multinomial probabilities when T, the total number of elements in the sample space, is extremely large. The distinction between these different distributions becomes practically important only when samples are taken from relatively small sets of potential units for observation.

In order to compare the binomial and hypergeometric distributions, consider the case where there are only two classes. It can be shown that the mean and variance of the hypergeometric distribution when $K = 2$ (assuming class 1 represents success and class 2 represents failure, and X is the number of successes in N trials) are

$$E(X) = N\left(\frac{T_1}{T}\right) \quad \text{and} \quad \text{var}(X) = \left(\frac{T - N}{T}\right)\left[N\left(\frac{T_1}{T}\right)\left(1 - \frac{T_1}{T}\right)\right].$$

$$(4.13.2^*)$$

Recall that the mean and variance of the binomial distribution are

$$E(X) = Np \quad \text{and} \quad \text{var}(X) = Npq = Np(1 - p).$$

Note that the ratio T_1/T represents the proportion of successes in the entire sample space. If we were to take an element at random from the sample space, the probability that it would be a success is simply this ratio, T_1/T. In this sense, T_1/T in the hypergeometric distribution corresponds to p in the binomial distribution, and the similarity in means and variances becomes apparent. If we assume that $p = T_1/T$, the means are equal and the variances differ only by a factor of $(T - N)/T$, or $1 - (N/T)$. But N is the sample size, and T is the total number of elements in the entire sample space, which is sometimes called the population size. Since the two variances differ by a factor of $1 - (N/T)$, they become more identical as the ratio N/T approaches zero (that is, as the sample size becomes very small in relation to the population size). This agrees with the statement in the preceding paragraph, which essentially says that the distinction between the two distributions becomes practically important only when the population size is small enough (or the sample size is large enough) so that the sample represents a sizeable proportion of the population. Sampling from a Bernoulli process can be thought of as sampling from an infinite population, since p remains constant from trial to trial. Appropriately, then, the factor $(T - N)/T$ is called the *finite population correction*, allowing for the fact that the hypergeometric distribution represents the case in which we are sampling from a finite population. We will encounter this again when we discuss sampling theory in Chapter 11 (Volume II).

4.14 THE POISSON PROCESS AND DISTRIBUTION

The binomial distribution is the distribution of the random variable X, the number of successes, or occurrences of a particular event, in a sequence of trials from a stationary and independent Bernoulli process. This distribution has wide applicability, for there are many situations where a sequence of observations can be made and the assumptions underlying the Bernoulli process are satisfied or at least approximately satisfied. Frequently, however, we are not able to observe a finite sequence of trials. Instead, observations take place over a continuum, such as time. Suppose that we are observing the number of cars arriving at a toll booth in a given period of time, or the number of deaths occurring from a specific disease in a given period of time. Since the observations take place over time, it is difficult to think of these situations in terms of finite trials, although we are interested in the number of occurrences of a particular event, as in the Bernoulli process.

Suppose that we are interested in the number of occurrences of an event E in a given time period of length t. One way to look at this is to break the time period t into N equal intervals, each of length t/N, and to consider these as N independent, stationary trials from a Bernoulli process. From the binomial distribution, if we know p, the probability of an occurrence of E on any specific trial, we can calculate the probabilities for X, the number of occurrences of E in N trials. Furthermore, the expected number of occurrences is Np. Thus it looks as though the situation of interest (occurrences of an event over a time period of length t) can be treated as a Bernoulli process. However, there is one difficulty which has been ignored.

Since the event occurs at various points of time, what would prevent it from occurring twice or more in one of the "trials" of length t/N? Since the Bernoulli process considers only a dichotomous situation—that is, the occurrence or nonoccurrence of an event, there is no way to allow for the possibility of multiple occurrences within a single trial. This difficulty can be resolved to some extent by making N larger, thus dividing the period of length t into N equal intervals of smaller length. If N is chosen so that the probability of multiple occurrences in any single "trial" of length t/N is zero for all practical purposes, then the situation resembles that of a Bernoulli process.

As we make N larger and larger, the "trials" are shorter and shorter in terms of length of time, t/N. As a result, the probability of an occurrence in any single trial must become smaller and smaller. Consider the expected number of occurrences in a period of time t, and call this expectation $E(X)$. From the binomial distribution, $E(X) = Np$. The division of the time period into subperiods should not have any effect on the expected number of occurrences in a period of time t, so $E(X)$ should remain constant as N is varied. If this is true, it is clear that p must become smaller as N becomes larger, so that Np will remain constant. Now, if X is the number of occurrences of E in the given time period, we have, from the binomial distribution,

$$P(X = r \mid N, p) = \binom{N}{r} p^r q^{N-r} \qquad \text{for } r = 0, 1, \cdots, N.$$

It can be shown mathematically that if N is made larger and larger and p smaller and smaller in such a way that Np remains constant, then the distribution of X approaches the Poisson distribution (named after the French mathematician S. Poisson), which can be written as follows:

$$P(X = r \mid N, p) = \frac{e^{-Np}(Np)^r}{r!}. \qquad (4.14.1)$$

Since Np remains constant, it is customary to set $Np = \lambda$ (Greek lambda) and to write the Poisson distribution as follows:

$$P(X = r \mid \lambda) = \frac{e^{-\lambda}\lambda^r}{r!}, \qquad \text{where } r = 0, 1, 2, \cdots. \qquad (4.14.2^*)$$

Note that X can now take on any integral value, since there is no upper bound to the number of occurrences of an event in a given time period. The mathematical derivation of the Poisson distribution as a limiting form of the binomial distribution is beyond the scope of this book. However, given the functional form of the distribution as presented above, the mean and variance of the Poisson distribution can be derived in a manner similar to the derivation of the mean and variance of the binomial distribution in Section 4.6.

From the definition of expectation,

$$E(X) = \sum_r rP(r) = \sum_{r=0}^{\infty} r \frac{e^{-\lambda}\lambda^r}{r!}.$$

Canceling r from numerator and denominator and factoring out λ, we get

$$E(X) = \lambda \sum_{r=0}^{\infty} \frac{e^{-\lambda}\lambda^{r-1}}{(r-1)!}.$$

Since the expression being summed in the first expression for $E(X)$ is zero when $r = 0$, the limits of summation can be changed from 0 and ∞ to 1 and ∞. As a result, the sum in the second expression is simply the summation of all of the terms of a Poisson distribution with parameter λ (to see this, let $r' = r - 1$), and this must be one. But then

$$E(X) = \lambda. \tag{4.14.3*}$$

This is consistent with the argument above, in which it was noted that the expected number of occurrences in a period of time t is equal to Np, which is equal to λ.

The variance of the Poisson distribution is equal to the mean, λ. The derivation of the variance of the Poisson distribution parallels the derivation of the variance of the binomial distribution.

We have developed the Poisson distribution by considering a limiting form of the binomial distribution. It is possible to consider the occurrence of an event over a continuum as a process in itself, however, and thus to derive the Poisson distribution without any reference to the binomial. Suppose that we are interested in the number of occurrences of an event in a given period of time of length t. Furthermore, suppose that the occurrence or nonoccurrence of the event in any time interval is independent of its occurrence or nonoccurrence in any other time interval (the independence assumption). Also, suppose that the probability of an occurrence of the event in a time period of given length is the same no matter when the period begins and ends (the stationarity assumption). Then we say that **the occurrences of the event are generated by a stationary and independent Poisson process, and that the distribution of X, the number of occurrences in a time period of length t, is given by a Poisson distribution, where λ corresponds to the intensity of the process.** The Poisson process has a single parameter, λ, which is called the **intensity of the process** because it represents the expected number of occurrences in a time period of length t. For example, λ might be the number of cars per minute arriving at a toll booth, or the number of deaths per year attributed to a specific disease.

Since it is tedious to calculate Poisson probabilities by hand due to such factors as $e^{-\lambda}$, λ^r, and $r!$, tables of Poisson probabilities are available. In these tables, probabilities are given for various values of r and λ. From Table VI in the back of the book, we see that the Poisson probabilities for $\lambda = 1$ are as follows:

r	$P(X = r \mid \lambda = 1)$
0	.3679
1	.3679
2	.1839
3	.0613
4	.0153
5	.0031
6	.0005
7	.0001
>7	.0000

From these probabilities, cumulative probabilities can be determined. If $\lambda = 1$, for example, $P(X \le 2) = .3679 + .3679 + .1839 = .9197$. The PMF and CDF of the Poisson distribution with $\lambda = 1$ are presented in Figure 4.14.1. In general, the Poisson distribution is positively skewed,

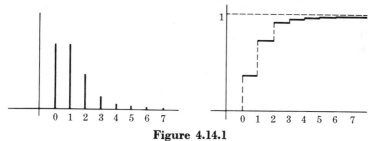

Figure 4.14.1

although it is more nearly symmetrical as λ becomes larger. Graphs of Poisson PMF's with $\lambda = 2, 5$, and 10 are presented in Figure 4.14.2. Note

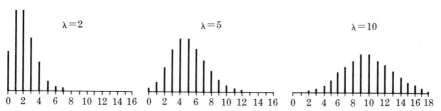

Figure 4.14.2

from these graphs that if λ is an integer, the Poisson distribution has two modes, located at $X = \lambda - 1$ and $X = \lambda$. It can be shown that if λ is *not* an integer, there is a single mode located at the integer value which is between $\lambda - 1$ and λ.

4.15   THE POISSON APPROXIMATION TO THE BINOMIAL

The calculation of probabilities is a tedious task for the binomial distribution as well as for the Poisson distribution. Unless one has access to special tables, the usual tables of the binomial distribution are much more limited than tables of other distributions (that is, only a few values of N and p are considered). Because of this, it is sometimes convenient to use other distributions to determine approximations to binomial probabilities. Both the Poisson distribution and the normal distribution provide good approximations to the binomial, although the conditions under which the approximation is good differ in the two cases. The normal distribution is continuous, and we will not discuss its use as an approximation for the binomial until later in this chapter, when we cover continuous distributions.

The relationship between the Bernoulli and Poisson processes, and hence between the binomial and Poisson distributions, was discussed in the previous section. We noted that the Poisson distribution could be derived as a limiting form of the binomial if the following three assumptions were simultaneously satisfied:

(1) N becomes large (that is, $N \rightarrow \infty$),
(2) p becomes small (that is, $p \rightarrow 0$),
and   (3) Np remains constant.

Under these conditions, the binomial distribution with parameters N and p can be approximated by the Poisson distribution with parameter $\lambda = Np$. This means that the Poisson distribution provides a good approximation to the binomial distribution if p is close to zero and N is large. Since p and q can be interchanged by simply interchanging the definitions of "success" and "failure," the Poisson is also a good approximation when p is close to one and N is large.

For an example of the use of the Poisson distribution as an approximation to the binomial distribution, consider the case in which $N = 10$ and $p = .10$. The Poisson parameter for the approximation is then $\lambda = Np = 10(.10) = 1$. The binomial distribution and the Poisson approximation are:

r	Binomial $P(X = r \mid N = 10, p = .10)$	Poisson $P(X = r \mid \lambda = 1)$
0	.3487	.3679
1	.3874	.3679
2	.1937	.1839
3	.0574	.0613
4	.0112	.0153
5	.0015	.0031
6	.0001	.0005
7	.0000	.0001
8	.0000	.0000
9	.0000	.0000
10	.0000	.0000

The two distributions agree reasonably well. If more precision is desired, a possible rule of thumb is that the Poisson is a good approximation to the binomial if $N/p > 500$ (this should give accuracy to at least two decimal places). In the example, $N/p = 100$, and yet the approximation does not seem to be too bad. In situations where the Poisson is not a good approximation, the normal distribution may be, as we will see in Section 4.22.

This section illustrates the fact that it is often possible to use one distribution to approximate another. This is particularly valuable when it is difficult to make probability calculations and when no tables are available for the original distribution. Of course, the use of *any* distribution is usually an approximation of sorts, for the assumptions underlying the distribution are seldom completely satisfied.

4.16 SUMMARY OF SPECIAL DISCRETE DISTRIBUTIONS

In the preceding sections a number of commonly encountered discrete families of distributions have been presented: the binomial, Pascal, geometric, hypergeometric, multinomial, and Poisson distributions. For each of these families, we discussed the conditions under which the distribution might arise, the functional form of the probability function $P(x)$, the parameters of the probability function, the mean and variance of the distribution, and the shape of the distribution for different values of the parameter(s). We have also seen how the distributions are related: the binomial, Pascal, and geometric distributions deal with the Bernoulli process; the hypergeometric distribution deals with samples from a finite population, and can be closely approximated by the binomial when the sample size is small relative to the population size; the multinomial distribution represents an extension of the binomial to the case in which there are more than two possible outcomes on each trial; and the Poisson distribution deals with observations made over a continuum, such as time, rather than in a finite sequence of trials—the Poisson can be used to approximate binomial probabilities if the number of trials is large and the parameter p is near zero or one.

In later chapters statistical techniques will be developed for the investigation of samples from these special distributions. These techniques include the estimation of values of the parameters, the testing of hypotheses concerning the parameters or concerning the distribution itself, and the making of decisions when there is uncertainty which can be expressed in probabilistic form with the aid of these distributions. At this point it might be instructive to suggest some practical situations in which these distributions might be applied.

There are numerous situations in which the assumptions of a Bernoulli process seem to be met, and in these situations the binomial, Pascal, and geometric distributions are applicable. One example, as we have seen, is the area of quality control, where we are interested in the items produced by a manufacturing process and, more specifically, in the percentage of defective items. If the probability of an item being defective remains the same over time, and the probability of one item being defective is independent of the past history of defectives and nondefectives, the manufacturing process can be thought of as a Bernoulli process. We might be interested in ascertaining the percentage of defectives; if this is too large the process might be considered "out of control," and adjustments would be necessary.

A gambler might also find the Bernoulli process useful, since the successive trials of many games of chance (for example, roulette) are independent and the chances of winning remain constant from trial to trial. A gambler is usually quite concerned about the possibility of a "losing streak" of such proportions that he would lose all of his capital. This is the classic problem of "gambler's ruin," which has been studied by many mathematicians. It turns out that the chances of being "wiped out" at the gambling tables depend upon the gambler's beginning capital, the stakes of the game, the probability of winning on any single trial, and the number of trials he plays. If the game is unfair to him, that is, if his expected gain on any single trial is negative, then if he plays for an indefinitely long period of time, he is sure to go broke. Problems of this general nature involve the theory of runs. A run, in terms of a Bernoulli process, is a sequence of consecutive successes or consecutive failures. The gamber is concerned about long runs of failures. The theory of runs is also applicable in many other situations.

It may be used in investigating the assumptions of a Bernoulli process: if a sample of data from a dichotomous process includes quite a few long runs of successes or failures, the assumptions of stationarity and independence may seem doubtful.

There are many other applications involving Bernoulli processes. A medical researcher might be interested in the probability that a certain medical treatment will successfully "cure" a patient's illness. A professor might be interested in the probability that a student will earn a certain grade on a multiple-choice examination just by pure guesswork. A marketing manager might be interested in the probability that a consumer will buy a particular brand of a product. The examples are endless, and you can no doubt think of some which are directly related to your particular area of interest.

Applications of the multinomial distribution are similar to the above applications except that they deal with situations where there are more than two possible outcomes at each trial. If a manufactured item can possibly have zero, one, two, or three defects, then the quality control manager might be interested in the probability of each of these possibilities. If a missile is shot at a target, it might completely destroy, partially destroy, or completely miss the target. The forecasting of election results involves the multinomial distribution if there are more than two candidates for a given elective office. The field of genetics deals with probabilities concerning various characteristics of the offspring of a particular pair of parents. Of course, the multinomial distribution is more difficult to deal with than the binomial distribution, so the set of possible outcomes is often grouped into two classes and the binomial distribution is utilized to make probability statements concerning these two classes. The binomial is also often used to approximate the hypergeometric distribution. In some applications, however, the sample size is reasonably large relative to the population size, and the hypergeometric distribution itself must be used. The hypergeometric distribution is applicable to situations involving sampling without replacement from a finite population. Most sample surveys (opinion polls, questionnaires, and so on) are of this nature. Another example of a hypergeometric application is the estimation of the size of an animal population from recapture data.

The Poisson process is applicable when observations are made over a continuum, such as time, and the assumptions of stationarity and independence are satisfied. The occurrence of demands for service is often Poisson-distributed. Examples of this are the number of cars passing a toll booth in a given time period, the number of customers entering a store in a given time period, and so on. These variables are of interest in determining the number of toll booths to construct, the number of clerks to hire, or the number of items to stock in a store. These topics have been widely studied under the names of *queuing theory* and *inventory theory*. Other random variables which have been shown to obey a Poisson process are the number of misprints in a book and the radioactive disintegration of certain elements.

In any application, it is first necessary to determine which distribution or distributions might be applicable to the situation at hand. In doing this, it is important that the assumptions be carefully investigated, for the probability statements resulting from a given distribution are only as valid as the assumptions underlying the distribution.

4.17 SPECIAL CONTINUOUS DISTRIBUTIONS

Just as there are commonly encountered discrete distributions, there are families of continuous distributions that have wide applicability, especially with regard to the development of statistical techniques. In the following sections a number of such distributions will be discussed, the most important of which is the normal distribution. You should keep in mind that a continuous probability distribution represents an idealization, as we pointed out in Section 3.7. In many situations, however, these idealizations are reasonable approximations and are much easier to deal with than the corresponding complicated discrete models.

4.18 THE NORMAL DISTRIBUTION

Heretofore, we have considered only discrete distributions. Now we consider a type of distribution having a domain which is *all* of the real numbers. This is the so-called "normal" or "Gaussian" distribution. The normal distribution is but one of a vast number of mathematical functions one might invent for a distribution; it is purely theoretical. At the very outset let it be clear that, like the binomial or other probability distributions, the normal distribution is not a fact of nature that one actually observes to be exactly true. Rather, the normal distribution is a theory about what might be true of the relation between intervals of values and probabilities for some variable. There is nothing magical about the normal distribution; it happens to be only one of a number of theoretical distributions that have been studied and found useful as an idealized mathematical concept. Normal distributions do not really exist, however, and in applied situations the closest we can come to finding a normal distribution will never quite correspond to the requirements of the mathematical rule. Many concepts in mathematics and science that are never quite true give good practical results nevertheless, and so does the normal distribution.

Like most theoretical functions, a normal distribution is completely specified only by its mathematical rule. Quite often, the distribution is symbolized by a graph of the functional relation generated by that rule, and the general picture that a normal distribution presents is the familiar bell-shaped curve of Figure 4.18.1. The horizontal axis represents all the different values of X, and the vertical axis their densities $f(x)$. The normal distribution is continuous for all values of X between $-\infty$ and ∞, and each conceivable nonzero *interval* of real numbers has a probability other than zero. For this reason, the curve is shown as never quite touching the horizontal axis; the tails of the curve show decreasing probability densities as values grow extreme in either direction from the mode, but any interval representing *any* degree of deviation from central tendency is possible in this theoretical distribution. The normal distribution is symmetric and unimodal, and the mean, median, and mode are all equal. Bear in mind that since the normal distribution is continuous, the height of the curve shows the probability *density* for each X value. However, as for any continuous distribution, the area cut off beneath the curve by any interval is a probability, and the entire area under the curve is one.

The student is warned not to let the graph of any distribution lead him into jumping to conclusions about the kind of distribution represented. Although the normal distribution graphs as a bell-shaped curve, just any bell-shaped curve is not necessarily a normal distribution. The kind of distribution, the family to which it belongs, depends absolutely on the function rule, and on nothing else. Only if each possible value of X is paired with a density in the way provided by the normal function rule can one say that the distribution is normal. Other rules may give similar pictures or probabilities that are close to their normal counterparts but, by definition, these distributions are not exactly normal.

The mathematical rule for a normal density function is as follows:

$$f(x) \;=\; \frac{1}{\sqrt{2\pi\sigma^2}}\, e^{-(x-\mu)^2/2\sigma^2}. \qquad (4.18.1^*)$$

This rule pairs a probability density $f(x)$ with each and every possible

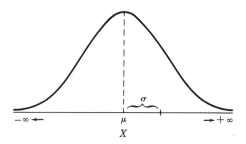

Figure 4.18.1

value of X. This rule looks somewhat forbidding to the mathematically uninitiated, but actually it is not. The π, the e, and the 2, of course, are simply positive numbers acting as mathematical constants. The "working" part of the rule is the exponent

$$-\frac{(x - \mu)^2}{2\sigma^2} \qquad (4.18.2)$$

where the particular value of the variable X appears, along with the two parameters μ and σ^2.

The more that the value of X differs from μ, the larger will the quantity in the numerator of this exponent be. However, this deviation of x from μ enters as a squared quantity, so that *two* different x values showing the same absolute deviation from μ have the same probability density according to this rule. This dictates the symmetry of the normal distribution. Furthermore, the exponent as a whole has a negative sign, meaning that the larger the absolute deviation of x from μ, the smaller will the density assigned to x be by this rule. This dictates that either tail of the distribution shows decreasing density, since the wider the departure of x values from μ, the lower will be the height of this function's curve. However, a little thought will convince you that no real and finite value of X can possibly make the density itself negative or exactly zero: the normal function curve never touches the x-axis, indicating that any interval of numbers will have a nonzero probability. On the other hand, any number, such as e, raised to the zero power is 1; when x exactly equals μ, the density is

$$\frac{1}{\sqrt{2\pi\sigma^2}}, \qquad (4.18.3)$$

which is the *largest* density value any x may have. This fact implies that the function curve must be unimodal, with maximum density at the mean, μ. Note that, unlike a probability, a density value such as expression (4.18.1) can be greater than 1.

It is very important to notice that the precise density value assigned to any x by this rule cannot be found unless the two parameters μ and σ are specified. The parameter μ can be any finite number, and σ can be any finite *positive* number. Thus, like the binomial and the other discrete distributions that we have studied, the normal distribution rule actually specifies a *family* of distributions. Although each distribution in the family has a density value paired with each x by this same general rule, the *particular* density that is paired with a given x value differs with different assignments of μ and σ. Thus, normal distributions may differ in their means (Figure 4.18.2); in their standard deviations (Figure 4.18.3); or in both means and standard deviations (Figure 4.18.4). Nevertheless, given the mean and standard deviation of the distribution, the rule for finding the probability density of any value of the variable is the same.

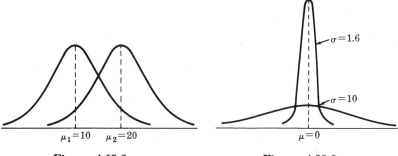

Figure 4.18.2   Figure 4.18.3

Since the normal distribution is really a family of distributions, statisticians constructing tables of probabilities found by the normal rule find it convenient to think of the variable in terms of a standardized random variable. That is to say, if the random variable z is standardized, so that $\mu = 0$ and $\sigma = 1$, then the rule (4.18.1) becomes simpler:

$$f(z) = \frac{1}{\sqrt{2\pi}}\, e^{-z^2/2}. \qquad (4.18.4^*)$$

For **standardized normal variables,** the density depends only on the *absolute* value of z; since both z and $-z$ give the same value z^2, they both have the same density. The higher the z in absolute value, the less the associated density. The standardized form of the distribution makes it possible to use one table of densities for any normal distribution, regardless of its particular parameters.

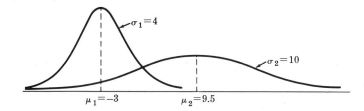

Figure 4.18.4

4.19 CUMULATIVE PROBABILITIES AND AREAS FOR THE NORMAL DISTRIBUTION

In Section 3.5 it was pointed out that when a distribution is continuous, the probability that X takes on some exact value x is, in effect, zero. For this reason, *all probability statements we will make using a normal distribution will be in terms either of cumulative probabilities or the probabilities of intervals.*

The cumulative probability

$$F(a) = P(X \leq a)$$

can be thought of as the area under the normal curve in the interval bounded by $-\infty$ and the value a (Figure 4.19.1).

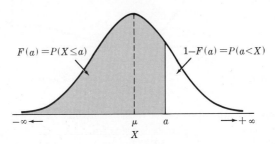

$$F(a) = P(X \leq a) \qquad 1-F(a) = P(a<X)$$

Figure 4.19.1

In terms of the density function, this area is given by

$$F(a) = \int_{-\infty}^{a} f(x)\,dx, \qquad (4.19.1)$$

where $f(x)$ is the normal density function.

These cumulative probabilities can be used to find the probability of any interval. For example, suppose that in some normal distribution we want to find the probability that X lies between 5 and 10, given some exact values for μ and σ. This is the probability represented by the area shown in Figure 4.19.2. The cumulative probability *up to and including* 10, or $F(10)$, minus the cumulative probability up to and including 5, or $F(5)$, gives the probability in the interval:

$$P(5 \leq X \leq 10) = F(10) - F(5).$$

In the same way, the probability of any other interval with limits a and b can be found from cumulative probabilities:

$$P(a \leq X \leq b) = F(b) - F(a).$$

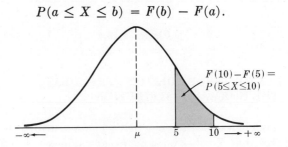

$$F(10)-F(5) = P(5 \leq X \leq 10)$$

Figure 4.19.2

4.20   THE USE OF TABLES OF THE NORMAL DISTRIBUTION

Most tables of the normal distribution give the cumulative probabilities for various *standardized* values. That is, for a given z value the table provides the cumulative probability *up to and including that standardized value* in a normal distribution. For example,

$$F(2) = P(z \leq 2),$$

$$F(-1) = P(z \leq -1),$$

and so on.

If the cumulative probability is to be found for a positive z, then Table I in the Appendix can be used directly. For example, suppose that a normal distribution is known to have a mean of 50 and a standard deviation of 5. What is the cumulative probability of a score of 57.5? The corresponding standardized value is

$$z = \frac{57.5 - 50}{5} = 1.5.$$

A look at Table I shows that for 1.5 in the first column, the corresponding cumulative probability in the column labeled $F(z)$ is .933, approximately. This is the probability of observing a value *less than or equal to* 57.5 in this particular distribution. The cumulative probability for any other positive z value can be found in the same way.

If the z value is negative, this procedure is changed somewhat. Since the normal distribution is symmetric, the density associated with any z is the same as for the corresponding value with a negative sign, or $-z$. However, the cumulative probability for a negative standardized value is 1 minus the cumulative probability for the z value with a positive sign:

$$F(-z) = 1 - F(z)$$

where z is the positive standardized score of the same magnitude. Figure 4.20.1 will clarify this point. The shaded area on the left is the cumulative

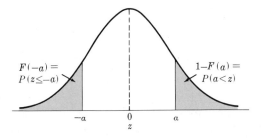

Figure 4.20.1

probability for a value $z = -a$. The table gives only the area falling below the point $z = a$, the positive number of the same absolute value as $-a$. However, the shaded area on the *right*, which is $1 - F(a)$, is the same as the shaded area on the *left*, $F(-a)$.

For example, suppose that in a distribution with mean equal to 107 and standard deviation of 70, we want the cumulative probability of the score 100. In standardized form,

$$z = \frac{100 - 107}{70} = -.1.$$

We look in the table for z equal to *positive* .1, which has a cumulative probability of approximately .5389. The cumulative probability of a z equal to $-.1$ must be approximately

$$F(-.1) = 1 - .5389 = .4611.$$

We may also answer questions about the probabilities of various intervals from the table. For example, what proportion of cases in a normal distribution must lie within one standard deviation of the mean? This is the same as the probability of the interval $(-1 \leq z \leq 1)$. The table shows that $F(1) = .8413$, and we know that $F(-1)$ must be equal to $1 - .8413$ or .1587.

Thus,

$$P(-1 \leq z \leq 1) = F(1) - F(-1) = .8413 - .1587 = .6826.$$

About 68 percent of all cases in a normal distribution must lie within one standard deviation of the mean.

To find the proportion lying between 1 and 2 standard deviations *above* the mean, we take

$$P(1 \leq z \leq 2) = F(2) - F(1).$$

From the table, these numbers are .9772 and .8413, so that

$$P(1 \leq z \leq 2) = .9772 - .8413 = .1359.$$

About 13.6 percent of cases in a normal distribution lie in the interval between 1σ and 2σ above the mean. By the symmetry of the distribution we know immediately that

$$P(-2 \leq z \leq -1) = .1359$$

as well.

Beyond what number must only 5 percent of all standardized scores fall? That is, we want a number b such that the following statement is true for a normal distribution:

$$P(b < z) = .05.$$

This is equivalent to saying that

$$P(z \le b) = .95.$$

A look at the table shows that roughly .95 of all observations must have z scores at or below 1.65. Thus,

$$P(1.65 < z) = .05, \text{ approximately.}$$

If intervals are mutually exclusive, their probabilities can be added by the "or" rule to find the probability that X falls into either interval. For example, we want to know

$$P(z < -2.58 \text{ or } 2.58 < z),$$

which is the same as $\qquad P(2.58 < |z|).$

For the first of these intervals, we find

$$F(2.58) = .995,$$

so that $\qquad F(-2.58) = 1 - .995 = .005.$

For the other interval, the probability is also

$$P(2.58 < z) = 1 - F(2.58) = .005.$$

The two intervals are mutually exclusive events, and so

$$P(z < -2.58 \text{ or } 2.58 < z) = .005 + .005 = .01.$$

Put in the other way,

$$P(2.58 < |z|) = .01,$$

the probability that z exceeds 2.58 in absolute value is about one in one hundred. This probability is given by the extreme tails of the distribution in the graph of Figure 4.20.2.

It is interesting to compare the probability just found with that given by the Tchebycheff inequality for *any* symmetric, unimodal distribution (Section 3.27):

$$P(2.58 \le |z|) \le \frac{4}{9}\left[\frac{1}{(2.58)^2}\right].$$

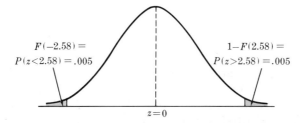

$$F(-2.58) = P(z < 2.58) = .005 \qquad\qquad 1 - F(2.58) = P(z > 2.58) = .005$$

$$z = 0$$

Figure 4.20.2

On working out the number on the right, we get about .067. That is, for *any* unimodal symmetric distribution the probability of a score deviating 2.58 or more standard deviations from expectation is less than about seven in one hundred. By specifying that the distribution is *normal*, we pin this probability down to the exact value .01. The more one assumes or knows about the population distribution, the stronger is the statement he can make about the probability of any degree of deviation from expectation.

4.21 THE IMPORTANCE OF THE NORMAL DISTRIBUTION

The normal is by far the most used distribution in inferential statistics. There are at least four very good reasons why this is true.

A population may be assumed to follow a normal law because of what is known or presumed to be true of the measurements themselves. There are two rather broad instances in which the random variable is a measurement of some kind, and the distribution is conceived as normal. The first is when one is considering a hypothetical distribution of "errors," such as errors in reading a dial, in discriminating between stimuli, or in test performance. Any observation can be assumed to represent a "true" component plus an error. Each error has a magnitude, and this number is thought of as a reflection of "pure chance," the result of a vast constellation of circumstances operative at the moment. Any factor influencing performance at the moment contributes a tiny amount to the size of the error and to its direction. Furthermore, such errors are appropriately considered to push the observed measurement up or down with equal likelihood, to be independent of each other over samplings, and to "cancel out" in the long run. Thus, in theory, it makes sense that errors of measurement or errors of discrimination follow something like the normal rule.

In other instances, the distribution of *true magnitudes* of some trait may be thought of as normal. For example, human heights form approximately a normal distribution, and so, we believe, does human intelligence. Indeed, there are many examples, especially in biology, where the distribution of measurements of a natural trait seems to follow something closely resembling a normal rule. No obtained distribution is ever exactly normal, of course, but some distributions of measurements have relative frequencies very close to normal probabilities. However, the view that the distribution of almost any trait of living things should be normal, although prominent in the nineteenth century, has been discredited. The normal distribution is not, in any sense, "nature's rule," and all sorts of distributions having little resemblance to the normal distribution occur in all fields.

It may be convenient, on mathematical grounds alone, to assume a normally distributed population. Mathematical statisticians do not devote so much attention to normal distributions just because they think bell-shaped curves are pretty! The truth of the matter is that the normal function has important mathematical properties shared by no other theoretical distribution. Assuming a normal distribution gives the statistician an extremely rich set of mathematical consequences that he can use in developing methods. Very many problems in mathematical statistics either are solved, or can be solved, *only* in terms of a normal population distribution. We will find this true especially when we come to methods for making inferences about a population variance. The normal distribution is the "parent" of several other important theoretical distributions that figure in statistics. In some practical applications the methods developed using normal theory work quite well even when this assumption is not met, despite the fact that the problem can be given a *formal* solution only when a normal population distribution is assumed. In other instances, there just is not any simple way to solve the problem when the normal rule does not hold for the population, at least approximately.

The normal distribution may serve as a good approximation to a number of other theoretical distributions having probabilities that are either laborious or impossible to work out exactly. For example, in the following section it will be shown that the normal distribution can be used to approximate binomial probabilities under some circumstances, even though the normal distribution is continuous and the binomial is discrete. For very large N, binomial probabilities are troublesome to work out exactly; on the other hand, intervals based on the normal function give probabilities that can be used as though they were binomial probabilities.

There is a very intimate connection between the size of sample N and the extent to which a sampling distribution approaches the normal form. Many sampling distributions based on large N can be approximated by the normal distribution even though the population distribution itself is definitely not normal. This is the extremely important principle that we will call the **central limit theorem.** The normal distribution is the *limiting form* for large N for a very large variety of sampling distributions. This is one of the most remarkable and useful principles to come out of theoretical statistics, and we will discuss it in more detail in Chapter 5.

Thus, you see, it is no accident that the normal distribution is the workhorse of inferential statistics. The assumption of normal population distributions or the use of the normal distribution as an approximation device is not as arbitrary as it sometimes appears; this distribution is part of the very fabric of inferential statistics. These features of the normal distribution will be illustrated in the succeeding sections. It is important to repeat at this point the caution that the normal distribution, although it has wide applicability, is by no means a "universal nature's rule." Whenever a normal distribution is used in statistical inference or decision, its applicability should be somehow justified.

4.22 THE NORMAL APPROXIMATION TO THE BINOMIAL

One of the interpretations given to the normal distribution is that it is the limiting form of a binomial distribution as $N \to \infty$ for a fixed p. If one draws samples of N from a Bernoulli distribution with the probabilities of the two categories being p and q, respectively, he should expect a binomial distribution of number of successes. Suppose that first he does this for all possible samples of size 10. Next he draws all possible samples with $N = 10,000$, then repeats the sampling with an N of 10,000,000, and so on. For each sample size, there will be a different binomial distribution he should observe when *all possible* samples of that size have been drawn.

How does the binomial distribution change with increasing sample size? In the first place, the actual range of the possible number of successes grows larger, since the whole numbers 0 through N form a larger set with each increase in N. Second, the expectation, Np, is larger for each increase in N for a fixed p, and the standard deviation $\sqrt{Npq}$ also increases with N. Finally, the probability associated with any given *exact* value of X tends to *decrease* with *increase* in N.

However, suppose that for any value of N the number of successes X is put into standardized form:

$$z = \frac{X - Np}{\sqrt{Npq}} = \frac{X - E(X)}{\sigma} . \qquad (4.22.1)$$

Then regardless of the size of N, the mean of the z values for a binomial distribution will be 0, and the standard deviation 1. The probability of any z value is the same as for the corresponding X. Any z interval can be given a probability in the particular binomial distribution, and the same interval can also be given a probability for a normal distribution. We can compare these two probabilities for any interval; if the two probabilities are quite close over all intervals, the normal distribution gives a good approximation to the binomial probabilities. On making this comparison of binomial and normal interval probabilities, we should find that the larger the sample size N the better is the fit between the two kinds of probabilities. **As N grows infinitely large, the normal and binormal probabilities become identical for any interval.**

For example, imagine that $p = .5$, and $N = 5$. Here the expectation is $(5)(.5)$, or 2.5, and the standard deviation is $\sqrt{5(.5)(.5)} = \sqrt{1.25}$ or about 1.12. The binomial distribution for these values of p and N is given below:

r	$P(r)$
5	.0312
4	.1563
3	.3125
2	.3125
1	.1563
0	.0312

For the moment, let us pretend that this is actually a continuous distribution, and that the binomial probability $P(X = r)$ is actually the probability associated with an interval of width 1 centered at r:

$$P(r - \tfrac{1}{2} \leq X \leq r + \tfrac{1}{2}).$$

Now we shall convert X into a standardized random variable z, by taking

$$z = \frac{X - E(X)}{\sigma} = \frac{X - 2.5}{1.12}.$$

But then each X-interval corresponds to a z-interval, and the probability of the X-interval can be stated in terms of the z-interval:

$$P(r - \tfrac{1}{2} \leq X \leq r + \tfrac{1}{2}) = P\left(\frac{r - \tfrac{1}{2} - 2.5}{1.12} \leq \frac{X - 2.5}{1.12} \leq \frac{r + \tfrac{1}{2} - 2.5}{1.12}\right)$$

$$= P\left(\frac{r - 2.5 - \tfrac{1}{2}}{1.12} \leq z \leq \frac{r - 2.5 + \tfrac{1}{2}}{1.12}\right). \quad (4.22.2)$$

Using this formula and assuming that z is normally distributed with mean 0 and variance 1, we can determine probabilities for these intervals from tables of the normal distribution:

r	z-interval	Binomial probability	Normal approximation
5	1.79 to 2.68	.0312	.0367
4	.89 to 1.79	.1563	.1500
3	.00 to .89	.3125	.3132
2	− .89 to .00	.3125	.3132
1	−1.79 to − .89	.1563	.1500
0	−2.68 to −1.79	.0312	.0367

The probabilities for the normal approximation were determined from (4.22.2), with one exception. Since z ranges from $-\infty$ to $+\infty$, the normal probabilities for the extreme intervals (corresponding to $r = 5$ and $r = 0$) were computed by assuming that the extreme limit was infinite, so that the normal probabilities would sum to one. Thus, the first probability is actually $P(z \geq 1.79)$ rather than $P(1.79 \leq z \leq 2.68)$, and the last probability is $P(z \leq -1.79)$ rather than $P(-2.68 \leq z \leq -1.79)$.

Observe how closely the normal probabilities approximate their binomial counterparts; each normal probability is within .01 of the corresponding binomial probability. The difference in the probabilities is smallest for the middle intervals and is largest for the extremes. Thus, even when N is only 5, the normal distribution gives a respectable approximation to binomial probabilities for $p = .5$.

Now suppose that for this same example, N had been 15. Here the expectation is 7.5, and the standard deviation is $\sqrt{(15)(.25)} = 1.94$. The distribution, with z-intervals and both binomial and normal probabilities, is shown in Table 4.22.1.

Table 4.22.1

r	z-interval	Binomial probability	Normal approximation
15	3.608 to 4.124	.00003	.0002
14	3.092 to 3.608	.0005	.0012
13	2.577 to 3.092	.0032	.0036
12	2.061 to 2.577	.0139	.0147
11	1.546 to 2.061	.0416	.0409
10	1.030 to 1.546	.0916	.0909
9	.515 to 1.030	.1527	.1501
8	.000 to .515	.1964	.1984
7	$-$.515 to .000	.1964	.1984
6	-1.030 to $-$.515	.1527	.1501
5	-1.546 to -1.030	.0916	.0909
4	-2.061 to -1.546	.0416	.0409
3	-2.577 to -2.061	.0139	.0147
2	-3.092 to -2.577	.0032	.0036
1	-3.608 to -3.092	.0005	.0012
0	-4.124 to -3.608	.00003	.0002

For this distribution the normal approximation gives an even better fit to the exact binomial probabilities; the average absolute difference in probability over the intervals is about .001, whereas when N was only 5, the average absolute difference was about .004. In general, as N is made larger, the fit between normal and binomial probabilities grows increasingly good. In the limit, when N approaches infinite size, the binomial probabilities are exactly the same as the normal probabilities for any interval, and thus one can say that the normal distribution is the limit of the binomial. This is true *regardless of the value of p.*

On the other hand, for any finite N, the more p departs from .5, the less well does the normal distribution approximate the binomial. When p is not exactly equal to .5, the binomial distribution is somewhat skewed for any finite sample size N, and for this reason the normal probabilities will tend to fit less well than for $p = .5$, which always gives a symmetric distribution. For example, notice that for $N = 5$ and $p = .3$, illustrated below, the correspondence is not so good as before:

r	Binomial probability	Normal approximation
5	.00243	.0017
4	.02835	.0239
3	.13230	.1379
2	.30870	.3365
1	.36015	.3365
0	.16807	.1635

Here the normal probabilities are still reasonably close to the binomial, but the fit is not as good as for $p = .5$.

Given a sufficiently large N, normal probabilities may always be used to approximate binomial probabilities irrespective of the value of p, the true probability of a success. The more that p departs in either direction from .5, the less accurate is this approximation for any given N. In practical situations where the sampling distribution is binomial, the normal approximation may ordinarily be used safely if the *smaller* of Np, the number of successes expected, or Nq, the expected number of failures, is 10 or more. Otherwise, binomial probabilities can either be calculated directly or found from tables that are readily available. Obviously, this rule requires a larger sample size the smaller the value of either p or q, and if one must use the normal approximation to make an inference about the value of p, he is wise to plan on a relatively large sample.

The graphs presented in Figures 4.22.1 and 4.22.2 give some idea of the speed with which the binomial distribution approaches its normal approximation. In Figure 4.22.1, graphs are shown for $p = .10$ and sample sizes of 10, 20, and 50. Because $p = .10$, the binomial distribution is positively skewed, but the skewness decreases as N gets larger. This skewness is the

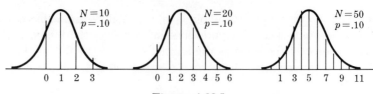

Figure 4.22.1

reason that for any given N, the normal approximation is less accurate as p approaches 0 or 1. In Figure 4.22.2, graphs are shown for $p = .50$ and sample sizes of 10, 20, and 50. Since $p = .50$, the binomial distribution is symmetric, and thus for a given sample size the normal approximation is better in this case than in Figure 4.22.1. As N gets larger, the normal approximation becomes more and more accurate.

In a more advanced study of the normal distribution we would show that as N becomes infinite with fixed p the mathematical rule for a binomial distribution becomes identical with the rule for a normal density function, with $\mu = Np$ and $\sigma = \sqrt{Npq}$. Unfortunately, this sort of demonstration is beyond our managing at this point.

In any practical use of the normal approximation to binomial probabilities, it is important to remember that here we regard the binomial distribution as though it were continuous, and actually find the normal probability associated with an interval with real lower limit of $(X - .5)$ and real upper limit of $(X + .5)$, where X is any given number of successes. Thus, in order to find the normal approximation to the binomial probability of X, we take

$$P\left(\frac{X - Np - .5}{\sqrt{Npq}} \le z \le \frac{X - Np + .5}{\sqrt{Npq}}\right) \qquad (4.22.3)$$

by use of the standardized normal tables.

As noted above, the normal distribution need not be particularly good as an approximation to the binomial if either p or q is quite small, and if either p or q is extremely small the normal approximation may not be satisfactory even when N is quite large. In these circumstances another theoretical distribution, the Poisson, provides a better approximation to binomial probabilities, as we saw in Section 4.15.

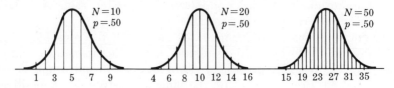

Figure 4.22.2

4.23   THE THEORY OF THE NORMAL DISTRIBUTION OF ERROR

The fact that the limiting form of the binomial is the normal distribution actually provides a rationale for thinking of random error as distributed in this normal way. Consider an object measured over and over again independently and in exactly the same way. Imagine that the value S obtained on any occasion is a sum of two independent parts:

$$S = T + e.$$

That is, the obtained score S is a sum of a constant *true* part T plus a random and independent *error* component. However, the error portion can also be thought of as a sum:

$$e = g[e_1 + e_2 + e_3 + \cdots + e_N].$$

Here, e_1 is a random variable that can take on only two values

$e_1 = 1$, when factor 1 is operating

$e_1 = -1$, when factor 1 is not operating.

The g is merely a constant, reflecting the "weight" of the error in S. Similarly, the other random errors are attributable to different factors, and take on only the values 1 and -1. Now imagine a vast number of influences at work at the moment of any measurement. Each of these factors operates independently of each of the others, and whether any factor exerts an influence at any given moment is purely a chance matter. If you want to be a little anthropomorphic in your thinking about this, visualize old Dame Fortune tossing a vast number of coins on any occasion, and from the result of each deciding on the pattern of factors that will operate. When coin one comes up heads, e_1 gets value 1, and if tails, value -1, and the same principle determines the value for every other error portion of the observed score.

Now under this conception, the number of factors operating at the moment one observes S is a number of successes in N independent trials of a Bernoulli experiment. If this number of successes is X, then

$$e = gX - g[N - X] = g[2X - N].$$

The value of e is exactly determined by X, the number of "influences" in operation at the moment, and the probability associated with any value of e must be the same as for the corresponding value of X. **If N is very large, then the distribution of e must approach a normal distribution.** Furthermore, if any factor is equally likely to operate or not operate at a given moment, so that $p = 1/2$, then

$$E(e) = g[2E(X) - N],$$

so that, since $E(X) = Np = N/2$,

$$E(e) = 0.$$

In the long run, over all possible measurement occasions the errors all "cancel out." This makes it true that

$$E(S) = E(T) + E(e) = T,$$

the long-run expectation of a measurement S is the true value T, provided that error really behaves in this random way as an additive component of any score.

This is a highly simplified version of the argument for the normal distribution of errors in measurement. Much more sophisticated rationales can be invented, but they all partake of this general idea. Moreover, the same kind of reasoning is sometimes used to explain why distributions of natural traits, such as height, weight, and size of head, follow a more or less normal rule. Here, the mean of some population is thought of as the "true" value, or the "norm." However, associated with each individual is some departure from the norm, or error, representing the culmination of all the billions of chance factors that operate on him, quite independently of other individuals. Then by regarding these factors as generating a binomial distribution, we can deduce that the whole population should take on a form like the hypothetical normal distribution. However, this is only a theory about how errors might operate and there is no reason at all why errors must behave in the simple additive way assumed here. If they do not, then the distribution need not be normal in form at all.

4.24 THE EXPONENTIAL DISTRIBUTION

Although the normal distribution is the most important of the continuous distributions, there are nevertheless many instances in which the normal distribution is not applicable. Consider the Poisson process discussed in Section 4.14. The Poisson process is often applicable when we are interested in occurrences of an event over time. In particular, the Poisson distribution is the distribution of the number of occurrences of an event in a given time period of length t, and the single parameter of the Poisson distribution is λ, the intensity of the process. Suppose that instead of the number of occurrences in a given time period, we were interested in the amount of time until the *first* occurrence. For example, we might be interested in the amount of time until the first incoming telephone call, or the amount of time until the first breakdown of a new piece of equipment. Since we assume that the Poisson process is stationary and independent, it is equivalent to say that we are interested in the amount of time *between* occurrences, such as the amount of time between incoming calls or between breakdowns.

Let the random variable X represent the amount of time until the *first* occurrence of a given event E, where the occurrences of E are governed by a stationary and independent Poisson process with intensity λ per time period t. It is important to specify the unit of time, t, to which λ refers. Suppose that we were interested in the probability of a specified number of occurrences in a period of length s, where s is not necessarily equal to t. Consider the ratio s/t, and call this ratio T. If t is a unit of time, then T is the number of units of time contained in s. Because of the independence and stationarity of the process, if the intensity per unit of time t is λ, then the intensity per T units of time (that is, in a period of length s) is λT. For example, an intensity of 6 telephone calls per hour is equivalent to an intensity of 18 calls per three hours $(T = 3)$ or 3 calls per half-hour $(T = 1/2)$ or 1 call per 10 minutes $(T = 1/6)$. Using the above definitions of t and T, we can then generalize the Poisson distribution as follows: **If occurrences of an event are generated by a stationary and independent Poisson process with intensity λ per unit of time t, and we are interested in the number of occurrences in T units of length t of time, then the intensity per T units is equal to λT, and the corresponding Poisson distribution is**

$$P(r \text{ occurrences in } T \text{ units} \mid \lambda) = \frac{e^{-\lambda T}(\lambda T)^r}{r!} \quad \text{for} \quad r = 0, 1, 2, \cdots$$

$$(4.24.1^*)$$

Recalling that X is the amount of time until the *first* occurrence, consider the event $X > T$. If $X > T$, then there must have been no failures in the first T units of time. From the above Poisson distribution,

$$P(0 \text{ occurrences in time } T \mid \lambda) = \frac{e^{-\lambda T}(\lambda T)^0}{0!} = e^{-\lambda T}.$$

Thus,

$$P(X > T) = P(0 \text{ occurrences in time } T) = e^{-\lambda T},$$

and

$$P(X \leq T) = 1 - P(X > T) = 1 - e^{-\lambda T},$$

where it is assumed that $T \geq 0$. This is simply the cumulative distribution function of X, which we can rewrite in functional form as

$$F(x) = \begin{cases} 1 - e^{-\lambda x} & \text{if } x \geq 0, \\ 0 & \text{if } x < 0. \end{cases}$$

$$(4.24.2)$$

Since $F(x)$ is a continuous function, the density function $f(X)$ is equal to the derivative of $F(x)$:

$$f(x) = \frac{d}{dx} F(x) = \begin{cases} \lambda e^{-\lambda x} & \text{if } x \geq 0, \\ 0 & \text{if } x < 0. \end{cases}$$

$$(4.24.3^*)$$

This distribution is called the exponential distribution, and it is the distribution of the amount of time until the first occurrence of an event, or, equivalently, the distribution of the amount of time between occurrences of an event, where the occurrences are governed by a Poisson process. Note that the relationship between the exponential and the Poisson distributions is analogous to the relationship between the geometric and the binomial distributions. The Poisson and binomial distributions concern the number of occurrences of an event. The exponential and geometric distributions concern the amount of time and the number of trials, respectively, until the first occurrence.

It can be shown that the mean and variance of the exponential distribution are

$$E(X) = \frac{1}{\lambda} \quad \text{and} \quad \text{var}(X) = \frac{1}{\lambda^2}. \qquad (4.24.4)$$

Since the derivation of $E(X)$ and var (X) require the technique of integration by parts, we will not concern ourselves with the derivation. Note again the parallel between the exponential and the geometric distributions. The mean of the exponential distribution is $1/\lambda$ and the mean of the geometric distribution is $1/p$.

An example of graphs of exponential density functions with $\lambda = \frac{1}{2}$ and $\lambda = 2$ is presented in Figure 4.24.1. Notice that the exponential distribution is positively skewed, with the mode occurring at the smallest possible value, zero.

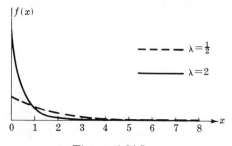

Figure 4.24.1

4.25   THE UNIFORM DISTRIBUTION

Many of the examples presented in Chapter 3 concerning continuous distributions dealt with the uniform distribution on the unit interval from zero to one. The density function was shown to be

$$f(x) = \begin{cases} 1 & \text{if } 0 \le x \le 1, \\ 0 & \text{elsewhere.} \end{cases}$$

It is possible to consider a more general uniform distribution, one which is defined on an interval from a to b rather than from zero to one (of course, one possibility is that $a = 0$ and $b = 1$), with the obvious restriction that $a < b$. Since the distribution is "uniform," the density function must be constant over the interval (a, b):

$$f(x) = \begin{cases} k & \text{if } a \le x \le b, \\ 0 & \text{elsewhere.} \end{cases} \qquad (4.25.1)$$

The only problem, then, is the determination of k. We know that $k \ge 0$, since $f(x)$ cannot be negative, and that the total probability must be equal to one:

$$\int_{-\infty}^{\infty} f(x) \, dx = 1.$$

But

$$\int_{-\infty}^{\infty} f(x) \, dx = \int_{a}^{b} k \, dx = k \int_{a}^{b} dx = k \, (b - a),$$

so

$$k = \frac{1}{b - a}.$$

The density function of the uniform distribution on the interval from a to b, where $a < b$, is thus

$$f(x) = \begin{cases} \dfrac{1}{b - a} & \text{if } a \le x \le b, \\[2ex] 0 & \text{elsewhere.} \end{cases} \qquad (4.25.2^*)$$

The corresponding cumulative distribution function is equal to

$$F(x) = \int_{-\infty}^{x} f(x) \, dx = \begin{cases} 1 & \text{if } x > b, \\[2ex] \displaystyle\int_{a}^{x} \frac{1}{b - a} \, dx = \frac{x - a}{b - a} & \text{if } a \le x \le b, \\[2ex] 0 & \text{if } x < a. \end{cases} \qquad (4.25.3)$$

Graphs of the PDF and CDF of the uniform distribution are presented in Figure 4.25.1.

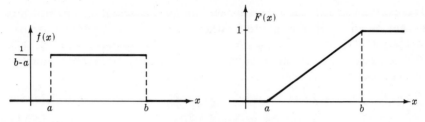

Figure 4.25.1

The mean and variance of the uniform distribution can be easily calculated:

$$E(X) = \int_{-\infty}^{\infty} xf(x)\, dx = \int_{a}^{b} \frac{x}{b-a}\, dx = \frac{1}{b-a} \int_{a}^{b} x\, dx$$

$$= \frac{b^2 - a^2}{2(b-a)} = \frac{(b-a)(b+a)}{2(b-a)}$$

$$= \frac{b+a}{2}, \tag{4.25.4}$$

and

$$E(X^2) = \int_{-\infty}^{\infty} x^2 f(x)\, dx = \int_{a}^{b} \frac{x^2}{b-a}\, dx = \frac{1}{b-a} \int_{a}^{b} x^2\, dx$$

$$= \frac{b^3 - a^3}{3(b-a)} = \frac{(b-a)(b^2 + ab + a^2)}{3(b-a)}$$

$$= \frac{b^2 + ab + a^2}{3}, \tag{4.25.5}$$

so

$$\operatorname{var}(X) = E(X^2) - [E(X)]^2 = \frac{b^2 + ab + a^2}{3} - \left(\frac{b+a}{2}\right)^2$$

$$= \frac{4b^2 + 4ab + 4a^2 - 3b^2 - 6ab - 3a^2}{12}$$

$$= \frac{(b-a)^2}{12}. \tag{4.25.6}$$

Note that the mean, $E(X) = (b+a)/2$, seems intuitively reasonable, for it is the midpoint of the interval from a to b. The distribution is "uniform," so we would expect the mean to be located at the midpoint of the interval.

4.26   THE BETA DISTRIBUTION

The final continuous distribution to be considered in this chapter is the **beta distribution.** The mathematical rule for the beta distribution is

$$f(x) = \begin{cases} \dfrac{(N-1)!}{(r-1)!(N-r-1)!}\, x^{r-1}(1-x)^{N-r-1} & \text{if } 0 \le p \le 1, \\[2ex] 0 & \text{elsewhere.} \end{cases}$$

$$(4.26.1^*)$$

The beta distribution has two parameters, r and N, and the mean and variance can be shown to be

$$E(X) = \frac{r}{N} \text{ and } \text{var}(X) = \frac{(r/N)(1-r/N)}{N+1} = \frac{r(N-r)}{N^2(N+1)}. \quad (4.26.2)$$

We will not be concerned with the derivation of the above formulas, but will instead attempt to give some intuitive justification for the introduction of the beta distribution.

The density function of the beta distribution is very similar to the probability mass function of the binomial distribution. If we let $x = p$, the beta density is of the form

$$f(p) = \begin{cases} \dfrac{(N-1)!}{(r-1)!(N-r-1)!}\, p^{r-1}(1-p)^{N-r-1} & \text{if } 0 \le p \le 1, \\[2ex] 0 & \text{elsewhere.} \end{cases}$$

$$(4.26.3^*)$$

This is similar to the binomial probability function, with $N-1$ substituted for N, $r-1$ substituted for r, and $N-r-1$ substituted for $N-r$. But the beta distribution is continuous, whereas the binomial distribution is discrete. How can this be? The explanation is quite simple; in the binomial distribution, the random variable is X, the number of successes, and X can obviously only take on whole number values. Thus, the binomial distribution is discrete. In the beta distribution, the random variable is p, which in terms of a Bernoulli process is the probability of a "success" on any single Bernoulli trial. But the only limitation on probabilities in general is that they must be between zero and one. Thus p can take on an infinite number of possible values (all real numbers between zero and one), and hence the beta distribution is continuous.

Now that the problem of discrete versus continuous distributions has been cleared up, can the beta distribution be given an interpretation in terms of the Bernoulli process? We will informally (and briefly) present such an interpretation; a more formal discussion is given in Chapter 8. In all previous distributions involving the Bernoulli process (binomial, Pascal, geometric), the probability p was a parameter of the distribution. In order to calculate probabilities with these distributions, then, p must be given (or assumed to take on a certain value). From a given p, we can calculate probabilities concerning sample outcomes from the process. If, on the other hand, we have observed some sample results but we do not know p, then the beta distribution is applicable. As we shall see in Chapter 8, the beta distribution represents our knowledge about p, given that we have observed a sample of r successes in N trials from the Bernoulli process of interest. Suppose that we observed $r = 4$ successes in $N = 10$ trials from a Bernoulli process. The expected value of p is then $r/N = .40$, which seems reasonable, and the variance of p is

$$r(N - r)/N^2(N + 1) = 4(6)/100(11) = .0218.$$

Furthermore, probabilities can be calculated for any intervals:

$$P(a \leq p \leq b) = \int_a^b f(p)\, dp$$

$$= \int_a^b \frac{(N - 1)!}{(r - 1)!(N - r - 1)!} p^{r-1}(1 - p)^{N-r-1}\, dp.$$

The shape of the beta distribution depends on r and N. If $r = N/2$ (referring to the Bernoulli process interpretation, this implies that the number of successes equals the number of failures in the sample from the Bernoulli

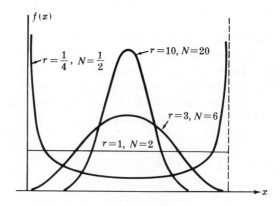

Figure 4.26.1

process), the distribution is symmetric. If $r < N/2$, the distribution is positively skewed, while if $r > N/2$, the distribution is negatively skewed. Finally, if $r > 1$ and $N > 2$, the distribution is unimodal with the mode equal to $(r - 1)/(N - 2)$; if $r \leq 1$ or $N \leq 2$, the distribution is either unimodal with mode at zero or one, **U**-shaped with modes at zero *and* one, or simply the uniform distribution on the unit interval (the last case holds if $r = 1$ and $N = 2$). Thus, the family of beta distributions can take on a wide variety of shapes as r and N are varied. Some examples with $r/N = \frac{1}{2}$ are presented in Figure 4.26.1 and some examples with $r/N = 1/20$ are presented in Figure 4.26.2.*

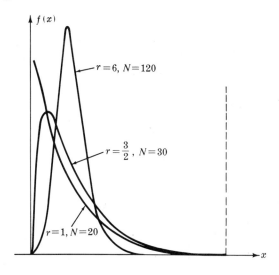

Figure 4.26.2

4.27 SUMMARY OF SPECIAL CONTINUOUS DISTRIBUTIONS

Four important families of continuous distributions have been presented in this chapter: the normal, exponential, uniform, and beta distributions. There are numerous other useful continuous families of distributions, some of which will be introduced later as sampling distributions (the t, χ^2, and F distributions) and some of which are beyond the scope of this book. It should be emphasized again that continuous distributions are a theoretical idealization and should be recognized as such. In some cases it is possible

* *Note*: Figures 4.26.1 and 4.26.2 are based on Figures 7.1 and 7.2 in Raiffa and Schlaifer, *Applied Statistical Decision Theory* (Boston: Graduate School of Business Administration, Harvard University, 1961).

to relate continuous distributions to real processes, as the exponential is related to the Poisson process and represents the amount of time until the first occurrence of an event. A related distribution which we did not discuss is the gamma distribution, which is the distribution of the amount of time until the Xth occurrence of an event in a Poisson process. Also, the beta distribution can be related to the Bernoulli process. However, the normal distribution can be given no such interpretation; it is strictly a theoretical function which happens to describe adequately the probability distribution of a large number of real-world variables. This necessitates a slightly different approach to the validation of the normal probability model in any application. In previous cases it was important to check the assumptions underlying a particular distribution (for example, stationarity and independence). Since this cannot be done with the normal distribution unless it is being used to approximate some other distribution, such as the binomial, it is necessary to compare sample data with the theoretical model in order to see if the model provides a "good fit" to the data. In the next chapter we will discuss the presentation of sample data in frequency distributions, which are analogous to probability distributions. In later chapters, we will discuss statistical techniques for investigating the "goodness of fit" of a theoretical model to a sample of data.

Because they are usually much easier to work with than discrete distributions, which may become quite complex (in a computational sense), continuous distributions are extremely useful in the development of statistical techniques. This should become clear as the basic statistical theory is presented in the next few chapters. Numerous applications of continuous distributions will be discussed in these chapters, and as a result we will not discuss specific applications of the special continuous families of distributions at this time.

EXERCISES

1. Explain the terms "stationary" and "independent" as applied to a series of Bernoulli trials.

2. A multiple choice examination is so constructed that, in principle, the probability of a correct choice for any item by guessing alone is $\frac{1}{4}$. If the test consists of 10 items, what is the probability that a student will have *exactly* 5 correct answers if he is just guessing?

3. The probability that an item produced by a certain production process is defective is .05. (a) Assuming stationarity and independence, what is the probability that a lot of 10 items from the process contains no defectives? (b) What is the probability that the first defective item is the fifth item produced?

4. Suppose that the probability that a duck hunter will successfully hit a duck is .25 on any given shot. He decides to hunt until he has hit 4 birds. (a) What is the expected number of shots that he will have to take? (b) What is the probability that he will take more than 10 but fewer than 15 shots? (c) If he only has enough ammunition for 20 shots, what is the probability that he will run out of ammunition before he hits 4 ducks?

5. Suppose that the probability that a particular football team wins a certain game is p, and that this probability remains constant from game to game.
   (a) If the outcomes of the games played by the team are independent of each other, for what value of p would the team have at least an even chance of winning all ten games on its schedule?
   (b) Suppose that p is .70. What is the probability that the team loses no more than two of its ten games?
   (c) If p is .70, what is the probability that the second loss occurs in the 8th game?
   (d) Suppose that p is .70 for home games and .50 for away games. The team plays five games at home and five games away from home. Now what is the probability that the team will lose exactly two games of the ten? What is the expected number of wins?

6. If the expected number of successes in N Bernoulli trials is 3 and the variance of the number of successes is 2.1, find N and p.

7. The probability that a person who is exposed to a certain contagious disease will catch it is .20.
   (a) Find the probability that the 20th person exposed to this disease is the 5th one to catch it.
   (b) If 20 persons are exposed to the disease, what is the expected number of persons who will catch the disease? What is the probability that *exactly* 5 catch it? What is the probability that *more than* 5 catch it?
   (c) If 1000 persons are exposed to the disease, what is the probability that at least 250 catch it?
   (d) What assumptions did you make in answering parts (a)–(c)?

8. Compare the binomial, Pascal, and geometric distributions. Is it possible to determine binomial probabilities by using the Pascal and geometric distributions? The event "no successes in ten trials," which is a binomial event, is related to what event involving the geometric distribution?

9. A manufacturer of parts for automobiles knows that the proportion of defective parts in the past has been about 10 percent. However, today he has taken a random sample of 10 parts and found that 2 were defective. What is the probability of his getting two or more defective parts by chance in such a sample if the probability of a defective part is only .10? Would you say that his finding suggests strongly that more defective parts than usual are being turned out?

10. A pair of dice are rolled for a total of four trials. If the dice are fair, find the probability that the dice total to 7 on exactly 3 of the four trials. Find the

probability that the dice total to 7 on at most 2 of the 4 trials. What is the expected number of trials (out of four) that the dice will total to 7?

11. In the primary elections of a certain state, any candidate for governor must carry at least 55 percent of the counties in order to win the nomination outright, without the necessity for a run-off election. A public opinion poll run on a sample of ten counties indicates that the candidate will carry only four of the ten. Assuming that the ten counties are a random sample of a large number of counties in the state, how probable is this result if the candidate will actually carry 55 percent of the counties? How probable is this result if he will actually carry 45 percent? Which of the two figures, 55 percent or 45 percent, is the better bet in the light of the evidence? Why?

12. Compute the binomial probabilities, the Poisson approximation, and the normal approximation for the following values of N and p:

   (a) $N = 10, p = .5$ (b) $N = 10, p = .05$
   (c) $N = 20, p = .5$ (d) $N = 20, p = .05.$

   Comment on the relative merit of the two approximations in each case.

13. Under what conditions can we approximate a binomial distribution by using a Poisson distribution, and under what conditions can we approximate a binomial distribution by using a normal distribution? Justify your answers with a brief intuitive discussion.

14. For any given match, the probabilities that a soccer team will win, tie, or lose are .3, .4, and .3. If the team plays ten matches and the outcomes are assumed to be independent, what is the probability that they will win 5, tie 3, and lose 2?

15. What is the probability that in 24 tosses of a fair die, each face will occur exactly 4 times? What is the probability that in 6 tosses, each face will occur exactly once?

16. Just as the binomial distribution is a special case of a much more general principle in algebra (the binomial expansion), so the multinomial distribution is a special case of another quite general mathematical principle. See if you can state the general rule, and then use this rule to prove that the sum of all probabilities in multinomial distribution must sum to 1.00. Why are no tables or graphs of multinomial distributions shown in the text?

17. Explain the differences among the binomial, multinomial, and hypergeometric distributions. When is the binomial a good approximation to the hypergeometric?

18. A university committee consists of six faculty members and four students. If a subcommittee of four persons is to be chosen randomly from the committee, what is the probability that it will consist of *at least* two faculty members? Compute this probability from the hypergeometric distribution *and* from the binomial approximation to the hypergeometric. How good is the approximation?

19. If twelve cards are to be drawn at random without replacement from a standard 52-card deck, what is the probability that there will be exactly three cards of each suit?

20. One of the key assumptions in the derivation of the binomial distribution is that the various trials are independent. Is this same assumption made in the calculation of probabilities for poker hands or in deriving the hypergeometric rule? Explain.

21. An instructor in an elementary laboratory class in psychology knows that out of 50 students, 12 are freshmen, 25 are sophomores, 11 are juniors, and only 2 are seniors. He assigns these students to particular laboratory sections at random, ten to a section. What is the probability that a particular section consists of 1 freshman, 1 sophomore and 8 juniors? (Assume that this is the first section of 10 students assigned.) What is the probability that the first section consists only of sophomores, *and* that the second section consists of juniors and the two seniors?

22. Suppose that cars arrive at a toll booth according to a Poisson process with intensity 5 per minute. What is the probability that no cars will arrive in a particular minute? What is the probability that more than 8 cars will arrive in a particular minute? What is the probability that exactly 9 cars will arrive in a given two minute period, and what is the expected number of arrivals during that period? Comment on the applicability of the assumptions of the Poisson process for this example.

23. Suppose that incoming telephone calls behave according to a Poisson process with intensity 12 per hour. What is the probability that more than 15 calls will occur in any given one hour period? If the person receiving the calls takes a fifteen minute coffee break, what is the probability that no calls will come in during the break? (Calculate this probability in two different ways, once using the Poisson distribution and once using the exponential distribution.)

24. Consider the occurrence of misprints in a book, and suppose that they occur at the rate of 2 per page. Assuming a Poisson process, what is the expected number of pages until the first misprint? What is the probability that the first misprint will *not* occur in the first page? Comment on the applicability of the Poisson assumptions to the occurrence of misprints or typing errors.

25. Prove that the variance of the Poisson distribution is equal to λ.

26. Think of two applications in your area of interest which involve the Bernoulli process and two which involve the Poisson process.

27. Accidents occur on a particular stretch of highway at the rate of 3 per week. What is the probability that there will be no accidents in a given week? What is the probability that there will be no accidents for two consecutive weeks? Compute these probabilities (a) using the Poisson distribution, (b) using the exponential distribution.

28. Comment on the following statement: "The exponential distribution is to the Poisson distribution as the geometric distribution is to the binomial distribution." In general, what is the difference between a Bernoulli process and a Poisson process?

29. If the median of an exponential distribution is 2, find λ.

30. If the first quartile of an exponential distribution is 2, find λ.

31. Suppose that the instructor in your History of Babylonia course informs you that on the final examination two students received grades of 60 and 30 respectively, and that the standardized scores corresponding to these grades are 1 and -1, respectively.
    (a) What is the mean and the standard deviation of the scores?
    (b) If the scores are approximately normally distributed, what proportion of the scores should lie between 25 and 65?
    (c) If the scores are approximately normally distributed, between what two scores do the *middle* 50 percent of cases lie?

32. In this problem, and those to follow, the notation $N(\mu, \sigma)$ will be used to denote a normal density function with mean μ and standard deviation σ, where μ and σ are particular values. Thus $N(0, 1)$ indicates a standardized normal distribution having a mean of 0 and a standard deviation of 1, so that z is a standardized normal variable. For the following events, make a sketch of the regions under the normal curve associated with each event, and use the tables to compute the probability of each such event.
    (a) $P(0 < z \leq 1.96)$, $P(-1.65 \leq z \leq 0)$, $P(-.63 < z \leq 1.02)$
    (b) $P(1.3 \leq z)$, $P(.45 \leq z)$, $P(-.45 < z \leq 1.30)$
    (c) $P(-.32 \leq z)$, $P(z \leq .75)$, $P(-.32 < z \leq .75)$
    (d) Let A be the event $(-.67 \leq z \leq .67)$ and let B be the event $(.26 \leq z \leq 1.30)$. Find (to two significant places) the probabilities of the following events:

$$(A \cap B), \quad (A \cap \bar{B}), \quad (\bar{A} \cap B) \cup (A \cap \bar{B}), \quad (A \cup B).$$

33. If X is distributed as $N(3, 4)$, use the table to find:
    (a) $P(1 \leq X \leq 7)$, $P(X \leq 5)$, $P(-2 \leq X \leq 1.5)$
    (b) a number a such that $P((X - 2) < a) = .95$
    (c) the .05, .25, .67, and .99 fractiles of the distribution.

34. If the .35 fractile of a normal distribution is 105 and the .85 fractile is 120, find the mean and standard deviation of the distribution.

35. Let us assume that at a certain university the actual time that a graduate student spends working toward his Ph.D. degree is approximately normally distributed with a mean of 200 weeks and a standard deviation of 50 weeks. Given 1000 students beginning graduate work toward the Ph.D. at the same time, what number should still be working toward the degree after 300 weeks? Find the number of weeks within which the *middle* 50 percent of students should complete their degrees. (Here we will simply ignore drop-outs and failures.)

36. The mean score of 1000 students on a national aptitude test is 100 and the standard deviation is 15. We will assume that the distribution of such scores is well-approximated by a normal distribution. Find how many students have scores:
    (a) between 120 and 155 inclusive
    (b) greater than 95
    (c) less than or equal to 175
    (d) 5 or more points below the score cutting off the lowest 25 percent
    (e) scores 10 or more points above the score cutting off the lowest 75 percent
    (f) between 70 and 120, inclusive.

37. A sample of four independent observations is taken from a normal population with mean μ and variance σ^2.
    (a) What is the probability that *all four* sample observations will fall between the limits $\mu + \sigma$ and $\mu - \sigma$?
    (b) What is the probability that *all four* sample values will fall between the limits $\mu + 1.96\sigma$ and $\mu - 1.96\sigma$?
    (c) What is the probability that *exactly one* value will be greater than $\mu + \sigma$?
    (d) What is the probability that two sample values will fall below $\mu - 2\sigma$ and the other two sample values will fall above $\mu + 2\sigma$?

38. Why is the normal distribution of such importance in statistical work? Try to think of some applications in your area of interest which might involve the normal distribution.

39. Under what conditions would the continuous uniform distribution be applicable? It is also possible to consider a *discrete* uniform distribution in which all of the possible values of the random variable have the same probability. How might this distribution relate to the discussion in Section 2.6?

40. If the random variable is uniformly distributed on the interval from a to b with mean 6 and variance 3, find a and b.

41. Find the mean and variance of the beta distribution with parameters $r = 2$ and $N = 6$, and graph the density function. Do the same for the following beta distributions:
    (a) $r = 4, N = 6$ (b) $r = 4, N = 12$ (c) $r = 8, N = 12$
    Explain how the different values of r and N affect the shape and location of these four distributions.

42. If the mean and variance of a beta distribution are 2/3 and 1/72, respectively, find r and N.

5

FREQUENCY
AND SAMPLING
DISTRIBUTIONS

The preceding three chapters involved probability theory, and we are now ready to take up the first major set of concepts in statistics, utilizing what we have learned about probability theory. First of all, the idea of a frequency distribution for sets of observations will be introduced, together with some of the mechanics for constructing distributions of data. The idea of a frequency distribution is parallel to the idea of a probability distribution; the former concerns observed data, and the latter concerns a theoretical state of affairs. Summary measures of frequency distributions will then be discussed, followed by the concept of a sampling distribution. The material in this chapter will be of great value in the development of the theory of statistical inference and decision in the following chapters.

5.1 RANDOM SAMPLING

Given some simple experiment and the sample space of elementary events, the set of outcomes of N separate trials is a **sample.** When the sample is drawn *with replacement*, the same elementary event can occur more than once. When the sample is drawn *without replacement*, the same elementary event can occur no more than once in a given sample. Ordinarily, probability and statistics deal with **random samples.** The word "random" has been used rather loosely in the foregoing discussion, and now the time has come to give it a more restricted meaning in connection with random samples.

It has already been suggested that probability calculations can be made quite simple when the elementary events in a sample space space have equal probabilities. The theory of statistics deals with samples of size N from a specified sample space, and here, too, great simplification is introduced if each distinct sample of a particular size can be assumed to have equal probability of selection. For this reason, the elementary theory of statistics is based on the idea of simple random sampling:

A method of drawing samples such that each and every distinct sample of the same size N has exactly the same probability of being selected is called simple random sampling.

In other words, simple random sampling corresponds to the situation where all possible samples of the same size have exactly the same probability. For us, sampling "at random" will always mean simple random sampling, as defined above. This does not mean, however, that the theory of statistics does not apply to situations where samples have unequal probabilities of occurrence; the theory and methods can be extended to any sampling scheme where the probabilities of the various samples are *known*, even though they are unequal. Sampling methods other than simple random sampling can be quite valuable, particularly in areas such as survey sampling. We will discuss some of these alternative sampling schemes in Chapter 11 (Volume II).

We shall also have many occasions to require *independent random sampling*. You may recall that independence was formally defined in Section 2.26. Intuitively, we think of the term as follows: elementary events are sampled independently when the occurrence of one elementary event has *absolutely no connection* with the probability of occurrence of the same or another elementary event on a subsequent trial. A series of tosses of a coin can be thought of as a random and independent sampling of events; what happens on one toss has no conceivable connection with what happens on any other. Each trial of a simple experiment is a sampling of elementary events, and the trials are independent when no connection exists between the particular outcomes of different trials. Recall that independence of trials is an important assumption for the Bernoulli and Poisson processes.

Take care to notice that our assuming samples of size N to be equally probable does not imply that *events* must be equally probable. Thus, in the marble example in Section 2.17, samples of N marbles were assumed equally probable, but the probabilities of the *colors* of the marbles were not equal. The assumption of equal probabilities is simply a way of saying that any elementary event has just as good an opportunity as any other to serve as a sample on any given experimental trial.

5.2 RANDOM NUMBER TABLES

In practical situations, there are a number of schemes for sampling the elementary events randomly and independently. One of the most common is by use of a table of random numbers. Random number tables consist of many pages filled with digits, from 0 through 9. These tables have been composed in such a way that each digit is approximately equally likely to occur in any spot in the table and there is no systematic connection between the occurrence of any digit in the sequence and any other. Most books on the design of experiments contain pages of random numbers, and very extensive tables of random numbers may be found in a book prepared by the RAND Corporation; Table VIII in the Appendix was taken from this book.

In order to use these tables for random sampling, one must list the possible elementary events (distinct units that might be observed). Each unit is then given a different number. The table of random numbers is entered by choosing some starting point at random and reading to the right or left, up or down. If the sample space contains only ten or fewer individual points, then the digits 0 through 9 themselves are used; if there are one hundred or fewer members of the sample space, digits are grouped by pairs, 00 through 99, and so on. Then the first N nonrepeated random numbers or groups of numbers are chosen, where N is the required sample size. The individual units corresponding to these numbers comprise the sample, which can be regarded as chosen at random.

This method is subject to two drawbacks. In the first place, unless one has an infinitely large table of random numbers, the one he uses must contain either some differences in probability or some dependence among the digits. This makes a sample chosen with any given table only approximately random. Second, and much more important, is the fact that a *listing* of the members of the sample space (all potential observation-units) must be possible. Except in some very restricted situations this is exceedingly difficult, or even impossible, to do. What usually happens is that selection is made from some sample space much smaller than the one the statistician would really like to use. For example, he might like to make statements about all people of a certain age, but the only group he can really sample randomly is college sophomores in a certain locale. Then one of two things must occur; either the statistician assumes that the sample space actually used is itself a random sample from the larger space, or he restricts his inferences to the group that he can sample randomly. Regardless of which course he takes, if he is going to make probability statements the sample must be drawn at random from *some* specific sample space of potential observations.

Quite apart from their utility for drawing samples, random number tables are interesting as another instance of the notion of "randomness" that is idealized in the theory of probability. A random process is one in which only chance factors determine the particular outcome of any single trial. The possible outcomes are known in advance, but not the exact outcome to be realized on any given trial. Nevertheless, built into the process is some regularity, so that each class of outcome (or event) can reasonably be assigned a probability. Perhaps the simplest example of a random phenomenon is the result of tossing a coin. Only chance factors determine whether heads or tails will come up on a particular toss, but a fair coin is so constructed that there is just as much physical reason for heads to appear as for tails, and so we have the justification we need in order to say that, in the long run, heads will appear just as often as tails. This property of the coin is idealized when we say that the probability of heads is .50. A fair die is constructed in such a way that each of the six sides should have the same physical opportunity to appear on a given throw, although unknown factors dictate which side actually does appear on that throw; we idealize this property of the die when we say that the probability of, say, three spots is 1/6. One can construct a random number table by use of a ten-sided die, each side labeled with a number from 0 through 9. Then any sequence of tosses gives a sequence of random numbers. In practice, of course, random numbers are generated electronically, but the general idea is the same; the physical process is such that no single number is favored for any particular outcome, and chance alone actually dictates the number that does occur in any place in a sequence. When we use random numbers to select samples, we are using this property of randomness, which was inherent in the process by which the numbers themselves were generated. The point is that the ideas of randomness and of probability are not just misty abstractions; we know how to create processes and devise operations that are approximately "random" and we know how to manufacture events with particular probabilities, at least with probabilities that are approximately known.

In summary, the calculation of probabilities requires that the basic outcomes of observation, or elementary events, occur with probabilities that are known. Random sampling methods are ways of observing sequences or sets of N such elementary events so that the probabilities associated with the various possible sample results actually *are* known. In simple random sampling each potential observation unit or elementary event has the same probability of appearing in a sample of N observations.

The student may be a bit puzzled as we proceed with the insistence on random sampling with *replacement*, when it is perfectly obvious that in most sampling N different subjects or observation units are chosen and each unit is *not* replaced before the next is sampled. This is done in sampling in business, the social sciences, the behavior sciences, and related fields because the number of potential observations in the sample space is extremely large. The selection and nonreplacement of a unit does virtually nothing to the probabilities of events. In such situations, sampling without replacement can be treated exactly like sampling with replacement. On the other hand, as we saw in Section 4.13 and as we shall see in Chapters 6 and 11, when the number of elementary events in the sample space is relatively small, sampling without replacement requires special procedures, since the probabilities of events are altered appreciably with each new observation made. This does no real violence to the notion of simple random sampling, however; instead of requiring that each elementary event be equally likely to occur on each trial, we require that *all samples of N possible outcomes be equally likely to occur*. However, in most elementary applications of statistics, sampling is thought of as being with replacement.

5.3 MEASUREMENT SCALES

Prior to any concerns he may have with summarizing his data, or comparing it with some theoretical state of affairs, the statistician must first measure the phenomena he studies. It therefore seems appropriate to mention the process of measurement at this point and to indicate briefly the role measurement considerations will play in the remainder of this text. We will not even attempt to undertake a thorough discussion of this topic, but we will try to suggest its relevance to statistics.

Whenever the statistician makes observations of any kind, he employs some scheme for classifying and recording what he observes. Any phenomenon or "thing" will have many distinguishable characteristics or attributes, but *the statistician must first single out those properties relevant to the question being studied*. For example, suppose that a marketing manager is interested in consumers' brand preferences. Individuals differ in ways as diverse as body temperature, size of head, weight, color of hair, income, occupation, name, and so on ad infinitum. All these properties are ignored by the marketing manager as immaterial to his immediate purpose, which is determining which brand each individual prefers.

When the statistician has singled out the property or properties he wants to study, he applies his classifying scheme to each observation. Such a scheme is essential if he wants to record, summarize, and communicate what he observes. At its very simplest, *this scheme is a rule for arranging observations into equivalence classes, so that observations falling into the same set are thought of as qualitatively the same and those in different classes as qualitatively different in some respect*. In general, *each observation is placed in one and only one class, making the classes mutually exclusive and exhaustive*.

The process of grouping individual observations into qualitative classes is measurement at its most primitive level. Sometimes this is called **categorical or nominal scaling.** The set of equivalence classes itself is called a **nominal scale.** There are many areas in science where the best one can do is to group observations into classes, each given a distinguishing symbol or name. For example, taxonomy in biology consists of grouping living things into phyla, genera, species, and so on, which are simply sets or classes.

Psychiatric nosology contains classes such as schizophrenic, manic-depressive, paretic, and so on; individual patients are classified on the basis of symptoms and the history of their disease. Classification is also prevalent in business and economics: investments are classified in finance (and classes of investments may be further broken down into subclassifications, as with AA bonds, A bonds, and so on); costs may be classified as fixed or variable costs, or may be classified in numerous other ways.

Notice that even this kind of measurement can be thought of in terms of a function; the domain of this function is a set of individual observations, and the range is the set of equivalence classes (the nominal scale). The measurement procedure is like a function rule that permits the assignment of each individual observation to one and only one qualitative class. The class paired with an individual informs us of some qualitative property that he exhibits.

The word "measurement" is usually reserved, however, for the situation where each observation is assigned a *number*; this number reflects a magnitude of some *quantitative* property. A function exists associating one and only one number with each observation, and the measurement procedure fills the role of a rule for this function.

There are at least three kinds of numerical measurement that can be distinguished; these are often called ordinal scaling, interval scaling, and ratio scaling. A really accurate description of each kind of measurement is beyond the scope of this book, but we can gain some idea of these distinctions by using the function terminology once again.

Imagine a set of objects $\mathcal{O}$ and any pair of objects in that set, say o_1 and o_2. Now suppose that there is some property that all objects in the set show, such as value, or weight, or length, or intelligence, or motivation. Furthermore, let us suppose that each object o has a certain amount or degree of that property. In principle we could assign a number, $t(o)$, to any object, standing for the amount that o actually "has" of that characteristic.

Ideally, in order to measure an object o, we would like to determine this number $t(o)$ directly. However, this is not always (or even usually) possible, and so what we do is devise a procedure for pairing each object with another number, $m(o)$, that can be called its *numerical measurement*. The measurement procedure we use constitutes a function rule, telling how to give an object o its $m(o)$ value. But just any measurement procedure will not do; *we want the various $m(o)$ values at least to reflect the different $t(o)$ values for different degrees of the property*. A measurement rule would be nonsense if it gave numbers having no connection at all with the amounts of some property different objects possess. Even though we may never be able to determine the $t(o)$ values exactly, we would like to find $m(o)$ numbers that will at least be related to them in a systematic way.

The measurement numbers must be good reflections of the true quantities, so that information about magnitudes actually is contained in our numerical measurements.

Measurement operations or procedures differ in the information that the numerical measurements themselves provide about the true magnitudes. Some ways of measuring permit us to make very strong statements about what the differences or ratios among the true magnitudes must be, and thus about the actual differences in, or proportional amounts of, some property that different objects possess. On the other hand, some measurement operations permit only the roughest inferences to be made about true magnitudes from the measurement numbers themselves.

Suppose that we had a measurement procedure or rule for assigning a number $m(o)$ to each object o, and suppose that the following statements were true:

1. For any pair of objects o_1 and o_2,
   $$m(o_1) \neq m(o_2) \text{ only if } t(o_1) \neq t(o_2).$$
2. $m(o_1) > m(o_2)$ only if $t(o_1) > t(o_2)$.

In other words, by this rule we can at least say that if two measurements are unequal the true magnitudes are unequal, and if one measurement is larger than another, then one magnitude exceeds another. Any measurement procedure for which both statements 1 and 2 are true is an example of **ordinal scaling,** or measurement at the **ordinal level.**

For example, suppose that the objects in question were minerals of various kinds. Each mineral has a certain degree of *hardness*, represented by the quantity $t(o)$. We have no way to know these quantities directly, and so we devise the following measurement rule: take each pair of minerals and find if one *scratches* the other. Presumably, the harder mineral will scratch the softer in each case. When this has been done for each pair of minerals, give the mineral scratched by everything some number, the mineral scratched by all but the first a higher number, and so on, until the mineral scratching all others but scratched by nothing else gets the highest number of all. In each pair the "scratcher" gets a higher number than the "scratchee." (Here we are assuming tacitly that "*a* scratches *b*" and "*b* scratches *c*" implies that "*a* scratches *c*," and that we will get a *simple ordering* of "what scratches what.")

This measurement procedure gives an example of *ordinal* scaling. The possible numerical measurements themselves might be some set of numbers, such as $(1, 2, 3, 4, \cdots)$ or $(10, 17, 24, \cdots)$ or even other symbols having some conventional order, as $(A, B, C, \cdots)$. In any case, the assignment of numbers or symbols to objects is a form of ranking, showing which is "more" something. In the example, if one mineral gets a higher number than another then we can say that the first mineral is harder than the second. Notice, however, that this is really all we can say about the degree or amount of hardness each possesses. If we find two objects where $m(o_1) = 3$ and $m(o_2) = 2$ by this procedure, then it is perfectly true that the difference between the *numbers* is $3 - 2 = 1$, but we have no basis at all for saying that $t(o_1) - t(o_2) = 1$. The difference in hardness between the two minerals could be any positive number. In short, *although the numbers standing for ordinal measurements may be manipulated by arithmetic, the answer cannot necessarily be interpreted as a statement about the true magnitudes of objects, nor about the true amounts of some property.*

However, other measurement procedures give functions pairing objects with numbers where much stronger statements can be made about the true magnitudes from the numerical measurements. Suppose that the following statement, in addition to statements 1 and 2, were true:

3.　　$t(o) = x$ if and only if $m(o) = ax + b$, where $a > 0$.

That is, the measurement number $m(o)$ is some *linear function* of the true magnitude x (the rule for a linear function is x multiplied by some constant a and added to some constant b). When the statements 1, 2, and 3 are all true, the measurement operation is called **interval scaling,** or measurement at the **interval-scale level.**

Much stronger inferences about magnitudes can be made from interval-scale measurements than from ordinal measurements. In particular, we can say something precise about *differences* in objects in terms of magnitude. Consider two objects, o_1 and o_2, once again. Then if we find

$$m(o_1) - m(o_2) = 4,$$

we can conclude that　　$t(o_1) - t(o_2) = 4a,$

the difference between the magnitudes of the two objects is 4 units, where a is simply some constant changing measurement units into "real" units, whatever they may be.

For example, finding temperature in Fahrenheit units is measurement on an interval scale. If object o_1 has a reading of 180° and o_2 has 160°, the difference $(180 - 160)$ times a constant actually *is* the difference in "temperature-magnitude" between the objects. It is perfectly meaningful to say that the first object has twenty units more temperature than the second. However, even stronger statements can be made. If $m(o_3) = 140°$, so that

$$m(o_1) - m(o_3) = 40°$$

and　　$$m(o_1) - m(o_2) = 20°,$$

then　　$$\frac{40°}{20°} = \frac{t(o_1) - t(o_3)}{t(o_1) - t(o_2)} = 2.$$

One can say that in true amount of "temperature," object o_3 is twice as different from o_1 as o_2 is from o_1.

Given some measurement operation yielding an interval scale of some property, then *any other measurement operation for the same property also gives an interval scale, provided that the second way of measuring yields numbers that are a linear function of the first:* thus, let some second measurement operation give numbers $g(o)$ to each object o, and let it be true that

$$g(o) = c[m(o)] + d, \quad c > 0,$$

where c and d are constants. However, by statement 3,

$$m(o) = ax + b,$$

so that $$g(o) = c[ax + b] + d = cax + bc + d.$$

The number ca and the number $bc + d$ are constants, and so the $g(o)$ values satisfy statement 3 as well, showing that this second measurement procedure gives an interval scale. A familiar example of this principle is temperature measurement in Fahrenheit and in centigrade degrees; each is an interval-scale measurement procedure, and the reading on one scale is a function of the reading on the other. If the $m(o)$ are Fahrenheit readings, and the $g(o)$ are centigrade, then it will always be true that for any three objects

$$\frac{m(o_1) - m(o_3)}{m(o_1) - m(o_2)} = \frac{g(o_1) - g(o_3)}{g(o_1) - g(o_2)}.$$

Hence, if we conclude from a Fahrenheit thermometer that o_3 is twice as different from o_1 as o_2 is from o_1, we would reach exactly the same conclusion from a centigrade thermometer.

When measurement is at the interval-scale level, any of the ordinary operations of arithmetic may be applied to the differences between numerical measurements, and the result interpreted as a statement about magnitudes of the underlying property. The important part is this interpretation of a numerical result as a quantitative statement about the property shown by the objects. This is not generally possible for ordinal-scale measurement numbers, but it *can* be done for *differences* between interval-scale numbers. In very simple language, one can do arithmetic to his heart's content on any set of numbers, but his results are not necessarily true statements about amounts of some property objects possess unless interval-scale requirements are met by the procedure for obtaining those numbers.

Interval scaling is about the best one can do in most scientific work, and even this level of measurement is all too rare in business and the social and behavioral sciences. However, especially in the physical sciences, it is some-

times possible to find measurement operations making the following statement true:

4. $t(o) = x$ if and only if $m(o) = ax$, where $a > 0$.

When the measurement operation defines a function such that statements
1 through 4 are all true, then measurement is said to be at the **ratio-scale
level.** For such scales, *ratios* of numerical measurements can be interpreted
directly as ratios of magnitudes of objects:

$$\frac{m(o_1)}{m(o_2)} = \frac{t(o_1)}{t(o_2)} .$$

For example, the usual procedure for finding the length of objects provides
a ratio scale. If one object has a measurement value of 10 feet, and another
a value of 20 feet, then it is quite legitimate to say that the second object
has twice as much length as the first. Notice that this is not a statement one
ordinarily makes about the *temperatures* of objects (on an interval scale);
if the first object has a temperature reading of 10° and the second 20°, we
cannot necessarily say that the second has twice the temperature of the first.
Only when scaling is at the ratio level can the full force of ordinary arithmetic be applied directly to the measurements themselves, and the results
reinterpreted as statements about magnitudes of objects.

There are any number of examples that could be adduced to illustrate
the differences among these levels of measurement and other possible
intermediate levels as well. Our immediate concern, however, is to see the
implications of different levels of measurement for probability and statistics.

Most of the statistical techniques introduced in this book are designed
for numerical observations, presumably arising from application of measurement techniques at the interval-scale level. This does not mean that these
techniques cannot be applied to numerical data where the interval-scale
level of measurement is not met; these techniques can be applied to *any*
numerical data where the purely *statistical* assumptions are met. It does
mean, however, that caution must be exercised in the interpretation of
these statistical results in terms of some property the statistician intended
to measure. The statistical conclusion may be quite valid for these data
as numbers, but the statistician must think quite seriously about the further
interpretations he gives to the result.

Of course, it is completely obvious that interval scales simply cannot be
attained for some things. It may be that an ordering of objects on some basis
is the very best the statistician can do. In this case, ordinal methods like
those in Chapter 12 are helpful; each of these techniques is designed specifically for ordinal data. Even when measurement is only nominal, statistical
methods can be applied; some of these are also discussed in Chapter 12.

The most common situation is, however, one where numerical measurements are made, but where the interval-scale assumptions are hard to justify. In these cases, the user of the statistical techniques should understand the limitations imposed by his measurement device (not by statistics); the road from objects to numbers may be easy, but the return trip from numbers to properties of objects is not.

5.4 FREQUENCY DISTRIBUTIONS

As we have just seen, even in the simplest instances of measurement the statistician makes some N observations and classifies them into a set of qualitative measurement classes. These qualitative classes are mutually exclusive and exhaustive, so that each observation falls into one and only one class.

As a summarization of his observations, the statistician often reports the various possible classes, together with the number of observations falling into each. This may be done by a simple listing of classes, each paired with its frequency number, the number of cases observed in that category. The same information may be displayed as a graph, perhaps with the different classes represented by points or segments of a horizontal axis, the frequency shown by a point or a vertical bar above each class. Regardless of how this information is displayed, such a listing of classes and their frequencies is called a *frequency distribution. Any representation of the relation between a set of mutually exclusive and exhaustive measurement classes and the frequency of each is a frequency distribution.*

Notice that a frequency distribution is a function; each of a set of classes is paired with a number, its frequency. Thus, in principle, a frequency distribution of real or theoretical data can be shown by any of the three ways one specifies any function: an explicit listing of class and frequency pairs, graphically, or by the statement of a rule for pairing a class of observations with its frequency. In describing actual data, the first two methods are almost always used alone, but there are circumstances where the statistician wants to describe some theoretical frequency distribution, and in this case the mathematical rule for the function sometimes is stated. It should be clear by now that a frequency distribution is analogous to a probability distribution; the former represents the relationship between events (in terms of classes) and their observed frequencies in a given sample, and the latter represents the relationship between events and their probabilities. A frequency distribution, just as a probability distribution, can be specified by a listing, in graphical form, or in functional form.

The set of measurement classes may correspond to a nominal, an ordinal, an interval, or a ratio scale. Although the various possible classes will be qualitative in some instances and number classes in others, a frequency distribution can always be constructed, provided that each and every observation goes into one and only one class. Thus, for example, suppose that some N native United States citizens are observed, and the state in which each was born is noted. This is like nominal measurement, where the measurement classes consist of the fifty states (and the District of Columbia) into which subjects could be categorized. On the other hand, the subjects might be students in a course and graded according to A, B, C, and so on. Here the frequency distribution would show how many got A, B, and so on. Notice that in this case the measurement actually is ordinal; A is better than B, B better than C. Nevertheless, the frequency distribution is constructed in the same way, except that the various classes are displayed in their proper order.

Even when measurement is at the interval- or ratio-scale level, one actually reports frequency distributions in this general way. Suppose that N college students were each weighed, and the weights noted to the nearest pound. The set of weight classes into which students are placed would perhaps consist of fifty or sixty different numbers; a weight class is the set of students getting the same weight number. There may be only one, or even no students, in a particular weight class. Nevertheless, the frequency distribution would show the pairing of each possible class with some number from zero to N, its frequency.

As we shall see, when the number of possible classes is very large, or even potentially infinite, the task of constructing a frequency distribution is made easier by a process of combining classes. The principle is, however, always the same: a display of the function relating classes and their frequencies.

5.5 FREQUENCY DISTRIBUTIONS WITH A SMALL NUMBER OF MEASUREMENT CLASSES

The reasons for dealing with frequency distributions rather than raw data are not hard to see. Raw data almost always take the form of some sort of listing of pairs, each consisting of an object and its measurement class or number. Consider a study done on a group of twenty-five consumers in order to determine their brand preference. The consumers were classified as preferring brand A, B, C, or D. Table 5.5.1 lists the brand preferences of these twenty-five consumers. This set of pairs does contain all the relevant information, but if there are a great many objects being measured such a listing is not only laborious but also very confusing to anyone who is trying to get a picture of the set of observations as a whole. If we are interested

Table 5.5.1

Consumer	Brand preferred
1	A
2	B
3	A
4	A
5	A
6	C
7	D
8	A
9	A
10	A
11	D
12	B
13	D
14	B
15	A
16	B
17	D
18	B
19	D
20	A
21	B
22	B
23	A
24	A
25	D

only in the "pattern" of brand preference in the group, the numbers assigned to the persons are irrelevant, and all that is necessary is the number of individuals having each brand preference.

These data condense into the following frequency distribution, where f is the frequency:

Class	f
A	11
B	7
C	1
D	6
	—
	$25 = N$

In another study, 2000 families in some city were interviewed about their preferences in art, music, literature, recreation, and so forth. After each interview, the family was characterized as having "highbrow," "middlebrow," or "lowbrow" tastes. This we can consider as a case of ordinal-scale measurement, since these three "taste" categories do seem to be ordered: "highbrow" and "lowbrow" should be more different from each other than "middlebrow" is from either. A listing of families in terms of these classes would be confusing, if not grounds for libel suits. However, just as before, we can construct a frequency distribution summarizing these data:

Class	f
Highbrow	50
Middlebrow	990
Lowbrow	960
	$2000 = N$

Notice how much more clearly the characteristics of the group as a whole emerge from a frequency distribution such as this than they possibly could from a listing of 2000 names each paired with a rating.

It is appropriate to point out that a frequency distribution provides clarity at the expense of some information in the data. It is not possible to know from the frequency distribution alone whether the family of John Jones at 2193 Spring Street is high-, middle-, or low-browed. Such information about *particular* objects is sacrificed in a frequency distribution to gain a picture of the group of measurements *as a whole*. This is true of all descriptive statistics; we want clear pictures of large numbers of measurements, and we can do this only by losing detail about particular objects. The process of weeding out particular qualities of the object that happen to be irrelevant to our purpose, begun whenever we measure, is continued when we summarize a set of measurements.

5.6   GROUPED DISTRIBUTIONS

In the distributions shown above, there were a few "natural" categories into which all the data fit. It often happens, however, that data are measured in some way giving a great many categories into which a given observation might fall. Indeed, the number of potential categories of description often vastly exceeds the number of cases observed, so that little or no economy of description would be gained by constructing a distribution showing each *possible* class. The most usual such situation occurs when the data are measured in numerical terms. For instance, in the measurement of height or weight, there are, in principle, an infinite number of numerical classes into which an observation might fall: all the positive real numbers. Even when the measurements result in fewer than an infinite number of possibilities, it is still quite common for the number of available categories to be very large.

For this reason, it is necessary to form **grouped** frequency distributions. Here the domain of the frequency function is not each possible measurement category, but rather intervals or groupings of categories.

For example, consider a number of people who have been given an intelligence test. The data of interest are the numerical IQ scores. Immediately, we run into a problem of procedure. It could very well turn out that for, say, 150 individuals, the IQ score would be different for each case, and thus a frequency distribution using each different IQ number as a measurement class would not condense our data any more than a simple listing.

The solution to this problem lies in grouping the possible measurement classes into new classes, called **class intervals,** each including a range of score possibilities. Proceeding in this way, if the IQs 105, 104, 100, 101, and 102 should turn up in our data, instead of listing each in a different class, we might put them all into a single class, 100–105, along with any other IQ measurements that fall between these limits. Similarly, we group other sets of numbers into class intervals. On our doing this, one way that the frequency distribution might look for a group of 150 persons is shown in Table 5.6.1.

Forming class intervals has enabled us to condense the data so that a simple statement of the frequency distribution can be made in terms of only a few classes. In this distribution, the various class intervals are shown in order

Table 5.6.1

Class interval	Midpoint	f
124–129	126.5	8
118–123	120.5	0
112–117	114.5	10
106–111	108.5	20
100–105	102.5	65
94– 99	96.5	22
88– 93	90.5	23
82– 87	84.5	2
		$150 = N$

in the first column on the left. The extreme right column lists the frequencies, each paired with the class in that row of the table. The sum of the frequencies must be N, the total number of observations. The middle column contains the midpoint of each class interval; these will be discussed in Section 5.9.

5.7 CLASS INTERVAL SIZE AND CLASS LIMITS

The first problem in constructing a grouped frequency distribution is deciding upon a class interval size. Just how large shall the range of numbers included in a class be? The class interval size for a grouped distribution is the difference between the largest and the smallest number that may go into any class interval. We will let the symbol i denote this interval size.

In the example above, as in most examples to follow, the size i is the same for each class interval. For our example, $i = 6$, meaning that the largest and the smallest score going into a class interval differ by six units. Take the class interval labeled 100–105. It may seem that the smallest number going into this interval is 100, and the largest 105; this is not, however, true. Actually, the numbers 99.5, 99.6, 99.7 would also go in this interval, should they occur in the data. Also, the numbers 105.1, 105.2 are included. On the other hand, 99.2 or 105.8 would be excluded. The interval actually includes any number *greater than or equal to* 99.5 *and less than* 105.5. These are called the **real limits** of the interval, in contrast to those actually listed, which are the **apparent limits.** Thus, the real limits of the interval 100–105 are

$$99.5 \text{ to } 105.5$$

and $\qquad\qquad i = 105.5 - 99.5 = 6.$

In general,

Real lower limit = apparent lower limit minus .5(unit difference)
Real upper limit = apparent upper limit plus .5(unit difference).

The term "unit difference" demands some explanation. In measuring something we usually find some limitation on the accuracy of the measurement, and seldom can one measure with *any* desired degree of accuracy. For this reason, measurement in numerical terms is always rounded, either during the measurement operation itself, or after the measurement has taken place. In measuring weight, for example, we obtain accuracy only within the nearest pound, or the nearest one tenth of a pound, or one hundredth of a pound, and so on. If one were constructing a frequency distribution where weight had been rounded to the nearest pound, then a unit difference is one pound, and the real limits of an interval such as 150–190 would be 149.5–190.5. The i here would be 41 pounds.

On the other hand, suppose that weights were accurate to the nearest one tenth of a pound; then the unit difference would be .1 pound, and the real class interval limits would be

$$149.05-190.05.$$

The way that rounding has been carried out will have a real bearing on how the successive class intervals will be constructed. Suppose that we wanted to construct the distribution of annual income for a group of American men, where the income figures have been rounded to the nearest thousand dollars. We have decided that the apparent upper limit of the top interval shall be 25,000 dollars, and that i should equal 5,000. Then the classes would have the following limits:

Real	Apparent
20,500–25,500	21,000–25,000
15,500–20,500	16,000–20,000
10,500–15,500	11,000–15,000
5,500–10,500	6,000–10,000
500– 5,500	1,000– 5,000

Here, one half of a unit difference is 500 dollars.

Now suppose that each man's income has been rounded to the nearest one hundred dollars. Once again, with $i = \$5,000$ and with the top apparent limit $25,000, we form class intervals, but this time the class limits are:

Real	Apparent
20,050–25,050	20,100–25,000
15,050–20,050	15,100–20,000
10,050–15,050	10,100–15,000
5,050–10,050	5,100–10,000
50– 5,050	100– 5,000

This difference in unit for the two distributions could make a difference in the "picture" we get of the data. For example, a man making exactly $20,100 would fall into the top interval in the second distribution, but into the second from top interval in the first. It is important to decide upon the accuracy represented in the data before the real limits for the class intervals are chosen. Is every digit recorded in the raw data regarded as significant, or will the data be rounded to the "nearest" unit?

5.8 INTERVAL SIZE AND THE NUMBER OF CLASS INTERVALS

One can use any number for i in setting up a distribution. However, it should be obvious that there is very little point in choosing i smaller than the unit difference (for example, letting $i = .1$ pound when the data are in nearest whole pounds), or in choosing i so large that all observations fall into the same class interval. Furthermore, i is usually chosen to be a *whole number* of units, whatever the unit may be. Even within these restrictions, there is considerable flexibility of choice. For example, the data of Section 5.6 could have been put into a distribution with $i = 3$, as shown in Table 5.8.1. Notice that this distribution with $i = 3$ is somewhat different from the distribution with $i = 6$, even though they are based on the same set of data. For one thing, here there are sixteen class intervals, whereas there were eight before. This second distribution also gives somewhat more detail than the first about the original set of data. We now know, for example, that there was no one in the group of persons who got an IQ score between the real limits of 83.5 and 86.5, whereas we could not have told this from the first distribution. We can also tell that more cases showed IQs

Table 5.8.1

Class interval	Midpoint	f
126–128	127	5
123–125	124	3
120–122	121	0
117–119	118	0
114–116	115	7
111–113	112	3
108–110	109	8
105–107	106	13
102–104	103	44
99–101	100	20
96– 98	97	10
93– 95	94	12
90– 92	91	14
87– 89	88	9
84– 86	85	0
81– 83	82	2
		$150 = N$

between 101.5 and 104.5 than between 98.5 and 101.5; conceivably this could be a fact of some importance. On the other hand, the first distribution gives a simpler and, in a sense, a neater picture of the group than does the second.

The decision facing the maker of frequency distributions is, "Shall I use a small class-interval size and thus get more detail, or shall I use a large class-interval size and get more condensation of the data?" There is no fixed answer to this question; it depends on the kind of data and the uses to which they will be put. However, a convention does exist which says that for most purposes *ten to twenty class intervals* give a good balance between condensation and necessary detail, and in practice one usually chooses his class-interval size to make about that number of intervals.

5.9   MIDPOINTS OF CLASS INTERVALS

There is one more feature of frequency distributions that has not yet been discussed. When one puts a set of measurements into a frequency distribution, he loses the power to say exactly what the original numbers were. For the example of Section 5.8, if a person in the group of 150 cases has an IQ of 102, he is simply counted as one of the 44 individuals who make up the frequency for the interval 102–104. You cannot tell from the distribution exactly what the IQs of those 44 individuals were, but only that they fell within the limits 102–104. What do we call the IQs of these 44 individuals? We call them all 103, the midpoint of the interval. **The midpoint of any class interval is that number which would fall exactly halfway among the possible numbers included in the interval.** A moment's thought will convince you that 103 falls halfway between the real limits of 101.5 and 104.5. In a like fashion, all 5 cases falling into the class interval 126–128 will be called by the midpoint 127; all 3 cases in the interval 123–125 will be called 124, and so on for each of the other class intervals.

The real limits can also be defined in terms of the midpoint:

$$\text{Real limits} = \text{midpoint} \pm .5i.$$

Thus, given only the midpoints of the distribution, one can find the class limits.

In the computations using grouped distributions the midpoint is used to substitute for each raw score in the interval. For this reason, it is convenient to choose an odd number for i whenever possible; this makes the midpoint a whole number of units and simplifies computations.

5.10   FREQUENCY DISTRIBUTIONS WITH OPEN OR UNEQUAL CLASS INTERVALS

Quite often the data are such that it is not possible to make a frequency distribution with intervals of a constant size. This most commonly occurs when exact values are not known for some of the observations. For example, suppose that we are measuring temperature, using a thermometer which only goes up to 100 degrees Fahrenheit, and some of the observations are greater than this upper limit. We cannot give exact values for these observations, although we can say that they are greater than 100, that is, that they fall in the top class of the frequency distribution, in an interval called "100 or more." This interval is open, since there is no way to determine its upper real limit. Furthermore, there is no way to give such an open interval a midpoint.

It is perfectly correct to show open intervals at either end of a frequency distribution. However, no computations involving the midpoints can be carried out from this distribution, unless the cases in the open intervals are dropped from the sample.

In other instances, it may be that there are extreme scores in a distribution that are widely separated from the bulk of the cases. When this is true, a class-interval size that is small enough to show "detail" in the more concentrated part of the distribution will eventuate in very many classes with zero frequency before the extreme scores are included. Enlarging the class-interval size will reduce the number of unnecessary intervals, but will also sacrifice detail. When this situation arises, it is often wise to have a varying class-interval size, so that the class intervals are narrow where the detail is desired, but rather broad toward the extreme or extremes of the distribution. In this case, class-interval limits and midpoints can be found in the usual way, provided that one takes care to notice where the interval size changes. Furthermore, midpoints of these intervals can be used for further computations using the methods given in Sections 5.16 and 5.17. For example, suppose that we were measuring daily changes in security prices. In the more concentrated part of the distribution (that is, the part near zero), we might be interested in narrow class intervals. Observations more than, say, two units (dollars in this example) from zero might be infrequent enough that we would like wider class intervals for these observations.

5.11   GRAPHS OF DISTRIBUTIONS: HISTOGRAMS

Now we direct our attention for a time to the problem of putting a frequency distribution into graphic form. Often a graph shows that the old bromide, "a picture is worth a thousand words," is really true. If the purpose is to provide an easily grasped summary of data, nothing is so effective as a graph of a distribution.

There are undoubtedly all sorts of ways any frequency distribution might be graphed. A glance at any national news magazine will show many examples of graphs which have been made striking by some ingenious artist. However, we shall deal with only three "garden varieties": the histogram, the frequency polygon, and the cumulative frequency polygon.

The histogram, which we discussed in terms of probability distributions in Section 3.6, is really a version of the familiar "bar graph." In a histogram each class or class interval is represented by a portion of the horizontal axis. Then over each class interval is erected a bar or rectangle equal in height to the frequency for that particular class or class interval. The histograms representing the frequency distributions in Sections 5.5 and 5.8 are shown in Figures 5.11.1–5.11.3. It is customary to label both the horizontal and the vertical axes as shown in these figures, and to give a label to the graph as a whole. In the case where class intervals are employed

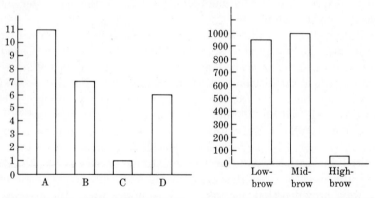

Figure 5.11.1. Brand Preferences of Twenty-Five Consumers

Figure 5.11.2. Taste Ratings for Two Thousand American Families

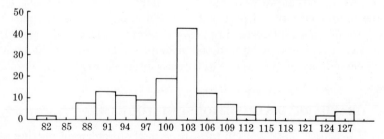

Figure 5.11.3. Intelligence Quotients of a Sample of One Hundred and Fifty Subjects

to group numerical measurements, a saving of time and space may be accomplished by labeling the class intervals by their midpoints (as in Figure 5.11.3) rather than by their apparent or real limits. With categorical measurements, this problem does not arise, of course, since each class can only be given its name or symbol.

If you will look at the two histograms shown in Figures 5.11.4 and 5.11.5, you will get very different first impressions, even though they represent the same frequency distribution. These two figures illustrate the effect that proportion can have on the viewer's first impressions of graphed distributions. Various conventions about graphs of distributions exist in different fields, of course. In fact, those who make a business of using statistics for persuasion often choose a proportion designed to give a certain impression. However, one convention is to arrange a graph so that the vertical axis is about three-fourths as long as the horizontal. This usually will give a clear and esthetically pleasing picture of the distribution.

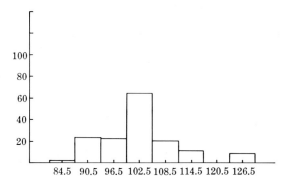

Figure 5.11.4

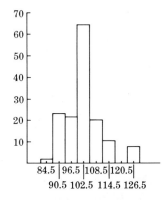

Figure 5.11.5

5.12 FREQUENCY POLYGONS

The histogram is a useful way to picture any sort of frequency distribution, regardless of the scale of measurement of the data. However, a second kind of graph is often used to show frequency distributions, particularly those that are based upon numerical data: this is the frequency polygon. In order to construct the frequency polygon from a frequency distribution one proceeds exactly as though one were making a histogram; that is, the horizontal axis is marked off into class intervals and the vertical into numbers representing frequencies. However, instead of using a bar to show the frequency for each class, a point on the graph corresponding to the midpoint of the interval and the frequency of the interval is found. These points are then joined by straight lines, each being connected to the point immediately succeeding and the point immediately following, as in Figure 5.12.1.

The frequency polygon is especially useful when there are a great many potential class intervals, and it thus finds its chief use with numerical data that could, in principle, represent a continuous variable. For a distribution based on a very large number of numerical measurements on a potentially continuous scale, if we maintained the same proportions in our graph, but employed a much greater number of class intervals, the frequency polygon would provide a picture more like a smooth curve, as shown in Figures 5.12.2–5.12.4, based on several thousand cases. Although the function

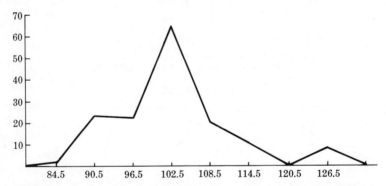

Figure 5.12.1. Intelligence Quotients of a Sample of One Hundred and Fifty Subjects

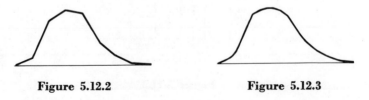

Figure 5.12.2   **Figure 5.12.3**

Figure 5.12.4

rule describing the frequency polygon may be extremely complicated to state, the function rule for a smooth curve approximately the same as that of the distribution may be relatively easy to find. For this reason, a frequency polygon based upon interval- or ratio-scale measurements with a relatively large number of classes is sometimes "smoothed" by creating a curve that approximates the shape of the frequency polygon. This new curve may have a rule that is simple to state and that may serve as an approximation of the rule describing the frequency distribution itself. Graphically, the smooth curve may be a good approximation to the frequency polygon, just as a probability density function can sometimes be thought of as a good approximation to the probability mass function of a discrete random variable (Section 3.6).

5.13 CUMULATIVE FREQUENCY DISTRIBUTIONS

For some purposes it is convenient to make a different arrangement of data into a distribution, called a cumulative frequency distribution, which is analogous to a cumulative distribution function (Section 3.8). Instead of showing the relation that exists between a class interval and its frequency, **a cumulative distribution shows the relation between a class interval and the frequency at or below its real upper limit.** In other words, the cumulative frequency shows how many cases fall into *this interval or below.* Thus, in the distribution in Section 3.5, the lowest interval shows two cases, and its cumulative frequency is 2. Now the next class interval has a frequency of 23, so that its cumulative frequency is 23 + 2, or 25. The third interval has a frequency of 22, and its cumulative frequency is 22 + 23 + 2, or 47, and so on. The cumulative frequency of the very top interval must always be N, since all cases must lie either in the top interval or below. The cumulative frequency distribution for the example of Section 5.6 is given at the top of page 266.

Cumulative frequency distributions are often graphed as polygons. The axes of the graph are laid out just as for a frequency polygon, except that the class intervals are labeled by their *real upper limits* rather than by their midpoints. The numbers on the vertical axis are now cumulative frequencies,

Class interval	Cumulative frequency
124–129	150
118–123	142
112–117	142
106–111	132
100–105	112
94– 99	47
88– 93	25
82– 87	2

of course. The graph of the example from Section 5.6 is shown in Figure 5.13.1. Cumulative frequency distributions always have more or less this characteristic **S** shape, although the slope and the size of the "tails" on the **S** will vary greatly from distribution to distribution. Like a frequency polygon, a cumulative frequency polygon is sometimes smoothed into a curve as similar as possible to the polygon. You sometimes encounter a reference to such a smoothed cumulative distribution curve under the term **ogive.**

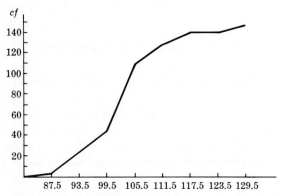

Figure 5.13.1

5.14   JOINT FREQUENCY DISTRIBUTIONS
AND INFERENCES ABOUT ASSOCIATION

It should be apparent that actual data can be put into the form of a *joint frequency distribution* in exactly the same way that events are shown in a joint probability distribution. The only difference is that the observed frequency is shown for each joint event or joint interval. When the events of both attributes are qualitative, then the joint distribution is often called a **contingency table.** For numerical scores, it is often convenient to show occurrence of a joint event by a point on a graph, where the X and Y axes symbolize the two variables. However, for the same reason that numerical distributions are grouped into class intervals, it may be more convenient to break each axis into class intervals to show the frequency of occurrence of the joint intervals in the data. Such bivariate frequency distributions are often called **scatter plots** or **scatter diagrams.**

The principle embodied in Bernoulli's theorem holds for such bivariate (or even multivariate) frequency distributions: given random samples from a specific joint or bivariate distribution of events, the obtained relative frequencies of the various joint events must, in the long run, equal the probabilities for such joint events. Thus, any sample's joint distribution provides an estimate of the joint probability distribution, just as for single variate distributions.

Can the independence of events be judged from a sample joint distribution? Suppose that some N cases were sampled in all. A joint frequency distribution could be constructed. In any cell the obtained relative frequency of a joint event $X \cap Y$, or

$$\text{Relative frequency } (X \cap Y) = \frac{\text{freq. } (X \cap Y)}{N}, \qquad (5.14.1)$$

is an estimate of $P(X \cap Y)$. Because of sampling error, this relative frequency in the sample need not be equal exactly to $P(X \cap Y)$. Nevertheless, in the long run, as the sample size grows indefinitely large, the probability approaches 1.00 that the joint relative frequency will agree very closely with the true joint probability. Similarly, the probability is large that the relative frequency associated with any X will agree closely with $P(X)$, and the same is true of Y. **If the variables are independent, then we should expect that in the long run**

$$\frac{\text{freq. } (X \cap Y)}{N} = \left(\frac{\text{freq. } (X)}{N} \right) \left(\frac{\text{freq. } (Y)}{N} \right),$$

the relative frequency of any joint event $(X \cap Y)$ should equal the relative frequency of X times the relative frequency of Y for those particular values of X and Y.

Provided that X and Y are independent, the best estimate we can make of what to *expect* in a cell, given only the marginal frequencies, is

$$\text{Expected frequency } (X \cap Y) = \frac{[\text{freq. } (X)] \, [\text{freq. } (Y)]}{N}. \qquad (5.14.2)$$

Now suppose that the statistician draws a sample and actually constructs the table of joint frequencies. By the rule just given, he can estimate what the expected frequency of each cell should be, provided that X and Y were independent. Any departure of any cell's frequency from this expected number is a small piece of evidence that X and Y are not independent, and he may even say that X and Y are *apparently associated*. However, for any sample, it is quite possible for X and Y apparently to be associated, even though the variables are really independent. Conversely, finding relative frequencies that *exactly* conform to expectation about independent variables does not necessarily mean that X and Y are independent. Unequivocal statements about independence or association can never be made from sample results alone; these statements can be made with complete confidence only if the probabilities are known. The statistician *never* has enough evidence to say for sure, unless his set of observations includes the entire sample space.

On the other hand, the more the joint frequencies depart from the frequencies to be expected for independent X and Y, the better is the bet that X and Y are associated. Big departures from expectation should be unusual for relatively large sample size and for truly independent random variables. Of course, we still cannot say for sure, even for very large departures, that X and Y are associated. In Chapter 12, however, we will discuss statistical techniques for investigating the question of independence in a contingency table.

5.15 STATISTICAL RELATIONS AND ASSOCIATION

Most applications of statistics to experimental data are concerned with association among two or more variables. Regardless of what the statistician studies, he can be sure that *for any observation showing a given value on one variable, there will be a whole distribution of values that might occur on some other variable*. That is, given the value of X, there is some conditional distribution of Y. Almost never will it be possible for one and only one Y value to occur for each possible X; instead there will be any number of different Y values that one might observe, each with some probability.

This is what we mean when we say that any set of data shows *variability*. No matter how we restrict the sample space, making it true that all observations represent the same event when looked at in one way, the individuals sampled will "scatter" over many events when regarded in some other way. In the physical sciences this variability can often be made so small as to be negligible; given X, different possibilities for Y still exist, but different observations give Y values so nearly alike as to be practically identical. This is seldom the case with observations in business, the social sciences, and the behavioral sciences.

For this reason, it is useful to think of each value of X as paired with some distribution of Y values, and to discuss this pairing of X values with Y distributions as a *statistical relation*. When each and every X value possible is paired with exactly the same Y distribution, then the variables X and Y are statistically independent, and one says that there is *no* statistical relationship. On the other hand, when different conditional distributions of Y are paired with different values of X, then one says that a statistical relationship *does* exist. In short, statisticians are very interested in comparing distributions to see if they are identical or different in some way; the reason for this interest becomes clear when differences among distributions are viewed as suggesting a statistical relationship or association between variables.

A comparison of obtained Y distributions, conditional on specific X values, or different qualitative categories, gives clues to association between variables. Nevertheless, even though differences among frequency distributions may suggest association, the goodness of this evidence must be evaluated in terms of probability. In particular, the experimenter knows that sample distributions need not be like the true probability distributions, and, unfortunately, association is defined in terms of the true state of affairs for all such observations. The theory of statistical inference gives the ingredients for this judgment about association from samples; we will develop this theory in later chapters.

5.16 MEASURES OF CENTRAL TENDENCY FOR FREQUENCY DISTRIBUTIONS

Any frequency distribution is a summarization of data, but for many purposes it is necessary to summarize still further. Rather than compare entire frequency distributions with each other or with hypothetical distributions, it is generally more efficient to compare only certain characteristics of distributions. As with probability distributions, the two most important characteristics are central tendency and dispersion. In this section we discuss measures of central tendency for frequency distributions, and in the following section we will discuss measures of dispersion.

The measures of central tendency considered for probability distributions have counterparts for frequency distributions. The three most important measures are the sample mean, the sample median, and the sample mode. **The sample mean of a sample of observations is determined by taking the simple arithmetic average of the values;** if the sample size is N and the observed values are $X_1, X_2, \cdots, X_N$, then the **sample mean M** is

$$M = \frac{\sum_{i=1}^{N} X_i}{N}.$$

(5.16.1*)

The definition and computation of the arithmetic mean for raw data is simple enough, but the situation is slightly more complicated when one wishes to find the mean of a grouped distribution of scores. You will recall (Section 5.9) that when a distribution is grouped into class intervals the midpoint x of each class interval was taken to represent the score of each of the cases in the interval. Thus, in an interval 59–73 with midpoint 66 and frequency 16, the sum of the scores of the 16 cases falling into this interval is taken to be 66 summed 16 times, or $(66)(16) = xf$. Similarly, when all of the scores in any interval are assumed the same, their sum is the midpoint of that particular interval times the frequency for that interval, or xf. Then the sum of *all* of the scores in the distribution is taken to be the sum of the values of x times f over all of the respective intervals, and thus the mean is found from

$$M = \frac{\Sigma xf}{N}.$$ (5.16.2)

(Note that here x is *any* midpoint, f is the frequency corresponding to that interval, and the sum is taken over *all intervals*.)

For example, consider the distribution shown in Table 5.16.1. The frequency of each class interval is multiplied by its midpoint, and these are then summed and divided by N to given the distribution mean.

The mean calculated from a distribution with grouped class intervals need not agree exactly with that calculated from raw scores. Information

Table 5.16.1

Class	x	f	xf
74–78	76	10	760
69–73	71	18	1278
64–68	66	16	1056
59–63	61	16	976
54–58	56	11	616
49–53	51	27	1377
44–48	46	17	782
39–43	41	49	2009
34–38	36	22	792
29–33	31	6	186
24–28	26	8	208
		200	$10{,}040 = \Sigma xf$

$$M = \frac{\Sigma xf}{N} = \frac{10{,}040}{200} = 50.2$$

is lost and a certain amount of inaccuracy introduced when scores are grouped and treated as though each corresponded to the midpoint of some interval. The coarser the grouping, in general, the more likely is the distribution mean to differ from the raw-score mean. For most practical work the rule of ten to twenty class intervals gives relatively good agreement, however. Nevertheless, it is useful to think of the mean calculated from any given distribution as the mean of that *particular* distribution, a particular set of groupings with their associated frequencies.

The median for a set of observed data is ordinarily defined in slightly different ways depending upon whether N is odd or even, and upon whether raw data or a grouped frequency distribution is to be described. For a set of raw scores, **when N is odd and the N values are arranged in numerical order, the median corresponds to the $((N + 1)/2)$th value in order; when N is even and the N values are arranged in numerical order, the median is defined as the number midway between the $(N/2)$th value and the $((N/2) + 1)$th value.** Then either for odd or even N it will be true that exactly as many values fall above as fall below the median.

On the other hand, this way of computing a median is not usually applicable if the data have been arranged into a grouped frequency distribution. For any such grouped distribution, the median is defined as the point at or below which exactly 50 percent of the observed values fall. Consequently the first step in finding the median of a grouped distribution is to construct the *cumulative* frequency distribution. This is illustrated in Table 5.16.2.

Table 5.16.2

Class	f	cf
74–78	10	200
69–73	18	190
64–68	16	172
59–63	16	156
54–58	11	140
49–53	27	129
44–48	17	102
39–43	49	85
34–38	22	36
29–33	6	14
24–28	8	8
	——	
	200	

The last column given shows the cumulative frequencies for the class intervals. Since, by definition, the median will be that point in the distribution at or below which 50 percent of the cases fall, the cumulative frequency *at* the median score should be $.50N$. Thus, the cumulative frequency is $.50(200) = 100$ at the median for this example. Where would such a score fall? It can be seen that it could not fall in any interval below the real limit of 43.5, since only 85 cases fall at or below that point. However, it does fall below the real limit 48.5, since 102 cases fall at or below that point. Hence, we have located the median as being in the class interval with real limits 43.5 and 48.5.

We must still ascertain the value that corresponds to the median, and this is where the process of interpolation comes in. The median is somewhere in the interval 44–48. We assume that the 17 cases in that interval are evenly scattered over the interval width of 5 units, as in Figure 5.16.1.

The median score, which is greater than or equal to exactly 100 cases, must then exceed not only the 85 cases below 43.5, but also equal or exceed 15 cases above 43.5. In other words, the median is 15/17 of the way *up* the interval from the lower limit. Next, what score is exactly 15/17 of the way between 43.5 and 48.5? Since the difference between these two limits is 5, the class interval size, we take $(15/17)5$, or 4.4, as the amount that must be added to the lower real limit to find the median score, so that the median must be $43.5 + 4.4$ or 47.9.

A little computational formula that summarizes the steps just described is given by:

$$\text{Median} = \text{lower real limit} + i \frac{(.50N - cf \text{ below lower limit})}{f \text{ in interval}}, \quad (5.16.3)$$

where the lower real limit used belongs to the interval containing the median, and the *cf* refers to the cumulative frequency *up to* the lower limit of the interval. For the example, we find

$$\text{Median} = 43.5 + 5 \frac{[.50(200) - 85]}{17} = 47.9.$$

If the median can fall only in some class interval with nonzero frequency, this method of interpolation gives a unique value. However, if the interval

Figure 5.16.1

frequency happens to be zero, then actually any point in the interval serves equally well as the distribution median; here one usually takes the midpoint as the median.

Even when the raw data are available, it is often worthwhile to define and compute the median as for a grouped frequency distribution. In the first place, for a sizable N, ordering all of the cases by their score magnitudes can be a considerable chore, and it may be simpler to construct a grouped distribution. Secondly, a troublesome problem arises when two or more cases in the raw data are tied in order at the median position, and here it often makes sense to calculate the median by interpolation as for a grouped distribution.

In principle, a median may be found for any distribution in which the variable represents an interval or even an ordinal scale; it is not, in general, applied to a distribution in which the measurement classes are purely categorical, since such classes are unordered.

The **mode** is certainly the most easily computed and the simplest to interpret of all the measures of central tendency. It is merely the midpoint or class name of the most frequent measurement class. If a case were drawn at random from the N observations, then that case is more likely to fall in the **modal class** than any other. So it is that in the graph of any distribution, the modal class shows the highest "peak" or "hump" in the graph. The mode may be used to describe any distribution, regardless of whether the events are categorized or numerical.

On the other hand, there are some disadvantages to the mode. One is that there may be more than one modal class. It is perfectly possible for two or more measurement classes to show frequencies that are equal to each other and higher than the frequency shown by any other class. In this case, there is ambiguity about which class gives *the* mode of the distribution, as two or more values are "most popular." Fortunately, this does not happen very often; even though there are two or more "humps" in the graph of a distribution, the class having the highest frequency or probability is still taken as *the* mode. Nevertheless, in such a distribution, where one measurement class is not clearly most popular, the mode loses its effectiveness as a characterization of the distribution as a whole.

Another disadvantage is that the mode is very sensitive to the size and number of class intervals employed when events are numerical; the value of the modes may be made to "jump around" considerably by changing the class intervals for a distribution.

Finally, the mode of a sample distribution forms a very undependable source of information about the mode of the underlying probability distribution. For these reasons, our use of the mode will be restricted to situations where: (1) the data are truly nominal scale in nature; (2) only the simplest, most easily computed, measure of central tendency is needed.

The median is considerably less sensitive to the distribution's grouping into class intervals than is the mode. Furthermore, when one is making inferences about a large "population" of potential observations from a sample, the median is generally more useful and informative than the mode, although the median itself is not ordinarily so useful as the mean, particularly when our purpose is to make inferences beyond the sample. The median has mathematical properties making it difficult to work with, whereas the mean is mathematically tractable. For this and other reasons, mathematical statistics has taken the mean as the focus of most of its inferential methods, and the median is relatively unimportant in inferential statistics. Nevertheless, as a description of a given set of data, the median is extremely useful in communicating the typical score.

Comparisons between the sample mean, sample median, and sample mode are analogous to comparisons between the mean, median, and mode of a probability distribution. Such comparisons were discussed in some detail in Sections 3.15–3.20, so we will not repeat them here. It should be noted that there are other measures of central tendency, such as the sample mid-range (the average of the highest and the lowest observations), the sample mid-interquartile range (the average of the .25 fractile and the .75 fractile of the frequency distribution), and sample trimmed means (sample means calculated after some of the extreme observations have been discarded). In general, these other measures, while they may be useful in certain situations, are not as valuable as the three measures we have concentrated on in this section. A final point of interest is that we have considered the sample mean, median, and mode as descriptive measures of central tendency of sets of data; in Chapter 6 we will consider them as "estimates" of population parameters, and we will see that the sample mean is the most useful measure in this sense.

5.17   MEASURES OF DISPERSION FOR FREQUENCY DISTRIBUTIONS

As we pointed out in Section 3.21, an important feature (in addition to central tendency) of a distribution is dispersion, or spread, or variability. For probability distributions, we generally measure this dispersion with the variance or its square root, the standard deviation. Recall that the variance is the expectation of the squared deviations about the mean, $E(X - \mu)^2$. Now **we define the sample variance to be the average of the squared deviations of the sample values from the sample mean:**

$$s^2 = \frac{\sum_{i=1}^{N} (X_i - M)^2}{N} .$$

(5.17.1*)

Note that s^2 will be zero if and only if each and every observation has the same value. The more the observations differ from each other and from M, the larger s^2 will be.

The sample variance is defined in Equation (5.17.1) *as though* the raw score of each of N cases were known, and can be computed by this formula when the raw data have not been grouped into a distribution. However, when data are in a grouped distribution, an equivalent definition of the variance is given by

$$s^2 = \frac{\Sigma(x - M)^2 \text{freq.} (x)}{N}, \tag{5.17.2}$$

where x is, as usual, the midpoint of an interval. Here, for each interval, the deviation of the midpoint from the mean is squared, and multiplied by the frequency for that interval. When this has been done for each interval, the average of these products is the variance, s^2. Just as with the mean, the value of s^2 calculated for a grouped distribution need not agree exactly with that based on the raw scores; nevertheless, if a relatively large number of class intervals is used these two values should agree very closely.

Since the sample variance involves squared terms, it has the disadvantage of not being in the same units as the observations and the sample mean. This problem can be resolved by taking the square root of the sample variance, which is the sample standard deviation, s. The sample variance and sample standard deviation can be computed from a set of data by formulas (5.17.1) and (5.17.2). This entails finding the mean, subtracting it successively from each score, squaring each result, adding the squared deviations together, and dividing by N to find the variance. Obviously, this is relatively laborious for a sizable number of cases. However, it is possible to simplify the computations by a few algebraic manipulations of the original formulas. This may be shown as follows: for any single deviation, expanding the square gives

$$(X_i - M)^2 = X_i^2 - 2X_iM + M^2,$$

so that, on averaging these squares, we have

$$\sum_i \frac{(X - M)^2}{N} = \sum_i \frac{(X_i^2 - 2X_iM + M^2)}{N}$$

$$= \sum_i \frac{X_i^2}{N} - 2\sum_i \frac{X_iM}{N} + \sum_i \frac{M^2}{N}.$$

However, wherever M appears it is a constant over the sum, and so

$$\sum_i \frac{(X_i - M)^2}{N} = \sum_i \frac{X_i^2}{N} - 2M \sum_i \frac{X_i}{N} + \frac{NM^2}{N},$$

or
$$s^2 = \sum_i \frac{X_i^2}{N} - 2M^2 + M^2 = \frac{\sum_i X_i^2}{N} - M^2. \qquad (5.17.3^*)$$

Finally,

$$s = \sqrt{\frac{\sum_i X_i^2}{N} - M^2}. \qquad (5.17.4^*)$$

These last formulas give a way to calculate the indices s^2 and s with some savings in steps. These formulas will be referred to as the "raw-score computing forms" for the variance and standard deviation. For an example of the use of these forms, study the example shown in Table 5.17.1, based on a sample of size seven. These two methods must always agree exactly, as in the example. Another convenient formula for calculating s^2 is

$$s^2 = \frac{N \sum X_i^2 - (\sum X_i)^2}{N^2}.$$

It is also possible to find similar computing forms for grouped distributions. Starting with the definition of the variance of a grouped distribution,

$$s^2 = \frac{\Sigma(x - M)^2 \text{ freq. } (x)}{N},$$

Table 5.17.1

Scores X_i	Deviation method		Raw-score method X_i^2
	$d_i = (X_i - M)$	d_i^2	
11	6	36	121
10	5	25	100
9	4	16	81
8	3	9	64
6	1	1	36
−4	−9	81	16
−5	−10	100	25
$35 = \sum_i X_i$	0	$268 = \sum_i d_i^2$	$443 = \sum_i X_i^2$

$M = 5$

$$s^2 = \frac{\sum_i d_i^2}{N} = \frac{268}{7} = 38.28 \qquad s^2 = \frac{\sum_i X_i^2}{N} - M^2$$

$$= \frac{443}{7} - 25 = 38.28$$

$$s = \sqrt{38.28} = 6.19$$

we could develop the following computing forms:

$$s^2 = \frac{\Sigma x^2 \text{ freq. } (x)}{N} - M^2 \tag{5.17.5}$$

$$= \frac{\Sigma x^2 \text{ freq. } (x)}{N} - \left[\frac{\Sigma x \text{ freq. } (x)}{N} \right]^2. \tag{5.17.6}$$

Then, as usual,

$$s = \sqrt{s^2}. \tag{5.17.7}$$

This little derivation is left to you as an exercise.

An example of the computation of the mean and standard deviation from a grouped frequency distribution is given in Table 5.17.2.

Why is the sample variance calculated in terms of squared deviations from the sample mean, M, instead of from some other measure of central tendency? Recall that the variance of a probability distribution is calculated in terms of squared deviations from μ because the expected squared deviation is smallest when the deviations are taken from μ. Similarly, for the sample variance, the average squared deviation is smallest when the deviations are taken from M. To show this, consider

$$s_C^2 = \sum_i \frac{(X_i - C)^2}{N}.$$

Table 5.17.2

Class	x	f	xf	x^2	x^2f
46–50	48	6	286	2304	13814
41–45	43	8	338	1849	14792
36–40	38	10	380	1444	14440
31–35	33	5	165	1089	5445
26–30	28	3	84	784	2252
21–25	23	1	23	529	529
		33	$1276 = \Sigma xf$		$51282 = \Sigma x^2 f$

$$M = \frac{\Sigma xf}{N} = \frac{1276}{33} = 38.67$$

$$s^2 = \frac{51282}{33} - 1495.36 \qquad s = \sqrt{58.64} = 7.66$$

$$= 58.64$$

Adding and subtracting M for each score would not change the value of s_C^2 at all. However, if we did so, we could expand each squared deviation as follows:

$$(X_i - C)^2 = (X_i - M + M - C)^2$$
$$= (X_i - M)^2 + 2(X_i - M)(M - C) + (M - C)^2.$$

Substituting into the expression for s_C^2, and distributing the summation, we have

$$s_C^2 = \sum_i \frac{(X_i - M)^2}{N} + 2 \sum_i \frac{(X_i - M)(M - C)}{N} + \sum_i \frac{(M - C)^2}{N}.$$

Notice that $(M - C)$ is the same for every score summed, so that

$$s_C^2 = \sum_i \frac{(X_i - M)^2}{N} + 2(M - C) \sum_i \frac{(X_i - M)}{N} + (M - C)^2.$$

However, the first term on the right above is simply s^2, the variance about the mean, and the second term is zero, since

$$\sum_i \frac{(X_i - M)}{N} = \sum_j \frac{X_i}{N} - \sum_i \frac{M}{N} = M - \frac{1}{N} \sum_i M = M - \frac{NM}{N}$$
$$= M - M = 0.$$

On making these substitutions, we find

$$s_C^2 = s^2 + (M - C)^2.$$

Since $(M - C)^2$ is a squared number it can be only positive or zero, and so s_C^2 must be greater than or equal to s^2. The value of s_C^2 can be equal to s^2 only when M and C are the same. In short, we have shown that the sample variance calculated about the sample mean will always be smaller than about any other point.

Similarly, if we were to consider the average *absolute* deviation rather than the absolute squared deviation, we would find that the average absolute deviation is smallest when the deviations are taken from the sample median:

$$\text{Sample A.D.} = \sum_i \frac{|X_i - Md|}{N}. \tag{5.17.8}$$

In addition to the sample variance and the sample absolute deviation, there are other measures of dispersion, such as the range (the difference between the highest and lowest observations) and the interquartile range (the difference between the .75 fractile and the .25 fractile of the frequency distribution). In general, however, the sample variance is the most useful of these measures of dispersion. It should be noted at this point that in Chapter 6 we will have cause to use a "modified" sample variance, the modification being a change in the denominator from N to $N - 1$:

$$\hat{s}^2 = \frac{\sum (X_i - M)^2}{N - 1}.$$ (5.17.9*)

To differentiate this "modified" sample variance from s^2, we shall denote it by $\hat{s}^2$. Strictly speaking, this is not an "average" squared deviation; we will see in Chapter 6 why it is sometimes considered more useful than s^2. Incidentally, computing forms can be derived for $\hat{s}^2$ as well as for s^2:

$$\hat{s}^2 = \frac{N\Sigma X_i^2 - (\Sigma X_i)^2}{N(N - 1)},$$ (5.17.10)

or

$$\hat{s}^2 = \frac{\sum_i X_i^2}{N - 1} - \left(\frac{N}{N - 1}\right)M^2.$$ (5.17.11)

5.18 POPULATIONS, PARAMETERS, AND STATISTICS

Formerly we have called the entire set of elementary events the *sample space*, since this term is useful and current in probability theory. However, it is often more common to find the word **population** used to mean the totality of potential units for observation; these potential units for observation are very often real or hypothetical sets of people or objects, and *population* provides a very appropriate alternative to *sample space* in such instances. Nevertheless, whenever the term "population" is used in the following, we shall mean only the sample space of elementary events from which samples are drawn.

Given a population of potential observations, we shall think of the particular numerical score assigned to any particular unit observation as a value of a random variable; the distribution of this random variable is the **population distribution.** This distribution will have some mathematical form, with a mean μ, a variance σ^2, and all the other characteristic features of any distribution. If you like, you can usually think of the population distribution as a frequency distribution based upon some large but finite number of cases. However, population distributions are almost always discussed as though they were theoretical probability distributions; the process of random sampling of single units with replacement insures that the long-run relative frequency of any value of the random variable is the same as the probability of that value. Later we shall have occasion to idealize the population distribution and treat it as though the random variable were continuous. This is virtually impossible to be true of real-world observations, but we shall assume that it is "true enough" as an approximation to the population state of affairs.

Population values such as μ and σ^2 will be called **parameters of the population** (or sometimes, true values). Strictly speaking, a parameter is a value entering as an arbitrary constant in the particular function rule for a probability distribution, although the term is used more loosely to mean any value summarizing the population distribution. Just as parameters are characteristics of populations, so are **statistics** summary properties of samples. The sample mean M, the sample variance s^2, and so on, are examples of statistics. The main business of this chapter is to show the relation of statistics, particularly the sample mean and variance, to their parameter counterparts, the population mean and variance.

5.19 SAMPLING DISTRIBUTIONS

In actual practice, random samples seldom consist of single observations. Almost always some N observations are drawn at random from the same population. Since this is true, we must designate another kind of theoretical distribution, called a **sampling distribution.**

A sampling distribution is a theoretical probability distribution that shows the functional relation between the possible values of some summary characteristic of N cases drawn at random and the probability (density) associated with each value of the summary characteristic over all possible samples of size N from a particular population.

In general, the sampling distribution of values for a sample statistic will not be the same as the population distribution. We shall see, however, that *the sampling distribution always depends in some specifiable way upon the population distribution.*

We have already used sampling distributions in the preceding chapters. For example, a binomial distribution is a sampling distribution. Recall that a binomial distribution is based on a two-category population, or a Bernoulli process. A sample of N cases is drawn at random with replacement from such a process, and the number (or proportion) of successes is calculated for each sample. Then the binomial distribution is the sampling distribution showing the relation between each possible sample result and the theoretical probability of occurrence.

Another example of a sampling distribution already introduced is the multinomial distribution. Here, the population has several event categories. A sample of size N is drawn at random with replacement and summarized according to how many of each measurement class occur. The multinomial distribution gives the probability of each possible sample.

Other examples of sampling distributions will now be given. A most important distribution we shall employ is the sampling distribution of the mean. Here, samples of N cases are drawn at random from some population and each observation is measured numerically. The population distribution has some mean μ, and for each sample drawn the sample mean M is calculated. **The theoretical distribution that relates the possible values of the sample mean to the probability (density) of each over all possible samples of size N is called the sampling distribution of the mean.** Furthermore, there is a population variance, σ^2. For each sample of size N drawn, the sample variance s^2 is found. The theoretical distribution of sample variances in relation to the probability of each is the sampling distribution of the variance. By the same token, *the sampling distribution of any summary characteristic* (mode, median, range, etc.) *of samples of N cases may be found, given the population distribution and the sample size N. This fact is all-important for the theory of statistics, as it provides ways to make inferences about population parameters on the basis of random samples, in terms of the probability of a sample's value arising by chance from a certain population.* Much of the material in the remainder of this book will deal with various ways of using theoretical sampling distributions to make such inferences.

5.20   CHARACTERISTICS OF SINGLE-VARIATE SAMPLING DISTRIBUTIONS

A sampling distribution is a theoretical probability distribution, and like any such distribution, is a statement of the functional relation between the values or intervals of values of some random variable and probabilities. Sampling distributions differ from population distributions in that *the random variable is always the value of some statistic based on a sample of N cases,* such as the sample mean, the sample variance, the sample median, etc. Thus, a plot of a sampling distribution, such as that in Figure 5.20.1, always

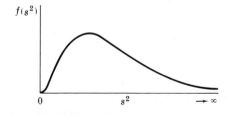

Figure 5.20.1. A Theoretical Sampling Distribution of s^2 for $N = 7$

has for the abscissa (or horizontal axis) the different sample statistic values that might occur. Figure 5.20.1, for example, shows a theoretical distribution for sample variances for all possible samples of size 7 drawn from a particular population. Any point on the horizontal axis is a possible value of a sample variance, and the height of the curve on the vertical axis gives the probability density $f(s^2)$, for that particular value.

Like population distributions, sampling distributions may be either continuous or discrete. The binomial distribution is discrete, although in applied problems it is sometimes treated as though it were continuous. Most of the commonly encountered sampling distributions based on a continuous population distribution will be continuous.

It should be clear that for any given population sampled at random there will be any number of different sampling distributions, depending upon the particular statistic used to describe the sample. Thus there will be a theoretical sampling distribution of means, a different distribution that is the theoretical sampling distribution of variances, still another distribution that is the sampling distribution of medians, and so on, for as many ways as a statistic can be computed for a sample of N cases. Obviously, we should seldom choose to describe a sample in *some* ways, such as describing the entire sample by the square root of the score of the third case observed; even so, the theoretical distributions of even such esoteric statistics can be worked out.

It is a good idea to reiterate at this point; so far three distinct kinds of distributions have been introduced. The first was the **population distribution,** which is a theoretical distribution describing the probability associated with various values of a random variable. The second kind of distribution with which we dealt was the **frequency distribution,** or the **sample distribution.** This is a distribution summarizing a set of data by describing the frequencies associated with various classes, based on a randomly chosen subset of a population. Finally, there was the **sampling distribution.** This is a theoretical probability distribution which relates various values or intervals of values of some sample statistic to their probabilities of occurrence over all possible samples of size N given a specific population distribution. In order to know any sampling distribution exactly, we must specify the population distribution.

When the sample size is one and the statistic is simply the observed value of the random variable, the sampling distribution will be exactly like the population distribution. In general, however, the sampling distribution will *not* be identical to the population distribution. The sampling distribution must be known before the probability of the occurrence of a particular sample statistic value or interval of values can be calculated. It is very important to bear these distinctions in mind, since all three kinds of distributions will be considered in the sections to follow.

5.21   THE MEAN AND VARIANCE OF A SAMPLING DISTRIBUTION

Except for a few statistical curiosities that need not concern us here, any random variable will have a determinable mean and variance. Since a sample statistic is a random variable, the mean and variance of any sampling distribution are defined in the usual way. That is, let G be any sample statistic; then if the sampling distribution of G is discrete, its expectation or mean is

$$E(G) \; = \; \mu_G \; = \; \Sigma g P(g).$$

If the variable G is continuous, then

$$E(G) \; = \; \mu_G \; = \; \int_{-\infty}^{\infty} g f(g) \, dg,$$

just as for any other continuous random variable.

In the same way, the variance of a sampling distribution for some statistic G can be defined:

$$\sigma_G^2 \; = \; E(G - \mu_G)^2, \tag{5.21.3*}$$

or

$$\sigma_G^2 \; = \; E(G^2) - [E(G)]^2. \tag{5.21.4*}$$

The variance of the statistic G gives a measure of the dispersion of particular sample values about the average value of G over all possible samples of size N. The standard deviation σ_G of the sampling distribution reflects the extent to which sample G values tend to be *unlike* the expectation, or are *in error*. To aid in distinguishing the standard deviation of a sampling distribution from the standard deviation of a population distribution, a standard deviation such as σ_G is usually called the *"standard error"* of the statistic G. When one speaks of the standard error of the mean, he is referring to the standard deviation of the distribution of possible sample means for all possible samples of size N drawn from a specified population. Similar meanings hold for the standard error of the median, the standard error of the standard deviation, the standard error of the range, and so on.

For an example of the determination of the mean and variance of a sampling distribution, consider the sampling distribution of the sample mean, M. Here

$$G \; = \; M \; = \; \frac{\sum_{i=1}^{N} X_i}{N}.$$

Using the algebra of expectations, we find that

$$E(M) = E \frac{\sum\limits_{i=1}^{N} X_i}{N} = \frac{1}{N} E(\sum\limits_{i=1}^{N} X_i) = \frac{1}{N} \sum\limits_{i=1}^{N} E(X_i).$$

But any $E(X_i)$ is equal to μ by definition, for the observations are assumed to have been taken at random from the same population. Thus,

$$E(M) = \frac{1}{N} \sum\limits_{i=1}^{N} \mu = \frac{N\mu}{N} = \mu. \qquad (5.21.5^*)$$

This statement can be interpreted as follows: **the mean of the sampling distribution of means is the same as the population mean.** However, it is not true that the sampling distribution of the mean will be the same as the population distribution; in fact, these distributions will ordinarily be quite different, depending particularly on sample size. In the first place, the variance of the sampling distribution of means will not be the same as σ^2, but instead will be smaller than σ^2 for samples of size 2 or larger.

Intuitively is seems quite reasonable that the larger the sample size the more confident we may be that the sample mean is a close estimate of μ. Now we can put that intuition on a firm basis by looking into the effect of sample size on the variance and standard deviation of the distribution of sample means. Using the rules concerning variances presented in Section 3.21,

$$\text{var } (M) = \text{var } \frac{\sum\limits_{i=1}^{N} X_i}{N} = \frac{1}{N^2} \text{var } (\sum\limits_{i=1}^{N} X_i) = \frac{1}{N^2} \sum\limits_{i=1}^{N} \text{var } (X_i).$$

But any var (X_i) is equal to σ^2 by definition. Therefore,

$$\text{var } (M) = \frac{1}{N^2} \sum\limits_{i=1}^{N} \sigma^2 = \frac{N\sigma^2}{N^2} = \frac{\sigma^2}{N}.$$

The variance of the sampling distribution of means for independent samples of size N is always the population variance divided by the sample size:

$$\sigma_M{}^2 = \frac{\sigma^2}{N}. \qquad (5.21.6^*)$$

It follows directly that the standard error of the sample mean is given by

$$\sigma_M = \sqrt{\sigma_M{}^2} = \frac{\sigma}{\sqrt{N}}. \qquad (5.21.7^*)$$

This is a most important fact, and it gives direct support to our feeling that large samples produce better estimators of the population mean than do small samples. When the sample size is only 1, then the variance of the sampling distribution is exactly the same as the population variance. If, however, the sample mean is based on two cases, $N = 2$, then the sampling variance is only $1/2$ as large as σ^2. Ten cases give a sampling distribution with variance only $1/10$ of σ^2, $N = 500$ gives a sampling distribution with variance $1/500$ of σ^2, and so on. If the sample size approaches infinity, then σ^2/N approaches zero. **If the sample is large enough to embrace the entire population, there is no difference between the sample mean and μ.**

In general, the larger the sample size, the more probable it is that the sample mean comes arbitrarily close to the population mean. This fact is often called **the law of large numbers,** and it is closely allied both to Bernoulli's theorem and the Tchebycheff inequality.

Given the Tchebycheff inequality and the variance of the mean, it is easy to see that the law of large numbers must be true. From the Tchebycheff inequality (3.27.1) it is true that for any random variable,

$$P(\,|\,X - \mu\,|\, < k) \geq 1 - \frac{\sigma^2}{k^2}, \qquad k > 0. \qquad (5.21.8)$$

Let the random variable be M, and the variance be $\sigma_M{}^2$. Then

$$P(\,|\,M - \mu\,|\, < k) \geq 1 - \frac{\sigma_M{}^2}{k^2}. \qquad (5.21.9^*)$$

When N becomes very large, $\sigma_M{}^2$ approaches zero. Thus, regardless of how small k is, the probability approaches 1 that the value of M will be within k units of μ when sample size grows large.

5.22 STATISTICAL PROPERTIES OF NORMAL POPULATION DISTRIBUTIONS—INDEPENDENCE OF SAMPLE MEAN AND VARIANCE

As suggested earlier, the normal distribution has mathematical properties that are most important for theoretical statistics. For the moment, we are going to discuss only two of these general properties, both of which will be useful to know in later sections. The first has to do with the independence of the sample mean and variance, and the second concerns the distribution of combinations of random variables.

Any sample consisting of N independent observations of the same random variable provides both a sample mean M and a sample variance s^2 (or a modified sample variance $\hat{s}^2$). These two values (M, s^2) obtained from any sample can be thought of as a joint event. But are these two estimates independent? That is, is the conditional distribution of s^2 given M the same as the marginal distribution of s^2, and is the distribution of M given s^2 like the marginal distribution of M?

The answer to these questions is provided by the following important principle:

Given random and independent observations, the sample mean
M and the sample variance (either $\hat{s}^2$ or s^2) are independent
if and only if the population distribution is normal.

The information contained in the sample mean in no way dictates the value of the sample variance, and vice versa, when a normal population is sampled. Furthermore, **unless the population actually is normal, these two sample statistics are not independent across samples.**

This is a most important principle, since a great many problems concerned with a population mean can be solved only if one knows something about the value of the population variance. At least an estimate of the population variance is required before particular sorts of inferences about the value of μ can be made. Unless the estimate of the population variance is statistically independent of the estimate of μ made from the sample, no simple way to make these inferences may exist. This question will be considered in more detail in Chapters 6 and 7. For the moment, suffice it to say that this principle is one of the main reasons for statisticians' assuming normal population distributions.

5.23 DISTRIBUTIONS OF LINEAR COMBINATIONS OF VARIABLES

So far we have emphasized the sample mean as a description of a particular set of data and as an estimator of the corresponding population mean. However, in many situations the sample mean and the mean of the population do not really tell the statistician what he wishes to find out from the data. It well may be that other ways of weighting and summing the scores or means obtained in one or more samples will answer particular questions about the phenomena under study. Therefore, we need to know something about the sampling distributions of weighted combinations of sample data.

Such weighted sums of sample scores are thought of as values of **linear combinations of random variables.** As a very simple case of a linear combination, imagine two distinct random variables, labeled X_1 and X_2 to show that their values need not be the same. We draw samples of two observations at a time, obtaining one value of X_1 and one of X_2. Then we combine these two numbers into a new value Y in some way such as

$$Y = 3X_1 - 2X_2.$$

We continue to sample X_1, X_2 pairs of values, and each time we combine the results in the same way, turning each pair of values into a single combined value. This gives a new random variable Y. The range of possible values of Y depends, as you can see, on the ranges of X_1 and X_2. Furthermore, over all possible such samples the probabilities of the various Y values must depend on the *joint* probabilities of (X_1, X_2) pairs. Over all possible samples of (X_1, X_2) pairs drawn at random there is some probability distribution of Y.

In general, given any n random variables, $(X_1, \cdots, X_n)$, a *linear combination* of values of those variables is a weighted sum,

$$Y = c_1X_1 + c_2X_2 + \cdots + c_nX_n = \sum_{i=1}^{n} c_iX_i, \qquad (5.23.1^*)$$

where $(c_1, \cdots, c_n)$ is any set of n real numbers (not all zero) used as weights.

In short, each sample produces a value for each of n random variables. These are weighted and summed in the same way for each sample to produce a new random variable Y, the linear combination.

The weights in a linear combination can be any set of n real numbers, provided that at least one weight is not zero. For example, for two random variables, a linear combination might be

$$Y = X_1 + X_2,$$

where $c_1 = 1$ and $c_2 = 1$; or

$$Y = 7X_1 - 10X_2,$$

where $c_1 = 7$ and $c_2 = -10$; or

$$Y = -(4/5)X_1 + (7/8)X_2,$$

with $c_1 = -4/5$, $c_2 = 7/8$, and so on.

Perhaps values for six random variables are represented in a sample, and the linear combination chosen is

$$Y = 5X_1 - 3X_2 + 7X_3 + (3/4)X_4 - .8X_5 + 0X_6.$$

Here, $(c_1, \cdots, c_6) = (5, -3, 7, 3/4, -.8, 0)$.

One very familiar linear combination of random variables is the mean of a random sample: given that X_1 is the value of the first case, X_2 that of the second, and so on, then

$$M = \frac{1}{N}X_1 + \frac{1}{N}X_2 + \cdots + \frac{1}{N}X_N = \frac{\Sigma X_i}{N},$$

where each value obtained is weighted by $1/N$ and the sum taken. This linear combination is also a random variable, and its distribution is the sampling distribution of the mean.

Now for an important principle involving the normal distribution:

Given some n independent random variables $X_1, \cdots, X_n$, each normally distributed, then any linear combination of these variables is also a normally distributed random variable.

In other words, if one takes samples, obtaining n independent values in each, and if the random variable represented by each value has a normal distribution, then any weighted sum of those values gives another normally distributed random variable. Thus, if X_1 and X_2 are independent and have normal distributions, then so does the variable Y, where

$$Y = 3X_1 - 2X_2,$$

or any other variable obtained by weighting and summing X_1 and X_2 values in some constant way over all samples.

An immediate and useful consequence of this principle is this:

Given random samples of N independent observations each drawn from a normal population, then the distribution of the sample means is normal, irrespective of the size of N.

We have just seen that the sample mean of N cases is a linear combination, where the score of case 1, representing random variable 1, is given the weight $1/N$, and so on for all other scores. Hence, if the N observations are independent of each other and are drawn from a normal population, the sampling distribution of the mean is exactly normal, regardless of now large or small N may be. As we saw in Section 5.21, the standard error of this distribution, σ_M, will be smaller the larger the N, and so the particular probabilities of various sample values will differ with N. On the other hand, the distribution *rule*, given μ and σ_M, will be that for the normal function.

We are going to make a great deal of use of this principle. When we can assume that a population with known μ and σ has a normal distribution a major problem in statistical inference is solved, since we can then give a probability to a sample mean's falling into any interval also in terms of a normal distribution. This will be illustrated in the next chapter.

5.24   MEANS AND VARIANCES OF LINEAR COMBINATIONS

While on the subject of linear combinations, we should look into the question of the mean and variance of the sampling distribution of a linear combination of random variables. These two principles do not depend upon the normal distribution, however. **Given n independent random variables and some linear combination,**

$$Y = c_1 X_1 + \cdots + c_n X_n = \sum_{i=1}^{n} c_i X_i,$$

the expected value of Y over all random samples is

$$E(Y) = E(\Sigma c_i X_i) = \Sigma E(c_i X_i) = \Sigma c_i E(X_i)$$
$$= c_1 E(X_1) + c_2 E(X_2) + \cdots + c_n E(X_n). \quad (5.24.1^*)$$

The expectation of any linear combination is the same linear combination of the expectations.

This follows quite simply from the algebra of expectations, since the expectation of a sum is always a sum of expectations, and the expectation of a constant times a variable is the constant times the expectation.

A special case of this principle gives us the following:

Given n independent sample values from the same distribution with mean μ, then the expectation of any linear combination of those sample values is

$$E(Y) = \mu(c_1 + c_2 + \cdots + c_n) = \mu \sum_{i=1}^{n} c_i. \quad (5.24.2^*)$$

This principle accords with our earlier result that the expectation of the sample mean is the population mean:

$$E(M) = E\left(\frac{1}{N} X_1 + \cdots + \frac{1}{N} X_N\right) = \mu\left(\frac{1}{N} + \frac{1}{N} + \cdots \frac{1}{N}\right) = \mu.$$

However, suppose that samples of two cases were drawn, and the linear combination were

$$Y = X_1 - X_2 .$$

In this instance it would not be true that the expectation of this combination equals μ; instead we would have $E(Y) = \mu(1 - 1) = 0$.

As a concrete example, suppose that samples of two cases were observed, one case being a male drawn from the population of all males, with $\mu_1 = 25$, and the other a female drawn from the female population, with $\mu_2 = 27$. The score of the male on an examination is represented by X_1 and that of the female by X_2. For each sample, the difference between the male's and the female's scores is taken:

$$Y = X_1 - X_2.$$

Then, over all such samples, the expectation is

$$E(Y) = E(X_1) - E(X_2) = \mu_1 - \mu_2 = -2.$$

Furthermore, if the scores of the two cases are independent, and if both populations have a normal distribution, the sample difference Y is normally distributed as well.

The variance of the sampling distribution of any linear combination can also be found by the following rule:

Given n independent random variables with variances σ_1^2, σ_2^2, ..., σ_n^2, respectively, and the linear combination

$$Y = c_1X_1 + c_2X_2 + \ldots + c_nX_n = \sum_{i=1}^{n} c_iX_i,$$

then the distribution of Y has variance given by

$$\sigma_Y^2 = \text{var}\,(\Sigma c_iX_i) = \Sigma \,\text{var}\,(c_iX_i) = \Sigma c_i^2 \,\text{var}\,(X_i) \qquad (5.24.3^*)$$

$$= c_1^2\sigma_1^2 + c_2^2\sigma_2^2 + \cdots + c_n^2\sigma_n^2.$$

The variance of a linear combination of independent random variables is a weighted sum of their separate variances, each weight being the *square* of the original weight given the variable.

Thus, in the preceding example, if the population of males has a variance of 10, and that of females a variance of 12, the variance of the difference, $X_1 - X_2$, is

$$\sigma_Y^2 = (1)^2\sigma_1^2 + (-1)^2\sigma_2^2 = 10 + 12 = 22.$$

As a more complicated example, suppose that three independent observations at a time are made, drawn respectively from populations represented by the three random variables

$$X_1, \text{ with } \mu_1 = 20, \ \sigma_1^2 = 5$$

$$X_2, \text{ with } \mu_2 = 16, \ \sigma_2^2 = 9$$

$$X_3, \text{ with } \mu_3 = 25, \ \sigma_3^2 = 7.$$

If all three random variables are normally distributed, what will the distribution of the linear combination

$$Y = \frac{X_1}{2} + \frac{X_2}{2} - X_3$$

be like? In the first place, we know from the principle of Section 5.23 that Y will be normally distributed. Second, the mean of the distribution of Y will be

$$E(Y) = \frac{\mu_1}{2} + \frac{\mu_2}{2} - \mu_3 = 10 + 8 - 25 = -7.$$

The variance of Y must be

$$\sigma_Y{}^2 = \left(\frac{1}{2}\right)^2 \sigma_1{}^2 + \left(\frac{1}{2}\right)^2 \sigma_2{}^2 + (-1)^2 \sigma_3{}^2$$

$$= \frac{5 + 9}{4} + 7 = 10.5.$$

Under these conditions, the principles of Sections 5.23 and 5.24 allow us to specify the distribution of Y completely, and we could proceed to evaluate the probability that an obtained value of Y falls into any given interval.

The variance of the sampling distribution of the mean, found in Section 5.21 to be σ^2/N, is a special case of Equation (5.24.3). Here, samples of N independent observations are made from the same distribution, with some true mean μ and with some true variance σ^2. The sample mean itself is, as we have seen, a linear combination of the scores, each weighted by $1/N$. Then, by the application of the principle,

$$\sigma_M{}^2 = \left(\frac{1}{N}\right)^2 \sigma^2 + \left(\frac{1}{N}\right)^2 \sigma^2 + \cdots + \left(\frac{1}{N}\right)^2 \sigma^2$$

$$= \frac{N}{N^2} \sigma^2 = \frac{\sigma^2}{N}.$$

You are probably wondering why anyone would be interested in linear combinations other than the sample mean. Actually, forming linear combinations of the data is one way of answering particular questions about what goes on in an experiment based on two or more samples. Each item of information about different experimental treatments that the experimenter gains from his data usually corresponds to one way of weighting and summing means of the various samples. The linear combination applied to the set of sample means gives an estimate of the same combination applied to the corresponding set of population means. Thus, the study of the sampling distribution of linear combinations forms an important part of the theory of sampling.

The most important point that should emerge from this discussion of linear combinations is that, given normal population distributions, the sampling distribution of the mean or of any other linear combination of scores or of sample means must be normal. Thus, the assumption of a normal distribution for the population settles the question of the exact form of the sampling distribution for any linear combination of scores in the sample data, including the mean. This is true quite irrespective of the sample size. Do not lose sight of the fact that the size of the sample does, however, determine how small the variance of the sampling distribution will be.

5.25   THE CENTRAL LIMIT THEOREM

It is quite common for the statistician to be concerned with populations where the distribution should *definitely not be normal*. We may know this either from empirical evidence about the distribution, or because some theoretical issue makes it impossible for the random variable to be normally distributed. An illustration is a distribution of the annual income of individuals, which is generally severely positively skewed. The normal distribution is symmetric, so the assumption of a normal distribution does not make sense in such instances.

Nonetheless, very often an inference must be made about the mean of such a population. To do this effectively the statistician needs to know the sampling distribution of the mean, and to know this exactly, he has to be able to specify the particular form of the population distribution. However, if we had enough evidence to permit this, we would likely have an extremely good estimate of the population mean in the first place, and would not need any other statistical methods!

The way out of this apparent impasse is provided by the central limit theorem, which can be given an approximate statement as follows:

If a population has a finite variance σ^2 and mean μ, then the distribution of sample means from samples of N independent observations approaches a normal distribution with variance σ^2/N and mean μ as sample N increases. When N is very large, the sampling distribution of M is approximately normal.

Absolutely nothing is said in this theorem about the form of the population distribution. Regardless of the population distribution, if sample size N is large enough, the normal distribution is a good approximation to the sampling distribution of the mean. This is the heart of the theorem, since, as we have already seen, the sampling distribution will have a variance σ^2/N and mean μ for any N.

The sense of the central limit theorem is illustrated by Figures 5.25.1 to 5.25.4. The solid curve in Figure 5.25.1 is a very skewed population distribution in standard form, and the dotted curve is a standardized normal distribution. The other figures show the standardized form of the sampling distribution of means for samples of size 2, 4, and 10 respectively from this population, together with the corresponding standardized normal distributions.

Even with these relatively small sample sizes it is obvious that each increase in sample size gives a sampling distribution more nearly symmetric and tending more toward the normal distribution with the same mean and

Figure 5.25.1. A Negatively Skewed Population Distribution with a Comparable Normal Distribution

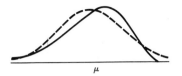

Figure 5.25.2. A Sampling Distribution of M for $N = 2$ with a Comparable Normal Distribution

variance. This symmetry increases with increasing sample size until, in the limit, the normal distribution is reached.

It must be emphasized that in most instances the tendency for the sampling distribution of M to be like the normal distribution is very strong, even for samples of moderate size. Naturally, the more similar to a normal distribution the original population distribution, the more nearly will the sampling distribution of M be like the normal distribution for any given sample size. However, even extremely skewed or other nonnormal distributions may yield sampling distributions of M that can be approximated quite well by the normal distribution for samples of at least moderate size. In the examples shown in Figures 5.25.1 to 5.25.4, the correspondence between the exact probabilities of intervals in the sampling distribution and intervals in the normal distribution is fairly good even for samples of only 10 observations; for a rough approximation, the normal distribution probabilities might be useful even here in some statistical work. In a great many instances, a sample size of thirty or more is considered large enough to permit a satisfactory use of normal probabilities to approximate the unknown exact probabilities associated with the sampling distribution of M. Thus, even though the central limit theorem is actually a statement about what happens *in the limit* as N approaches an infinite value, the principle at work is so strong that in many instances the theorem is practically useful even for moderately large samples.

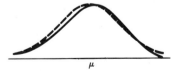

Figure 5.25.3. A Sampling Distribution of M for $N = 4$ with a Comparable Normal Distribution

Figure 5.25.4. A Sampling Distribution of M for $N = 10$ with a Comparable Normal Distribution

The mathematical proof of this theorem is extremely advanced, and many eminent mathematicians over the centuries contributed to its development before it was finally proved in full generality. However, some intuitive feel for why it should be true can be gained from the following example. Here we will actually work out the sampling distribution of the mean for a special and very simple population distribution.

Imagine a random variable with the following distribution:

x	$P(x)$
5	1/3
4	1/3
3	1/3

The μ of this little distribution is 4. However, instead of dealing directly with this random variable, let us consider the deviation d from μ, or

$$d = (X - \mu).$$

The distribution of d has a form identical to the distribution of X itself, although the mean d is zero. Since X can assume only three values, each equally probable, there are three possible deviation values of d, also equally probable.

Suppose that sample observations were taken two at a time, independently and at random. The deviation value of the first observation made is d_1 and that of the second d_2. Corresponding to each joint event (d_1, d_2) there is a mean deviation, $(d_1 + d_2)/2$, which is equal to $M - \mu$. The following table shows the mean deviation value that is produced by each possible joint event:

d_2

	-1	0	1	
1	0	.5	1	
0	−.5	0	5	
−1	−1	−.5	0	
	-1	0	1	d_1

Since X_1 and X_2 are independent, then d_1 and d_2 are also independent, according to Section 3.11. The probability associated with each cell in the table is thus $P(d_1)P(d_2)$, or 1/9.

Now let us find the probabilities of the various values of mean d. The value -1 can occur in only one way, and so its probability is 1/9. On the other hand, $-.5$ can occur in two mutually exclusive ways, giving a probability of 2/9. In this way we can find the other probabilities, and form the distribution of the mean d and the mean X values, as follows:

M	$Mean\ d$	$P(M) = P(mean\ d)$
5.0	1.0	1/9
4.5	.5	2/9
4.0	0	3/9
3.5	$-.5$	2/9
3.0	-1.0	1/9

where $M = $ mean $d + \mu$. Now notice that whereas in the original distribution there was no distinct mode, in this sampling distribution there *is* a distinct mode at 4, with exact symmetry about this point. What causes the sampling distribution to differ from the population distribution in this way? Look at the table of mean deviations corresponding to joint events once again: there are simply more possible ways for a sample to show a small than an extreme deviation from μ. There are more ways for a sample joint event to occur where the deviations tend to "cancel out" than where deviations tend to cumulate in the same direction.

Exactly the same idea could be illustrated for any discrete population distribution, whatever its form. The small and "middling" deviations of M from μ always have a numerical advantage over the more extreme deviations. This superiority in number of possibilities for a small mean deviation increases as N is made larger. Regardless of how skewed or otherwise irregular a population distribution is, by increasing sample size one can make the advantage given to small deviations so big that it will overcome any initial advantage given to extreme deviations by the original population distribution. The numerical advantage given to relatively small deviations from μ effectively swamps any initial advantage given to other deviations by the form of the population distribution, as N becomes large.

Perhaps this simple example gives you some feel for why it is that the sampling distribution of the mean approaches a unimodal, symmetric form. Of course, this is a far cry from showing that the sampling distribution must approach a *normal* distribution, which is the heart of the central limit theorem. Nevertheless, the basic operation of chance embodied in this theorem is of this general nature: for large N it is much easier to get a small deviation of M from μ than a large one.

5.26 THE CENTRAL LIMIT THEOREM AND LINEAR COMBINATIONS OF MEANS

As suggested above, it is very common for an experimenter to be interested in the means of several samples put into some kind of weighted sum, or linear combination. The central limit theorem applies to such weighted combinations as well:

Given K independent samples, containing $N_1, N_2, \ldots, N_K$ independent observations respectively, then the sampling distribution of any linear combination of means of those samples approaches a normal distribution as the size of each sample grows large.

It is easy to see that the central limit theorem applies to the sampling distribution of each mean separately as the sample size increases. In the limit, each mean has a normal sampling distribution. Then by the principle described in Section 5.23, any linear combination of normal random variables is normal, so that the sampling distribution of the linear combination of means approaches a normal distribution as the sample sizes all grow large.

This is but one of the most elementary extensions of the central limit theorem. It can be shown that a great many of the sampling distributions of all sorts of sample characteristic also approach a normal distribution with increasing sample size. However, in elementary work we will have most occasion to use the principle as it applies to means.

5.27 STANDARDIZED RANDOM VARIABLES CORRESPONDING TO SAMPLE MEANS

The standard error of the sample mean is

$$\sigma_M = \sqrt{\sigma_M{}^2} = \frac{\sigma}{\sqrt{N}}, \qquad (5.27.1^*)$$

so that when a sample mean is put into standardized form, we have

$$z_M = \frac{M - \mu}{\sigma_M} = \frac{M - \mu}{\sigma/\sqrt{N}}. \qquad (5.27.2^*)$$

The larger the standardized value of a mean, the *less likely* one is to observe a mean this much or more deviant from μ. Any given degree of departure of M from μ corresponds to a larger absolute standardized value, and hence a less probable result, the larger the sample size N.

For example, suppose that in sampling from some population a sample mean was found differing by 10 points from the population mean. Suppose that σ were 5 and the sample size were 2. Then the standard value z_M would be

$$z_M = \frac{10}{5/\sqrt{2}} = 2.8,$$

disregarding sign. The Tchebycheff inequality tells us that means this deviant or more so from expectation can occur with probability no greater than about $1/(2.8)^2$ or .13.

Suppose, however, that the sample size had been 200. Here

$$z_M = \frac{10}{5/\sqrt{200}} = 28.28,$$

so that a sample M deviating this much or more from expectation could occur only with probability no greater than about $1/(28.28)^2$ or .0013. An extent of deviation of M from the true mean that could occur relatively often for small samples is rare when the sample size is large. Notice that knowing the standard error of the sample mean is essential if we are going to judge the agreement between a sample and a population mean in terms of the probability of occurrence for a given extent of deviation. For the moment we are assuming that this information is simply given to us.

5.28   THE USES OF FREQUENCY AND SAMPLING DISTRIBUTIONS

Frequency distributions are useful primarily in a descriptive sense. That is, they describe the information contained in a particular sample result, in a particular set of observations. When the statistician must report the results of a sample, he generally presents a frequency distribution along with some summary measures such as the sample mean and sample variance. In performing this task, the statistician should attempt, insofar as possible, to report the data impartially. There are various ways to specify a frequency distribution and various summary measures of such distributions; these different presentations of the results may give entirely different impressions to the reader. A simple example of this was presented in Section 5.11, in which a slight change in the scales on the axes of a graph led to an entirely different impression. The moral is simple: it is very easy to "lie" with statistics, so it is a good idea to be somewhat wary when presented with the results of a statistical study. Conversely, it is important that the statistician attempt to present the data from such a study as impartially as possible. This means, for example, that when reporting the results of a study of family incomes in a certain community, the statistician should report the median income as well as the mean income, since the sample mean is greatly affected by a few very high incomes.

In contrast to frequency distributions, sampling distributions are of prime importance in the theory of statistical inference and decision rather than as descriptive methods. When an estimate of a population quantity is desired, we will investigate qualities of competing estimators which are based on their sampling distributions. When the reasonableness of a hypothesis, or a theory, is of interest, inferences about the hypothesis will be based directly on sampling distributions. Finally, in statistical decision theory, sampling distributions are an important input to the decision-making process. In all of these statistical procedures, we will make extensive use of summary measures (means and standard errors) of sampling distributions, probabilities associated with intervals of sample values, and the entire sampling distribution, which we will usually attempt to specify in functional form.

It should be pointed out that the discussion of sampling distributions in this chapter is incomplete. For example, we have said nothing about the distribution of the sample mean when we do not know the population variance, partly because a discussion of this distribution requires the introduction of the t distribution, which we have put off until the next chapter. Similarly, we have not discussed the distribution of the sample variance because this involves the χ^2 (chi-square) distribution, which will also be introduced in the next chapter. Other sampling distributions of interest include the distribution of the difference of two sample means (either from independent samples or dependent samples), the distribution of the ratio of two sample variances, the distribution of the median, the distribution of the sample correlation coefficient, and so on. In the following chapters, the most important of these sampling distributions will be presented (insofar as they are within the scope of and can be presented at the mathematical level of this book), along with the implications which they hold for statistical inference and decision.

5.29   TO WHAT POPULATIONS DO OUR INFERENCES REFER?

Most statisticians who use inferential statistics rely on the model of simple random sampling. Yet how does one go about getting such a "truly" random sample? It is not easy to do, unless, as in all probability sampling, each and every potential member of the population may somehow be listed. Then, by means of a device such as a random number table, individual members may be assigned to the sample with approximately equal probabilities.

How does one know the population to which the statistical inferences drawn from a sample apply? If random sampling is to be assumed, *the population is defined by the sample and the manner in which it is drawn. The only population to which the inferences strictly apply is that in which individuals have equal likelihood of appearing in the sample.* It should be obvious that simple random samples from one population may not be random samples of another population. For example, suppose that some one wishes to sample American college students. He obtains a directory of college students from a Midwestern university and, using a random number table, takes a sample of these students. He is not, however, justified in calling this a random sample of the population of American college students, although he may be justified in calling this a random sample of students at *that* university. *The population is defined not by what he said, but rather by what he did to get the sample.* For any sample, one should always ask the question, "What is the set of potential cases that could have appeared in my sample with equal probability?" If there is some well-defined set of cases which fits this qualification, then inferences may be made to that population. However, if there is some population whose members could not have been represented in the sample with equal probability, then inferences do not *necessarily* apply to that population when methods based on simple random sampling are used. Any generalization beyond the population actually sampled at random must rest on extrastatistical considerations.

From a mathematical-statistical point of view the assumption of random sampling makes it possible to determine the sampling distribution of a particular statistic given some particular population distribution. If the various values of the statistic can arise from samples having undetermined or unknown probabilities of occurrence, then the statistician has no way to determine the sampling distribution of that statistic. On the other hand, if each and every distinct sample of N observations has equal probability of occurrence, as in simple random sampling, then the sampling distribution of any given statistic is relatively easy to study, given, of course, the population situation. It is a sad fact that if one knows nothing about the probability of occurrence for particular samples of units for observation, he can use very little of the machinery of statistical inference. This is why the assumption of random sampling is not to be taken lightly. All the techniques and theory that we will discuss apply to random samples, and do not necessarily hold for any data collected in any way. In practical situations, the statistician may be hard put to show that a given sample is "truly" random, but he must realize that he is acting *as if* this were a random sample from some well-defined population when he applies methods of statistical inference. Unless this assumption is at least reasonable, the results of inferential methods mean very little, and these methods might as well be omitted. Data that are not the product of random sampling may yield extremely valuable conclusions in their own right, but there is usually little to be gained from the application of inferential methods to such data. Certainly, the application of some statistical method does not somehow magically make a sample random, and the conclusions therefore valid. Inferential methods apply to random samples, and there is no guarantee of their validity in other circumstances.

EXERCISES

1. Why are random samples of such importance in statistics?

2. If one has a fair coin, he can generate his own table of random numbers in the following manner: let $x_1 = 1$ if the coin comes up heads on the first toss, and $x_1 = 0$ if it comes up tails. Similarly, let $x_2 = 1$ if it comes up heads on the second toss and let $x_2 = 0$ if it comes up tails. Define x_3 and x_4 similarly for the third and fourth toss. Now toss the coin four times, and find

$$y = x_1 + 2x_2 + 4x_3 + 8x_4.$$

List the possible values of y and the probabilities for the different values. Would you call y a random number? Suppose that you simply ignored values of y of 10 or greater. Would the resulting numbers be random digits?

3. Distinguish between independent random sampling and nonindependent random sampling. By considering a simple experiment such as the drawing of a series of cards from a well-shuffled deck of cards, give examples of both independent and dependent random sampling. How does this distinction relate to the binomial and hypergeometric distributions?

4. In many institutions of higher education, a student's record is evaluated by a grade point average. For example, 4 points might be given for each semester hour of A, 3 for each hour of B, 2 for each hour of C, 1 for each hour of D, and 0 for each hour of F. These numerical values are multiplied by the number of hours of each grade earned, summed, and divided by the total number of hours to arrive at a grade point average. Thus, a grade point average of near 4.00 indicates a fine record and a grade point average below 2.00 indicates a poor record. In light of the discussion of measurement scales given in the text, discuss the merits of this procedure.

5. What level of measurement would you expect to attain when measuring the following variables?
   (a) temperature
   (b) income
   (c) sex
   (d) height
   (e) brand preference
   (f) academic achievement
   (g) color of hair
   (h) occupation.

6. In your own words, carefully distinguish between a probability distribution, a frequency distribution, and a sampling distribution.

7. What is one assuming about a set of data grouped into a frequency distribution when he: (a) considers the real, rather than the apparent, limits to be the class boundaries; and (b) uses the midpoint of any interval to represent all of the scores grouped into that interval.

8. Comment on the following statement: "The process of data-gathering and reporting often involves the loss or deliberate sacrifice of some potential information."

9. Make up two examples from your area of interest in which a frequency distribution would very likely require the employment of unequal or open class intervals.

10. In one week, thirty patients were admitted to a large state hospital. A list of the patients classified by sex and by tentative diagnosis follows. For each sex form a frequency distribution by type of diagnosis and display this information via a pair of histograms superimposed on the same graph.

Patient	Sex	Diagnosis	Patient	Sex	Diagnosis
1	M	Senile dementia	16	M	Senile dementia
2	F	Schizophrenia	17	F	Senile dementia
3	M	Schizophrenia	18	M	Manic-depressive
4	M	Manic-depressive	19	F	Schizophrenia
5	F	Senile dementia	20	M	Paretic
6	M	Senile dementia	21	F	Schizophrenia
7	M	Schizophrenia	22	M	Schizophrenia
8	F	Manic-depressive	23	F	Unclassified
9	F	Schizophrenia	24	F	Schizophrenia
10	F	Unclassified	25	F	Schizophrenia
11	F	Schizophrenia	26	F	Unclassified
12	F	Senile dementia	27	M	Senile dementia
13	M	Schizophrenia	28	M	Senile dementia
14	M	Manic-depressive	29	F	Schizophrenia
15	M	Schizophrenia	30	M	Paretic

11. The diameters of parts produced by a certain production process are given below. Starting at 0 and using $i = .0625$, construct a frequency distribution for these data. Repeat the procedure with $i = .1250$ and comment on any differences in the two distributions.

.581	.630	.460	.511	.351	.180	.450	.240
.630	.684	.500	.554	.380	.195	.489	.261
.460	.500	.365	.405	.278	.143	.357	.190
.511	.554	.405	.449	.308	.158	.396	.211
.351	.380	.278	.308	.212	.109	.272	.145
.180	.195	.143	.158	.109	.056	.139	.074
.450	.489	.357	.396	.272	.139	.349	.186
.240	.261	.190	.211	.450	.074	.189	.099

12. For the two frequency distributions obtained in Exercise 11, construct the corresponding frequency polygons and cumulative frequency polygons.

13. In a learning experiment, cats were given successive trials at learning a complicated maze. A given cat was trained until he could traverse the maze without error five times in a row. If the cat could not go through the maze five times successfully after 30 trials, training was stopped. The numbers of trials it took each of 36 cats to learn this maze to the criterion is given below. Construct a frequency distribution for these data.

Cat	Trials	Cat	Trials	Cat	Trials
1	5	13	30+	25	16
2	12	14	7	26	5
3	7	15	30+	27	30+
4	15	16	9	28	9
5	9	17	30+	29	30+
6	10	18	10	30	19
7	30+	19	13	31	8
8	30+	20	15	32	20
9	8	21	30+	33	30+
10	30+	22	6	34	17
11	6	23	30+	35	30+
12	30+	24	11	36	22

How would you give a qualitative description of the performance of these animals in the maze on the basis of what this distribution shows?

14. Suppose that thirty-two persons participated in a series of gambles and that the net gain for each person is shown below. Construct a histogram, a frequency polygon, and a cumulative frequency polygon for this set of data.

$2.74	−$1.05	$ .25	−$1.39
4.09	− 1.56	.37	− 2.08
.95	− .36	.09	− .48
2.29	− .88	.21	− 1.17
2.27	− .87	.22	− 1.16
.87	.33	− .08	.44
.20	− .08	.02	− .10
1.16	.44	− .10	.59

15. The incomes of the families in a particular neighborhood are shown below. Construct a histogram with $i = 1000$, another histogram with $i = 5000$, and another histogram with $i = 333$ to represent this data. Compare the three histograms and discuss.

5,500	12,500	8,600	11,000
11,000	8,600	31,000	6,500
8,700	4,700	7,100	7,300
9,200	9,700	11,300	10,200
6,700	8,800	10,600	14,300
15,400	10,100	13,000	16,700
9,900	9,200	9,400	9,500
13,200	9,500	9,800	10,600

16. A box of loose change is known to contain 100 coins of the following denominations: 35 pennies, 50 nickels, and 15 quarters. You draw a random sample of ten coins, with replacement. What is the frequency distribution of coins which you should *expect* to get? What is the probability of obtaining this frequency distribution exactly?

17. Suppose that a group of 1000 college students were classified by sex (700 males, 300 females) and by class in school (400 freshmen, 300 sophomores, 200 juniors, and 100 seniors). If sex is independent of class in school, construct a contingency table showing all possible joint frequencies (such as: the frequency of freshman males). Compare this with the frequencies which were actually observed:

Class in School

	Freshman	*Sophomore*	*Junior*	*Senior*
Male	300	250	100	50
Female	100	50	100	50

18. For the data in Exercise 11, compute the sample mean
    (a) directly from the data
    (b) from the grouped frequency distribution with $i = .0625$
    (c) from the grouped frequency distribution with $i = .1250$.
    Explain any differences in your three answers.

19. For the data in Exercise 11, compute the sample median
    (a) directly from the data
    (b) from the grouped frequency distribution with $i = .0625$
    (c) from the grouped frequency distribution with $i = .1250$.
    Explain any differences in your three answers. Also, compute the mode according to (b) and (c) and compare the three sample means (from Exercise 18), the three sample medians, and the two sample modes as measures of central tendency.

20. When one finds the median of a grouped distribution by interpolation, what is he assuming about the distribution of values *within* a given class interval? How does this assumption show up in the cumulative frequency polygon?

21. How does the value of the sample mean computed from a grouped distribution depend on the particular choice of intervals? Does the error introduced by treating each score as equal in value to the midpoint of its interval have any systematic connection with the extent to which the distribution is symmetric?

22. The scores for twelve students on a particular examination were as follows:

$$
\begin{array}{cccc}
18 & 15 & 19 & 27 \\
13 & 30 & 24 & 11 \\
5 & 16 & 17 & 20
\end{array}
$$

Compute the mean, median, standard deviation, and average absolute deviation.

23. Construct the standardized scores for the twelve students whose raw scores are given in Exercise 22, and then show that the mean of these standardized scores is 0 and the standard deviation of the standardized scores is 1.

24. In a group of 50 boys and 50 girls of high-school age, the number of calories consumed on a particular day by each person can be summarized by the following two frequency distributions:

Class interval	Males	Females
5000–5499	1	1
4500–4999	2	0
4000–4499	4	2
3500–3999	16	5
3000–3499	12	10
2500–2999	7	20
2000–2499	5	8
1500–1999	2	2
1000–1499	1	2
	—	—
	50	50

(a) compute the sample mean, median, and mode for each of the two frequency distributions

(b) compute the sample variance for each of the two frequency distributions.

25. For the data in Exercise 15, find the mean income and the median income. Which do you think is a better measure of central tendency? Why? Also, find the sample variance and standard deviation [*Hint:* To make this easier, first transform the data so that you will not have to work with such large numbers.]

26. Show that

$$
s^2 = \frac{N \Sigma X_i^2 - (\Sigma X_i)^2}{N^2} .
$$

27. Starting with Equation (5.17.9), derive the computational formulas (5.17.10) and (5.17.11).

28. Given the following very simple population distribution, construct the theoretical sampling distribution for the sample mean based on three independent observations from the population.

x	$P(X = x)$
9	1/6
6	1/2
3	1/3

[*Hint*: Enumerate the possible combinations of events for samples of size 3, and then translate these into values of the sample mean.]

29. From the population distribution given in Exercise 28, find the sampling distribution of the sample *median* based on three independent observations from the population. Compare the standard error of the sample median with the standard error of the sample mean.

30. For the population distribution given in Exercise 28, find the expectation and the variance of the *sample mean* from a sample of size three
    (a) directly from the *sampling* distribution of the sample mean
    (b) from the mean and variance of the *population* distribution.

31. For the population distribution given in Exercise 28, find the sampling distribution of the *sample variance* for a sample of three independent observations by enumerating the possible combinations of events for samples of size three and translating these into values of the sample variance.

32. If X is normally distributed with mean 50 and variance 100, Y is normally distributed with mean 80 and variance 25, Z is normally distributed with mean 100 and variance 144, and X, Y, and Z are independent, find the distribution of:

    (a) $X - Y$ (c) $6Z + 4X - 30Y$
    (b) $X + Y - Z$ (d) $(X + Y + Z)/3$.

33. The argument for the normal theory of error, presented in Section 4.23 of the text, was based on the normal approximation to the binomial, assuming the action of a vast number of "influences" that might affect any given numerical observation. However, can one make a rationale for the normal distribution of error, once again imagining the number of influences to be very large, but without depending directly on the binomial assumption?

34. To demonstrate the central limit theorem, draw 100 samples of size five from a random number table and calculate the sample mean for each of the 100 samples. Construct a frequency distribution of sample means. Do the same for 100 samples of size ten and compare the two frequency distributions. Does the central limit theorem appear to be working?

35. To demonstrate the central limit theorem under a skewed population distribution, follow the same procedure as in Exercise 34 but perform the following transformation before computing sample means:

Random number	X
0, 1, 2, 3	0
4, 5, 6	1
7, 8	2
9	3

36. Suppose that X is normally distributed with mean 150 and variance 400. If a sample of 16 independent trials is randomly chosen from the X population, find
    (a) $P(M \geq 150)$
    (b) $P(145 < M < 149)$
    (c) the mean and variance of M.

37. Answer (a)–(c) in Exercise 36 if the sample size is 400, and compare your answers to these obtained when $N = 16$. In each case, how would your answers be affected if the assumption of normality was not applicable?

38. Suppose that samples are taken from two normal populations. The first population has mean 80 and variance 81, and the second population has mean 100 and variance 100. If $N_1 = 9$ is the size of the sample from the first population and $N_2 = 10$ is the size of the sample from the second population, find
    (a) the distribution of $X_1 + X_2$
    (b) the distribution of $X_1 - X_2$
    (c) $P(170 \leq X_1 + X_2 \leq 185)$
    (d) the distribution of M_1, the sample mean of the sample from the first population
    (e) the distribution of M_2, the sample mean of the sample from the second population
    (f) the distribution of $M_1 + M_2$
    (g) the distribution of $M_1 - M_2$
    (h) $P(M_1 - M_2 > -25)$.

39. If a sample of size N_1 is taken from a normal population with mean μ_1 and variance σ_1^2, and a sample of size N_2 is taken from a normal population with mean μ_2 and variance σ_2^2, what is the distribution of the difference between the two sample means, $M_1 - M_2$? What if the populations were not normal but the sample sizes were both large?

6

ESTIMATION

In this chapter we take up the first concepts of statistical inference and decision. In Chapter 5 summary measures of frequency distributions were considered, along with their sampling distributions. Now we will consider the use of such sample statistics as estimates of population quantities. The process of generalizing from the sample to the population is the process of statistical inference. There are two types of estimation: (1) point estimation, which consists of the use of a single sample statistic to determine a single value, which is to be used as an estimate of a population quantity, and (2) interval estimation, which involves the determination of an interval of values within which the population quantity must lie with a given "confidence." We will first discuss point estimation and then interval estimation.

6.1 SAMPLE STATISTICS AS ESTIMATORS

Each population parameter has its parallel in some sample statistic; the population mean μ has its sample counterpart in M, the variance σ^2 in the sample variance s^2, the population proportion p in the sample proportion P, and so on. The value of a sample statistic must contain evidence about the value of the corresponding population value, and a central problem in inferential statistics is the use of sample statistics as *estimators* of population values. The use of a sample value to infer the population value is called **point estimation,** since a single value (or point in the "space" of all possible values) is taken as the estimate.

A word regarding terminology: a sample statistic is an **estimator** of a population quantity, and a specific value of a sample statistic, computed from a particular sample, is an **estimate** of a population quantity. Thus, an estimator is a random variable, and we can talk of its probability distribution (which is a *sampling distribution*, since the estimator is a sample statistic), its expectation, and so on; an estimate is a specific value of this random variable.

How does one go from the sample value to an inference about the population parameter? The fact that the sample represents only a small subset of observations drawn from a much larger set of potential observations makes it risky to say that any estimate is exactly like the population value. As a matter of fact, they very probably will not be the same, as all sorts of different factors of which we are in ignorance may make the sample a poor representation of the population. Such factors we lump together under the general heading of *chance* or *random effects*. Nevertheless, in the long run such samples drawn at random should reflect the population characteristics. This is the reason that we insist on random samples, so that "in-the-long-run" statements may be made in terms of principles such as Bernoulli's theorem and the Tchebycheff inequality. However, practical action can seldom wait for "in the long run"; things must be decided here and now in the face of limited evidence. We need to know how to use the available evidence in the best possible ways to infer the characteristics of the population. Therefore general principles for estimation have been formulated.

There are a number of properties that have been put forth as desirable for a statistic to have in order to be a good estimator of a population parameter. Four of these properties are *unbiasedness, consistency, efficiency,* and *sufficiency.* Although few statistics satisfy all of these desirable properties as estimators, each criterion is an important property for an estimator to possess if possible. These four criteria for good estimators will be discussed in the next four sections.

6.2 UNBIASED ESTIMATORS

Suppose that one is interested in estimating the value of population parameter θ (Greek theta), and he is considering the use of some sample statistic G as an estimate of the value of θ. Then an estimate of the parameter θ made from the sample statistic G is said to be an **unbiased estimate** if

$$E(G) = \theta. \tag{6.2.1*}$$

That is, the sample quantity G is unbiased as an estimator of θ if the expectation of G is θ; in the long run, G averaged over all possible random samples is exactly θ.

For example, consider the mean of a sample as an estimator of the mean of the population. Here

$$G = M = \frac{\sum_i X_i}{N}$$

and
$$\theta = \mu.$$

Is the sample mean an unbiased estimator of the population mean μ? Yes, since we proved in Section 5.21 that

$$E(M) = \mu. \tag{6.2.2*}$$

The mean of a random sample is an unbiased estimate of the population mean μ.

In exactly the same way, for samples from a Bernoulli process with fixed sample size N, it can be shown that the *sample proportion P of cases in a given category is an unbiased estimator of the population proportion p* [this is really just a special case of (6.2.2)]:

$$E(P) = p. \tag{6.2.3*}$$

Since $P = R/N$, $E(P) = E(R/N)$. But N is fixed, so

$$E(P) = \frac{1}{N} E(R).$$

Now, $E(R)$ is just the mean of a binomial distribution, and is equal to Np. Thus,

$$E(P) = \frac{Np}{N} = p,$$

and P is an unbiased estimator of p.

What would an example of a *biased* estimator be like? An example is provided by the sample variance. The sample variance s^2 is a biased estimator of the population variance σ^2, since

$$E(s^2) \neq \sigma^2. \tag{6.2.4*}$$

This can be demonstrated as follows: the expectation of a sample variance is

$$E(s^2) = E\left(\frac{\sum_i X_i^2}{N} - M^2\right) = E\left(\frac{\sum_i X_i^2}{N}\right) - E(M^2).$$

Let us consider the two terms on the extreme right separately. By the algebra of expectations,

$$E\left(\frac{\sum_i X_i^2}{N}\right) = \frac{\sum_i E(X_i^2)}{N}.$$

From the definition of the variance of the population,

$$\sigma^2 = E(X_i^2) - \mu^2,$$

so that $\qquad\qquad E(X_i^2) = \sigma^2 + \mu^2$

for any observation i. Thus,

$$E\left(\frac{\sum_i X_i^2}{N}\right) = \frac{\sum_i (\sigma^2 + \mu^2)}{N} = \sigma^2 + \mu^2. \qquad (6.2.5)$$

Now the variance of the sampling distribution of means is, from Equations (5.21.4) and (6.2.2),

$$\sigma_M^2 = E(M^2) - \mu^2,$$

so that $\qquad\qquad E(M^2) = \sigma_M^2 + \mu^2. \qquad (6.2.6)$

Putting these two results [(6.2.5) and (6.2.6)] together, we have

$$E(s^2) = \sigma^2 - \sigma_M^2 \qquad (6.2.7^*)$$

(the expectation of the sample variance is the *difference* between the population variance σ^2 and the variance of the sampling distribution of means σ_M^2). In general this difference will not be the same as σ^2, since ordinarily the variance σ_M^2 will not be zero, and so the sample variance is biased as an estimator of the population variance. In particular, the sample variance is, on the average, smaller than the population variance σ^2. Since [from Equation (5.21.6)]

$$\sigma_M^2 = \sigma^2/N,$$

the expectation of the sample variance can be written in the form

$$E(s^2) = \sigma^2 - \frac{\sigma^2}{N} = \left(\frac{N-1}{N}\right)\sigma^2, \qquad (6.2.8^*)$$

so that on the average the sample variance is *too small* by a factor of $\dfrac{N-1}{N}$.

Since this is true, a way emerges for correcting the variance of a sample to make it an unbiased estimator. **An unbiased estimator of the variance based on any sample of N independent cases is**

$$\hat{s}^2 = \frac{N}{N-1} s^2. \qquad (6.2.9^*)$$

Note that $\hat{s}^2$ is the "modified" sample variance introduced in Chapter 5. The caret or "hat" ($\wedge$) will always be placed over the symbol for a sample variance when it is an unbiased estimator, to distinguish it from the ordinary sample variance. It is simple to show that $\hat{s}^2$ is indeed unbiased:

$$E(\hat{s}^2) = \frac{N}{N-1} E(s^2) = \frac{N}{(N-1)} \frac{(N-1)}{N} \sigma^2 = \sigma^2. \quad (6.2.10^*)$$

Quite often it is convenient to calculate the unbiased variance estimator $\hat{s}^2$ directly, without the intermediate step of calculating s^2. This is done by either formula (5.17.10) or (5.17.11).

Even though we will be using the square root of $\hat{s}^2$, or $\hat{s}$, to estimate the population σ, it should be noted that $\hat{s}$ is not *itself* an unbiased estimator of σ, and that

$$E(\hat{s}) \neq \sigma$$

in general. The correction factor used to make $\hat{s}$ an unbiased estimator of σ depends upon the form of the population distribution; thus, for the normal distribution, an unbiased estimator for large N is provided by

$$\text{Unbiased estimator of } \sigma = \left[1 + \frac{1}{4(N-1)} \right] \hat{s}. \quad (6.2.11)$$

Furthermore, special tables exist for correcting the estimate of σ for relatively small samples from such populations. However, the problem of estimating σ from $\hat{s}$ is bypassed, in part, by the methods we will use in making inferences about σ^2, and, if the sample size is reasonably large, the amount of bias in $\hat{s}$ as an estimator of σ ordinarily is rather small. For these reasons we will not trouble to correct for the bias in $\hat{s}$ found from the *unbiased* estimator $\hat{s}^2$ of σ^2.

Some modern statistics texts completely abandon the idea of the sample variance s^2 as used here, and introduce only the unbiased estimator $\hat{s}^2$ as *the* variance of a sample. However, this is apt to be confusing in some work, and so we will follow the older practice of distinguishing between the sample variance s^2 as a descriptive statistic and as a biased estimator of σ^2, and $\hat{s}^2$ as an unbiased estimator of σ^2.

6.3 CONSISTENCY

For an estimator to be a good one, **the sample estimator should have a higher probability of being close to the population value θ the larger the sample size N.** Statistics that have this property are called consistent estimators. More formally, the statistic G is a **consistent estimator** of θ if for any arbitrary positive number ϵ,

$$P(|G - \theta| < \epsilon) \to 1, \text{ as } N \to \infty. \quad (6.3.1^*)$$

The probability that G is within a certain distance ϵ of the parameter θ

approaches 1 as the sample size N approaches infinity, however small the size of the positive number ϵ.

It is not difficult to show that the sample mean M is a consistent estimator of the population mean μ. From the law of large numbers, Equation (5.21.9), it is true that

$$P(|M - \mu| < k) \geq 1 - \frac{\sigma_M^2}{k^2}. \tag{6.3.2}$$

But we know that

$$\sigma_M^2 = \frac{\sigma^2}{N},$$

and thus $\quad\quad P(|M - \mu| < k) \geq 1 - \frac{\sigma^2}{Nk^2}.$

When N becomes very large, σ^2/Nk^2 approaches zero, and therefore

$$P(|M - \mu| < k) \to 1 \quad \text{as} \quad N \to \infty. \tag{6.3.3*}$$

Thus, M *is a consistent estimator of* μ, since this corresponds to Equation (6.3.1) with $G = M$, $\theta = \mu$, and $\epsilon = k$.

The above proof suggests a sufficient (but not necessary) condition for an estimator to be consistent: if G is an unbiased estimator of θ, and var $(G) \to 0$ as $N \to \infty$, then G is a consistent estimator of θ. To see that this must be true, write the Tchebycheff inequality (3.27.1) as follows:

$$P(|G - E(G)| < \epsilon) \geq 1 - \frac{\sigma_G^2}{\epsilon^2}. \tag{6.3.4}$$

But G is unbiased, so $E(G) = \theta$. Also, $\sigma_G^2 \to 0$ as $N \to \infty$, so as N becomes large, σ_G^2/ϵ^2 approaches zero, and the probability on the left-hand side thus must approach 1, in which case G is a consistent estimator of θ:

$$P(|G - \theta| < \epsilon) \to 1 \text{ as } N \to \infty. \tag{6.3.5}$$

This shows that the conditions given above are *sufficient* conditions for G to be consistent. To show that they are not *necessary* conditions, we point out that the sample variance s^2 is a consistent estimator of σ^2 even though it is *not* an unbiased estimator of σ^2, thus violating the first condition.

The sample mean, the sample variance and many other statistics are consistent estimators, as they tend in likelihood to get closer to the true population value the larger the sample. It is, however, possible to create sample statistics that are not consistent estimators. For example, suppose that one wished to estimate the population mean, and he decided to take the value of the second observation made in any sample as the estimator. Would this be a consistent estimator? No, since the second observation's score would not tend in likelihood to get any closer to the mean value of the population the larger the sample. According to the criterion of consistency, this would not be a good way to estimate the mean. (Would this be an unbiased estimator of the mean, however?)

It should be noted that an important distinction between unbiasedness and consistency is that the former is a fixed-sample property (if an estimator is unbiased, it is unbiased for any fixed sample size), while the latter is an *asymptotic* property (that is, it is concerned only with what happens as the sample size becomes very large). Whenever the sample size is quite small, of course, asymptotic properties such as consistency are not of much interest.

6.4   RELATIVE EFFICIENCY

A third criterion for evaluating a statistic G as an estimator of a parameter θ is that G be efficient relative to other statistics that might be used to estimate θ. You will recall that the standard deviation (or standard error, as it is called when the distribution it describes is a sampling distribution) represents the extent of the difference that chance factors tend to create between a sample estimate and a true parameter value. Good estimators should have sampling distributions with small standard errors, given the N of the sample.

Suppose that there were two different sample statistics G and H calculated from the same data, and that these two statistics were each unbiased estimators of the same population parameter θ. The theoretical sampling distribution of G determined for samples of size N has a standard error σ_G. There is also a theoretical sampling distribution of H having a standard error σ_H. Each of these two standard errors reflects the tendency of the sample statistic to deviate by chance from the population value. Then, for any N, the **efficiency** of the statistic G relative to the statistic H, both as estimators of θ, is given by

$$\frac{\sigma_H{}^2}{\sigma_G{}^2} = \text{efficiency of } G \text{ relative to } H. \qquad (6.4.1^*)$$

The more efficient estimator has the smaller standard error, so that the ratio is greater than 1.00 when the variance of the more efficient estimator appears in the denominator. Other things, such as sample size, being equal, relatively inefficient statistics have relatively larger standard errors than other estimators of the same parameter.

For example, in a unimodal symmetric distribution the population mean and the population median both have the same value, $Md = \mu$. We wish to estimate this value μ by drawing a random sample of N cases. We could use either the sample mean (statistic G, let us say) or we could use the sample median (statistic H). In any given sample, these values will very likely not be the same. Which should we use to estimate μ? The answer involves efficiency. If the standard error of the mean is represented by σ_M and the standard error of the distribution of sample medians is represented by σ_{Md}, then it is true that for unimodal symmetric population distributions σ_{Md} is greater than σ_M for a given sample size $N > 2$. For such populations, and for a reasonably large N, the magnitude of error in any given estimate of the population mean from the sample mean is likely to be less than in an estimate of μ from the sample median. This says that for such situations

$$\frac{\sigma_{Md}^2}{\sigma_M^2} > 1.00, \tag{6.4.2}$$

the relative efficiency of the mean will be greater than 1.00, indicating that the mean is a more efficient estimator than the median for symmetric unimodal populations. For a specific example, if we assume that the population is normally distributed, then, approximately,

$$\sigma_{Md}^2 = \frac{\pi \sigma^2}{2N}.$$

In this case,

$$\frac{\sigma_{Md}^2}{\sigma_M^2} = \frac{\pi \sigma^2 / 2N}{\sigma^2 / N} = \frac{\pi}{2} = 1.57.$$

This is one reason that so much of statistical inference deals with the mean rather than the median; the median is less efficient than the mean as an estimator of μ for a symmetric unimodal population distribution, and such distributions are very often assumed in statistical work.

For a normal distribution, then, the sample mean is a more efficient estimator of μ than the sample median. Furthermore, it can be shown that the sample mean is the most efficient estimator of μ; that is, of all unbiased estimators of μ, the sample mean has the smallest standard error. As a result, the sample mean is called the **minimum variance unbiased estimator** of μ.

As presented in this section, the concept of relative efficiency is restricted to unbiased estimators. A more general concept is that of minimum mean-square error. If G is an estimator of θ, the mean-square error of G is

$$\text{MSE } (G) = E(G - \theta)^2. \tag{6.4.3}$$

Observe that if G is unbiased, the mean-square error is identical to the variance of G, since $E(G) = \theta$. If we define the *bias* of an estimator G to be

$$B(G) = E(G) - \theta, \tag{6.4.4}$$

then the mean-square error can be written as the sum of the variance and the square of the bias:

$$\text{MSE } (G) = \text{var } (G) + [B(G)]^2. \qquad (6.4.5)$$

In some situations, a certain biased estimator may have a smaller MSE than any unbiased estimator, in which case it might be preferred to any unbiased estimator.

6.5 SUFFICIENCY

A concept of great importance in the theory of estimation is that of a sufficient estimator. A statistic G is said to be a **sufficient** estimator of the parameter θ if G contains *all* of the information available in the data about the value of θ. A sufficient statistic G is a "best" estimator of θ in the sense that G cannot be improved by considering any other aspects of the sample data not already included in the statistic G itself.

A slightly more formal definition of sufficiency may be given as follows: consider the estimator G of θ once again, and then let H be any other sample statistic. We can consider the *conditional* distribution of H given the value of G. Then if the conditional distribution of H given G does not depend in any way on the value of θ, G is a sufficient statistic.

An example of a sufficient estimator is the sample proportion P when samples are drawn at random from a Bernoulli process. The sampling distribution of P can be found from the binomial rule in this situation, of course. If we are estimating the population value p, then P is a sufficient statistic, since there is no information we could add to P to make it a better estimator of p. For a normal distribution, the sample mean M is a sufficient estimator of μ, the population mean; for this situation the sample mean "wraps up" all of the available information about μ in the sample. This means that in order to estimate μ from a normal distribution, we need not concern ourselves with the set of individual sample results, $(x_1, x_2, \cdots, x_N)$; all that is needed is the sample mean, M, and the sample size, N. Observe that the entire set of sample results is sufficient, but that M and N form a smaller set which is also sufficient. That is, M and N summarize the results of the sample without any loss of information, thereby enabling the statistician to convey all of the available information from the sample in reasonably simple form.

Sufficient estimators do not always exist, and hypothetical situations can be constructed where no way to find a sufficient estimator of a given parameter can be found. Nevertheless, sufficient estimators, when they do exist, are very important, since if one can find a sufficient estimator it is ordinarily possible to find an unbiased and efficient estimator based on the sufficient statistic (like sufficient estimators, unbiased and efficient estimators do not always exist).

In most of the work to follow, the estimators will be the sample mean and the corrected sample variance, both of which will fulfill these criteria for good estimators in the particular situations where they will be used. This does not imply, however, that other estimators failing to meet one or more of these criteria are useless. In particular, special situations exist where other methods of estimating central tendency may be better than either mean or median, and variance estimates other than s^2 or $\hat{s}^2$ are called for. Other statistics are useful on occasion, but we will focus most attention on the mean and the corrected variance, since they occupy a central place in the "classical" statistical methods we will be treating in this and the next chapter. When we consider the decision-theoretic approach to statistics in Chapters 8 and 9, a somewhat different view of the estimation problem will be presented; estimation will be thought of as a decision-making problem, and the properties presented in the preceding sections will no longer be of primary interest.

6.6 METHODS FOR DETERMINING GOOD ESTIMATORS

The preceding sections have dealt with some desirable properties for point estimators: unbiasedness, consistency, relative efficiency, and sufficiency. If we have a particular statistic G and would like to use it as an estimator of a parameter θ, we can attempt to find out which, if any, of the four desirable properties G satisfies (or we may even look at other properties such as minimum mean-square error). Unless G is a reasonably simple statistic (that is, unless it can be expressed as a fairly simple mathematical function of the sample values), it may be very difficult to see if it possesses any of the desirable properties. First, the sampling distribution of G may not be known, and the mean and standard error of G may depend on the form of this distribution. Secondly, even if the sampling distribution is known, it may be difficult to determine the mean and variance or to determine the limiting properties of the sampling distribution as $N \to \infty$ (which may be necessary to investigate asymptotic properties such as consistency).

Suppose that these problems were not encountered with a particular statistic G, and G was found to be unbiased and consistent. This information is of some value, but it certainly is not enough information to indicate that G is the "best" estimator that we could find in this situation. For example, if we consider a symmetric unimodal population, the sample median is an unbiased and consistent estimator of the population mean, μ. The sample mean, on the other hand, is not only unbiased and consistent, but is a more efficient estimator of μ than the sample median, as Section 6.4 indicated. How do we know that there is not yet another estimator of μ that is even more efficient, or more desirable in some other way, than M? In other words, if we want to estimate θ, and we have investigated the properties of several alternative estimators of θ, how do we know that there is not yet *another* estimator, unbeknownst to us, which is a "better" estimator than those we have already considered?

The question posed above suggests that some methods for the determination of estimators are needed. If methods for the determination of "good" estimators could be developed, we would be saved some of the trouble of trying to find estimators, determining their properties, and worrying about the possible existence of "better" estimators. Two such methods, the method of maximum likelihood and the method of moments, are discussed in the next two sections. A third method, involving decision theory, will be discussed in Chapter 9. Finally, a fourth method, concerned primarily with a specific class of problems dealing with curve fitting, will be presented in Chapter 10.

Before we discuss these methods, it should be pointed out that according to the criteria we have developed so far (that is, the desirable properties of estimators), it is generally not possible to find a "best" estimator for any particular situation. This is why the terms "good," "better," and "best," when applied to estimators, have been enclosed in quotation marks. Usually there are several alternative estimators possessing some of the desirable properties, but the choice among them is not at all clear-cut. For example, consider two estimators G and H. Suppose that G is unbiased and H is not unbiased, but that H has a smaller mean-square error than G. Assume further that it is not possible to modify H to allow for its bias because the bias cannot be expressed in simple form. If you are quite concerned about the amount of bias in H because of the features of the problem at hand, you may prefer G. If, on the other hand, a small mean-square error is important to you, you may prefer H. The former condition might prevail if you have to make an estimate over and over again for different sets of data and you would like to be correct "on the average." The latter condition might prevail if you have a single set of data and you would like to be as "close" as possible to the true value of the population parameter, even if this means that you will have to accept a biased estimator. If you have a large sample, you might be most concerned about consistency, for this is a large-sample property. At any rate, neither estimator, G nor H, is clearly "better" than the other. The methods to be presented will not determine "best" estimators—they will determine estimators with certain desirable properties, and it is up to the statistician to decide if these properties are satisfactory for the problem at hand. Many factors enter into this decision, some of which are not even directly related to the properties (such as the time that would be required to find alternative estimators). With this in mind, we will discuss the method (or principle) of maximum likelihood and the method of moments.

6.7 THE PRINCIPLE OF MAXIMUM LIKELIHOOD

Basically, the problem of the person using statistics is, "Given several possible population situations, any one of which might be true, how shall I bet on the basis of the evidence so as to be as confident as possible of being right?" Everyone who uses statistical inference is faced with this question. He knows that any number of things might be true of the population. Fortunately, he can "snoop" on the population to a certain extent by taking a sample, but he knows that the evidence he gains might be faulty. Since this evidence is all there is, the person must use it nevertheless to form an opinion or make up his mind. How does one decide on the basis of evidence that might possibly be erroneous? The principle of maximum likelihood gives a general strategy for such decisions and may be paraphrased as follows.

Suppose that a random variable X has a distribution that depends only upon some population parameter θ. The form of the density function will be assumed known, but not the value of θ. A sample of N independent observations is drawn, producing the set of values $(X_1, X_2, \cdots, X_N)$. Let

$$L(X_1, \cdots, X_N \mid \theta)$$

represent the likelihood or probability (density) of this particular sample result *given* θ. For each possible value of θ, the likelihood of the sample result might be different. Then, **the principle of maximum likelihood requires us to choose as our estimate that possible value of θ making $L(X_1, \ldots, X_N \mid \theta)$ take on its largest value.**

In effect this principle says that when faced with several parameter values, any of which might be the true one for a population, the best "bet" is that parameter value which would have made the sample actually obtained have the highest probability. When in doubt, place your bet on that parameter value which would have made the obtained result most likely.

This principle may be illustrated very simply for the binomial distribution. Suppose that a sample is drawn at random from a population of college graduates. Each graduate is classified either as a "liberal arts major" or as a graduate with some other major field. Now three possibilities or hypotheses are entertained about the proportion of college graduates from liberal arts colleges. Hypothesis 1 states that .5 of all graduates are from such colleges, hypothesis 2 states that .4 of the graduates are from such schools, and hypothesis 3 states that .6 are liberal arts graduates.

Now suppose that some 15 college graduates are drawn at random and classified according to "liberal arts degree" versus "other degree." The result is that 9 out of the 15 hold a liberal arts degree (we are assuming that no one in the population holds more than one degree). What decision about the three available hypotheses should we reach? This is a simple binomial sampling problem, so that the probability of each sample result may be calculated for each of the three hypothetical values of the parameter p: $p = .5$, $p = .4$, or $p = .6$.

If $p = .5$, the probability of a sample result such as that obtained would be

$$\binom{15}{9}(.5)^9(.5)^6 = .153.$$

For $p = .4$, the probability becomes

$$\binom{15}{9}(.4)^9(.6)^6 = .061.$$

Finally, if p were .6, the probability of 9 successes out of 15 would be

$$\binom{15}{9}(.6)^9(.4)^6 = .207.$$

The use of the principle of maximum likelihood to decide among these three possibilities leads to the choice of hypothesis 3, that .6 is the population proportion, since this is the parameter value among the possibilities considered that would have made the obtained sample result most likely a priori.

It should be apparent that this approach could be made much more general. Surely we would like to be able to consider values of p other than .5, .4, and .6. The parameter p of a Bernoulli process can take on any value between 0 and 1, inclusive. For any value of p in this interval, the likelihood function can be written as

$$L(X_1, X_2, \cdots, X_N \mid p) = \binom{N}{r}p^r(1 - p)^{N-r}. \qquad (6.7.1^*)$$

Now the task is to find the value of p for which the likelihood function is maximized. For this we shall use the calculus (the reader not interested in the mathematical details can safely skip the derivations). First, since the logarithm of a positive argument is a monotonically increasing function of that argument, maximizing the likelihood function is equivalent to maximizing the logarithm of the likelihood function, which in this case is

$$\log L = \log\binom{N}{r} + r \log p + (N - r) \log (1 - p).$$

To maximize this function with respect to p, we differentiate with respect to p and set the result equal to zero (assuming that the logarithms are natural logarithms, that is, that they are taken to the base e):

$$\frac{d}{dp} \log L = \frac{r}{p} - \frac{N - r}{1 - p} = 0.$$

Thus,

$$\frac{r(1 - p) - (N - r)p}{p(1 - p)} = 0,$$

so

$$r(1 - p) - (N - r)p = 0,$$

$$r - rp - Np + rp = 0,$$

and hence

$$p = r/N. \tag{6.7.2*}$$

In general, then, the value of p which maximizes the likelihood function is the sample P, or r/N. We call this the **maximum-likelihood estimator** of p. In the example of college graduates, the maximum likelihood estimate based on the sample of size 15 would simply be equal to 9/15. Among all possible values of p, the value 9/15 makes the occurrence of the actual result have the greatest likelihood (check a few values of p to satisfy yourself that this is true). Formally, to show that $p = r/N$ is a maximum, note that the second derivative of $\log L$ with respect to p is negative.

For a slightly more difficult example, suppose that we are sampling from a normally distributed population with known variance σ^2 and that we wish to find the maximum-likelihood estimator of the population mean, μ, on the basis of a sample of size N from the population: $(X_1, X_2, \cdots, X_N)$. The density of each X_i is, by Equation (4.18.1),

$$f(X_i \mid \mu) = \frac{1}{\sqrt{2\pi\sigma^2}} \exp\left[-(X_i - \mu)^2/2\sigma^2\right]. \tag{6.7.3}$$

Since the trials are assumed to be independent, the likelihood function is simply the product of the N density functions:

$$L(X_1, X_2, \cdots, X_N \mid \mu) = f(X_1 \mid \mu)f(X_2 \mid \mu) \cdots f(X_N \mid \mu)$$

$$= \left(\frac{1}{\sqrt{2\pi\sigma^2}} \exp\left[-(X_1 - \mu)^2/2\sigma^2\right]\right)$$

$$\cdots \left(\frac{1}{\sqrt{2\pi\sigma^2}} \exp\left[-(X_N - \mu)^2/2\sigma^2\right]\right)$$

$$= \left(\frac{1}{\sqrt{2\pi\sigma^2}}\right)^N \exp\left[-\sum_{i=1}^{N} (X_i - \mu)^2/2\sigma^2\right]. \tag{6.7.4*}$$

Now, to maximize $L(X_1, \cdots, X_n \mid \mu,)$ we once again take the logarithm and use the calculus to maximize $\log L$:

$$\log L = N \log \left(\frac{1}{\sqrt{2\pi\sigma^2}} \right) - \frac{\sum\limits_{i=1}^{N} (X_i - \mu)^2}{2\sigma^2}.$$

$$\frac{d}{d\mu} \log L = -\frac{1}{2\sigma^2} \sum_{i=1}^{N} 2(X_i - \mu)(-1) = \frac{\sum(X_i - \mu)}{\sigma^2}.$$

Setting this equal to zero, we find that

$$\sum_{i=1}^{N} (X_i - \mu) = 0,$$

or

$$\sum_{i=1}^{N} X_i - \sum_{i=1}^{N} \mu = 0,$$

or

$$\sum X_i - N\mu = 0,$$

which implies that

$$\mu = \frac{\sum X_i}{N}. \tag{6.7.5*}$$

Thus, for populations having a normal distribution, the sample mean M is a maximum-likelihood estimator of μ. The reader can check the second order conditions to satisfy himself that this does indeed produce the maximum value for the likelihood function.

The principle of maximum likelihood is quite useful in theoretical statistics, since general methods exist for finding the value of θ that maximizes the likelihood of a particular sample result. The principle gains its greatest importance from the fact that if a *sufficient* estimator of a parameter exists, the maximum-likelihood estimator is based on this sufficient statistic. In addition, maximum-likelihood estimators possess certain desirable large-sample properties.

Finally, maximum-likelihood estimators possess an *invariance property*. This means that if G is a maximum-likelihood estimator of θ and the function $h(\theta)$ is a function which has an inverse [that is, there is a one-to-one relationship between values of θ and the corresponding values of $h(\theta)$], then $h(G)$ is a maximum-likelihood estimator of $h(\theta)$. For example, if a sample of size N is taken from a normally distributed population, it can be shown that the sample variance s^2 is a maximum-likelihood estimator of the population variance σ^2. The property of invariance implies that s, the sample standard deviation, is also a maximum-likelihood estimator of σ, the population standard deviation. Note that this example shows that maximum-likelihood estimators are not always unbiased, since s^2 is not unbiased (Section 6.2). Thus, the principle of maximum likelihood provides a fairly routine way of finding estimators having "good" properties.

However, the principle of maximum likelihood is introduced here not only because of its importance in estimation, but also because of the general point of view it represents about inference. This point of view is that *true population situations should be those making our empirical results likely; if a theoretical situation makes our obtained data have very low likelihood of occurrence, then doubt is cast on the truth of the theoretical situation.* Theoretical propositions are believable to the extent that they accord with actual observation. If a particular result should be very unlikely given a certain theoretical state of affairs, and we do get this result nevertheless, then we are led back to examine our theory. Good theoretical statements accord with observation by giving predictions having high probabilities of being observed.

Naturally, the results of a single experiment, or even of any number of experiments, cannot prove or disprove a theory. Replications, variants, different ways of measuring the phenomena, must all be brought into play. Even then, proof or disproof is never absolute. Nevertheless, the principle of maximum likelihood is in the spirit of empirical science, and it runs throughout the methods of statistical inference.

6.8 THE METHOD OF MOMENTS

A second method for determining point estimators is the method of moments, which is not too widely used, primarily because it does not in general produce estimators with all of the desirable properties possessed by maximum-likelihood estimators. Nevertheless, it is occasionally useful, particularly when the principle of maximum likelihood is difficult to apply. The method of moments essentially amounts to equating sample moments and population moments and solving the resulting equations for the parameter(s) of interest. Formally, the kth population moment (about the origin) of the random variable X is defined as $E(X^k)$, which can be determined by using the usual rules of expectation:

$$E(X^k) = \begin{cases} \sum x^k P(X = x) & \text{if } X \text{ is discrete,} \\ \int_{-\infty}^{\infty} x^k f(x) \, dx & \text{if } X \text{ is continuous.} \end{cases} \tag{6.8.1*}$$

The kth population moment is sometimes denoted by μ_k. If a sample of size N is taken from the population, and the values $X_1, X_2, \cdots, X_N$ are

THE METHOD OF MOMENTS

observed, the kth sample moment is defined as M_k, where

$$M_k = \frac{\sum_{i=1}^{N} X_i^{k}}{N}. \tag{6.8.2*}$$

Given these definitions of M_k and μ_k, then, the method of moments consists of equating M_1 and μ_1, M_2 and μ_2, and so on, until enough equations are obtained to permit a solution in terms of θ, the parameter being estimated. Since the number of equations required varies with the form of the distribution of X and with the specific parameter(s) of interest, it is best to demonstrate the method of moments with some examples.

The simplest example of the application of the method is in the case in which we wish to estimate a population moment, μ_k. It is obvious that the estimator will simply be the corresponding sample moment, M_k. In this way, for example, the sample mean is determined to be the estimator of the population mean according to the method of moments. Similarly, M_2 is the estimator of μ_2. Given these two estimators, we can determine an estimator of the population variance, σ^2. Specifically, we will show that the method of moments determines the sample variance s^2 as the estimator of σ^2. First, note that

$$\sigma^2 = E(X^2) - [E(X)]^2$$
$$= \mu_2 - (\mu_1)^2.$$

Using the method of moments, we set $\mu_1 = M_1$ and $\mu_2 = M_2$:

$$\text{est. } \sigma^2 = M_2 - M_1^2$$

From our definition of sample moment,

$$M_2 = \frac{\sum X_i^2}{N},$$

so

$$\text{est. } \sigma^2 = \frac{\sum X_i^2}{N} - M^2.$$

But, as we have seen, the right-hand side of this equation is equal to s^2, and thus s^2 is the estimator of σ^2 determined by the method of moments. It should be pointed out that if we extended the notion of the method of moments to allow consideration of moments about the mean as well as moments about the origin, the derivation just presented would be unnecessary. Defining the kth population moment about the mean as $E(X - \mu)^k$, and the kth sample moment about the mean as

$$\sum_{i=1}^{N} (X_i - M)^k / N,$$

we find that if $k = 2$ we get σ^2 and s^2, respectively. Notice that the population moments are taken about the population mean and the sample moments are taken about the sample mean.

As another example of the application of the method of moments, consider the binomial distribution. To estimate the parameter p, note that $E(X) = Np$ and $M = r$. Equating these and solving for p, we get

$$\text{est. } p = \frac{r}{N}.$$

For most of the distributions we have studied, it is only necessary to look at the first one or two moments in order to determine an estimator. An advantage of the method of moments, then, is that it is not too difficult in most cases to find an estimator. If we desire to estimate simultaneously several parameters or if the formulas for the population moments are complicated or not known, the method of moments will not be so easy to apply. In these situations, however, it is generally at least as difficult, if not more so, to find a maximum-likelihood estimator (for example, it may be quite difficult, or even impossible, to differentiate L or $\log L$). In addition, while the method of moments seems intuitively satisfying (one would hope that a sample moment would provide a reasonably "good" estimator of the corresponding population moment), it does not provide as much of a "guarantee" of desirable properties as does the principle of maximum likelihood. As a result, it is not usually used if it is feasible to use the principle of maximum likelihood instead. In reasonably uncomplicated situations, though, the two methods often result in identical or virtually identical estimators. In estimating the parameter p of a Bernoulli process, this is the case, as we have seen, and it is also true in estimating μ and σ^2 of a normal distribution or λ of a Poisson process, to mention but a few examples.

6.9 PARAMETER ESTIMATES BASED ON POOLED SAMPLES

Sometimes it happens that one has several independent samples, and that each provides an estimate of the same parameter or set of parameters. The most usual situation occurs when one is estimating μ or σ^2 or both. When this happens there is a real advantage in pooling the sample values to get an unbiased estimate. These pooled estimates are actually weighted averages of the estimates from the different samples. The big advantage lies in the fact that the sampling error will tend to be smaller for the pooled estimate than that for any single sample's value taken alone.

Suppose that there were two independent samples, based on N_1 and N_2 observations respectively. From each sample a mean M is calculated, each of which estimates the same value μ. Then the pooled estimator of the mean is

$$\text{Pooled } M = \text{est. } \mu = \frac{N_1 M_1 + N_2 M_2}{N_1 + N_2}. \tag{6.9.1}$$

It is easy to see that this must give an unbiased estimator of μ:

$$E\left[\frac{N_1 M_1 + N_2 M_2}{N_1 + N_2}\right] = \frac{N_1 E(M_1) + N_2 E(M_2)}{N_1 + N_2}$$

$$= \frac{\mu(N_1 + N_2)}{N_1 + N_2} = \mu.$$

For some J different and independent samples, where any given sample j gives an estimate M_j of μ, the pooled estimator is

$$\text{est. } \mu = \frac{\sum_j N_j M_j}{\sum_j N_j}. \tag{6.9.2}$$

The fact that the pooled estimator is likely to be better than the single sample estimators taken alone is shown by the standard error of the pooled mean. For two independent samples, each composed of independent observations drawn from the same population, the standard error of the *pooled* mean is

$$\sigma_M = \frac{\sigma}{\sqrt{N_1 + N_2}}, \tag{6.9.3}$$

which *must* be smaller than the standard error either of M_1 or of M_2. Similarly, for any number J of independent samples, with $N_1, N_2, \cdots, N_J$ as the respective sample sizes,

$$\sigma_M = \frac{\sigma}{\sqrt{\sum_j N_j}}. \tag{6.9.4*}$$

Naturally, these standard errors refer to the situation where the respective samples all come from the *same* population with mean μ and variance σ^2.

In the same general way it is possible to find pooled estimators of σ^2. For two independent samples,

$$\text{est. } \sigma^2 = \frac{N_1 s_1^2 + N_2 s_2^2}{N_1 + N_2 - 2}, \tag{6.9.5}$$

which is an unbiased estimator of σ^2.

For any number J of independent samples, then, the corresponding unbiased estimator of σ^2 is

$$\text{est. } \sigma^2 = \frac{\sum_j N_j s_j^2}{(\sum_j N_j) - J}.$$ (6.9.6)

We will make use of this pooling principle in Chapter 7.

6.10 SAMPLING FROM FINITE POPULATIONS

It has been mentioned repeatedly that most sampling problems deal with populations so large that the fact that samples are taken without replacement of single cases can safely be ignored. However, it may happen that the population under study is not only finite, but relatively small, so that the process of sampling without replacement has a real effect on the sampling distribution.

Even in this situation the sample mean is still an unbiased estimator regardless of the size of the population sampled. Hence no change in procedure for estimation of the mean is needed.

However, for finite populations the sample variance is biased as an estimator of σ^2 in a way somewhat different from the former, infinite-population, situation. When samples are drawn *without replacement* of individuals, the unbiased estimator of σ^2 is

$$\text{est. of } \sigma^2 = \frac{(N)(T-1)}{(N-1)(T)} s^2,$$ (6.10.1)

where T is the *total number* of elements in the population.

Another difference from the infinite-population situation comes with the variance of the sampling distribution of means. For a population with T cases in all, from which samples of size N are drawn, the sampling variance of the sample mean is

$$\sigma_M^2 = \left(\frac{T-N}{T-1}\right) \frac{\sigma^2}{N}.$$ (6.10.2)

The variance of the mean tends to be *somewhat smaller* for a fixed value of N when sampling is from a finite population than when it is from an infinite population. Note that here the size of σ_M^2 depends both upon T, the total number in the population, and upon N, the sample size. An unbiased estimator of the variance of the mean is thus given by

$$\text{est. } \sigma_M^2 = \left(\frac{T-N}{T-1}\right) \frac{\hat{s}^2}{N}.$$ (6.10.3)

The square root of this value gives the estimate of σ_M for sampling from a finite population.

Sometimes the number $C = N/T$ is called the "sampling fraction," and when T is at least moderately large, the variance of the mean, $\sigma_M{}^2$, is approximately

$$\frac{\sigma^2}{N}(1 - C). \qquad (6.10.4)$$

When T is so large that C is virtually zero, the sampling variance is essentially the same as for the infinite-population situation. Note that $(1 - C)$ is what we called the "finite-population correction" in Section 4.13.

6.11 INTERVAL ESTIMATION—CONFIDENCE INTERVALS

In estimating the mean μ of a normal population, the sample mean M has been shown to be a maximum-likelihood estimator and a method of moments estimator. M also satisfies the four desirable properties of point estimators listed in Sections 6.2–6.5. Thus, the sample mean is a good estimator of μ. Nevertheless, it is clear that sample mean ordinarily will not be equal to the true population value because of sampling error. It is necessary to qualify our estimate in some way to indicate the general magnitude of this error. Usually this is done by showing a **confidence interval,** which is **an estimated range of values with a given probability of covering the true population value.** When there is a large degree of sampling error the confidence interval estimated from any sample will be large; the range of values likely to cover the population mean is wide. On the other hand, if sampling error is small the true value is likely to be covered by a small estimated range of values; in this case one can feel confident that he has "trapped" the true value within a small range of estimation for any sample.

Recall that if the population being sampled from is normal, then the sample mean M is normal, with $E(M) = \mu$ and var $(M) = \sigma^2/N$. The corresponding standardized random variable is then

$$z_M = \frac{M - \mu}{\sigma/\sqrt{N}}.$$

Now, since M is normally distributed, z_M is a standard normal random variable, and, from the tables of the normal distribution, we can determine probabilities for various intervals. For example,

$$P(-1.96 \leq z_M \leq +1.96) = .95.$$

But, on substituting $(M - \mu)/(\sigma/\sqrt{N})$ for z_M and algebraically manipulating within the parentheses, we find that:
The probability is approximately .95 that

$$-1.96\sigma_M \leq \mu - M \leq +1.96\sigma_M. \qquad (6.11.1^*)$$

That is, in very nearly 95 percent of all samples of size N from the population in question, the sample mean must lie within 1.96 standard errors to either side of the true mean. You can check this for yourself from Table I.

However, the statement above is still true if we alter the inequality by adding M to each term: **The probability is approximately .95 that**

$$M - 1.96\sigma_M \leq \mu \leq M + 1.96\sigma_M. \qquad (6.11.2^*)$$

That is, over all possible samples, the probability is about .95 that the range between $M - 1.96\sigma_M$ and $M + 1.96\sigma_M$ will include the true mean, μ.

This range of values between $M \pm 1.96\sigma_M$ is called **95 percent confidence interval for** μ. The two boundaries of the interval, or $M + 1.96\sigma_M$ and $M - 1.96\sigma_M$, are called **the 95 percent confidence limits.**

How can we interpret the probability statement (6.11.2) and similar statements involving confidence limits? Under the long-run frequency interpretation of probability, μ is a fixed value rather than a random variable. Thus, the probability statement is really not about μ, but about *samples*. The population either does or does not have a μ equal to a given number, and according to the long-run frequency interpretation of probability it does not make sense to say "the probability is such and such that the true mean takes on the value x." Before any sample mean is observed, however, the sample mean M *is* a random variable. In theory, one could consider all possible samples, compute M for all samples, and determine a 95 percent confidence interval for each and every sample. The confidence limits for any particular sample depend on the value of M in that sample. In short, over all possible samples there will be many possible 95 percent confidence intervals. Some of these confidence intervals will represent the event "covers the true mean," and others will not. If one such confidence interval were sampled at random, then the probability is .95 that it covers the true mean. The long-run frequency interpretation of a confidence interval, then, is as follows: if samples were drawn repeatedly under identical conditions, and if the 95 percent confidence interval were computed for each of these samples, then in the long run 95 percent of these confidence intervals would "cover," or include, the true value of μ.

It should be pointed out that according to the subjective interpretation of probability, it is possible to consider μ as a random variable and to interpret Equation (6.11.2) as a probability statement about μ. Prior to observing the sample, both M and μ are random variables; after observing the sample and computing M, μ is still a random variable and it is possible to make probability statements about μ. This interpretation will be discussed in more detail in Chapters 8 and 9.

In determining confidence intervals, there is no reason to limit ourselves to 95 percent confidence intervals. **A 99 percent confidence interval for the mean is given by**

$$M - 2.58\sigma_M \leq \mu \leq M + 2.58\sigma_M. \qquad (6.11.3^*)$$

Notice that the 99 percent confidence interval is larger than the 95 percent interval. More possible values are included, and thus one can assert with more confidence that the true mean is covered. A 100 percent confidence interval would be

$$M - \infty\sigma_M < \mu < M + \infty\sigma_M.$$

Here, one can be completely confident that the true value is covered, since the range includes all possible values of μ. On the other hand, a 0 percent interval is simply M; for a continuous sampling distribution of M, the probability is zero that M exactly equals μ.

In the particular situation where the population σ is known, an analogy to the idea of finding confidence intervals for the mean is tossing rings of a certain size at a post. The size of the confidence interval is like the diameter of a ring, and random sampling is like random tosses at the post. The predetermined size of the ring determines the chances that the ring will cover the post on a given try, just as the size of the confidence interval governs the probability that the true value μ will be covered by the range of values in the estimated interval. Reducing the size of σ_M by increasing N is like improving our aim, since we are more likely to cover the true value with an interval of a given size in terms of σ_M. Later when we deal with situations where the population σ is unknown, the analogy becomes that of tossing rings with a certain *distribution* of sizes at a post; here, the probability that the ring covers the post on a given try depends on the distribution of ring-sizes used, just as the probability that the confidence interval covers the true mean depends on the distribution of estimated confidence intervals over random samples. This idea will be elaborated upon later in this chapter.

It is interesting to note that, in principle, the 95 percent, or 99 percent, or any other confidence interval for a mean might be defined in other ways. For example, for a normal sampling distribution of M it is true that the probability is approximately .95 that

$$-1.75\sigma_M \leq \mu - M \leq 2.33\sigma_M.$$

This implies that a 95 percent confidence interval might possibly be found with limits

$$M - 1.75\sigma_M \quad \text{and} \quad M + 2.33\sigma_M.$$

In a similar way, still another 95 percent confidence interval might be found having limits

$$M - 2.06\sigma_M \quad \text{and} \quad M + 1.88\sigma_M,$$

and so on. In principle, 95 percent confidence limits for μ might be found in any number of ways, since any number of ways exist to find areas in a normal distribution equivalent to probabilities of .95.

If so many different ways of defining the 95 percent confidence interval are possible, what is the advantage of defining this confidence interval to have limits

$$M - 1.96\sigma_M \quad \text{and} \quad M + 1.96\sigma_M?$$

The answer is that, in the particular case of the mean, this way of defining the confidence interval gives the *shortest* possible range of values such that the probability statement holds. Naturally, there is an advantage in pinning the population parameter within the narrowest possible range with a given probability, and this dictates the form of the confidence interval we use. A secondary consideration, as we shall see in Chapter 7, is that this way of defining the confidence interval corresponds to the standard two-tailed test of the mean, so that a parameter value falling outside the 95 percent confidence interval for the mean for a particular sample would be rejected as a hypothesis (at the .05 level, two-tailed) in the light of this sample. A final consideration is that this way of defining the confidence interval results in an interval centered about M, which has been shown to be a "good" point estimator of μ.

As an example of a confidence interval for the mean when σ_M is known, consider a sample of 50 cases giving a mean of 143. The population standard deviation, σ, is 35. The 95 percent confidence interval will be found. First of all,

$$\sigma_M = 35/\sqrt{50} = 4.9.$$

Then the desired interval is

$$143 - (1.96)(4.9) \leq \mu \leq 143 + (1.96)(4.9),$$

or

$$133.39 \leq \mu \leq 152.60.$$

In succeeding sections methods for finding confidence intervals will be given not only for means, but also for proportions, for differences between means, and for variances.

Although it has been assumed in this section that the population is normally distributed, the same general approach to interval estimation can be used in other cases as well. The sampling distribution of M must be known, however, so that the probabilities such as the following can be calculated:

$$P\left(-a \le \frac{M - \mu}{\sigma/\sqrt{N}} \le a\right). \tag{6.11.4}$$

Performing some simple algebraic manipulations within the parentheses, we have

$$P\left(-a \le \frac{M - \mu}{\sigma/\sqrt{N}} \le a\right) = P\left(-a \frac{\sigma}{\sqrt{N}} \le M - \mu \le a \frac{\sigma}{\sqrt{N}}\right)$$

$$= P\left(-a \frac{\sigma}{\sqrt{N}} \le \mu - M \le a \frac{\sigma}{\sqrt{N}}\right)$$

$$= P\left(M - a \frac{\sigma}{\sqrt{N}} \le \mu \le M + a \frac{\sigma}{\sqrt{N}}\right).$$

Now, if

$$P\left(-a \le \frac{M - \mu}{\sigma/\sqrt{N}} \le a\right) = 1 - \alpha,$$

we say that the interval

$$M \pm a \frac{\sigma}{\sqrt{N}} \tag{6.11.5*}$$

is a $100(1 - \alpha)$ **percent confidence interval for** μ**.** The $(1 - \alpha)$ terminology is related to hypothesis testing, and will become clear in the next chapter.

Note that no assumptions have been made about the sampling distribution of M in the above analysis. In order to determine a $100(1 - \alpha)$ percent confidence interval for μ, it is necessary to find the value a such that the probability given by Equation (6.11.4) is equal to $(1 - \alpha)$. If M is normally distributed (which may be true even if the population is *not* normally distributed, because of the central limit theorem), a can be determined from the tables of the standard normal distribution. If M is not normally distributed, then other tables (or other methods, if tables are not available) will be needed. One other important point is that it has been assumed that the population standard deviation σ is known. If σ is not known, we may need to use the t tables, as we will see in Section 6.18. A final caution is that if the distribution of M is not symmetric, then the lower and upper limits in Equation (6.11.4) may not have the same absolute value. In general, the upper limit should be denoted by b rather than by a in order to show that it is not necessarily the negative of the lower limit.

6.12   APPROXIMATE CONFIDENCE INTERVALS FOR PROPORTIONS

Forming an exact confidence interval for a proportion is a considerably more complicated business than it is for a single mean. For small samples, this is because the distribution we are dealing with is a discrete distribution rather than a continuous distribution. For large samples (large enough so that the normal approximation is applicable), it is because the standard error of a proportion P depends on the true population parameter p (recall that the standard error of the mean does not depend upon μ). Thus, since a sample only provides us with P, an *estimate* of p, the true standard error is unknown unless p is known. This could obviously not be the case, for if p were known we would certainly not be attempting to determine a confidence interval for p!

When N is small, confidence intervals for p can be determined directly from tables of the binomial distribution. An understandable discussion of this procedure would require several examples, so we will confine ourselves to the large-sample case. When N is large, so that the sampling distribution of the proportion is approximately normal, the $100(1 - \alpha)$ percent confidence limits for a proportion are given approximately by

$$\frac{N}{N + z^2}\left[P + \frac{z^2}{2N} \pm z\sqrt{\frac{PQ}{N} + \frac{z^2}{4N^2}}\right], \qquad (6.12.1)$$

where z is the standard score in a normal distribution cutting off the *upper* $\alpha/2$ proportion of values. These limits define the $100(1 - \alpha)$ percent confidence interval for p.

This rather mysterious-looking form for finding the confidence limits for a proportion can be justified as follows: since the sampling distribution of P for large samples can be assumed approximately normal, the appropriate confidence limits should be approximately

$$P - z\sqrt{\frac{(pq)}{N}} \quad \text{and} \quad P + z\sqrt{\frac{(pq)}{N}}\,.$$

The possible values of p corresponding to the limits thus satisfy the relation

$$\frac{N(P - p)^2}{(p - p^2)} = z^2.$$

Working out this quadratic equation in terms of p, we have

$$(N + z^2)p^2 - (2NP + z^2)p + NP^2 = 0.$$

Solving for p by the rule for a quadratic equation,

$$\frac{-b \pm \sqrt{b^2 - 4ac}}{2a}\,,$$

where

$$a = N + z^2,$$
$$b = -(2NP + z^2),$$

and

$$c = NP^2,$$

and arranging terms gives expression (6.12.1).

If N is very large, this confidence interval may be replaced by the simpler approximation

$$P - z\sqrt{\frac{PQ}{N}} \leq p \leq P + z\sqrt{\frac{PQ}{N}}. \qquad (6.12.2)$$

For example, suppose that $N = 100$ and $P = .4$, and that a 95.44 percent confidence interval for p was desired (this value was not selected arbitrarily—it corresponds to z-limits of ± 2, so the calculations are not too difficult). Using Equation (6.12.1), the confidence limits are .308 and .500; using Equation (6.12.2), the confidence limits are .302 and .498. You can verify that these limits are correct by using the formulas with $N = 100$, $P = .4$, and $z = 2$.

6.13 SAMPLE SIZE AND THE ACCURACY OF ESTIMATION OF THE MEAN

The width of any confidence interval for the mean μ depends upon σ_M, the standard error, and anything that makes σ_M proportionately smaller reduces the width of the interval. Thus, any increase in sample size, operating to reduce σ_M, makes the confidence interval shorter. For example, the σ value for some normal population is 45. When samples of size 9 are taken the 95 percent confidence interval based on any sample mean has limits

$$M - 1.96(15)$$

and

$$M + 1.96(15).$$

However, if samples of size 81 are taken, these limits come closer together:

$$M - 1.96(5)$$

and

$$M + 1.96(5),$$

so that the interval itself is smaller for any sample. Samples of 900 cases give a confidence interval which is relatively quite narrow: the limits are

$$M - 1.96(1.5)$$

and

$$M + 1.96(1.5).$$

Continuing in this way, if the sample size becomes extremely large, the width of the confidence interval approaches zero.

A practical result of this relation between the standard error of the mean and sample size is that *the population mean may be estimated within any desired degree of precision, given a large enough sample size*. This principle has already been introduced in terms of the Tchebycheff inequality (Section 3.27), but it can be stated even more strongly when the sampling distribution has some known form, such as the normal distribution. For instance, how many cases should one sample if he wants the probability to be .99 that his sample mean lies within $.1\sigma$ of the true mean? That is, the experimenter wants

$$P(|\,M - \mu\,| \le .1\sigma) = .99.$$

Assuming that the sampling distribution is nearly normal, which it should be if sample size is large enough, this is equivalent to requiring that the 99 percent confidence interval should have limits

$$M - .1\sigma$$

and
$$M + .1\sigma.$$

However, this is the same as saying that

$$.1\sigma = 2.58\sigma_M,$$

so that
$$.1\sigma = 2.58\,\frac{\sigma}{\sqrt{N}}\,.$$

Solving for N, we find

$$\sqrt{N} = 25.8,$$

$$N = 665.64.$$

In short, if the experimenter makes 666 independent observations, he can be sure that the probability of his estimate's being wrong by more than $.1\sigma$ is only one in one hundred. Notice that we do not have to say exactly what σ is in order to specify the desired accuracy in σ units and to find the required sample size.

If the experimenter is willing to take a somewhat larger chance of an absolute error exceeding $.1\sigma$, then the required N is smaller. If the probability of an error exceeding $.1\sigma$ in absolute magnitude is to be .05, then

$$.1\sigma = 1.96\sigma_M,$$

$$.1\sigma = 1.96\,\frac{\sigma}{\sqrt{N}},$$

so that $N = 384$.

This idea may be turned around to find the accuracy *almost* insured by any given sample size. Given that N is, say, 30, within how many σ units of the true mean will the sample estimate fall with probability .95? Here, letting k stand for the required number of σ units,

$$k\sigma = 1.96 \frac{\sigma}{\sqrt{N}},$$

or
$$k = \frac{1.96}{\sqrt{30}} = \frac{1.96}{5.48} = .358.$$

For $N = 30$, the chances are about 95 in 100 that our estimate of μ falls within about .36 population standard deviations of the true value, provided that the sampling distribution is approximately normal.

It is obvious that the accuracy of estimation in an experiment could be judged in this way. If the sample size is known, then one can judge how "off" the estimate of the mean is likely to be in terms of the population standard deviation. Furthermore, if one has some idea of the desirable accuracy of estimation, then he can set his sample size so as to attain that degree of accuracy with high probability. This is a very sensible and easy way to determine necessary sample size. Unfortunately, it does require the statistician to think to some extent about the population he is sampling and what he is going to do with the result. For the problem at hand, just how much error in estimation can he tolerate? We will be better able to answer this question in Chapter 9, when decision theory is presented. One of the important decisions facing the statistician is the determination of sample size, and the decision depends on the trade-off between costs, or losses, due to imprecise estimates, and the cost of sampling. For any contemplated sample, if the anticipated increase in precision is "worth more" to the statistician than the cost of taking the sample, then the sample should be taken. This idea will be elaborated on in Chapter 9.

6.14 THE PROBLEM OF UNKNOWN σ^2: THE t DISTRIBUTION

If we are sampling from a normally distributed population and the mean μ and the variance σ^2 of the population are known, then the standardized random variable

$$z_M = \frac{M - E(M)}{\sigma_M} = \frac{M - \mu}{\sigma/\sqrt{N}} \qquad (6.14.1^*)$$

has a normal distribution. As a result, we can use the facts we know about normal distributions in making inferences based on this sampling distribution. In many situations, however, the variance of the population is *not known*, in which case the standard error of the sample mean cannot be determined. One possible way to allow for this is to use an *estimator* of σ^2, namely the sample variance, s^2, which we have defined as follows:

$$s^2 = \frac{\sum(X_i - M)^2}{N}.$$

If this is done, the corresponding random variable becomes

$$t = \frac{M - E(M)}{\text{est. } \sigma_M} = \frac{M - \mu}{s/\sqrt{N-1}}. \qquad (6.14.2^*)$$

If we use $\hat{s}^2$ instead of s^2 as an estimator of σ^2, then we can write

$$t = \frac{M - \mu}{\hat{s}/\sqrt{N}}. \qquad (6.14.3^*)$$

There is an extremely important difference between the two ratios, z_M and t. For z_M, the numerator $(M - E(M))$ is a random variable, the value of which depends upon the particular sample drawn from a given population situation; on the other hand, the denominator is a constant, σ_M, which is the same regardless of the particular sample of size N we observe. Now contrast this ratio with the ratio t; just as before, the numerator of t is a random variable, but the denominator is also a random variable, since the particular value of s or $\hat{s}$—and hence the estimate of σ_M—is a sample quantity. Over several different samples, the same value of M must give us precisely the same value of z_M; however, over different samples, the same value of M will give us different t values. Similar intervals of t and z_M values should have different probabilities of occurrence. For this reason it is risky to use the ratio t as though it were z_M unless the sample size is large enough for the Central Limit Theorem to apply.

The solution to the problem of the nonequivalence of t and z_M rests on the study of t itself as a random variable. That is, suppose that the value of t were computed for each conceivable sample of N independent observations drawn from some normal population distribution with true mean μ. Each sample would have some t value,

$$t = \frac{M - E(M)}{\text{est. } \sigma_M} = \frac{M - \mu}{s/\sqrt{N-1}} = \frac{M - \mu}{\hat{s}/\sqrt{N}}. \qquad (6.14.4)$$

Over the different samples the value of t would vary, of course, and the different possible values would each have some probability (density). A random variable such as t is an example of a *test*-statistic, so called to distinguish it from an ordinary descriptive statistic or estimator, such as M or $\hat{s}^2$. The t value depends on other sample statistics, but is not itself an estimator of a population value. Nevertheless, such test-statistics have sampling distributions just as ordinary sample statistics do, and these sampling distributions have been studied extensively.

In order to find the exact distribution of t, one must assume that the basic population distribution is normal. The main reason for the necessity of this assumption is that only for a normal distribution will the basic random variables in numerator and denominator, sample M and s, be statistically independent; this is a use of the important fact mentioned in Section 5.22. Unless M and s are independent, the sampling distribution of t is extremely difficult to specify exactly. On the other hand, for the special case of normal populations, the distribution of the ratio t is quite well known. In order to learn what this distribution is like, let us take a look at the rule for the density function associated with this random variable.

The density function for t is given by the rule

$$f(t) = G(\nu) \left[1 + \frac{t^2}{\nu} \right]^{-(\nu+1)/2}. \qquad (6.14.5)$$

Here, $G(\nu)$ stands for a constant number which depends *only* on the parameter ν (Greek nu), and how this number is found need not really concern us. Let us focus our attention on the "working part" of the rule, which involves only ν and the value of t. This looks very different from the normal distribution function rule in Section 4.18. As with the normal function rule, however, a quick look at this mathematical expression tells us much about the distribution of t.

First of all, notice that the particular value of t enters this rule only as a squared quantity, showing that the distribution of sample t values must be symmetric, since a positive and a negative value having the same absolute size must be assigned the same probability (density) by this rule. Second, since all the constants in the function rule are positive numbers, and the entire term involving t is raised to a negative power, the largest possible density value is assigned to $t = 0$. Thus $t = 0$ is the distribution mode. If we inferred from the symmetry and unimodality of this distribution that the mean of t is also 0, we should be quite correct. In short, the t distribution is a unimodal, symmetric, bell-shaped distribution having a graphic form much like a normal distribution, even though the two function rules are quite dissimilar. Loosely speaking, the curve for a t distribution differs from the standardized normal in being "plumper" in extreme regions and "flatter" in the central region, as Figure 6.14.1 shows. (Note that both t and the standardized normal distribution have a mean of zero.)

The most important feature of the t distribution will appear if we return for a look at the function rule. Notice that the only unspecified constants in the rule are those represented in (6.14.5) by ν and $G(\nu)$, which depends only on ν. This is a **one-parameter distribution**: the single parameter is ν, called the **degrees of freedom.** *Ordinarily, in most applications of the t distribution to problems involving a single sample, ν is equal to $N - 1$, one less*

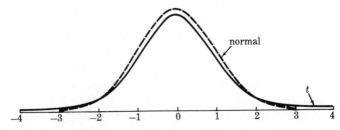

Figure 6.14.1. Distribution of t with $\nu = 4$, and Standardized Normal Distribution

than the number of independent observations in the sample. For samples of N independent observations from any normal population distribution, the exact distribution of sample t values depends only on the degrees of freedom, $N - 1$. Remember, however, that the value of $E(M)$ or μ must be specified when a t ratio is computed, although the true value of σ need not be known.

In principle, the value of ν can be any positive number, and it just happens that $\nu = N - 1$ is the value for the degrees of freedom for the particular t distributions we will use first. Later we will encounter problems calling for t distributions with other numbers of degrees of freedom. Like most theoretical distributions, the t distribution is actually a family of distributions, with general form determined by the function rule, but with particular probabilities dictated by the parameter ν. For any value of ν, the mean of the distribution of t is 0. For $\nu > 2$ the variance of the t distribution is $\nu/(\nu - 2)$, so that the smaller the value of ν the larger the variance. As ν becomes large the variance of the t distribution approaches 1.00, which is the variance of the standardized normal distribution.

Incidentally, the random variable t is often called "Student's t," and the distribution of t, "Student's distribution." This name comes from the statistician W. S. Gossett, who was the first to use this distribution in an important problem, and who first published his results in 1908 under the pen-name "Student."

6.15   THE t AND THE STANDARDIZED NORMAL DISTRIBUTION

As we have seen, the "shape" of the t distribution is not unlike that of the normal distribution. Just as for the standardized normal, the mean of the distribution of t is 0, although the variance of t is greater than 1.00. Given any extreme interval of fixed size on either tail of the t distribution, the probability associated with this interval in the t distribution is larger than that for the corresponding normal distribution of z_M. The smaller the value of ν, the larger is this discrepancy between t and normal probabilities at the extreme ends of each distribution. This reflects and partly explains the danger of using a t ratio as though it were a z ratio; extreme values of t are relatively more likely than comparable values of z_M. A small sample size corresponds to a small value of ν, or $N - 1$, and thus there is serious danger of underestimating the probability of an extreme deviation from expectation when sample size is small. This is apparent in the illustration (Figure 6.14.1) showing the distribution of t together with the standardized normal function.

Suppose that a sample of 5 observations is drawn, and from this sample we compute a ratio, t, using the estimate of σ_M from the sample. Furthermore, suppose that

$$t = \frac{M - E(M)}{\text{est. } \sigma_M} \geq 2.13.$$

That is, we obtain a value for t greater than or equal to 2.13. In the t distribution for samples of size 5 $(\nu = 4)$, this interval has probability of .05. That is, when sample size is 5, so that degrees of freedom are 4, the probability of obtaining a ratio in this interval of values is $1/20$. However, if the ratio is interpreted as a z_M variable, then the normal probability for this interval is .0166. Incorrectly considering a t ratio as a standardized normal variable leads one to underestimate the probability of values in extreme intervals, which, as we shall see in Chapter 7, are the intervals of greatest interest in significance tests.

On the other hand, notice what should happen to the distribution of t as ν becomes large (sample size grows large), as suggested both by Figure 6.14.1 and by the variance of a t distribution. *As the sample size N grows large, the distribution of t approaches the standardized normal distribution. For large numbers of degrees of freedom, the exact probabilities of intervals in the t distribution can be approximated closely by normal probabilities.*

The practical result of this convergence of the t and the normal probabilities is that the t ratio *can* be treated as a z_M ratio, provided that the sample size is substantial. The normal probabilities are quite close to—though not identical with—the exact t probabilities for large ν. On the other hand, when sample size is small the normal probabilities cannot safely be used, and instead one uses a special table based on the t distribution.

How large is "large enough" to permit use of the normal tables? If the population distribution is truly normal, a sample size of 30 or 40 permits a quite accurate use of the normal tables. Beyond this sample size, the normal probabilities are extremely close to the exact t probabilities. Of course, if a great deal of accuracy is desired, the t tables should be used even for larger sample sizes. For most purposes, an N of 30 or 40 should be large enough to permit the use of the normal tables.

Recall that the stipulation is made that the population distribution be *normal* when the t distribution is used, even when the normal approximations are substituted for the exact t-distribution probabilities. As we have already seen, for a normal population the distribution of sample means must be normal anyway; the difficulty with the use of a normal sampling distribution for small N comes solely from the fact that our estimate of the standard error is a random variable rather than a constant over samples, and this is the reason we must use the t distribution. Thus, the t distribution is related to the normal distribution in two distinct ways: the parent distribution must be normal if t probabilities are to be found exactly, and for sufficiently large N, the distribution of t approaches the normal sampling distribution in form.

It is apparent that the requirement that the population be normal limits the usefulness of the t distribution, since this is an assumption that we can seldom really justify in practical situations. Fortunately, when sample size is fairly large, and provided that the parent distribution is roughly unimodal and symmetric, the t distribution apparently still gives an adequate approximation to the exact (and often unknown) probabilities of intervals for t ratios under these circumstances. However, one should insist on a relatively *larger* sample size the *less* confident that he is that the normal rule holds for the population, if he plans to use the t distribution. In effect, if the sample size is large enough so that the normal probabilities are good approximations to the t probabilities anyway, then the form of the parent distribution is more or less irrelevant. However, often the sample size is so small that the t distribution must be used and here it *is* somewhat risky to make inferences from t ratios unless the population is more or less normally distributed.

6.16 TABLES OF THE t DISTRIBUTION

Unlike the table of the standardized normal function, which suffices for all possible normal distributions, tables of the t distribution must actually include many distributions, each depending on the value of ν, the degrees of freedom. Consequently, tables of t are usually given only in abbreviated form; otherwise, a whole volume would be required to show all the different t distributions one might need.

Table II in the Appendix shows selected percentage points of the distribution of t, in terms of the value of ν. Different ν values appear along the left-hand margin of the table. The top margin gives values of Q, which is $1 - P(t \leq a)$, one minus the cumulative probability that t is less than or equal to a specific value a, for a distribution within the given value for ν.

A cell of the table then shows the value of t cutting off the upper Q proportion of values in a distribution with v degrees of freedom.

This sounds rather complicated, but an example will clarify matters considerably: suppose that $N = 10$, and we want to know the value *beyond which* only 10 percent of all sample t values should lie. That is, for the distribution of t shown in Figure 6.16.1, we want the t value that cuts off the shaded area in the curve, the upper 10 percent.

First of all, since $N = 10$ and $v = N - 1 = 9$, we enter the table for the row marked 9. Now since we want the upper 10 percent, we find the column for which $Q = .1$. The corresponding cell in the table is the value of t we are looking for, $t = 1.383$. We can say that in a t distribution with 9 degrees of freedom, the probability is .10 that a t value *equals or exceeds* 1.383. Since the distribution is symmetric, we also know that the probability is .10 that a t value *equals or falls below* -1.383. If we wanted to know the probability that t equals or exceeds 1.383 in absolute value, then this must be .10 $+$ .10 $=$ .20, or $2Q$.

Suppose that in a sample of 21 cases, we get a t value of 1.98. We want to see if this value falls into the upper .05 of all values in the distribution. We enter row $v = 21 - 1 = 20$, and column $Q = .05$; the t value in the cell is 1.725. Our obtained value of t is larger than this, and so the obtained value does fall among the top 5 percent of all such values. On the other hand, suppose that the obtained t had been -3. Does this fall either in the top .001 or the bottom .001 of all such sample values? Again with $v = 20$, but this time with $Q = .001$, we find a t value of 3.552. This means that at or above 3.552 lie .001 of all sample values, and also at or below -3.552 lie .001 of all sample values. Hence our sample value does not fall into either of these intervals.

The very last row, marked ∞, shows the z values that cut off various areas in a normal distribution curve. If you trace down any given column, you find that as v gets larger the t value bounding the area specified by the column comes closer and closer to this normal deviate value, until finally, for an infinite sample size, the required value of t is the same as that for z.

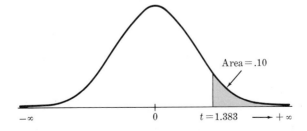

Figure 6.16.1

6.17 THE CONCEPT OF DEGREES OF FREEDOM

Before we proceed to the uses of the t distribution, it is well to examine the notion of degrees of freedom. The degrees of freedom parameter reflects the fact that a t ratio involves a sample standard deviation as the basis for estimating σ_M. Recall the basic definitions of the sample variance and the sample standard deviation:

$$s^2 = \frac{\sum (X - M)^2}{N}$$

and
$$s = \sqrt{\frac{\sum (X - M)^2}{N}}.$$

The sample variance and standard deviation are both based upon a sum of squared deviations from the sample mean. However, recall another fact of importance about deviations from a mean: in Section 5.17 it was shown that

$$\sum_i (X_i - M) = 0$$

(the sum of deviations about the mean must be zero).

These two facts have an important consequence: suppose that you are told that $N = 4$ in some sample, and that you are to guess the four deviations from the mean M. For the first deviation you can guess any number, and suppose you say

$$d_1 = 6.$$

Similarly, quite at will, you could assign values to two more deviations, say

$$d_2 = -9$$

and
$$d_3 = -7.$$

However, when you come to the fourth deviation value, you are *no longer free* to guess any number you please. The value of d_4 *must* be

$$d_4 = 0 - d_1 - d_2 - d_3,$$

or
$$d_4 = 0 - 6 + 9 + 7 = 10.$$

In short, given the values of any $N - 1$ deviations from the mean, which could be any set of numbers, the value of the last deviation is completely determined. Thus we say that there are $N - 1$ degrees of freedom for a sample variance, reflecting the fact that only $N - 1$ deviations are "free" to be any number, but that given these free values, the last deviation is completely determined. It is not the sample size per se that dictates the distribution of t, but rather the number of degrees of freedom in the variance (and standard deviation) estimate. We will encounter the concept of degrees of freedom again in Sections 6.21 and 6.28 in connection with the χ^2 and F distributions.

6.18 CONFIDENCE INTERVALS FOR THE MEAN WHEN σ^2 IS UNKNOWN

When we are sampling from a normal population and we do not know the variance of the population, the sampling distribution of M is not normal even though the population itself is normally distributed, as we saw in Section 6.14. Unless N is large enough for the central limit theorem to apply, the confidence limits cannot be determined from the tables of the normal distribution. When the sample size is so small that the normal approximation to the sampling distribution is not good, then the t distribution is used to establish confidence limits for the mean. For some fixed percentage representing the confidence level, $100(1 - \alpha)$ percent, the sample confidence limits depend upon three things: the sample value of M, the estimated standard error, or est. σ_M, and the number of degrees of freedom, ν. For some specified value of ν, then, the $100(1 - \alpha)$ percent confidence limits are found from

$$M - t_{(\alpha/2;\,\nu)}\ (\text{est. } \sigma_M)$$

and
$$M + t_{(\alpha/2;\,\nu)}\ (\text{est. } \sigma_M). \tag{6.18.1*}$$

Here, $t_{(\alpha/2;\,\nu)}$ represents the value of t that bounds the upper $\alpha/2$ proportion of cases in a t distribution with ν degrees of freedom. In Table II this is the value listed for $Q = \alpha/2$ and ν. Thus, if one wants the 99 percent confidence limits, $\alpha = .01$, and one looks in the table for $Q = .005$.

For example, imagine a study using 8 independent observations drawn from a normal population. The sample mean is 49 and the estimated standard error of the mean is 3.7. Now we want to find the 95 percent confidence limits. First of all, $\alpha = .05$, so that $Q = .025$. The value of ν is $N - 1$, or 7. The table shows a t value of 2.365 for $Q = .025$ and $\nu = 7$, so that $t_{(\alpha/2;\,\nu)} = 2.365$. The confidence limits are

$$49 - (2.365)(3.7) = 40.25$$

and
$$49 + (2.365)(3.7) = 57.75.$$

Over all random samples, the probability is .95 that the true value of μ is covered by an interval such as that between 40.25 and 57.75, the confidence interval calculated for this sample.

In summary, confidence limits are calculated in much the same way, and have the same general interpretation, when based on the t distribution as for the normal distribution. The essential difference is that values of t corresponding to $\alpha/2$ and ν must be used instead of normal z values.

In Equation (6.18.1), the estimate of σ_M is given either by $s/\sqrt{N-1}$ or by $\hat{s}/\sqrt{N}$. These two estimates are equivalent, and the $100(1-\alpha)$ percent confidence interval can then be written in the form

$$M \pm t_{(\alpha/2,\nu)} \frac{s}{\sqrt{N-1}} \qquad \text{or} \qquad M \pm t_{(\alpha/2,\nu)} \frac{\hat{s}}{\sqrt{N}}. \qquad (6.18.2^*)$$

6.19 THE SAMPLING DISTRIBUTION OF DIFFERENCES BETWEEN MEANS

Often in statistical work we are interested not just in a single mean, but in the difference between two means. For example, we may be interested in the difference in the production rates of two machines, or the difference in the IQ's of children from two different schools, or the difference in performance of two different types of securities, and so on. The comparison of two populations is of interest more often than the investigation of a single population, and the comparison of primary interest concerns the means of the populations. Suppose that two populations, 1 and 2, are of interest. We draw a sample of size N_1 from population 1 and an *independent* sample of size N_2 from population 2, and consider the difference between their means, $M_1 - M_2$. Now suppose that we keep on drawing pairs of independent samples of these sizes from these populations. For each pair of samples drawn, the difference $M_1 - M_2$ is recorded. What is the distribution of such sample *differences* that we should expect in the long run? In other words, what is the sampling distribution of the difference between two means?

You may already have anticipated the form of the sampling distribution of the difference between two means, since all the groundwork for this distribution has been laid in Section 5.23. The difference between sample means drawn from independent samples is actually a linear combination:

$$(1)M_1 + (-1)M_2.$$

Let us apply the results of Sections 5.23 and 5.24 to this problem. In the first place,

$$E(M_1 - M_2) = E(M_1) - E(M_2) = \mu_1 - \mu_2, \qquad (6.19.1^*)$$

which accords with principle (5.24.1) for any linear combination. Second, what is the standard error of the difference between two independent sample means? By principle (5.24.3),

$$\text{var. } (M_1 - M_2) = (1)^2 \sigma^2{}_{M_1} + (-1)^2 \sigma^2{}_{M_2}$$
$$= \sigma^2{}_{M_1} + \sigma^2{}_{M_2}. \tag{6.19.2*}$$

Hence, the standard error of the difference, $\sigma_{\text{diff.}}$, is

$$\sigma_{\text{diff.}} = \sqrt{\sigma^2{}_{M_1} + \sigma^2{}_{M_2}} = \sqrt{\frac{\sigma_1^2}{N_1} + \frac{\sigma_2^2}{N_2}}, \tag{6.19.3*}$$

provided that samples 1 and 2 are completely independent.

Notice that there is no requirement at all that the samples be of equal size. Regardless of the sample sizes, the expectation of the difference between two means is always the difference between their expectations, and the variance of the difference between two *independent* means is the *sum* of the separate sampling variances.

Furthermore, these statements about the mean and the standard error of a difference between means are true regardless of the form of the parent distributions. However, the form of the sampling distribution can also be specified under the following condition:

If the distribution for each of two populations is normal then the distribution of differences between sample means is normal.

This follows quite simply from the principle in Section 5.23 for linear combinations. When we can assume both populations normal, the form of the sampling distribution is known to be *exactly* normal.

On the other hand, one or both of the original distributions may not be normal; in this case the central limit theorem comes to our aid:

As both N_1 and N_2 grow infinitely large, the sampling distribution of the difference between means approaches a normal distribution, regardless of the form of the original distributions.

In short, when we are dealing with two very large samples, then the question of the form of the original distributions becomes irrelevant, and we can approximate the sampling distribution of the difference between means by a normal distribution.

In the above discussion, it was assumed not only that the two populations were normal, but also that their respective variances were known. If the first assumption still holds but only the *ratio* of σ_1^2 to σ_2^2 is known, then the sampling distribution of the difference in means is a t distribution. The mean of this sampling distribution is still $\mu_1 - \mu_2$, but since the variances are not known, an estimate of the standard error is needed. To simplify the determination of this estimate, one further assumption becomes necessary: the assumption that the variances of the two populations are equal, that $\sigma_1{}^2 = \sigma_2{}^2$.

Given these assumptions, then the distribution of t for a difference has the same form as for a single mean, except that the degrees of freedom are

$$\nu = N_1 - 1 + N_2 - 1 = N_1 + N_2 - 2.$$

When samples are drawn from populations with equal variance, then the estimated standard error of a difference takes a somewhat different form. First of all, when $\sigma_1 = \sigma_2 = \sigma$,

$$\sigma_{\text{diff.}} = \sqrt{\frac{\sigma^2}{N_1} + \frac{\sigma^2}{N_2}} = \sqrt{\sigma^2 \left(\frac{1}{N_1} + \frac{1}{N_2}\right)}. \qquad (6.19.4^*)$$

Now, as we showed in Section 6.9, when one has two or more estimates of the same parameter σ^2, the *pooled* estimate is actually better than either one taken separately. From Equation (6.9.5) it follows that

$$\text{est. } \sigma^2 = \frac{N_1 s_1{}^2 + N_2 s_2{}^2}{N_1 + N_2 - 2} \qquad (6.19.5)$$

is our best estimate of σ^2 based on the two samples. Hence

$$\begin{aligned} \text{est. } \sigma_{\text{diff.}} &= \sqrt{\text{est. } \sigma^2 \left(\frac{1}{N_1} + \frac{1}{N_2}\right)} \\ &= \sqrt{\left(\frac{N_1 s_1{}^2 + N_2 s_2{}^2}{N_1 + N_2 - 2}\right)\left(\frac{N_1 + N_2}{N_1 N_2}\right)}. \end{aligned} \qquad (6.19.6)$$

In order to justify the use of the t distribution in problems involving a difference between means, one must make two assumptions: the populations sampled are normal, and the population variances are homogeneous, σ^2 having the same value for each population. Formally, these two assumptions are essential if the t probabilities given by the table are to be exact. On the other hand, in practical situations these assumptions are sometimes violated with rather small effect on the conclusions.

The first assumption, that of a normal distribution in the populations, is apparently the less important of the two. So long as the sample size is even moderate for each group quite severe departures from normality seem to make little practical difference in the conclusions reached. Naturally, the results are more accurate the more nearly unimodal and symmetric the population distributions are, and thus if one suspects radical departures from a generally normal form then he should plan on larger samples.

On the other hand, the assumption of homogeneity of variances is more important. In older work it was often suggested that a separate test for homogeneity of variance be carried out before the t test itself, in order to see if this assumption were at all reasonable. However, the most modern authorities suggest that this is not really worth the trouble involved. In circumstances where they are needed most (small samples), the tests for homogeneity are poorest. Furthermore, for samples of equal size relatively big differences in the population variances seem to have relatively small consequencies for the conclusions derived from a t test. On the other hand, when the variances are quite unequal the use of different sample sizes can have serious effects on the conclusions. The moral should be plain: given the usual freedom about sample size in experimental work, *when in doubt use samples of the same size.*

However, sometimes it is not possible to obtain an equal number in each group. Then one way out of this problem is by the use of a correction in the value for degrees of freedom. This is useful when one cannot assume equal population variances and samples are of different size. In this situation, however, the separate standard errors are computed from each sample and the pooled estimate is not made. Then the corrected number of degrees of freedom is found from

$$\nu = \frac{(\text{est. } \sigma^2_{M_1} + \text{est. } \sigma^2_{M_2})^2}{(\text{est. } \sigma^2_{M_1})^2/(N_1 + 1) + (\text{est. } \sigma^2_{M_2})^2/(N_2 + 1)} - 2. \quad (6.19.7)$$

This need not result in a whole value for ν, in which case the use of the nearest whole value for ν is sufficiently accurate for most purposes. Of course, if the samples are large enough so that the normal approximation in applicable, then the normal tables can be used and it is not necessary to compute ν.

6.20 CONFIDENCE INTERVALS FOR DIFFERENCES IN MEANS

From the previous section, we can see that confidence intervals for differences in means can be determined either from the normal tables or from the t tables, depending on whether the sample size is large and on whether or not the variances of the two populations are known. In either case, of course, it is assumed that the two populations are normally distributed.

If σ_1^2 and σ_2^2 are known, then a $100(1 - \alpha)$ percent confidence interval for $\mu_1 - \mu_2$ is given by

$$(M_1 - M_2) \pm z \sqrt{\frac{\sigma_1^2}{N^1} + \frac{\sigma_2^2}{N_2}}, \quad (6.20.1^*)$$

where z is determined from the normal tables from our choice of $(1 - \alpha)$.

If σ_1^2 and σ_2^2 are. *not* known, but we can assume that they are equal, then we must use an estimate of the standard error of the difference between means, and the $100(1 - \alpha)$ percent confidence interval for $\mu_1 - \mu_2$ is given by

$$(M_1 - M_2) \pm t_{(\alpha/2;\, \nu)} \sqrt{\left(\frac{N_1 s_1^2 + N_2 s_2^2}{N_1 + N_2 - 2}\right)\left(\frac{N_1 + N_2}{N_1 N_2}\right)}, \quad (6.20.2^*)$$

where the degrees of freedom, ν, equals $N_1 + N_2 - 2$. Here t is determined from the t tables.

For example, suppose that we assume that $\sigma_1^2 = \sigma_2^2$ and that we take independent samples of size $N_1 = 5$ and $N_2 = 7$, with the following sample results:

$$M_1 = 18 \qquad M_2 = 20$$
$$s_1^2 = 4.8 \qquad s_2^2 = 5.0.$$

The estimated standard error of the difference is found by the pooling procedure given in the last section:

$$\text{est. } \sigma_{\text{diff.}} = \sqrt{\text{est. } \sigma^2 \left(\frac{1}{N_1} + \frac{1}{N_2}\right)}$$

$$= \sqrt{\frac{(5)(4.8) + (7)(5.0)}{10}\left(\frac{12}{35}\right)}$$

$$= \sqrt{2.02}$$

$$= 1.42.$$

Suppose we wanted a 99 percent confidence interval. The confidence limits are then

$$(M_1 - M_2) \pm t_{(.005;\, 10)}(\text{est. } \sigma_{\text{diff.}}),$$

or

$$-2 \pm (3.169)(1.42),$$

or approximately -6.5 and $+2.5$.

6.21   THE CHI-SQUARE DISTRIBUTION

Thus far we have been concerned with the sampling distribution of the sample mean M and of differences in sample means. In some instances the sampling distribution of the sample variance, s^2, is of interest. If the population from which we are sampling is normally distributed, the sampling distribution of s^2 can be expressed in terms of a χ^2 (chi-square) distribution. We will first discuss the theoretical development of the chi-square distribution and then show its relationship to the sampling distribution of s^2.

Suppose that there exists a population having a normal distribution of values of a random variable X. The mean of this distribution is $E(Y) = \mu$, and the variance is

$$E(Y - \mu)^2 = \sigma^2.$$

Now cases are sampled from this distribution *one* at a time, $N = 1$. For each case sampled the **squared standardized value**

$$z^2 = \frac{(Y - \mu)^2}{\sigma^2} \qquad (6.21.1)$$

is computed. Let us call this squared standardized value $\chi^2_{(1)}$, so that

$$\chi^2_{(1)} = z^2. \qquad (6.21.2)$$

Now we will look into the sampling distribution of this variable $\chi^2_{(1)}$.

First of all, what is the range of values that χ^2 might take on? The original normal variable Y ranges over all real numbers, and this is also the range of the standardized variable z. However, $\chi^2_{(1)}$ is always a squared quantity, and so its range must be all the *nonnegative* real numbers, from zero to a positive infinity. We can also infer something about the form of this distribution of $\chi^2_{(1)}$. The bulk of the cases (about 68 percent) in a normal distribution of standardized scores must lie between -1 and 1. Given a z between -1 and 1, the corresponding $\chi^2_{(1)}$ value lies between 0 and 1, so that the bulk of this sampling distribution will fall in the interval between 0 and 1. This implies that the form of the distribution of $\chi^2_{(1)}$ will be very skewed, with a high probability for a value in the interval from 0 to 1, and relatively low probability in the interval with lower bound 1 and approaching positive ∞ as its upper bound. The graph of the distribution of $\chi^2_{(1)}$ is represented in Figure 6.21.1.

Figure 6.21.1 pictures **the chi-square distribution with one degree of freedom. The distribution of the random variable $\chi^2_{(1)}$, where**

$$\chi^2_{(1)} = \frac{(Y - \mu)^2}{\sigma^2}, \qquad (6.21.3)$$

and Y is normally distributed with mean μ and variance σ^2, is a chi-square distribution with 1 degree of freedom.

Now let us go a little further. Suppose that samples of *two* cases are drawn independently and at random from a normal distribution. We find the squared standardized value corresponding to each observation:

$$z_1^2 = \frac{(Y_1 - \mu)^2}{\sigma^2}$$

$$z_2^2 = \frac{(Y_2 - \mu)^2}{\sigma^2}.$$

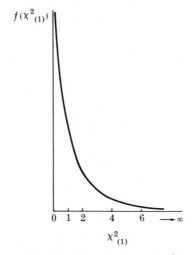

Figure 6.21.1. The Distribution of χ^2 for $\nu = 1$

The *sum* of these two squared standardized values is found and designated as the random variable $\chi^2_{(2)}$:

$$\chi^2_{(2)} = \frac{(Y_1 - \mu)^2}{\sigma^2} + \frac{(Y_2 - \mu)^2}{\sigma^2} = z_1{}^2 + z_2{}^2. \qquad (6.21.4)$$

If we look into the distribution of the random variable $\chi^2_{(2)}$ we find that the range of possible values extends over all nonnegative real numbers. However, since the random variable is based on *two* independent observations, the distribution is somewhat less skewed than for $\chi^2_{(1)}$; the probability is not so high that the sum of two squared standardized values should fall between 0 and 1. This is illustrated in Figure 6.21.2. This illustrates a **chi-square distribution with two degrees of freedom.**

Finally, suppose that we took N independent observations at random from a normal distribution with mean μ and variance σ^2 and defined

$$\chi^2_{(N)} = \frac{\sum\limits_{i}^{N} (Y_i - \mu)^2}{\sigma^2} = \sum_i z_i{}^2. \qquad (6.21.5^*)$$

The distribution of this random variable will have a form that depends upon the number of independent observations taken at one time. **In general, for N independent observations from a normal population, the sum of the squared standardized values for the observations has a chi-square distribution with N degrees of freedom.** Notice that the standardized values must be relative to the *population mean* and the *population standard deviation*.

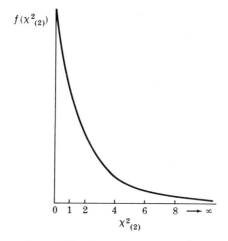

Figure 6.21.2. The Distribution of χ^2 for $\nu = 2$

Another way of expressing the definition of a chi-square variable is as follows: **a random variable has a chi-square distribution with N degrees of freedom if it has the same distribution as the sum of the squares of N independent random variables, each normally distributed, and each having expectation 0 and variance 1.** Chi-square distributions for larger numbers of degrees of freedom have the form illustrated in Figure 6.21.3.

The function rule assigning a probability density to each possible value of χ^2 is given by

$$f(\chi^2) = h(\nu) \exp(-\chi^2/2) (\chi^2)^{(\nu/2)-1} \qquad \text{for } \chi^2 \geq 0,$$

$$\nu > 0. \qquad (6.21.6^*)$$

As shown in Figures 6.21.1 through 6.21.3, the plot of this density function always presents the picture of a very positively skewed distribution, at least for relatively small values of ν, with a distinct mode at the point $\nu - 2$

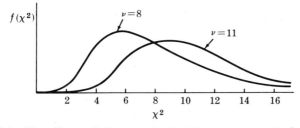

Figure 6.21.3. The General Form of the Distribution of χ^2 for Larger Numbers of Degrees of Freedom

for $\nu > 2$. However, as ν is increased, the form of the distribution appears to be less skewed to the right. In this function rule the value $h(\nu)$ is a constant depending only on the parameter ν. Unlike the normal and the t distributions, it is not so easy to infer the characteristics of the distribution of χ^2 from this function rule. However, the thing to notice is that there is only one value other than χ^2 that must be specified in order to find the density: this is the parameter ν. Like the t distribution, the distribution of χ^2 depends only on the degrees of freedom, the parameter ν. The family of chi-square distributions all follow this general rule, but the exact form of the distribution depends on the number of degrees of freedom, ν. In principle, the value of ν can be any positive number, but in the applications we will make of this distribution in this chapter, ν will depend only on the sample size, N.

From the original definition of χ^2, it is quite easy to infer what the mean of this distribution must be. For N independent observations we defined

$$\chi^2_{(N)} = \frac{\sum_{i}^{N} (Y_i - \mu)^2}{\sigma^2} = \sum_{i}^{N} z_i^2,$$

where the random variable Y is normal and z is the corresponding standardized random variable. If we take the expectation of $\chi^2_{(N)}$ over all possible samples of N independent observations, we find

$$E[\chi^2_{(N)}] = E\left(\sum_{i}^{N} z_i^2\right)$$

$$= \sum_{i}^{N} E(z_i^2). \tag{6.21.7}$$

However, for any i, $E(z_i^2) = 1$, by the definition of a standardized random variable. Thus,

$$E[\chi^2_{(N)}] = N. \tag{6.21.8*}$$

If a chi-square variable has a distribution with ν degrees of freedom, then the expectation of the distribution is simply ν. In our derivation, $\nu = N$, the number of independent observations, and so the mean of the distribution is the number of degrees of freedom, or N.

The variance of a chi-square distribution with ν degrees of freedom is always

$$\text{var. } [\chi^2_{(\nu)}] = 2\nu. \tag{6.21.9*}$$

Giving the degrees of freedom ν actually gives all the information one needs to specify the particular chi-square distribution completely.

6.22   TABLES OF THE CHI-SQUARE DISTRIBUTION

Like the t distribution, the particular distribution of χ^2 depends on the parameter ν, and it is difficult to give tables of the distribution for all values of ν that one might need. Thus, Table III in the Appendix, like Table II, is a condensed table, showing values of χ^2 that correspond to percentage points in various distributions specified by ν. The rows of Table III list various degrees of freedom ν and the column headings are probabilities Q, just as in the t table. The numbers in the body of the table give the values of χ^2 such that in a distribution with ν degrees of freedom, *the probability of a sample chi-square value this large or larger is Q* (Figure 6.22.1).

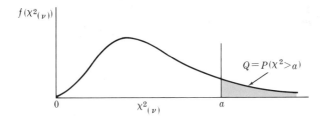

Figure 6.22.1

For example, suppose that we are dealing with a chi-square distribution with 5 degrees of freedom. Find the row ν equal to 5, and the column headed .05. The value in this row and column is approximately 11.071, showing that for 5 degrees of freedom, random samples showing a chi-square value of 11.071 *or more* should in the long run occur about five times in a hundred. As another example, look at the row labeled $\nu = 2$ and the column headed .10. The entry in this combination of row and column indicates that in a distribution with 2 degrees of freedom, chi-square values of 4.605 or more should occur with probability of about 1 in 10, under random sampling. Finally, look at the row for $\nu = 24$, and the column for .001; the cell entry indicates that chi-square values of 51.179 or more occur with probability of only about .001 in a distribution with 24 degrees of freedom.

6.23   THE ADDITION OF CHI-SQUARE VARIABLES

When several independent random variables each have a chi-square distribution, then the distribution of the *sum* of the variables is also known. This is a most important and useful property of variables of this kind, and it can be stated more precisely as follows:

If a random variable $\chi^2_{(\nu_1)}$ has a chi-square distribution with ν_1 degrees of freedom, and an independent random variable $\chi^2_{(\nu_2)}$ has a chi-square distribution with ν_2 degrees of freedom, then the new random variable formed from the sum of these variables,

$$\chi^2_{(\nu_1+\nu_2)} = \chi^2_{(\nu_1)} + \chi^2_{(\nu_2)}, \tag{6.23.1*}$$

has a chi-square distribution with $\nu_1 + \nu_2$ degrees of freedom.

In short, the new random variable formed by taking the sum of two independent chi-square variables is itself distributed as χ^2, with degrees of freedom equal to the sum of those for the original distributions. A little thought will convince you that this must be so; if $\chi^2_{(\nu_1)}$ is the sum of squares of ν_1 independent variables, each normally distributed with mean 0 and variance 1, and $\chi^2_{(\nu_2)}$ is based on the sum of ν_2 independent such squares, then $\chi^2_{(\nu_1+\nu_2)}$ must also be a sum of $\nu_1 + \nu_2$ squared values of independent normal variables each with mean 0 and variance 1. Therefore $\chi^2_{(\nu_1+\nu_2)}$ qualifies as a chi-square variable by definition.

6.24 THE DISTRIBUTION OF THE SAMPLE VARIANCE FROM A NORMAL POPULATION

In this section, the sampling distribution of the sample variance will be studied. It will be assumed that the population actually sampled is normal, and the results in this section apply, strictly speaking, only when this assumption is true.

Recall that the sample variance was defined by

$$s^2 = \frac{\sum_i^N (Y_i - M)^2}{N}.$$

Suppose, however, that we actually knew the value of μ. For any single deviation from μ, $(Y_i - \mu)$, it is true that

$$(Y_i - \mu) = (Y_i - M) + (M - \mu), \tag{6.24.1}$$

since adding and subtracting the same number to any given number does not change the value. Now, for any squared deviation from μ,

$$(Y_i - \mu)^2 = [(Y_i - M) + (M - \mu)]^2$$
$$= (Y_i - M)^2 + (M - \mu)^2 + 2(Y_i - M)(M - \mu). \tag{6.24.2}$$

Summing over all of the squared deviations from μ in the sample, we have

$$\sum_i^N (Y_i - \mu)^2 = \sum_i (Y_i - M)^2 + \sum_i (M - \mu)^2$$
$$+ 2 \sum_i (Y_i - M)(M - \mu). \quad (6.24.3)$$

The value of $(M - \mu)^2$ is constant over all individuals i, so that

$$\sum_i^N (M - \mu)^2 = N(M - \mu)^2.$$

Furthermore, since $(M - \mu)$ is the same for all i, then

$$\sum_i^N (Y_i - M)(M - \mu) = (M - \mu) \sum_i (Y_i - M)$$
$$= 0,$$

because the sum of the deviations about a sample mean must be 0. Hence,

$$\sum_i (Y_i - \mu)^2 = \sum_i (Y_i - M)^2 + N(M - \mu)^2. \quad (6.24.4^*)$$

If we divide the entire expression by σ^2, this becomes

$$\frac{\sum_i (Y_i - \mu)^2}{\sigma^2} = \frac{\sum_i (Y_i - M)^2}{\sigma^2} + \frac{N(M - \mu)^2}{\sigma^2}. \quad (6.24.5)$$

Now look at the term on the left in this expression: given N independent trials from a normal population distribution,

$$\frac{\sum_i (Y_i - \mu)^2}{\sigma^2} = \chi^2_{(N)}, \quad (6.24.6^*)$$

so that this sum of squared deviations from μ, divided by σ^2, qualifies as a chi-square variable with N degrees of freedom.

Furthermore, if the population distribution is normal, the sampling distribution of M must be normal, with mean μ and variance σ^2/N. Thus,

$$\frac{N(M - \mu)^2}{\sigma^2} = \frac{(M - \mu)^2}{(\sigma^2/N)} = \chi^2_{(1)}. \quad (6.24.7^*)$$

It follows that the last term in Equation (6.24.5) qualifies as a chi-square variable with 1 degree of freedom, since this is just z_M^2 from a normal distribution. Putting these two facts together, we have

$$\chi^2_{(N)} = \frac{\sum_i (Y_i - M)^2}{\sigma^2} + \chi^2_{(1)}, \quad (6.24.8)$$

or
$$\chi^2_{(N)} = \frac{Ns^2}{\sigma^2} + \chi^2_{(1)}. \tag{6.24.9*}$$

The term on the left is a chi-square variable with N degrees of freedom, and the term on the extreme right is a chi-square variable with 1 degree of freedom. From this fact it can be shown, by use of the addition property of chi-square variables and the fact that M and s^2 are independent in samples from a normally distributed population, that

$$\frac{Ns^2}{\sigma^2} = \chi^2_{(N-1)}, \tag{6.24.10*}$$

the ratio Ns^2/σ^2 is a chi-square variable with $N - 1$ degrees of freedom.

If, instead of s^2, we use the unbiased estimator $\hat{s}^2$, we find that

$$\frac{(N-1)\hat{s}^2}{\sigma^2} = \chi^2_{(N-1)}, \tag{6.24.11*}$$

meaning that the ratio of $(N - 1)\hat{s}^2$ to σ^2 is a random variable distributed as chi-square with $N - 1$ degrees of freedom. This fact is the basis for inferences about the variance of a single normally distributed population.

This same idea applies to the sampling distribution of any estimate of a variance, provided that the basic population sampled is normal. Thus, it is also true that when the pooled estimate of σ^2 is made from two independent samples of N_1 and N_2 cases,

$$\frac{(N_1 + N_2 - 2)\text{ est. }\sigma^2}{\sigma^2} = \chi^2_{(N_1+N_2-2)}. \tag{6.24.12*}$$

The pooled sample estimate of σ^2, multiplied by the degrees of freedom and divided by the true value of σ^2, is a chi-square variable with $\nu = N_1 + N_2 - 2$.

The sampling distribution of the unbiased estimate $\hat{s}^2$ is actually the distribution of the variable

$$\hat{s}^2 = \frac{\chi^2_{(\nu)}\sigma^2}{\nu}, \quad \text{or} \quad \frac{\text{est. }\sigma^2}{\sigma^2} = \frac{\chi^2_{(\nu)}}{\nu}, \tag{6.24.13*}$$

a chi-square variable multiplied by σ^2 and divided by degrees of freedom. However, inferences are usually made in terms of the distribution of $[(N - 1)\hat{s}^2]/\sigma^2$ since this is just a chi-square variable; thus, one can get by with only one chi-square table for this as well as for other uses of this distribution.

6.25   CONFIDENCE INTERVALS FOR THE VARIANCE AND STANDARD DEVIATION

Finding confidence intervals for the variance is quite simple, provided that the normal distribution rule holds for the population.

Suppose that we had a sample of some N independent observations, and we wanted the 95 percent confidence limits for σ^2. For samples from a normal distribution, it must be true that

$$P\left[\chi^2_{(N-1;\,.975)} \leq \frac{(N-1)\hat{s}^2}{\sigma^2} \leq \chi^2_{(N-1;\,.025)}\right] = .95, \qquad (6.25.1^*)$$

where $\chi^2_{(N-1;\,.975)}$ is the value in a chi-square distribution with $N-1$ degrees of freedom cutting off the upper .975 of sample values, and $\chi^2_{(N-1;\,.025)}$ is the value cutting off the upper .025 of sample values. This inequality can be manipulated to show that it is also true that

$$P\left[\frac{(N-1)\hat{s}^2}{\chi^2_{(N-1;\,.025)}} \leq \sigma^2 \leq \frac{(N-1)\hat{s}^2}{\chi^2_{(N-1;\,.975)}}\right] = .95. \qquad (6.25.2^*)$$

That is, the probability is .95 that the true value of σ^2 will be covered by an interval with limits found by

$$\frac{(N-1)\hat{s}^2}{\chi^2_{(N-1;\,.025)}}$$

and

$$\frac{(N-1)\hat{s}^2}{\chi^2_{(N-1;\,.975)}}.$$

Suppose, for example, that a sample of 15 cases is drawn from a normal distribution. We want the 95 percent confidence limits for σ^2. The value of $\hat{s}^2$ is 10; this is our best single estimate of σ^2, of course. For $\nu = 14$ and for $Q = .025$, $\chi^2_{(14;\,.025)}$ is found from Table III to be 26.12, and the value of $\chi^2_{(14;\,.975)}$ is 5.63. Thus the confidence limits for σ^2 are

$$\frac{(14)\,(10)}{26.12}, \text{ or } 5.36,$$

and

$$\frac{(14)\,(10)}{5.63}, \text{ or } 24.87.$$

We can say that the probability is .95 that an interval such as 5.36 to 24.87 covers the true value of σ^2.

In general, to find the $100(1 - \alpha)$ percent confidence limits, we take

$$\frac{(N - 1)\hat{s}^2}{\chi^2_{(N-1;\alpha/2)}}$$

and $$\frac{(N - 1)\hat{s}^2}{\chi^2_{(N-1;1-(\alpha/2))}}.$$ (6.25.3*)

These limits can always be turned into confidence limits for σ by replacing each limiting value by its square root.

On the other hand, this procedure for finding confidence limits for σ^2 contains a serious difficulty. When a confidence interval for μ is found by the methods of Section 6.11 or of Section 6.18, the confidence limits are taken to be equally distant from the obtained value of M. Since the normal sampling distribution of M is symmetric, and the distribution of t is also symmetric, these methods of constructing confidence intervals for M give *shortest* intervals. As mentioned in Section 6.11, it is perfectly possible to construct 95 percent, 99 percent, or any other confidence intervals for M in a variety of ways, but making the confidence limits equidistant from M gives the shortest of all possible such intervals. In other words, by the methods of 6.11 and 6.18, we "pin down" the value of μ within the smallest possible estimated range of values.

Unfortunately, this is not true when we find a confidence interval for σ^2 by the method just introduced. Here, the confidence limits are determined by values of χ^2 cutting off highest and lowest intervals having the same probability, $\alpha/2$, in the asymmetric chi-square distribution. This does make it true that the *failure* of the confidence interval to cover any hypothetical value of σ_0^2 is equivalent to the rejection of σ_0^2 as a hypothesis at the α level, two-tailed, as we shall see in Chapter 7. It does *not*, however, make it true that we have the shortest possible such interval of possible values of σ^2 for a given value of $(1 - \alpha)$, nor that we have the shortest possible interval of "good bets" about the value of σ^2. For this reason, the confidence interval for σ^2, although corresponding to procedures for testing hypotheses about σ^2, is not particularly useful when our interest lies in obtaining the best possible estimates of σ^2 from the data. More advanced methods do exist for finding this shortest possible confidence interval for σ^2, and useful tables exist for finding these intervals.

Regardless of whether confidence intervals are found by the method given here, or by a method giving shortest intervals, it is true that the larger the sample size, the narrower is the interval with a given probability of covering the true value of σ^2. For this reason, when we are estimating σ^2 for any purpose, the larger the sample size the more confident we can be of coming close to the value of σ^2 with our sample estimate $\hat{s}^2$.

6.26 THE IMPORTANCE OF THE NORMALITY ASSUMPTION IN INFERENCES ABOUT σ^2

In Section 6.15 it was pointed out that although the rationale for the t distribution demands the assumption of a normal population distribution, in practice the t distribution may be applied when the parent distribution is not normal, provided that N is at least moderately large. *This is not the case for inferences about the variance, however.* One runs a considerable risk of error in using the chi-square distribution either to test a hypothesis about a variance or to find confidence limits unless the population distribution is normal, or approximately so. The effect of the violation of the normality assumption is usually minor for large N, but can be quite serious when inferences are made about variances for moderate N. This principle is an important one and will be emphasized several times in succeeding sections.

6.27 THE NORMAL APPROXIMATION TO THE CHI-SQUARE DISTRIBUTION

As the number of degrees of freedom grows infinitely large, the distribution of χ^2 approaches the normal distribution. You should hardly find this principle surprising by now! Actually, the same mechanism is at work here as in the central limit theorem: for any ν greater than 1, the chi-square variable is equivalent to a *sum* of ν independent random variables. Thus, like the sampling distribution of the mean, M, given enough summed terms the sampling distribution of $(N-1)\hat{s}^2/\sigma^2$ approaches the normal form in spite of the fact that each component of the sum does not have a normal distribution when sampled alone.

This fact is of more than theoretical interest when very large samples are used. For very large ν, the probability for any interval of values of $\chi^2_{(\nu)}$ can be found from the normal standardized scores

$$z = \frac{\chi^2_{(\nu)} - \nu}{\sqrt{2\nu}}, \qquad (6.27.1^*)$$

since the mean and variance of a chi-square distribution are ν and 2ν respectively. However, this approximation is not good unless ν is extremely large. A somewhat better approximation procedure is to find the value of $\chi^2_{(\nu)}$ that cuts off the *upper* Q proportion of cases by taking

$$\chi^2_{(\nu;Q)} = \tfrac{1}{2}\{z_Q + \sqrt{2\nu - 1}\}^2. \qquad (6.27.2)$$

This can be used to find chi-square values for ν greater than 100, which are not given in Table III of the appendix. The necessary z_Q values are listed at the bottom of Table III and can also be found from Table I.

6.28   THE F DISTRIBUTION

It is rather rare to find a problem that centers on the value of a single population variance. Somewhat more common, however, is the situation where the variances of two populations are compared for equality. That is, we ask if the variability of population 1 is precisely equal to that of population 2, so that $\sigma_1^2 = \sigma_2^2$. The F distribution is used to make inferences of this sort, although we will also use this same distribution for quite a different kind of problem in Chapter 11 (Volume II).

Imagine two distinct populations, each showing a normal distribution of the variable Y. The means of the two populations may be different, but each population shows the same variance, σ^2. We draw two independent random samples; the first sample, from population 1, contains N_1 cases, and that from population 2 consists of N_2 cases. From each sample, an unbiased estimate of σ^2 is computed, based on $\hat{s}_1^2$ and $\hat{s}_2^2$ for each pair of samples. From the results of Section 6.24 we know that

$$\frac{(N_1 - 1)\hat{s}_1^2}{\sigma^2} = \chi^2_{(N-1)},$$

so that

$$\hat{s}_1^2 = \frac{\sigma^2 \chi^2_{(\nu_1)}}{\nu_1} .$$

In the same way, we know that

$$\hat{s}_2^2 = \frac{\sigma^2 \chi^2_{(\nu_2)}}{\nu_2} .$$

For each possible pair of samples, one from population 1, and one from population 2, we take the *ratio* of $\hat{s}_1^2$ to $\hat{s}_2^2$ and call this ratio the random variable F:

$$F = \frac{\hat{s}_1^2}{\hat{s}_2^2} = \frac{\text{est. } \sigma_1^2}{\text{est. } \sigma_2^2} . \qquad (6.28.1^*)$$

What sort of distribution should the random variable F actually have when the hypothesis that $\sigma_1^2 = \sigma_2^2$ is true? First of all notice that when we put the two variance estimates in ratio, this is actually the ratio of two independent chi-square variables, each divided by its degrees of freedom:

$$F = \frac{[\chi^2_{(\nu_1)}/\nu_1]}{[\chi^2_{(\nu_2)}/\nu_2]}, \qquad (6.28.2^*)$$

since

$$\frac{\hat{s}_1^2}{\hat{s}_2^2} = \frac{(\hat{s}_1^2/\sigma_1^2)}{(\hat{s}_2^2/\sigma_2^2)}$$

provided that $\sigma_1^2 = \sigma_2^2$ (that is to say, the hypothesis of the equality of the population variance is true).

Showing F as a ratio of two chi-square variables each divided by its ν is actually a way of defining the F variable:

A random variable formed from the ratio of two independent chi-square variables, each divided by its degrees of freedom, is said to be an F ratio, and to follow the rule for the F distribution.

When this definition is satisfied, meaning that both parent populations are normal, have the same variance, and that the samples drawn are independent, then the theoretical distribution of F values can be found.

The density function for F is much too complicated to allow us to gain much idea of the form of the distribution by looking at its mathematical expression. Suffice it to say that the density function for F depends only upon two parameters, ν_1 and ν_2, which can be thought of as *the degrees of freedom associated with the numerator and the denominator of the F ratio*. The range for F is *nonnegative real numbers*. The expectation of F is $\nu_2/(\nu_2 - 2)$. In general form, the distribution for any fixed ν_1 and ν_2 is nonsymmetric, although the particular "shape" of the function curve varies considerably with changes in ν_1 and ν_2.

Before we can apply this theoretical distribution to an actual problem of comparing two variances, the use of F tables must be discussed.

6.29 THE USE OF F TABLES

Since the distribution of F depends upon two parameters ν_1 and ν_2, it is even more difficult to present tables of F distributions than those of χ^2 or t. Tables of F are usually encountered only in drastically condensed form.

Such tables give only those values of F that cut off the *upper* proportion Q in an F distribution with ν_1 and ν_2 degrees of freedom. The only values of Q given are those that are commonly used as α in a test of significance (that is, the .05, the .025, and the .01 values). Table IV in the Appendix shows F values required for significance at the $\alpha = Q$ level, given ν_1 and ν_2. The columns of each table give values of ν_1, the degrees of freedom for the numerator, and the rows give values of ν_2, the degrees of freedom for the denominator. Each separate table represents one value of Q. The entries in the body of the table are the values of F cutting off the upper Q percent of the distribution, that is, the $(1 - Q)$ fractiles.

The use of Table IV can be illustrated in the following way: suppose that two independent samples are drawn, containing $N_1 = 10$ and $N_2 = 6$

cases, respectively. The degrees of freedom associated with the two variances are $\nu_1 = 10 - 1 = 9$ and $\nu_2 = 6 - 1 = 5$. For the ratio

$$F = \frac{\hat{s}_1^2}{\hat{s}_2^2},$$

the degrees of freedom must be 9 for the numerator and 5 for the denominator. Now suppose that the obtained $F = 7.00$. Does this fall into the upper .05 of values in an F distribution with 9 and 5 degrees of freedom? We turn to Table IV and find the page for .05. Then we look at the column for $\nu_1 = 9$ and the row for $\nu_2 = 5$. The tabled value is 4.77. Our obtained F value of 7.00 exceeds 4.77, and thus our sample result falls among the upper 5 percent in an F distribution.

Suppose that we were interested in the *lower* rather than the upper tail of an F distribution. Let $F_{(\nu_1;\,\nu_2)}$ stand for an F ratio with ν_1 and ν_2 degrees of freedom, and let $F_{(\nu_2;\,\nu_1)}$ be an F ratio with ν_2 degrees of freedom in the *numerator*, and ν_1 degrees of freedom in the *denominator*. The numbers of degrees of freedom in numerator and denominator are simply reversed for these two F ratios. Then it is true that, for any positive number C,

$$P[C \leq F_{(\nu_1;\,\nu_2)}] = P\left[F_{(\nu_2;\,\nu_1)} \leq \frac{1}{C}\right]; \qquad (6.29.1^*)$$

the probability that $F_{(\nu_1;\,\nu_2)}$ is greater than or equal to some number C is the same as the probability that the reciprocal of $F_{(\nu_1;\,\nu_2)}$ is less than or equal to the *reciprocal* of C. Practically, this means that the *value required for F on the lower tail of some particular distribution can always be found by finding the corresponding values required on the upper tail of a distribution with numerator and denominator degrees of freedom reversed, and then taking the reciprocal.*

For instance, we want the F value that cuts off the *bottom* .05 of sample values in a distribution with 7 and 10 degrees of freedom. We find this by first locating the value which cuts off the *top* .05 in a distribution with 10 and 7 degrees of freedom: this is 3.64 from Table IV. Then the F value we are looking for on the *lower* tail when 7 and 10 are the degrees of freedom is the reciprocal:

$$F_{(.95;\,7,10)} = \frac{1}{F_{(.05;\,10,7)}} = \frac{1}{3.64} = .27.$$

Any obtained F value that has 7 and 10 degrees of freedom, and that has a value of .27 *or less* represents a result occurring no more than .05 of the time in this distribution of possible results.

6.30   CONFIDENCE INTERVALS FOR RATIOS OF VARIANCES

Finding confidence intervals for ratios of variances is not difficult, requiring a direct application of the F distribution. If two populations are normally distributed and we take independent samples from the two populations of size N_1 and N_2, obtaining variance estimates $\hat{s}_1^2$ and $\hat{s}_2^2$, then the ratio

$$F = \frac{\hat{s}_1^2/\sigma_1^2}{\hat{s}_2^2/\sigma_2^2}$$

has an F distribution with $N_1 - 1$ and $N_2 - 1$ degrees of freedom, from Section 6.28. But, from the previous section, this implies that

$$P\left[\frac{1}{F_{(\alpha/2; N_2-1, N_1-1)}} \leq \frac{\hat{s}_1^2/\sigma_1^2}{\hat{s}_2^2/\sigma_2^2} \leq F_{(\alpha/2; N_1-1, N_2-1)}\right] = 1 - \alpha, \quad (6.30.1^*)$$

which is equivalent to

$$P\left[\frac{1}{F_{(\alpha/2; N_2-1, N_1-1)}} \cdot \frac{\hat{s}_2^2}{\hat{s}_1^2} \leq \frac{\sigma_2^2}{\sigma_1^2} \leq F_{(\alpha/2; N_1-1, N_2-1)} \cdot \frac{\hat{s}_2^2}{\hat{s}_1^2}\right]. \quad (6.30.2^*)$$

Thus, to find the $100(1 - \alpha)$ percent confidence limits for σ_2^2/σ_1^2, we take

$$\frac{1}{F_{(\alpha/2; N_2-1, N_1-1)}} \cdot \frac{\hat{s}_2^2}{\hat{s}_1^2}$$

and

$$F_{(\alpha/2; N_1-1, N_2-1)} \cdot \frac{\hat{s}_2^2}{\hat{s}_1^2}. \quad (6.30.3^*)$$

Suppose, for example, that a sample of size 10 is drawn from the first population, with $\hat{s}^2 = 40$, and that a sample of size 8 is drawn from the second population, with $\hat{s}^2 = 120$. Then

$$\frac{\hat{s}_2^2}{\hat{s}_1^2} = \frac{120}{40} = 3,$$

$$F_{(.025; 9, 7)} = 4.82,$$

and

$$\frac{1}{F_{(.025; 7, 9)}} = \frac{1}{4.20} = .24.$$

Thus, the confidence limits for σ_2^2/σ_1^2 are

$$(.24)(3) = \quad .62$$

and

$$(4.82)(3) = 14.46.$$

The probability is .95 that the interval from .62 to 14.46 contains the true value of the ratio of variances, $\sigma_2{}^2/\sigma_1{}^2$. Of course, by taking reciprocals, we could find a 95 percent confidence interval for the ratio $\sigma_1{}^2/\sigma_2{}^2$; it would be the interval from .07 to 1.61.

Despite the risk of needless repetition, it is well to emphasize the importance of the normal-distribution assumption in inferences about population variances. Neither the chi-square nor the F distributions can safely be used for *variance* hypotheses unless the population distribution is normal or the sample sizes are quite large. This warning does not necessarily extend to all uses of these two distributions, but it is valid when the primary question to be answered is about one or more variances.

6.31 RELATIONSHIPS AMONG THE THEORETICAL SAMPLING DISTRIBUTIONS

Now that the major sampling distributions have been introduced, some of the connections among these theoretical distributions can be examined in more detail. Over and over again, the binomial, Poisson, normal, t, chi-square, and F distributions have proved their utility in the solution of problems in statistical inference. Remember, however, that none of these distributions is empirical in the sense that someone has taken a vast number of samples and found that the sample values actually do occur with exactly the relative frequencies given by the function rule. Rather, it follows mathematically (read "logically") that if we are drawing random samples from certain kinds of populations, various sample statistics *must* have distributions given by the several function rules. Like any other theory, the theory of sampling distributions deals with "if-then" statements. This is why the assumptions we have introduced are important; if we wish to apply the theory of statistics to making inferences from samples, then we cannot expect the theory necessarily to provide us with correct results unless the conditions specified in the theory hold true. As we have seen, from a practical standpoint these assumptions may be violated to some extent in our use of these theoretical distributions as approximations, especially for large samples. However, these assumptions are quite important for the *mathematical* justification of our methods, in spite of the possible applicability of the methods to situations where the assumptions are not met.

Apart from the general requirement of random sampling of independent observations, the most usual assumption made in deriving sampling distributions is that the population distribution is normal. The chi-square, the t, and the F distributions all rest on this assumption. The normal distribution is, in a real sense, the "parent" distribution to these others. This is one of the main reasons for the importance of the normal distribution; the normal function rule not only provides probabilities that are often excellent approximations to other probability (density) functions, but it also has highly convenient mathematical properties for deriving other distribution functions based on normal populations.

The chi-square distribution rests directly upon the assumption that the population is normal. As you will recall from Section 6.21, the chi-square variable is basically a sum of squares of independent *normal* variables, each with mean 0 and with variance 1. At the elementary level, the problem of the distribution of the sample variance can be solved explicitly only for normal populations; this sampling distribution depends on the distribution of χ^2, which in turn rests on the assumption of a normal distribution of single observations. Furthermore, in the limit the distribution of χ^2 approaches a normal form.

There are close connections in theory between the F distribution and both the normal and chi-square distributions. Basically, the F variable is a ratio of two independent chi-square variables, each divided by its degrees of freedom. Since a chi-square variable is itself defined in terms of the normal distribution, then the F distribution also rests on the assumption of two (or more) normal populations.

The t distribution has links with the F, chi-square, and normal distributions. The t ratio for a single mean can be written as

$$t = \frac{(M - \mu)/\sigma}{\sqrt{(\hat{s}^2/N\sigma^2)}} = \frac{(M - \mu)/\sigma_M}{\sqrt{(\hat{s}^2/\sigma^2)}} \, . \qquad (6.31.1^*)$$

The numerator of the t ratio is obviously merely z_M, a standardized score in the normal sampling distributions of means. However, consider the term in the denominator. Since, from Section 6.24 we know that

$$\frac{(N - 1)\hat{s}^2}{\sigma^2} = \chi^2_{(N-1)}$$

so that

$$\frac{\hat{s}^2}{\sigma^2} = \frac{\chi^2_{(N-1)}}{N - 1},$$

it follows that

$$t = \frac{z_M}{\sqrt{\chi^2_{(N-1)}/(N - 1)}} \, . \qquad (6.31.2^*)$$

In general, a t variable is a standardized normal variable z_M in ratio to the square root of a chi-square variable divided by ν. Let us look at t^2 for a single mean in the light of this definition:

$$t^2 = \frac{z_M{}^2}{\chi^2/(N - 1)} \, . \qquad (6.31.3)$$

The numerator of t^2 is, by definition, a chi-square variable with 1 degree of freedom, and the denominator is a chi-square variable divided by its degrees of freedom, $\nu = N - 1$. Furthermore, these two chi-square variables are independent, by the principle stating that for a normal population, $\hat{s}^2$ is independent of M (Section 5.22). Thus, for a single mean, t^2 qualifies as an F ratio, with 1 and $N - 1$ degrees of freedom. In general,

$$t^2_{(\nu)} = F_{(1,\nu_2)}, \qquad \nu = \nu_2; \qquad (6.31.4^*)$$

the square of t with ν degrees of freedom is an F variable with 1 and ν degrees of freedom.

You can check this for yourself by examining the column for 1 and ν_2 degrees of freedom in the F table. In the table of F values required for significance at the .05 level, each of the entries in this column is simply the square of the entry in the t table for $\nu = \nu_2$ and $t_{.025}$. Similarly, in the table for $\alpha = .01$, each entry in the column for 1 and ν_2 degrees of freedom is the square of the corresponding entry for $t_{.005}$ in the distribution of t.

This relation between t^2 and F lets us illustrate the importance of the assumption that the two populations have the same true σ^2 when a difference between means is tested. For n cases in each of two independent groups, the value of z^2 represented by the numerator of t^2 must be

$$z^2_{\text{diff.}} = \frac{(M_1 - M_2)^2}{\sigma^2_{\text{diff.}}}. \qquad (6.31.5)$$

When $\mu_1 = \mu_2$, this is a chi-square variable with one degree of freedom. Furthermore, the square of the denominator term in the t ratio must correspond to

$$\frac{\text{est. } \sigma^2_{\text{diff.}}}{\sigma^2_{\text{diff.}}} \qquad (6.31.6)$$

for this to be a chi-square variable divided by its degrees of freedom. When each population has the same true variance, σ^2, the denominator term we compute is equivalent to such a variable, and the value t^2 is then the square of the ratio we actually find:

$$t^2 = \frac{(M_1 - M_2)^2}{\text{est. } \sigma^2_{\text{diff.}}} = \frac{(\chi^2_{(1)}/1)}{(\chi^2_{(\nu_2)}/\nu_2)} = F. \qquad (6.31.7)$$

The ratio one calculates, not knowing the true $\sigma^2_{\text{diff.}}$, actually is distributed exactly as t (or its square distributed as F) when the true variances are equal. However, if the values of the two variances are not the same, the ratio of the estimated $\sigma^2_{\text{diff.}}$ to the true $\sigma^2_{\text{diff.}}$ is equivalent to

$$\frac{\text{est. } \sigma^2_{\text{diff.}}}{\text{true } \sigma^2_{\text{diff.}}} = \frac{\chi_a^2\sigma_1^2 + \chi_b^2\sigma_2^2}{(n-1)(\sigma_1^2 + \sigma_2^2)}, \qquad (6.31.8)$$

where $\chi_a{}^2$ and $\chi_b{}^2$ symbolize two possibly different values of a chi-square variable with $n - 1$ degrees of freedom. This ratio is not necessarily distributed as a chi-square variable divided by its degrees of freedom. Thus, when variances are unequal for the two populations, the ratio we compute is not really distributed *exactly* like the random variable t, since the square of the ratio we compute cannot be equivalent to an F ratio. For this reason, a correction procedure, such as that in Section 6.19, is required when the variances are unequal and sample size is small.

The important relationships among these theoretical distributions are summarized in Table 6.31.1, showing how the distribution represented in the column depends for its derivation upon the distribution represented in the row.

Table 6.31.1

Distribution	Chi-square	F	t
Normal	Parent, and limiting form as $\nu \to \infty$. Defined as sum of normal and independent z^2 values.	Parent, making values in numerator and denominator independent χ^2/ν values.	Parent, and limiting form as $\nu \to \infty$. Numerator is normal z.
Chi-square		Variables in numerator and denominator are independent χ^2/ν.	Denominator is $\sqrt{\chi^2/\nu}$.
F			$t^2{}_{(\nu)} = F_{(1,\nu)}$

EXERCISES

1. Suppose that the random variable X is normally distributed with mean μ and variance σ^2. Consider the following two estimators of μ, based on a sample of four independent observations, X_1, X_2, X_3, and X_4:

$$G = (X_1 + X_2 + X_3 + X_4)/4$$

$$H = (4X_1 + 3X_2 + 2X_3 + X_4)/10.$$

(a) Is G unbiased?
(b) Is H unbiased?
(c) Find the variance of G and the variance of H.
(d) Which of the two estimators is more efficient? What is the relative efficiency?

2. A sample of N independent observations is taken from a normally distributed population, with $N > 2$. Denote the N observations by $X_1, X_2, \cdots, X_N$. You want to estimate the mean of the population, μ. For each of the three estimators G, H, and M, answer the following questions.

$$G = X_1, \qquad H = (X_1 + X_2)/2, \qquad M = (X_1 + X_2 + \cdots + X_N)/N.$$

(a) Is the estimator sufficient?
(b) Is the estimator unbiased?
(c) Is the estimator consistent?
(d) What is the variance of the estimator?

3. Show that for a sample of size N from a Poisson process, the sample mean is an unbiased estimator of both the mean and the variance of the process.

4. What is meant by the statement that "consistency, unlike the other three properties of estimators considered in the text, is a large-sample property"?

5. Using Equations (6.4.3) and (6.4.4), derive formula (6.4.5).

6. Using the definition of bias given by Equation (6.4.4), find the bias of s^2 as an estimator of σ^2.

7. If G is an unbiased estimator of θ, under what conditions would it be possible for G^2 to be an unbiased estimator of θ^2? Explain your answer.

8. Suppose that we have a sample of N independent observations from a certain population. We want to estimate the population mean, μ. Under what conditions is the linear combination

$$G = \sum_{i=1}^{N} a_i X_i$$

an unbiased estimator of μ? [*Hint*: Find the conditions which the a_i must satisfy for $E(G) = \mu$.] Also, find the variance of G.

9. Given that $E(X) = \lambda$ and var $(X) = \lambda$ for the Poisson distribution, show that sample means based on samples of N independent observations from a Poisson process are

(a) unbiased estimators of λ
(b) distributed with variance λ/N
(c) consistent estimators of λ
(d) relatively more efficient as estimators of λ than are means based on samples of size $N - 1$.

10. Consider a distribution of the random variable X such that the density function is given by

$$f(x) = \frac{2b(c - bx)}{c^2}, \qquad 0 \le x \le c/b.$$

It turns out that this distribution, with two positive parameters, b and c, gives

$$E(X) = \frac{c}{3b}$$

and

$$\sigma_X{}^2 = \frac{c^2}{18b^2}.$$

(a) Given a sample of size N, consisting of independent observations, would you say that sample M is a *sufficient* estimator of the parameter c?

(b) Suppose that it is known that $c = 3$. Is the sample mean for N observations then an unbiased estimator for the value of b? Why?

(c) When $c = 3$, it is true that $b = E(X)/(2\sigma_X{}^2)$. Would you say then that $M/(2\hat{s}^2)$ calculated from a sample of N is an unbiased estimator of b? Why or why not?

(d) When $b = 3$, is sample M a consistent estimator of c?

11. Given the following very simple population distribution, construct the theoretical sampling distribution for the sample mean based on three independent observations from the population.

x	$P(X = x)$
1	1/3
2	1/3
3	1/3

[*Hint*: Enumerate the possible combinations of events for samples of size 3, and then translate these into values of the sample mean.] Similarly, find the distribution of the sample median and the sample mid-range (one-half times the sum of the largest value in the sample and the smallest value in the sample). Using these sampling distributions,

(a) Show that the sample mean, the sample median, and the sample mid-range are all unbiased estimators of the population mean.

(b) Find the variance of the sample mean, the sample median, and the sample mid-range.

(c) Find the relative efficiency of the sample mean as compared with the sample median and the relative efficiency of the sample mean as compared with the sample mid-range.

12. For samples of three independent observations from the distribution given in Exercise 11, find (by enumeration) the sampling distribution of the sample variance (s^2), the modified sample variance ($\hat{s}^2$), and the sample range (the difference between the largest and smallest values in the sample).

(a) Is the sample range unbiased as an estimator of the population variance?

(b) Compute the mean-square errors of the three sample statistics (s^2, $\hat{s}^2$, and the sample range). Which of the three statistics has the smallest mean-square error as an estimator of the population variance?

13. Prove that the kth sample moment is an unbiased estimator of the kth population moment of a random variable (provided, of course, that the kth population moment exists).

14. Using the sampling distributions of the sample mean and sample median determined in Exercises 28 and 29 in Chapter 5, compare the efficiency of these two estimators of the population mean for the given population distribution.

15. For a sample of size N from an exponential distribution with parameter λ, find the maximum-likelihood estimator of λ. Suppose that λ is the intensity per hour of the arrival of cars at a toll booth. If a sample of size ten is taken, with the following amounts of time between arrivals, compute the maximum-likelihood estimate of λ.

$$4 \quad 3 \quad 6 \quad 1 \quad 1 \quad 4 \quad 2 \quad 6 \quad 1 \quad 3$$

If the cost per hour of maintaining the toll booth can be expressed as $\lambda^2 + 2\lambda + 5$, what is the maximum-likelihood estimate of the cost per hour of maintaining the toll booth?

16. On the basis of a sample of size N from a Poisson process, determine an estimator of the Poisson parameter λ by using (a) the principle of maximum likelihood, (b) the method of moments.

17. In the text, formula (6.9.1) gives a way to form a pooled estimate of the mean based on two nonoverlapping samples. Try to arrive at a similar expression that applies when the first sample, of size N_1, and the second sample, of size N_2, have exactly N_3 cases in common. [*Hint*: A Venn diagram may be helpful.]

18. If you have two independent sample proportions drawn from the same Bernoulli process, determine a pooled estimate of the Bernoulli parameter p which is unbiased. Denote the two sample proportions by P_1 and P_2, and assume that they are based on samples of size N_1 and N_2, respectively.

19. The random variable X is normally distributed with unknown mean μ and known variance $\sigma^2 = 225$. A sample of size $N = 9$ from X is observed, with the sample values 42, 56, 68, 56, 48, 36, 45, 71, and 64.
    (a) Compute the sample mean and sample variance.
    (b) Name three intuitively reasonable point estimators for μ, and determine the corresponding estimates from the above data.
    (c) Which of the three estimators given in (b) would you prefer? Why?
    (d) Find an 80 percent confidence interval for μ.
    (e) Find a 95 percent confidence interval for μ.

20. Suppose that in a certain population of men, the mean weight is μ and the variance is 400. If it is assumed that weight is normally distributed in this population, and if a random sample (with replacement) of 25 persons from the population yields a sample mean weight of 168, find 50 percent and 98 percent confidence intervals for the population mean weight, μ.

21. Based on historical data, the variability of the diameter of a part produced by a particular machine can be represented by a standard deviation of .1 inch. However, nothing is known about the population mean. A sample of size 100 is taken, and the sample mean diameter is 8.4 inches. Find 95 percent and 75 percent confidence intervals for the population mean, assuming that the diameters are normally distributed and the sample observations are independent.

22. An educator is interested in the mean IQ in a given population (say, the freshman class at a certain university). He cannot afford to test the entire population, but he is willing to assume that IQ is normally distributed with variance 121 in the population of interest. He wants to take a random sample of members of the population in order to determine a 90 percent confidence interval for mean IQ. How large a sample will he need if he wants the length of the confidence interval to be no greater than 5? If he wants the length to be no greater than 1?

23. Suppose that you find a 90 percent confidence interval for the parameter μ, based on a particular sample, and that the confidence limits are 50 and 60. In terms of the long-run frequency interpretation of probability, how should this confidence statement be interpreted? Might a different interpretation be given by an advocate of subjective, or personal, probability? Explain.

24. In the process of estimating the mean of a population, a statistician wants the probability to be .95 that his estimate will come within $.2\sigma$ of the true mean. What sample size should he use? If the sample size used is 100, within what distance of the true mean (in σ units) can the statistician be 95 percent sure that his sample mean falls?

25. Given a sample mean of 45, a population variance of 25, and a sample of size 30 drawn from a normal population, establish a 99 percent confidence interval for the population mean based on this sample. Can you find *another* pair of limits for which it will also be true that the probability of these limits covering the true mean is .99? If so, are these limits closer together or farther apart than the limits originally found? Does this suggest a general principle?

26. Suppose that we take a sample of size 1 from an exponential distribution with parameter λ. Can you find a 90 percent confidence interval for λ based on this single sample observation?

27. In determining a $(1 - \alpha)$ percent confidence interval for the mean of a normal population, assuming that the variance is known, how large a sample is needed to make the confidence interval 1/3 as large as it is when the sample size is N? What are the implications of this result?

28. Discuss the following statement: "The determination of sample size in an experiment involves a balancing of the cost of taking the sample and the desired precision of the results of the sample."

29. In a sample of 100 items from a production process, 4 were found to be defective. Assuming that the production process behaves like a Bernoulli process, determine 95 percent and 99.7 percent confidence intervals for p, the probability of obtaining a defective item on any single trial.

30. Why is it that the probability that $(M - \mu)/\sigma$ is greater than 1.65 is only .05 for samples of size 5 from a normal distribution, whereas the probability that $(M - \mu)/\text{est. } \sigma$ is greater than 1.65 is approximately .09 for samples of the same size from the same population?

31. Does it seem reasonable that the t distribution should have "fatter tals" than the normal distribution? Why?

32. In Exercise 19, assume that the variance of X is unknown and find 80 percent and 95 percent confidence intervals for μ. Compare these with the intervals computed in parts (d) and (e) of Exercise 19.

33. Suppose that daily changes in the Dow-Jones Average of Industrial Stocks are normally distributed and that the change on any given day is independent of the change on any other day. A random sample of 81 daily changes is obtained, with sample mean .20 and sample variance (s^2) 1.50. Find a 90 percent confidence interval for μ, the population mean. Suppose that a second random sample is obtained, with a sample size of 25, a sample mean of .15, and a sample variance of 1.20. Using the information from both samples, find a 90 percent confidence interval for μ. How reasonable are the assumptions in this problem?

34. An IQ test is given to a randomly selected group of 10 freshmen at a given university and also to a randomly selected group of 5 seniors at the same university. For the freshmen, the sample mean is 120 and the (unbiased) sample variance is 196. For the seniors, the sample mean is 128 and the (unbiased) sample variance is 121. Determine a 95 percent confidence interval for $\mu_S - \mu_F$, where μ_S is the population mean IQ for the freshmen and μ_F is the population mean IQ for the seniors (assume that the population variances are equal for the two populations).

35. Independent random samples are taken from the output of two machines on a production line. The weight of each item is of interest. From the first machine, a sample of size 36 is taken, with sample mean weight of 120 grams and a sample variance of 4 grams. From the second machine, a sample of size 64 is taken, with a sample mean weight of 130 grams and a sample variance of 5 grams. It is assumed that the weights of items from the first machine are normally distributed with mean μ_1 and variance σ^2, and that the weights of items from the second machine are normally distributed with mean μ_2 and variance σ^2 (that is, the variances are assumed to be equal). Find a 99 percent confidence interval for the difference in population means, $\mu_1 - \mu_2$.

36. Assuming that the variances of two normally distributed populations are equal, develop a formula for finding a confidence interval for the *sum* of the means. [*Hint*: Consider the sampling distribution of $\mu_1 + \mu_2$.] Use this to

determine a 99 percent confidence interval for the sum of the means in Exercise 35.

37. Show that the fact that $[N/(N-1)]s^2$ is an unbiased estimator of σ^2 and that Ns^2/σ^2 is a chi-square variable implies that the expectation of a chi-square random variable with $N-1$ degrees of freedom is simply $N-1$.

38. Under what circumstances might one be interested in the lower tail of a chi-square distribution?

39. For samples from normal populations, the variance of the sampling distribution of $\hat{s}^2$, the unbiased estimator of σ^2, is $2\,\sigma^4/(N-1)$. Prove this, using the fact that the mean of a chi-square random variable is equal to the number of degrees of freedom. [*Hint*: Remember that the variance of the sampling distribution of $\hat{s}^2$ is equal to $E[(\hat{s}^2)^2] - (\sigma^2)^2$.]

40. Suppose that a sample of size 8 is chosen randomly from a normally distributed population with mean 24 and variance 9. The standardized value corresponding to each observed value is computed by subtracting 24 and dividing the result by 3. Let T denote the sum of the squares of the 8 standardized values, and find
    (a) $P(T \geq 20.09)$
    (b) $P(2.73 < T < 5.07)$
    (c) $P(T > 2.18)$.

41. In Exercise 20, if s^2 is the sample variance of the sample of size 25, find
    (a) $P(s^2 \leq 198.40)$
    (b) $P(304.64 \leq s^2 \leq 687.68)$
    (c) $P(531.20 < s^2)$
    (d) $E(s^2)$.

42. Suppose that two independent samples from the same normal population were pooled to form the following estimator of the population variance:

$$G = \frac{N_1 s_1^2 + N_2 s_2^2}{N_1 + N_2 - 2}.$$

Discuss the sampling distribution of this pooled estimator of σ^2.

43. Would the sampling distribution in Exercise 42 change if the samples were drawn from normal populations with different means but with the same variance? Explain.

44. Using the data in Exercise 33, find a 90 percent confidence interval for σ^2 based on
    (a)   the first sample only
    (b)   the second sample only
    (c)   both samples.
    Repeat the process under the assumption that the sample sizes were 10 and 5, respectively.

45. In Exercise 34, find a 95 percent confidence interval for the ratio of the variance of the freshman population to the variance of the senior population (σ_F^2/σ_S^2).

46. Given the fact that the expected value of an F variable with ν_1 and ν_2 degrees of freedom is equal to $\nu_2/(\nu_2 - 2)$, for $\nu_2 > 2$, prove that the variance of a t variable with ν degrees of freedom is $\nu/(\nu - 2)$.

47. If X and Y are normally distributed and we let $Z = cX + d$ and $W = bY + g$, discuss the relationship between the F ratio for σ_X^2 and σ_Y^2 and the F ratio for σ_Z^2 and σ_W^2. Consider the following two cases:
    (a) $c = b$ and $d = g$
    (b) $c \neq b$ and $d \neq g$.
    How would the t distribution for the difference between means be changed by the transformation from X and Y to Z and W?

48. For the data in Exercise 35, find 90 percent and 98 percent confidence intervals for the ratio of the variances of weights from the two machines.

7

HYPOTHESIS
TESTING

7.1 STATISTICAL TESTS

Let us assume that the general motivation for inferential statistics is
clear to you by now: how do you say something about the population, given
only the sample evidence? We have seen that certain sample statistics are
capable of giving good estimates of particular population parameters;
ordinarily, these point estimates are the first and foremost statistical inferences one makes from his data. On the other hand, there is almost always
a dark side to the picture; virtually any estimate will be in error to some extent due to fluctuations of sampling, and so we discussed the concept of a
sampling distribution of a particular statistic based on a sample of size N.
In particular, we focused on the mean of the sampling distribution as the
expected value of the statistic and the standard deviation of the sampling
distribution as the standard error. Also, we found that sampling distributions of statistics such as M can often be regarded as normal, particularly
when N is large. Given the form of the population distribution together
with the appropriate parameter values such as μ and σ, then, we can find
the probability that the sample value for a statistic such as M will fall into
any given interval of values in the sampling distribution. Using this, we developed the concept of interval estimation. Confidence intervals give us
some indication of the precision of a particular point estimate. Point estimates and interval estimates, then, provide us with methods for saying
something about the population, given only the sample evidence.

A statistician who samples from a population is generally trying to decide,
or at least form an opinion on, something about the population. Quite
often he wants to decide if some hypothetical population situation appears

reasonable in the light of the sample evidence. Sometimes the problem is to judge which of several possible population situations is best supported by the evidence at hand. In either instance, one is trying to make up his mind from evidence. However, as we shall see, there are many possible ways of making this kind of decision, depending on what information in the sample is actually used, the various possible true population situations being compared, and especially the risk one is willing to take of being wrong in his decision.

The first problem to be faced in making a decision from data is that of choosing the relevant and appropriate statistic for the particular purpose. Obviously, different combinations of the sample data give different kinds and amounts of information about the population. Reaching a conclusion about some population characteristic requires effective use of the right information in the sample, and various statistics differ in their relevance to different questions about the population. Sometimes the best statistic to permit us to form a judgment about the population also happens to be the best estimate of the particular population parameter under study, but in other instances statistics may be used which are not estimates of any population parameter at all, such as χ^2 and F statistics.

Granting for the moment that we know the information we need from the sample and how to extract it, how do we judge the tenability of a particular hypothesis about the population? It seems reasonable that a tenable or "good" hypothesis about the population should be consistent with the obtained sample results. That is, the obtained value of the sample statistic of interest should be one that is reasonably likely in light of the hypothesis. Naturally, in any given sample, the obtained value of the statistic probably will not *exactly* equal the expected value dictated by the hypothesis, and there is a theoretical sampling distribution of values for the statistic that can be determined from the specified hypothesis. The question then becomes: is the deviation between the obtained value and the expected value large enough, in light of the sampling distribution, to make the truth of the hypothesis seem very unlikely?

A deviation between the obtained value of the statistic and its expected value under the hypothesis implies one of two things: either the hypothesis is right, and the difference in value between the statistic and its expectation is the product of chance, or the hypothesis is wrong and we were not led to expect the right thing of our sample statistic. In particular, if the sample value falls so far from expectation that it lies in an interval of values very improbable given the hypothesis, but this sort of sample result would be made probable if some alternative hypothesis were true, then doubt is cast on the original hypothesis and we reject it in favor of the alternative. Obviously, however, these notions of "disagree with expectation," "improbable," and "probable" are relative matters; when do we say that a sample result disagrees "enough" with expectation under some hypothesis to warrant rejection of the hypothesis itself?

One of the stickiest problems in comparing sample results with expectation given by some hypothesis is deciding what we mean by a sufficient kind and degree of disagreement. How and by how much must the sample statistic's value disagree with expectation before we decide that one hypothesis should be abandoned in favor of another? What we need is a decision-rule, a guide giving the conclusion we will reach depending on how the data turn out, and which can be formulated even before the data themselves are seen. However, there is literally no end to the number of decision-rules that one might formulate for a particular problem. Some of these ways of deciding might be very good in a particular circumstance, but not so good in others, and so we need to decide how to decide on the basis of what we wish to accomplish in a specific situation.

The process of comparing two hypotheses in the light of sample evidence is traditionally called a "statistical test" or a "significance test." The competing hypotheses are stated in terms of the form, or of one or more parameters, of the distribution for the population. Then the statistic containing the relevant information is chosen, having a sampling distribution dictated by the population distribution specified in the particular hypotheses. The hypothesis that dictates the particular sampling distribution against which the obtained sample value is compared is said to be "tested." The significance test is based on the sampling distribution of the statistic given that the particular hypothesis is true. If the observed value of the statistic falls into an interval representing a kind and degree of deviation from expectation which is improbable given the hypothesis, but which would be relatively probable given the other, alternative, hypothesis, then the first hypothesis is said to be rejected. The actual decision to entertain or to reject any hypothesis is thus based on whether or not the sample statistic falls into a particular region of values in the sampling distribution dictated by that hypothesis. However, the immediate statistical outcome of any significance test can be thought of as a *conditional probability*, the probability of a result this much or more deviant from expectation in a particular way given that the hypothesis tested is really true. The interpretation and use of this conditional probability depends on the purposes of the statistician and on the particular decision-rule he may have adopted.

A word of warning before we go further: although we will sometimes speak of testing a single hypothesis, in practice the statistician always acts as though he is deciding between *two* hypotheses. A great many issues connected with the use of significance tests can be understood only if this decision-making task is presupposed.

7.2   STATISTICAL HYPOTHESES

A statistical hypothesis is usually a statement about one or more population distributions, and specifically about one or more parameters of such population distributions. It is always a statement about the population, not about the sample. The statement is called a hypothesis because it refers to a situation that *might* be true. Statistical hypotheses are almost never equivalent to the hypotheses of science, which are usually statements about phenomena or their underlying bases. Quite commonly, statistical hypotheses grow out of or are implied by scientific hypotheses, but the two are seldom identical. The statistical hypothesis is usually a concrete description of one or more summary aspects of one or more populations; there is no implication of *why* populations have these characteristics.

We shall use the following scheme to indicate a statistical hypothesis: a letter H followed by a statement about parameters, the form of the distribution, or both, for one or more specific populations. For example, one hypothesis about a population could be written

H: population in question is normally distributed with $\mu = 48$ and
   $\sigma = 13$,

and another could be written

H: population in question is represented by a Bernoulli process
   with $p = .5$.

Hypotheses that completely specify a population distribution are known as **simple hypotheses.** In general, the sampling distribution of any statistic also is completely specified given a simple hypothesis and N, the sample size.

We will also encounter hypotheses such as:

H: population is normal with $\mu = 48$.

Here, the exact population distribution is not specified, since no requirement was put on σ, the population standard deviation. When the population distribution is not determined completely, the hypothesis is known as **composite.**

Hypotheses may also be classified by whether they specify *exact* parameter values or merely *a range* or *interval* of such values. For example, the hypothesis

$$H: \mu = 100$$

would be an exact hypothesis, although

$$H: \mu \geq 100$$

would not be exact.

7.3 ASSUMPTIONS IN HYPOTHESIS TESTING

In applied situations, very seldom will a hypothesis specify the sampling distribution completely. Simple hypotheses are just not of interest or are not available in most practical situations. On the other hand, it *is* necessary that the sampling distribution be completely specified before any precise probability statement can be made about the sample results. For this reason assumptions usually are made, which, taken together with the hypothesis itself, determine the relevant statistic and its sampling distribution and justify a test of the hypothesis. There assumptions differ from hypotheses in that they are rarely or never tested against the sample data, although there are procedures available for such testing, as we shall see in later chapters. They are assertions that are simply assumed true, or the evidence is collected in such a way that they must be true.

First and foremost among such assumptions is that the sample is random. In all discussion of significance tests to follow, we shall assume simple random sampling unless it is specifically indicated otherwise. Independence of two or more random variables is another assumption frequently made. At other points we shall assume that the population distribution is normal, that the variances of different populations are equal, and so on. Wherever such assumptions underlying a test are important, they will be pointed out. Note, by the way, that these assumptions are similar to the assumptions we made in constructing confidence intervals.

These assumptions are necessary for the formal justification of many of the methods to be discussed. It is not true, however, that these assumptions are always or even ever realized in practice. The results of any data analysis leading to statistical inference can always be prefaced by "If such and such assumptions are true, then . . . ." The effects of the violation of these assumptions on the conclusions reached can be very serious in some circumstances and only minor in others. In later sections the importance of the most common assumptions will be discussed (some, such as the assumption of normality, were discussed somewhat in Chapter 6). Nevertheless, these assumptions should always be kept in mind, along with the conditional character of any result subject to the assumptions being true.

7.4 TESTING A HYPOTHESIS IN THE LIGHT
OF SAMPLE EVIDENCE

Every test of a hypothesis involves the following features:

1. The hypothesis to be tested is stated, together with an alternative hypothesis.

2. Additional assumptions are made, permitting one to specify the sample statistic that is most relevant and appropriate to the test, as well as the sampling distribution of that statistic.

3. Given the sampling distribution of the test statistic when the hypothesis to be tested is true, a **region of rejection** is decided upon. This is an interval of possible values deviant from expectation if the hypothesis were true, but which more or less accord with expectation if the alternative were true. This region of rejection contains values relatively improbable of occurrence if the first hypothesis were true, but relatively probable given the alternative. The risk one is willing to take in rejecting the tested hypothesis *falsely* determines the size of the region of rejection, and the alternative hypothesis determines its location.

4. The sample itself is obtained. If the computed value of the test statistic falls into the region of rejection, then doubt is cast on the hypothesis, and it is said to be rejected in favor of the alternative. If the result falls out of the region of rejection, then the hypothesis is not rejected, and the experimenter may choose either to accept the hypothesis or to suspend judgment, depending on the circumstances. A sample result falling into the region of rejection is said to be **significant** or **to depart significantly from expectation** under the hypothesis.

As an example, suppose that a statistician entertains the hypothesis that the mean of some population is 75:

$$H_0 : \mu = 75.$$

This is actually a composite hypothesis, since nothing whatever is said about the form of the population distribution nor about other parameters such as the standard deviation. The statistician is, however, prepared to assume that irrespective of what is true about the mean, the population has a normal distribution, and that the standard deviation for the population is 10; for the moment, let us imagine that both of these assumptions are reasonable in the light of what the statistician already knows about his problem. In effect, the addition of these two assumptions to the original hypothesis gives us this hypothesis:

$$H_0 : \mu = 75, \ \sigma = 10, \ \text{normal distribution.}$$

Now, what of the alternative hypothesis? It might be that the alternate hypothesis is

$$H_1 : \mu \neq 75.$$

Note that this alternative happens to be inexact, and really asserts that some (unspecified) value of μ other than 75 is true. However, even if H_1 is true and H_0 false, the statistician will still assume that the distribution is normal with a standard deviation of 10.

It has already been decided that 25 independent observations will be made. Given this sample size N, the information in the hypothesis H_0, and the assumptions, the sampling distribution of the mean is known completely; this must be a normal distribution with a mean of 75 and a standard error of $10/\sqrt{25}$, or 2, if the hypothesis H_0 is true.

Now, in accordance with H_1, the statistician decides that he will reject this hypothesis H_0 only if his sample mean falls among the extremely deviant ones in either direction from the value 75; in fact, he will reject H_0 only if the probability is .05 or less for the occurrence of a sample mean as deviant or more so from 75. Thus, the region of rejection for this hypothesis contains only 5 percent of all possible sample results when the hypothesis H_0 is actually true. Such sample values depart widely from expectation and are relatively improbable given that H_0 is true; on the other hand, samples falling into these regions are relatively more probable if a situation covered by H_1 is true. These regions of rejection in the sampling distribution are shown in Figure 7.4.1.

From tables of the normal distribution, he finds that a z value of 1.96 shows a cumulative probability $F(1.96)$ of about .975. Hence only .025 of all sample means should lie *above* 1.96 standard errors from μ. Similarly, it must be true that about .025 of all sample means will lie below -1.96 standard errors from μ. In short, the region of rejection includes all sample means such that, in this sampling distribution,

$$1.96 \leq z_M$$

or
$$z_M \leq -1.96,$$

giving a total probability of $.025 + .025 = .05$ for the combined intervals. A value of the sample mean that lies exactly on the boundary of a region of rejection is called a **critical value** of M; the critical z_M values here are -1.96 and 1.96, so that the critical values of M are

$$-1.96\sigma_M + \mu = -1.96(2) + 75 = 71.08$$

and
$$1.96\sigma_M + \mu = 1.96(2) + 75 = 78.92.$$

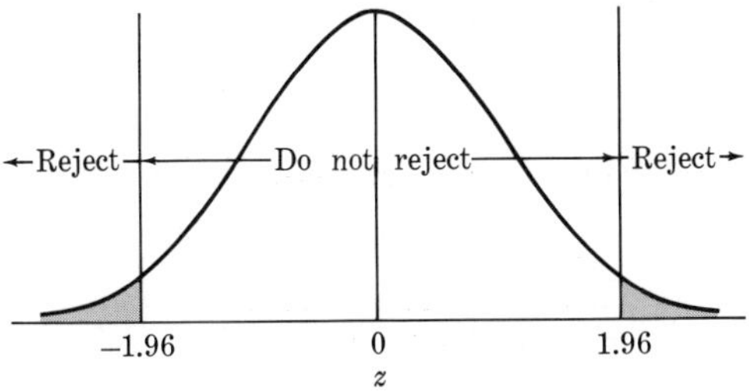

Figure 7.4.1

Now the sample is drawn and turns out to have a mean of 79. The corresponding z_M score is

$$z_M = \frac{79 - \mu}{\sigma_M} = \frac{79 - 75}{2} = 2.$$

Notice that this z_M value *does* fall into the region of rejection, since it is greater than a critical value of 1.96. Then the statistician says that the sample result is **significant beyond the 5 percent level.** Less than 5 percent of all samples should show results this deviant (or more so) from expectation under H_0, if H_0 is actually true. If the statistician has decided in advance that samples falling into this region show sufficient departure from expectation to be called *improbable* results, then doubt is cast on the truth of the hypothesis, and H_0 is said to be rejected.

In its essentials this simple problem exemplifies all significance tests. Naturally, the details vary with the particular problem. Hypotheses may be about variances, or any other characteristic feature of one or more populations. Other statistics may be appropriate to a test of the particular hypothesis. Other assumptions may be made to permit the specification of the sampling distribution of the relevant statistic. Other regions of rejection may be chosen in the sampling distribution. *But once these specifications are made, the statistician knows, even before he sees the sample, how he will make up his mind, if he must, in the light of what the sample shows.*

Here the region of rejection was simply specified without any particular explanation. In this step, however, lies a key problem of significance tests. In the choice of the region of rejection the experimenter is deciding how to decide from the data. It is perfectly obvious that there are very many ways that the experimenter could have chosen a rejection region. He could have decided to reject the hypothesis if the sample result fell beyond one standard error of the mean specified by the hypothesis, as in Figure 7.4.2, or if the sample result fell between 3 and 4 standard errors below the hypothetical mean, as in Figure 7.4.3, and so on, ad infinitum, for any interval or set of intervals.

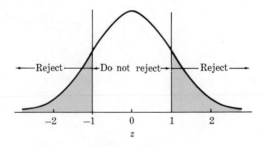

Figure 7.4.2

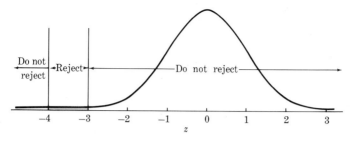

Figure 7.4.3

The regions of rejection in Figures 7.4.1 and 7.4.2 look very much like regions lying outside 95 percent and 68 percent confidence intervals for μ, respectively. They are *not*, however; confidence intervals for μ are centered around the *sample mean* M; these regions are centered around the *hypothesized population mean*. However, there *is* a relationship in these cases between a rejection region and a confidence interval. If, given the truth of H_0, the probability of a result falling in a rejection region is denoted by α, then the test is equivalent to the following: reject H_0 if the $100(1 - \alpha)$ percent confidence interval for μ does not contain the hypothesized value. In the example, the 95 percent confidence limits for μ are $M - 1.96\sigma_M$ and $M + 1.96\sigma_M$, or 75.08 and 82.92. Since the hypothesized mean, 75, does not lie within these limits, H_0 is rejected, and the sample result is said to be significant beyond the 5 percent level. In this case, $\alpha = .05$. The reader is cautioned, however, that this relationship between confidence intervals and rejection regions does not hold for all tests of hypotheses. We will see that in some cases the region of rejection lies in only one rather than both of the tails of the sampling distribution, in which case we can find no interpretation of the test in terms of "two-sided" confidence intervals such as those discussed in Chapter 6.

In the next section we are going to address ourselves to this problem of choosing a rejection region. This can be discussed most simply if we consider a choice between *two exact hypotheses*. Admittedly, this is an extremely artificial situation, and almost never does one actually have this kind of choice to make. Nevertheless, the terminology and procedures of hypothesis testing can best be understood by acting as if this were the statistician's problem.

7.5 CHOOSING A WAY TO DECIDE BETWEEN TWO EXACT HYPOTHESES

A statistician has two alternative theories about some economic behavior. According to Theory I, exactly .80 of all consumers should exhibit this behavior in a particular situation, whereas in Theory II, .40 should be the proportion doing this. The task of the statistician is to form a judgment about the two theories. Imagine, if you can, that the two theories exhaust the logical possibilities accounting for what he observes: either one or the other must be true.

These two theories suggest the two competing hypotheses

$$H_0: p = .80$$

and $\qquad H_1: p = .40.$

The subscripts 0 and 1 have no particular meaning here; they are merely indices to let us tell the hypotheses apart.

Some N consumers are to be drawn at random and independently for introduction into the particular situation of interest. These N consumers, then, are the "subjects" of an experiment. The occurrence of the behavior of interest for any subject is a "success," and for either hypothesis there is a binomial sampling distribution giving probabilities for various possible values of P, the sample *proportion* of successes. Of course, the probabilities of different sample P values differ under the two hypotheses, since $p = .8$ in one and only .4 in the other.

For the sake of simplicity, let us suppose that there are to be only ten cases in the sample. Then the sampling distributions of P found from the binomial distribution under the two hypotheses are as shown in Table 7.5.1. Thus, the statistician knows the probability of each possible sample result P, given the possible truth of either of the two hypotheses.

Table 7.5.1

P	$P(P)$ if $p = .8$	if $p = .4$
1.0	.107	.0001 (+)
.9	.268	.002
.8	.302	.011
.7	.201	.042
.6	.088	.111
.5	.026	.200
.4	.006	.251
.3	.001	.215
.2	.000 (+)	.121
.1	.0000 (+)	.040
.0	.0000 (+)	.006

How shall the statistician decide upon seeing the evidence? What he needs is a decision-rule, some way that lets him judge which of the two hypotheses is actually favored by the sample result. There are any number of decision-rules that could be formulated, but for the moment we will consider only the following three possibilities:

Decision-rule 1:   **If P is greater than or equal to .8, choose H_0; otherwise H_1.**

Decision-rule 2:   **If P is greater than or equal to .6, choose H_0; otherwise H_1.**

Decision-rule 3:   **If P falls between .2 and .8 inclusive, choose H_0; otherwise H_1.**

Regardless of which decision-rule is selected there are two ways that the statistician *could* be right, and two ways of his making an *error*. This is diagrammed below:

TRUE SITUATION

		H_0	H_1
	H_0	correct	error
DECISION			
	H_1	error	correct

If he decides H_0, and H_1 is actually true, an error is made. Furthermore, he can make an error if he decides on H_1 and H_0 is actually true. The other possibilities lead, of course, to correct decisions.

Given the sampling distributions under each of the two hypotheses, and given any decision-rule, we can find the *probabilities* of these two kinds of error. The statistician has no idea which hypothesis is true, and all he can go by is the occurrence of a P value leading him to reject one or the other of the hypotheses. Suppose that Decision-rule 1 were adopted. This rule requires that the occurrence of any P of .8 or more lead automatically to a decision that H_0 is true. What is the probability of observing such a sample result *if H_1 is true*? Notice that whenever this happens, there will be a wrong decision and so this probability is that of *one kind* of error. The binomial distribution for $p = .4$ shows a probability of about .013 for P greater than or equal to .8, and so the probability is 0.13 of *wrongly* deciding that H_0 is true.

In the same way, we can find the probability of the error made in choosing H_1 when H_0 is true. By Decision-rule 1, H_1 is chosen whenever P is less than or equal to .7; in the distribution under H_0, this interval of values has a probability of about .323. Thus the probabilities of error and correct decisions are, under Decision-rule 1,

TRUE SITUATION

		H_0	H_1
	H_0	.677	*.013*
DECISION			
	H_1	*.323*	.987

The probability of a correct decision is $1 - P(\text{error})$ for either of the possibly true situations (the two error probabilities appear in italics). Notice that the statistician is far more likely to make an erroneous judgment by using this rule when H_0 is true than for a true H_1.

Exactly the same procedure gives the erroneous and correct decision probabilities under Rule 2:

	H_0	H_1
H_0	.966	*.166*
H_1	*.034*	.834

By this second rule, the probability of error is relatively smaller when H_0 is true than it is for Rule 1. However, look what happens to the probability of the other error! This illustrates a general principle in the choice of decision-rules: *any change in a decision-rule that makes the probability of one kind of error smaller will ordinarily make the other error probability larger* (other things, such as sample size, being equal).

Before trying to choose between Rules 1 and 2, let us write down the probabilities for Decision-rule 3.

	H_0	H_1
H_0	.625	*.952*
H_1	*.375*	.048

Even on the face of it this decision-rule does not look sensible. The probability of error is large when H_0 is true, and the experimenter is almost sure to make an error if H_1 is the true situation! This illustrates that not all decision-rules are reasonable if the statistician has any concern at all with making an error; here, regardless of what is true the statistician has a larger chance of making an error using this rule than in using either Rules 1 or 2. Rules such as this are called **inadmissible** by decision theorists. We need confine our attention only to the two relatively "good" Rules 1 and 2.

There is a real problem in deciding between the Rules 1 and 2; Rule 1 is good for making error probability small when H_1 is true, but risky when H_0 represents the true state of affairs. On the other hand, Rule 2 makes for small chance of error when H_0 is true, but gives relatively large probability of error when H_1 holds. Obviously, any choice between these two rules must have something to do with the relative importance of the two kinds of errors. It might be that making an error is a minor matter when H_0 is true, but very serious given H_1. In this case Rule 1 is preferable. On the other hand, were error very serious given H_0, Rule 1 could be disastrous. In short, any rational way of choosing among rules must involve some notion of the loss involved in making an error.

7.6 ERRORS AND LOSSES

Decision theory, which grew out of problems of economic decision-making, is concerned with the possible outcomes of an action. Any decision maker has several courses of action open to him, and once he has chosen a course of action some event or chain of events is going to occur. However, the decision maker does not necessarily know what this outcome will be. Instead, there are various "states of the world" that might be true, and the outcome of any decision to take action depends both on what is decided and what is really true.

As a simple example, imagine a man interested in buying a sweater. His interest at the moment is in finding a sweater that will wash without shrinking. He locates a sweater that he likes, but he does not know if it is washable. For this particular sweater, two courses of action are open to him: "buy" or "do not buy." However, there are also two possibilities for the "real" nature of the sweater: either it won't shrink, or it will. The courses of action, the states of the world, and the different action-outcome possibilities are shown in the following table:

STATE OF THE WORLD

		Sweater will not shrink	*Sweater will shrink*
	Buy	(buy, good sweater)	(buy, sweater shrinks)
POSSIBLE ACTIONS	*Do not buy*	(do not buy, miss a good sweater)	(do not buy, miss a shrunken sweater)

In two of these action-outcome possibilities there is possibly a real economic gain to the buyer; the man does well if either the combination (buy, good sweater) or the combination (do not buy, miss a shrunken sweater) occurs. On the other hand, two possibilities for error exist as well: (buy, sweater shrinks), and (do not buy, miss a good sweater). In principle, each of these action-outcome possibilities has some value or utility for the buyer, and his choice to buy or not to buy should depend upon these relative values.

With only this to go on the buyer still has the problem of making up his mind. However, just as the statistician gathers data in order to "snoop" on the possible states of the world, the buyer decides that he will do a little snooping on his own, and he looks for the cleaning instructions tag on the sweater. He happens to know that only once in about one hundred times does a sweater factory label a garment "washable" when it is not. On the other hand, he knows that among washable sweaters, only nine in ten bear instructions that definitely say so. Thus, if he decides solely on the basis of what the tag says, he has the following probabilities of wrong and right decisions (assuming that this is a random sample from among one of the two sorts of sweaters):

	Will not shrink	*Will shrink*
Tag says washable (buy)	.90	.01
Tag does not say washable (do not buy)	.10	.99

Here, the chances are much smaller of making the mistake of buying a sweater that will shrink than of failing to buy a sweater that will not shrink. Is this a good way to decide? That depends upon the value to him of the various outcomes. If the sweater is very expensive and buying a sweater that shrinks means a big loss, then he probably should use this rule. On the other hand, if the man needs a sweater very badly and the price is low, so that much would be lost if this opportunity were mistakenly passed up, the rule is not so good. Here the various outcomes of the actions have a very real, and perhaps even a dollar-and-cents, value to the buyer. Not only the probabilities, but also the possible losses attached to the different outcomes, should figure in the buyer's choice of a way to decide.

In business situations, it is frequently true that a given action must be followed by one of a fairly small and specifiable set of outcomes. The decision maker does not know what the particular outcome of his action will be, but at least he is able to state the possible outcomes. The one that occurs may depend upon economic conditions in general, quality of product, competition, and so on. Furthermore, each of these possible outcomes to any action can be assigned a profit or loss value, at least in principle.

7.7 EXPECTED LOSS AS A CRITERION FOR CHOOSING A DECISION-RULE

Let us imagine each action-outcome combination as having some numerical value for the statistician. The exact nature of the value need not concern us here; it may reflect dollars, time, prestige, pleasure, or any of a variety of other things. Let us assume only that some function exists assigning one number to each action-outcome combination for a particular problem. Thus, the combination "decide H_0 and H_1 turns out to be true" is accompanied by the numerical value

$$L(H_0; H_1).$$

For our purposes in this chapter we need consider these values only as *losses*; more will be said about the values in Chapter 9. Each loss value is some positive number or zero given to an action-outcome combination. If the combination represents the best available action under the true circumstances, then the decision is *right* and the loss value is zero. On the other hand, if the action is less desirable than the best available action, then this is an *error* and the loss value is positive. (A more technical term for such a value is **opportunity loss**.)

Now let us return to the poor statistician who is faced with two hypotheses, and who was left in the lurch in Section 7.5. Suppose that the potential losses connected with the two erroneous decision-possibilities are

TRUE SITUATION

H_0 H_1

	H_0	H_1
H_0	0	5
H_1	10	0

POSSIBLE ACTIONS   *Decide*

so that
$$L(H_0; H_1) = 5$$
$$L(H_1; H_0) = 10.$$

Each possible decision rule D can be given an *expected loss value*,

$$\text{EL}(D, H_0) = \text{expected loss given } D \text{ and true } H_0,$$

for any specified true situation such as H_0. Here

$$\text{EL}(D, H_0) = L(H_1; H_0)P(\text{decide } H_1 \text{ by Rule } D \text{ given } H_0 \text{ actually true})$$
$$+ L(H_0; H_0)P(\text{decide } H_0 \text{ true given } H_0 \text{ true}).$$

Furthermore,

$$\text{EL}(D, H_1) = L(H_0; H_1)P(\text{decide } H_0 \text{ true given } H_1 \text{ actually true})$$
$$+ L(H_1; H_1)P(\text{decide } H_1 \text{ true given } H_1 \text{ true}).$$

For example, consider our Decision-rule 1 once again. Here

$$\text{EL}(D_1, H_0) = 10(.323) = 3.23,$$

which is just the loss associated with an error multiplied by its probability under D_1. When H_1 is true, then

$$\text{EL}(D_1, H_1) = 5(.013) = .065.$$

The two expected losses for D_1 can be written together as the first row in Table 7.7.1.

Notice that the expected loss using this rule is rather high when H_0 is true, but low when H_1 is true. The other rows of the table can be filled out using Decision-rules 2 and 3:

$$\text{EL}(D_2, H_0) = 10(.034) = .34,$$
$$\text{EL}(D_2, H_1) = 5(.166) = .83,$$
$$\text{EL}(D_3, H_0) = 10(.375) = 3.75,$$

and
$$\text{EL}(D_3, H_1) = 5(.952) = 4.76.$$

Table 7.7.1

TRUE

	H_0	H_1
D_1	3.23	.065
DECISION-RULE D_2	.34	.83
D_3	3.75	4.76

A good decision-rule is one giving low expected loss regardless of what the true situation turns out to be. Let us apply this criterion.

An inspection of this table shows, once again, the absurdity of Rule 3. Given either true situation (a column of Table 7.7.1), the expected loss for Rule 3 is greater than that for either Rule 1 or Rule 2, and hence this rule is far less desirable than either of the others. In general, any rule that gives a higher expected loss than another *regardless* of the true situation is disqualified (or is inadmissable).

Can we choose between Rules 1 and 2, however, on the basis of expected loss? A device within decision theory for choosing between rules is known as the "minimax" principle. Applying this principle leads us to choose that rule showing the *minimum maximum-expected-loss* over all possible true situations. For expected-loss tables such as 7.7.1, a minimax decision-rule is one having a largest entry in its row that is smaller than the largest entry in any other row of the table. Notice that in this situation, Rule 2 has .83 as the largest value in its row; this is smaller than the largest value for Rule 1, or 3.23. Thus, applying the minimax principle, we find that Rule 2 is the one to use. Using Rule 2, the largest amount that we expect long-run loss to be is smaller than for either Rule 1 or Rule 3.

This minimax criterion for choosing among decision-rules is used often in the theory of games and in decision theory, and it does give a way to compare decision-rules on the basis of their expected losses. However, as we will see in Chapter 9, this criterion is often criticized for being unduly conservative, and actually this way of choosing a decision-rule can produce results that go against our intuition in some problems. The minimax criterion seems especially to be out of the spirit of inference and decision-making, since it focuses only on the extreme potential loss in making errors, and not on the very large gains one can make *despite* the risk of being wrong on occasion. An alternative to the minimax criterion is given below.

7.8 SUBJECTIVE EXPECTED LOSS AS A CRITERION FOR CHOOSING AMONG DECISION-RULES

One of the considerations left completely out of our discussion so far is the fact that the statistician usually knows something (or at least believes something) about the hypotheses before the experiment takes place. The statistician does not operate in a vacuum; for any experimental problem many sources of prior information are open to him. There are related experiments (or even repetitions of the same experiment) done by others, there are theories that differ in their logical tightness and plausibility, there are the statistician's own informal observations of the world about him, and so on. The net result is that the statistician has some initial ideas about how credible each of the hypotheses is in the light of what he already knows. He is not certain about which one of the available hypotheses is true; otherwise, he would not trouble to do the experiment. However, if asked to bet about which is true, he may be able to give odds he considers fair for this bet. In short, there are prior beliefs, or opinions, or objective knowledge that make some of the possibilities for a true situation a better bet than others for the experimenter.

There is a school of thought among statisticians holding that these prior considerations must be brought into the choice of a decision-rule. As we noted in Chapter 2, it is possible to speak of the "personal probability" of the statistician reflecting the degree of prior confidence he has that any hypothesis is true *before* seeing the evidence. That is, imagine that we could assign numbers between 0 and 1 to hypotheses such as H_0 and H_1, standing for the statistician's personal probability for each. These numbers index his degree of belief in the respective hypotheses. Let the symbols

$$P(H_0) \text{ and } P(H_1)$$

stand for these numbers, where $P(H_0) + P(H_1) = 1.00$.

Suppose that the experimenter is totally in the dark about the true situation, knowing only that either H_0 or H_1 must be true. If he is willing to bet with 50–50 odds that either hypothesis will turn out true, then

$$P(H_0) = P(H_1) = 1/2.$$

On the other hand, suppose that you bet with the statistician about which will turn out eventually to be true, and he is willing to put up nine dollars to your one dollar on the bet that H_1 is true. In other words, he thinks that the correct odds are 9 to 1 against H_0 being true. In terms of personal probabilities, this is like saying

$$E(\$) = \$1P(H_1) - \$9P(H_0) = 0,$$

so that $P(H_1) = 9/10$ and $P(H_0) = 1/10,$

if from his point of view this bet is fair.

Given the personal probabilities, it is possible to compare decision-rules in still another way. We can form the "subjective expected loss" for each decision-rule:

$$EL(D) = EL(D, H_0)P(H_0) + EL(D, H_1)P(H_1).$$

This is just a weighted sum of the expected losses for a given rule, the weights being the personal probabilities. For example, for Decision-rule 1, Section 7.7,

$$EL(D_1) = 3.23(1/10) + (.065)(9/10)$$

$$= .382,$$

and for Rule 2, $EL(D_2) = .34(1/10) + .83(9/10)$

$$= .781.$$

Decision-rules may be compared by their subjective expected losses; the lower the subjective expected loss, the better the rule. In this case, given the loss values, and given the personal probabilities, Rule 1 is clearly superior to Rule 2. Even though error has a fairly high probability by this rule when H_0 is true, the statistician's own weighting of what he expects to be true of the situation tends to discount the likelihood of such an error.

7.9 CONVENTIONAL DECISION-RULES: TYPE I AND TYPE II ERRORS

In case you have forgotten again, we left our statistician long ago, lost in thought and trying to decide how to decide between his two hypotheses. Now we have come up with an answer. Or have we? *Provided* that the experimenter can furnish us with the loss-values of the possible errors, and *provided* that we can find out his personal probabilities, then the subjective expected loss gives an unequivocal way for him to pick a decision-rule. Unfortunately, these two provisions introduce a problem as big as before (a problem which we will attack in Chapters 8 and 9). Because of this, "classical" statisticians have adopted a convention regarding the selection of a decision rule. Before introducing this convention, it is necessary to introduce some new terminology.

In hypothesis testing, the statistician can make two kinds of errors: he can reject the hypothesis H_0 when it is in fact true, or he can accept H_0 when it is in fact false. We shall call the former a Type I error and the latter a Type II error. **A Type I error is that made when H_0 (the tested hypothesis) is falsely rejected, and a Type II error is that made when H_0 is falsely accepted.**

For any rejection region, we can determine the probability of each of the two kinds of error. We shall use the following notation:

$$\alpha = P(\text{Type I error}) \ = P(\text{Rejecting } H_0 \mid H_0 \text{ is true})$$

$$\beta = P(\text{Type II error}) = P(\text{Accepting } H_0 \mid H_0 \text{ is false}).$$

In "classical" statistics, given some hypothesis H_0 to be tested, the region of rejection is usually found as follows:

Convention

Set α, the probability of falsely rejecting H_0, equal to some small value. Then, in accordance with the alternative H_1, choose a region of rejection such that the probability of observing a sample value in that region is equal to α when H_0 is true. The obtained result is significant beyond the α level if the sample statistic falls within that region.

Ordinarily, the values of α used are .05 or .01, although, on occasion, larger or smaller values are employed. Even though only one hypothesis, H_0, may be exactly specified, and this determines the sampling distribution employed in the test, in choosing the region of rejection one acts as though there were two hypotheses, H_0 and H_1. The alternate hypothesis H_1 dictates which portion or tail of the sampling distribution contains the rejection region for H_0. In some problems, the region of rejection is contained in only one tail of the distribution, so that only extreme deviations in a given direction from expectation lead to rejection of H_0. In other problems, big deviations of either sign are candidates for the rejection region, so that the region of rejection lies in both tails of the sampling distribution.

The conventions about the permissible size of the α probability of Type I error actually grew out of a particular sort of experimental setting. Here it is known in advance that one kind of error is extremely important and is to be avoided. In this kind of experiment these conventional procedures do make sense when viewed from the decision-making point of view. Furthermore, designation of the hypothesis H_0 as the "null" hypothesis and the arbitrary setting of the level of α can best be understood within this context. An example frequently used to demonstrate this approach concerns the trial of an accused criminal. The two errors that can be made are: (a) set a guilty man free, and (b) convict an innocent man. According to the concepts of justice which have evolved in the United States, (b) is a much more serious error than (a). As a result, the hypothesis that the accused man is innocent is taken as the "null" hypothesis. (Historically, the hypothesis to be tested has been called the null hypothesis, primarily because it is a hypothesis that the statistician often sets out to "disprove"; H_0 is called the **null hypothesis** and H_1 is called the **alternative hypothesis.** This terminology serves no useful purpose, especially if hypothesis testing is viewed as a decision-making procedure, so for the most part we will avoid it.) Then α is the probability of convicting an innocent man, and the philosophy of the system of courts in the United States is to make α very small, even if this results in a fairly high β. This is clear from the many safeguards in the system which are designed to protect the accused, and from the statement: a man is presumed innocent until proven guilty (the word "proven" is used rather loosely here, implying that P(guilty) is quite high, but not necessarily equal to one). An interesting point with respect to this example is that courts in some countries appear to take the opposite approach, considering error (a) to be more serious and presuming a man guilty unless he is proven innocent. The choice of a "null" hypothesis, then, is very much dependent upon a set of values, which can be thought of as implying a set of losses.

As another example of an experimental setting where Type I error is clearly to be avoided, imagine that one is testing a new medicine, with the goal of deciding if the medicine is safe for the normal adult population. By "safe" we will mean that the medicine fails to produce a particular set of undesirable reactions on all but a very few normal adults. Now in this instance, deciding that the medicine is safe when actually it tends to produce reactions in a relatively large proportion of adults is certainly an error to be avoided. Such an error might be called "abhorrent" to the statistician and the interests he represents. Therefore, the hypothesis "medicine unsafe" or its statistical equivalent is cast in the role of the null hypothesis, H_0, and the value of α is chosen to be extremely small, so that the abhorrent Type I error is very unlikely to be committed. A great deal of evidence against the null hypothesis is required before H_0 is to be rejected. The statistician has complete control over Type I error, and regardless of any other feature his study of the medicine may have, he can be confident of taking very little risk of asserting that H_1, or "medicine safe" is true when actually H_0, or "medicine unsafe" is true.

In other words, the conventional practice of arbitrarily setting α at some very small level is based on the notion that one kind of error is extremely important and must be avoided if possible. This is quite reasonable in some contexts, such as the study of the safety of a new medicine or the guilt of an accused man. On the other hand, in many situations, there is no basis for assuming that one error is much more serious than the other. In these situations, the conventional approach is difficult to justify, and we would probably be better off to use the approach of the previous section. The introduction (formally) of losses to the problem provides us with a basis for selecting a decision rule; the conventional approach is strictly informal. The decision-theoretic approach to hypothesis testing eliminates the need for conventions, since it provides a way to select the optimal action (accept or reject) in *any* situation. It thus eliminates the need to concern ourselves with such designations as "null" hypotheses. Unfortunately, it does require the determination of losses. We will defer consideration of the full decision-theoretic approach to hypothesis testing until Chapter 9. In the remainder of this chapter, we will discuss testing under the assumption that α has been determined (although not necessarily small, such as in the convention). The student should keep in mind that the determination of α depends on losses. In the following section, an example is presented which demonstrates a situation in which the "classical" convention is somewhat unrealistic.

7.10   DECIDING BETWEEN TWO HYPOTHESES ABOUT A MEAN

Before we leave the problem of deciding between two exact hypotheses, one more example will be given, this time involving the mean of a population. Furthermore, in this example, something about losses will be known, so that the applicability of a conventional rule to the problem can be studied.

A personnel manager is assigned to study the possibility of using women in a certain job. Heretofore, only men have been used, and it is known that for the population of men on this job, the average performance score is 138, with a standard deviation of 20. If women perform exactly like men, then the population of women should have the same mean and standard deviation as the men. On the other hand, if women have a mean performance score of at least 142, a considerable profit will accrue to the management by employing them. From what the personnel manager already knows about women's performance in other situations, he believes that one of these two possibilities will turn out to be true.

Accordingly, the personnel manager frames two hypotheses, both of which deal with a population of women on this job:

$$H_0 : \mu = 138$$

$$H_1 : \mu = 142.$$

It is assumed that the population distribution in either situation will have a standard deviation of 20.

Now a sample of 100 women is to be drawn at random and put into this job situation on an experimental basis. The personnel manager feels that this sample is large enough to permit using the normal approximation to the sampling distribution of the mean.

The following conventional rule is used:

If the sample result falls among the highest 5 percent of means in a normal distribution given H_0, then reject H_0; otherwise, reject H_1.

This region of rejection for H_0 is shown in Figure 7.10.1.

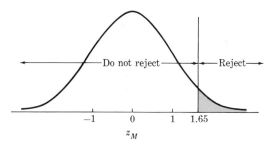

z_M

Figure 7.10.1

In this instance, there are once again two kinds of errors that can be made. The probabilities for each kind are labeled α and β respectively:

	TRUE	
	H_0	H_1
DECIDE H_0	$1 - \alpha$	β
DECIDE H_1	α	$1 - \beta$

In this situation with two *exact* hypotheses, it does make sense to talk about accepting or deciding in favor of H_0, since both probabilities can be known. In essence, the conventional rule says to fix α at .05, and the region of rejection chosen does just that.

However, notice that assigning the value of .05 to the α probability of error automatically fixes β as well. Given that H_0 is true, then the region of rejection must be bounded by a z_M score such that

$$F(z_M) = .95, \quad \text{or} \quad 1 - F(z_M) = .05.$$

The normal tables show that this z_M score is 1.65. In terms of a sample mean

$$z_M = \frac{M - 138}{\sigma_M} = \frac{M - 138}{2}.$$

Thus, the critical value of M is

$$M = 138 + 1.65\sigma_M = 138 + 3.30 = 141.30.$$

However, what would the z_M score for this critical mean be if H_1 were true?

$$z_M = \frac{141.3 - 142}{2} = -.35.$$

In a normal distribution, $F(-.35) = .36$, approximately, and so we can see that $\beta = .36$. Thus, the two error probabilities are

$$\alpha = .05$$

and

$$\beta = .36.$$

In short, the β probability of making an error is much greater than α. Our personnel manager is rather unlikely to decide on H_1 when H_0 is true, but has considerable chance of concluding H_0 when H_1 is true. Nevertheless, he can draw his sample at random, compute the mean, and decide by this rule; if the sample has a mean that is more than 1.65 standard errors above 138, then he concludes H_1 is true; otherwise H_0.

Suppose, however, that the possible losses to the management could be figured in this way: a decision that H_1 is true leads the management to hire women for the job, whereas women are not hired if the decision favors H_0. If the test indicates H_1 even though H_0 is true, the management is out the cost of the experiment, $10,000, but no other loss is incurred, since the women hired should perform like the men. On the other hand, if the second kind of error is made, and women are not hired even though they are really better than the men, the management is out not only the cost of the experiment but also the $80,000 net profit ($100,000 gross profit less $20,000 hiring and training costs) that might have been made in a year by hiring the women, resulting in a total loss of $90,000. Clearly, the two errors do not have the same consequences for the management. And yet, this decision-rule almost guarantees that the less important error will not be made, while permitting a large risk of making the expensive error.

This difficulty can be shown most clearly in terms of subjective expected loss. If the personal probabilities of the experimenter actually were

$$P(H_0) = P(H_1) = 1/2,$$

then

$$EL(D) = (10,000)(.05)(.5) + (90,000)(.36)(.5)$$

$$= \$16,250,$$

so that the subjective expected loss would be over sixteen thousand dollars.

On the other hand, suppose that another rule were adopted, with the critical value of M equal to 138.70. Then β is equal to .05 and α is equal to .36. In this circumstance,

$$\text{EL}(D) = (10,000)(.36)(.5) + (90,000)(.05)(.5)$$
$$= \$4050.$$

Obviously, this second rule is more desirable than the first if the expected loss is to be low. Still better rules exist if the probability β is made even smaller. In this situation, where losses can be considered, the adoption of an arbitrary decision-rule is not very wise.

On the other hand, if a sample does show a mean which is more than 1.65 standard errors away from 138, the experimenter can say that such samples are "improbable" given $\mu = 138$, regardless of whether or not this statement has any connection with a decision-rule he is actually going to use. If he *must* use the arbitrary rule, then the probability α is given by the probability of finding a sample mean in the rejection region for H_0. Nevertheless, the probability β may be so big that this rule is unreasonable for the actual decision.

7.11   THE OPTION OF SUSPENDING JUDGMENT

Must the statistician actually make a decision about what is true from his data? Naturally, choosing to suspend judgment and wait for more evidence is a decision to adopt a course of action. However, why cannot the statistician adopt this time-honored strategy of the scientist more often than he apparently does? In many applied situations the opportunity to suspend judgment is just not available to the statistician. He must decide here and now on the basis of what little evidence he has. Fortunately, however, statisticians often do have the privilege of waiting for more evidence.

The occurrence of a "significant" result in a significance test is not always a command to decide something. All that the significance test per se can give is a probability statement about obtaining such and such result if the given hypothesis is true. If one is using the rule represented by the rejection region, then this probability statement is also a statement of the probability of one kind of error. But here the direct contribution of statistics stops; the actual decision should depend on other factors, such as potential losses and personal probabilities, that are not a part of the formal mechanism of the "classical" statistical test. It does not make much sense to go about making decisions among hypotheses when one has no idea about the "goodness" of the decision-rule he uses. If the statistician has enough information about potential losses to be able to see that the conventional rule is not appropriate, then he should be able to figure out a better rule. On the other hand, if the situation in which he finds himself is so vague that he has no conception of the risk of loss through error, then how does he know that deciding among the hypotheses is better than suspending judgment regardless of the evidence? At least, suspending judgment is usually a conservative course of action.

Why, then, do statisticians bother with significance tests at all? *Regardless of what one is going to do with the information—change his opinion, adopt a course of action, or whatever—he needs to know relatively how probable is a result like that obtained, given a hypothetical true situation. Basically, a "classical" significance test gives this information, and that is all.* The conventions about significance level and regions of rejection can be regarded as ways of defining "improbable." The occurrence of a significant result in terms of these conventions is really a signal: "Here is a direction and degree of deviation which falls among those relatively unlikely to occur given that the tested hypothesis is true, but which is relatively more likely given the truth of some other hypothesis." If one should decide to reject the original hypothesis on the basis of a significant result, then at least he knows the probability of error in doing so. This does not mean that he must decide against the hypothesis simply because some conventional level of significance was met. Other options, such as suspending judgment, may actually be better actions under the circumstances regardless of the result of the conventional significance test. Even more emphatically, the occurrence of a nonsignificant result does not mean that you must accept the hypothesis as true. As we shall see, one often has not the foggiest idea of the error probability in saying that the tested hypothesis is true; here, making a decision to accept the tested hypothesis is absurd in the light of the unknown, and perhaps very large, probability of such an error. On occasion, however, the probability of such an error can be assessed, and here one can feel safe in asserting that the tested hypothesis is true when this error-probability is small. In other circumstances where this probability of error cannot be found, and the statistician is under no duress to make an immediate choice between hypotheses, then suspending judgment is a relatively safe way out. In still other circumstances the loss values of errors may be so small that it really does not make much difference what the experimenter decides. Until we discuss decision theory in more detail, then, the student is advised to think of "classical" hypothesis testing in the way that it has traditionally been used by most statisticians: a conventional signalling device saying, for a significant result, "Here is something relatively unlikely given the situation initially postulated, but which is rendered relatively much more likely under an alternative situation. The probability is remote of this being a chance occurrence under the initial hypothesis, and something interesting may be here."

7.12 THE POWER OF A STATISTICAL TEST

Suppose that we have a hypothesis, H_0, which we are interested in testing, and that it is an exact hypothesis concerning a particular parameter, which we shall call θ:

$$H_0 : \theta = \theta_0.$$

Furthermore, suppose that the value of α has been determined, and thus the rejection region is known. The probability of rejecting the hypothesis H_0 is then a function of θ. For any particular value of θ, say θ_1, we can compute

$$P(\text{reject } H_0 \mid \theta = \theta_1).$$

If $\theta_1 = \theta_0$, it is obvious that this probability is equal to α. On the other hand, if $\theta_1 \neq \theta_0$, the probability is equal to $1 - \beta$ if we think of a test of H_0 versus the alternative hypothesis $H_1 : \theta = \theta_1$. At any rate, we can compute the probability for any value θ_1, and it is known as the **power** of the statistical test of H_0 against the alternative H_1.

For example, consider the hypotheses in the preceding problem. When α was set equal to .05, then the hypothesis actually being tested was $H_0 : \mu = 138$. Suppose, however, that the true situation is represented by $H_1 : \mu = 142$. By the rule used, H_0 is rejected when a sample exceeds 141.3. This gives a probability of error β shown by the shaded region in Figure 7.12.1, if H_1 is true.

The unshaded area under the curve is $1 - \beta$, or the power of the test, given H_1. In this instance the power is about .64, so that the chances of correctly rejecting H_0 are about six in ten. Notice, however that *the power of a test of H_0 cannot be found until some true situation H_1 is specified.*

The power of a test of H_0 is not unlike the power of a microscope. It reflects the ability of a decision-rule to detect from evidence that the true situation differs from a hypothetical one. Just as a high-powered microscope lets us distinguish gaps in an apparently solid material that

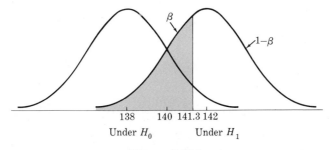

Figure 7.12.1

we would miss with low power or the naked eye, so does a high-powered test of H_0 almost insure us of detecting when H_0 is false. Pursuing the analogy further, any microscope will reveal "gaps" with more clarity the larger these gaps are; the larger the departure of H_0 from the true situation H_1, the more powerful is the test of H_0, other things being equal.

For example, suppose that the hypothesis to be tested is

$$H_0 : \mu = \mu_0,$$

but that the true hypothesis is

$$H_1 : \mu = \mu_1.$$

Here, μ_0 and μ_1 symbolize two different possible numerical values for μ. It will be assumed that under either hypothesis the sampling distribution of the mean is normal with σ known. Suppose now that the α probability of error is set at .05, and the rejection region is on the *right tail* of the distribution. The decision-rule is such that the critical value of M (the smallest M leading to the rejection of H_0) is $\mu_0 + 1.65\sigma_M$. The power of the test can then be figured for any possible true H_1.

First of all, suppose that μ_1 were actually $\mu_0 + \sigma_M$. Then in the distribution under H_1, the critical value of M has a z score given by

$$z_M = \frac{M - \mu_1}{\sigma_M}$$

$$= \frac{(\mu_0 + 1.65\sigma_M) - (\mu_0 + 1\sigma_M)}{\sigma_M} = .65.$$

The probability of a sample's falling into the region of rejection for H_0 is found from Table I to be .26. This probability is $1 - \beta$, the power of the test.

This may be clarified by Figure 7.12.2. The region of rejection for H_0

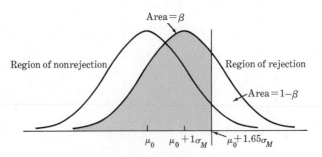

Figure 7.12.2

is the segment of the horizontal line to the right of $M = \mu_0 + 1.65\sigma_M$. The shaded portion under the curve for $\mu = \mu_0 + 1\sigma_M$ denotes β, the probability of *failing* to reject H_0, and the power is the unshaded part of the curve.

If another alternative hypothesis were true, say,

$$H_1 : \mu_1 = \mu_0 + 3\sigma_M,$$

then the power would be much larger. Here, relative to the sampling distribution based on the *true* mean, or $\mu_0 + 3\sigma_M$, the critical value of M would correspond to a z_M score of

$$z_M = \frac{(\mu_0 + 1.65\sigma_M) - (\mu_0 + 3\sigma_M)}{\sigma_M} = -1.35.$$

Table I shows us that above a z score of -1.35 lie .91 of sample means in a normal sampling distribution, so that the power is .91.

The power of the test for any true value of μ_1 can be found in the same way. Often, to show the relation of power to the true value of μ_1, so-called **power functions** or **power curves** are plotted. One such curve is given in Figure 7.12.3, where the horizontal axis gives the possible values of true μ_1 in terms of μ_0 and σ_M, and the vertical axis the value of $1 - \beta$, the power for that alternative. Notice that for this particular decision-rule, the power curve rises for increasing values of μ_1 and approaches 1.00 for very large values. On the other hand, for this decision-rule, when true μ_1 is less than μ_0 the power is very small, approaching 0 for decreasing μ_1 values. In any statistical test where the region of rejection is in the direction of the true value covered by H_1, the greater the discrepancy between the tested hypothesis and the true situation, the greater the power.

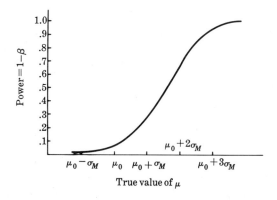

Figure 7.12.3. Power Curve for a One-Sided Test of a Mean, $\alpha = .05$

7.13   POWER AND THE SIZE OF α

Since β will ordinarily be small for large α, as we have seen in the discussion above, it follows that setting α larger makes for relatively more powerful tests of H_0. For example, the two power curves given below for the same decision-rule show that if α is set at .10 rather than at .05, the test with $\alpha = .10$ will be more powerful than that for $\alpha = .05$ over all possible true values under alternative H_1 (note Figure 7.13.1). Making the probability of error in rejecting H_0 larger has the effect of making the test more powerful, other things being equal.

In principle, if it is very costly to make the mistake of overlooking a true departure from H_0, but not very costly to reject H_0 falsely, one could (and perhaps should) make the test more powerful by setting the value of α at .10, .20, or more. This is not ordinarily done in scientific research (particularly in the physical sciences), however. There are at least two reasons why α is seldom taken to be greater than .05. In the first place, the problem of relative losses incurred by making errors is seldom faced in scientific research; hence conventions about the size of α are adopted. The other important reason is that given some fixed α the power of the test can be increased either by increasing sample size or by reducing the standard error of the test statistic in some other way. Of course, in many problems in business, the social sciences, and the behavioral sciences, it is feasible to assess losses due to erroneous decisions. Nevertheless, in these problems, the statistician may still be interested in increasing the power of a test by increasing sample size, as we see in the next section.

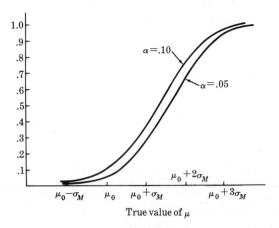

Figure 7.13.1. Power Curves for Two Tests with Equal Sample Sizes but with Different Values of α

7.14   THE EFFECT OF SAMPLE SIZE ON POWER

Given a population with true standard deviation σ, the standard error of the mean depends inversely upon the square root of sample size N. That is,

$$\sigma_M = \frac{\sigma}{\sqrt{N}} .$$

When N is large, then the standard error is smaller than when N is small. Provided that $1 - \beta > \alpha$ to begin with, *increasing the sample size increases the power of a given test of H_0 against a true alternative H_1.*

For example, suppose that

$$H_0 : \mu = 50$$

and

$$H_1 : \mu = 60,$$

where H_1 is true. Let us assume that true $\sigma = 20$. If samples of size 25 are taken, then

$$\sigma_M = \frac{20}{5} = 4.$$

Now the α probability for this test is fixed at .01, making the critical value of M be

$$M = \mu_0 + 2.33\sigma_M = 50 + (2.33)(4) = 59.32.$$

In the true sampling distribution (under H_1) this amounts to a z score of

$$z_M = \frac{59.32 - 60}{4} = \frac{-.68}{4} = -.17,$$

so that .57 of all sample means should fall into the rejection region for H_0. The power here is thus .57.

Now let the sample size be increased to 100. This changes the standard error of the mean to

$$\sigma_M = \frac{20}{10} = 2$$

and the critical value of M to

$$M = 50 + (2.33)(2) = 54.66.$$

The corresponding z score when $\mu = 60$ is

$$z_M = \frac{54.66 - 60}{2} = -2.67,$$

making the power now in excess of .99. With this sample size we would be almost certain to detect correctly that H_0 is false when this particular H_1 is true. With only 25 cases, we are quite likely not to do so.

The disadvantages of an arbitrary setting of α can thus be offset, in part, by the choice of a large sample size. Other things being equal and regardless of the size chosen for α, the test may be made powerful against any given alternative H_1 in the direction of the rejection region, provided that sample N can be made very large (see Figure 7.14.1).

Once again, however, it is not always feasible to obtain very large samples. Samples of substantial size are often costly for the statistician, if not in money, then in effort. Our ability to attain power through large samples only partly offsets the failure to choose a decision-rule according to costs, or losses, in such research; large samples may not really be necessary in some research using them, especially if the error thereby made improbable is actually not very important.

Even given a fixed sample size, the statistician has one more device for attaining power in tests of hypotheses. *Anything that makes σ, the population standard deviation, small will increase power, other things being equal.* This is one of the reasons for the careful control of conditions in good experimentation. By making conditions constant, the statistician rules out many of the factors that contribute to variation in his observations. Statistically this amounts to a relative reduction in the size of σ for some experimental population. When the statistician rules out some of the error variance from his observations he is *decreasing* the standard error of the mean, and thus *increasing* the power of the test against whatever hypothesis H_1 is true. Experiments in which the variability attributable to experimental or

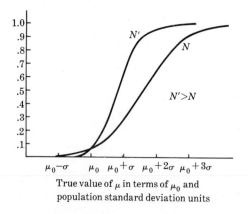

True value of μ in terms of μ_0 and
population standard deviation units

Figure 7.14.1. Power Curves for Two Tests with $\alpha = .05$ but with Different Sample Sizes, $N' > N$

sampling error is small are said to be **precise;** the result of such precision is that the statistician is quite likely to be able to detect when something of interest is happening. The application of experimental controls is like restricting inferences to populations with smaller values of σ^2 than otherwise, and thus control over error variation through careful experimentation implies powerful statistical tests. It follows that controlled experiments in which there is little "natural" variation in the materials observed can attain statistical power with relatively few observations, while those involving extremely variable material may require many observations to attain the same degree of power.

7.15 TESTING INEXACT HYPOTHESES

The primary notions of errors in inference and the power of a statistical test have just been illustrated for an extremely artificial situation, in which a decision must be made between two exact alternatives. Such situations are almost nonexistent in the real world. Instead, the statistician is far more likely to be called on to evaluate inexact hypotheses, each of which encompasses a whole range of possible true values. What relevance, then, does the discussion of the exact two-alternative case have to what statisticians actually do? The answer is that the statistician makes his inferences as though he were deciding between two exact alternatives, even though his interest lies in judging between inexact hypotheses. Thus, the mechanism we have been using for decisions between exact hypotheses is exactly the same as for any other set of alternatives.

An example should clarify this point. As a fairly plausible situation, imagine a production manager who must decide whether or not to replace a machine with a new machine. From past experience, the manager knows that the old machine produces at the mean rate of 100 units per minute, and the standard deviation of the rate per minute is 15. Furthermore, the distribution of values is approximately normal.

The new machine which is being considered is a newer model, and the manager has reason to suspect that it may produce at a higher rate than the old machine. In order to investigate this supposition, the manager obtains permission from the manufacturer of the new machine to try the machine on a trial basis. In doing so, he will obtain sample information which will aid him in making his decision. Basically, the question to be answered is "Does the new machine produce at a higher mean rate than the old machine?" We will assume that there is no reason to think that the standard deviation might be different for the new machine, nor the general form (approximately normal) of the distribution.

The answer to his question is tantamount to a decision between two *inexact* hypotheses:

$$H_0: \mu \leq 100$$

$$H_1: \mu > 100.$$

That is, the new machine has either a mean rate less than or equal to that of the old machine, or a mean rate greater than the old machine.

In choosing the decision-rule he will use, the manager decides to set α equal to .01. A result greater than 100 tends to favor H_1, and so he takes as his region of rejection all z values greater than or equal to 2.33, since this value cuts off the highest 1 percent of sample means in a normal sampling distribution.

What, however, is the hypothesis actually being tested? As written, H_0 is inexact, since it states a whole region of possible values for μ. One exact value is specified, however: this is $\mu = 100$. Actually, then, the hypothesis tested is $\mu = 100$ against some unspecified alternative greater than 100. In effect, the decision-rule can be put in the following form:

TRUE SITUATION

$\mu = 100$ $\mu > 100$

	$\mu = 100$	$\mu > 100$
$\mu = 100$	.99	(β)?
$\mu > 100$	.01	$(1 - \beta)$?

DECIDE (label at left of table)

The α probability of error can be specified in advance as .01, but β and the power are unknown, depending as they do upon the true situation. The experimenter has no real interest in the hypothesis that $\mu = 100$; he may even feel extremely confident that the true mean is not precisely 100. Nevertheless, this is a useful dummy hypothesis, in the sense that *if he can reject $\mu = 100$ with $\alpha = .01$, then he can reject any other hypothesis that $\mu < 100$ with $\alpha < .01$*. In other words, by this decision-rule, if he can be confident that he is not making an error in rejecting the hypothesis actually tested, then he can be even more confident in rejecting any other hypothesis covered by H_0.

But what does this do to β? Given some true mean μ_1 covered by H_1, *the power of the test of $\mu = 100$ is less than the power for any other hypothesis covered by H_0 with fixed α*. If he had chosen to test any other exact hypothesis embodied in H_0, such as $\mu = 90$, then neither α nor β would exceed those for $\mu = 100$. Testing the exact hypothesis with given α and β probabilities can be regarded as testing *all* hypotheses covered by H_0, with *at most* α and β probabilities of error (the β depending on the *true* mean, of course). This is illustrated in Figure 7.15.1.

Suppose now that a sample of 200 trials yields a mean of 103.5. The standard error of the mean is

$$\sigma_M = \frac{15}{\sqrt{200}} = 1.06.$$

Thus the z value corresponding to the sample mean is

$$z_M = \frac{103.5 - 100}{1.06} = 3.30.$$

Since this z value exceeds the critical value of 2.33, the hypothesis that $\mu = 100$ can be rejected with $\alpha < .01$, and any other hypothesis covered by H_0 can also be rejected with α less than .01. If the experimenter decides that the true mean is greater than 100 he could be wrong in this decision, but the probability of such an error is less than .01. Notice that we did not say that the production manager must decide in this way, but only that the α probability of error is slight should he make this decision.

On the other hand, suppose that the sample mean had been only 102. Here the z_M score would have been only 1.89, which is not large enough to place the sample in the region of rejection for H_0 if the probability α is to be .01. Under the decision-rule using $\alpha = .05$ this would lead to a rejection of the hypothesis, but not by the rule originally decided upon.

The same problem can be put on a much more realistic basis, particularly with regard to sample size, if the manager has an idea of some minimal value of μ he would like to be very sure to detect as a significant result if true. That is, suppose the experimenter in this problem knows that it is absolutely essential that he purchase the new machine if its mean rate

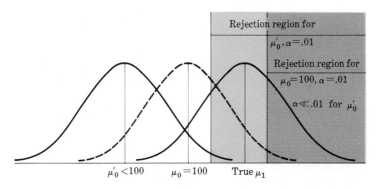

Rejection region for

$\mu_0', \alpha = .01$

Rejection region for

$\mu_0 = 100, \alpha = .01$

$\alpha \ll .01$ for μ_0'

$\mu_0' < 100$ $\mu_0 = 100$ True μ_1

Figure 7.15.1

of production is at least 103. This is important enough to him that he wants the power of the test to be at least .95 when μ is at least 103. Does his choice of a sample size and of a rejection region meet this qualification?

For $N = 200$, the standard error of the mean is 1.06. Consequently, the critical value of the mean is

$$\text{Critical } M = 2.33(1.06) + 100$$

$$= 102.47.$$

If the value of μ were actually 103, then in this sampling distribution under the true value of μ the critical value of M would correspond to

$$z_M = \frac{102.47 - 103}{1.06} = -.5.$$

In a normal distribution, 69 percent of all values must fall at or above $-.5$ standard errors from the mean. Hence, the power of this test against the true alternative $\mu = 103$ is only .69. If the statistician wants this power to be .95 for $\mu = 103$, then he must take a larger sample. In fact, the sample size required can be found from

$$-1.65 = \frac{M - 103}{\sigma/\sqrt{N}}$$

and

$$2.33 = \frac{M - 100}{\sigma/\sqrt{N}}.$$

Solving for N, we find that the required sample size is 396. Then for this sample size we can find the following probabilities of errors and correct decisions:

TRUE SITUATION

		$\mu \leq 100$	$100 < \mu < 103$	$103 \leq \mu$
	H_0	$1 - \alpha \geq .99$	$.99 < \beta < .05$	$.05 \geq \beta$
DECIDE				
	H_1	$\alpha \leq .01$	$.01 < 1 - \beta < .95$	$.95 \leq 1 - \beta$

In circumstances where there *is* such an important range of possible values of μ, representing a situation we want to be very sure to detect if true, then we can adjust the sample size or α value or both so as to make the power against such alternatives as great as necessary.

7.16 ONE-TAILED REJECTION REGIONS

In the example of Section 7.15 two inexact hypotheses were compared, having the form

$$H_0 : \mu \leq \mu_0$$

$$H_1 : \mu > \mu_0.$$

Here the entire range of possible values for the parameter under study (in this case, the population mean) was divided into two parts, that above and that below (or equal to) an exact value μ_0. The interest of the statistician lay in placing the true value of μ either above or below μ_0.

In this instance the appropriate region of rejection for H_0 consists of values of M relatively much larger than μ_0. Such values have a rather small probability of representing true means covered by the hypothesis H_0, but are more likely to represent true means in the range of H_1. Such a rejection region consisting of sample values in a particular direction from the expectation given by the exact value included in H_0 are called **directional** or **one-tailed** rejection regions. For the particular hypotheses compared in Section 7.15, the rejection region was one-tailed, since only the right or "high-value" tail of the sampling distribution under H_0 was considered in deciding between the hypotheses.

For some questions, the two inexact hypotheses are of the form

$$H_0 : \mu \geq \mu_0$$

$$H_1 : \mu < \mu_0.$$

Once again the region of rejection is one-tailed, but this time the lower or left tail of the sampling distribution contains the region of rejection for H_0. The choice of the particular rejection region thus depends both on α and the alternative hypothesis H_1.

Tests of hypotheses using one-tailed rejection regions are also called **directional.** The direction or sign of the value of the statistic (such as z_M) is important in directional tests since the sample result must show not only an extreme departure from expectation under H_0 but also a departure in the right direction to be considered strong evidence against H_0 and for H_1.

Directional hypotheses are implied when the basic question involves terms like "more than," "better than," "increased," "declined." The essential question to be answered by the data has a clear implication of a difference or change in a specific direction. For example, in the problem of Section 7.15 the production manager wanted specifically to know if the new machine produced at a higher rate than the old machine, indicating a directional hypothesis.

Many times, however, the statistician goes into a problem without a clearly defined notion of the direction of difference to expect if H_0 is false. He asks "Did something happen?" "Is there a difference?" or "Was there a change?" without any specification of expected direction. Next we will examine techniques for nondirectional hypothesis-testing.

7.17 TWO-TAILED TESTS OF HYPOTHESES

Imagine a study carried out on the "optical dominance" of human subjects. There is interest in whether or not the dominant eye and the dominant hand of a subject tend to be on the same or different sides. Subjects are to be tested for both kinds of dominance, and then classified as "same side" or "different side" in this respect. We will use the letters "S" and "D" to denote these two classes of subjects.

From knowledge he already has about the relative frequency of each kind of dominance the statistician reasons that if there actually is no tendency for eye dominance to be associated with hand dominance, then in a particular population of subjects he should expect 58 percent S and 42 percent D. However, he knows little about what to expect if there is some connection between the two kinds of dominance. To try to answer this question, our statistician draws a random sample of 100 subjects, each with a full set of eyes and hands, and classifies them.

The question at hand may be put into the form of two hypotheses, one exact and one inexact:

$$H_0 : p = .58$$

$$H_1 : p \neq .58.$$

The first hypothesis represents the possibility of no connection between the two kinds of dominance, and the inexact alternative is simply a statement that H_0 is not true, since H_1 does not specify an exact value of population p.

The α chosen is .05. The statistician then is faced with the choice of a rejection region for the hypothesis H_0. Either a very high percentage of S subjects in his sample or a very low percentage would tend to discount the credibility of the hypothesis that $p = .58$, and would lend support to H_1. Thus he wants to arrange his rejection region so that the H_0 will be rejected when extreme departures from expectation of *either* sign occur. This calls for a rejection region on *both* tails of the sampling distribution of sample P when $p = .58$. Since the sample size is relatively large, the binomial distribution of sample P may be approximated by a normal distribution, and the two rejection regions may be diagrammed as shown in Figure 7.17.1.

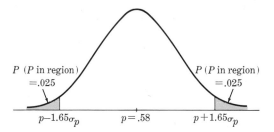

P (P in region)
$=.025$

P (P in region)
$=.025$

$p-1.65\sigma_p$ $p=.58$ $p+1.65\sigma_p$

Figure 7.17.1

Since the total probability of Type I error has been set at .05, the rejection region on the upper tail of the distribution will contain *the highest 2.5 percent,* and the lower tail *the lowest 2.5 percent of sample proportions,* given that $p = .05$. In short, each region should contain exactly the proportion $(1/2)\alpha$, or $\alpha/2$, of all samples under H_0. Consequently, either a very large or a very small sample P will lead to a rejection of H_0, and the total probability of Type I error is .05. The z score cutting off the upper rejection region is 1.96, and that for the lower is -1.96. Any sample giving a z beyond these two limits will lead to a rejection of H_0.

Since the basic sampling distribution is binomial, the standard error of P is

$$\sigma_P = \sqrt{\frac{pq}{N}} = \sqrt{\frac{(.58)(.42)}{100}}$$

$$= .049$$

when H_0 is true (Section 6.22). Using the normal approximation to the binomial we find

$$z = \frac{P - p}{\sigma_P}.$$

Now suppose that the proportion of S subjects is .69. Then

$$z = \frac{.69 - .58}{.049} = 2.24.$$

This value exceeds the critical z score of 1.96, and so the result is said to be significant beyond the 5 percent level. If the statistician concludes that eye and hand dominance do tend to be related then he runs a risk of less than .05 of being wrong.

Suppose, however, that the sample result had come out to be only .53. Then

$$z = \frac{.53 - .58}{.049} = -1.02,$$

making the result nonsignificant.

Although the example just concluded dealt with a hypothesis about a proportion, exactly the same procedure applies to two-tailed tests of means. An exact and an inexact hypothesis are posed:

$$H_0 : \mu = \mu_0$$

$$H_1 : \mu \neq \mu_0.$$

The exact hypothesis is tested by forming two regions of rejection, each region containing exactly $\alpha/2$ proportion of sample results when H_0 is true, and lying in the higher and lower tails of the distribution. A sample result falling into either of these regions of rejection is said to be significant at the α level. If the result is not significant, then judgment is suspended.

7.18 RELATIVE MERITS OF ONE- AND TWO-TAILED TESTS

In deciding whether a hypothesis should be tested with a one- or two-tailed rejection region, the primary concern of the statistician must be his original question. Is he looking for a directional difference between populations, or a difference only in kind or degree? By and large, most significance tests done in scientific research are nondirectional, simply because research questions tend to be framed this way. However, there are situations where one-tailed tests are clearly indicated by the question posed.

The powers of one- and two-tailed tests of the same hypothesis will be different, given the same α level and the same true alternative. If a one-tailed test is used, and the true alternative is in the direction of the rejection region, then the one-tailed test is more powerful than the two-tailed over all such possibly true values of μ. In a way, we get a little statistical credit in the one-tailed test for asking a more searching question.

Power curves for one- and two-tailed tests for means are compared in Figure 7.18.1. Sometimes, curves called **operating characteristic** (OC) **curves** or **error curves** are used by statisticians instead of power curves. Any one of the three types of curves can be derived from either of the other two, given a specific hypothesis testing situation and a choice of a rejection region. Recall that the power function is the probability of rejecting H_0 as a function of the parameter of interest. In the situation depicted in Figure 7.18.1, the power curve is

$$P(\text{reject } H_0 \mid \mu).$$

Similarly, the operating characteristic curve is

$$P(\text{accept } H_0 \mid \mu) = 1 - P(\text{reject } H_0 \mid \mu) = 1 - \text{power curve},$$

and the error curve is

$$P(\text{committing an error} \mid \mu).$$

Operating characteristic curves and error curves corresponding to the power curves in Figure 7.18.1 are presented in Figures 7.18.2 and 7.18.3. The relationships between the curves and the probabilities of error, α and β, can be summarized as follows:

	Region included in H_0	Region included in H_1
Power curve	α	$1 - \beta$
Operating characteristic curve	$1 - \alpha$	β
Error curve	α	β

Note that for a two-tailed test, the operating characteristic and error curves differ only at one point: the point corresponding to H_0. The operating characteristic curve is a smooth curve in this case, whereas the error curve has a "gap" at μ_0; at this point, the error curve takes on the value α, while the operating characteristic curve takes on the value $1 - \alpha$. From the graphs and the relationships between the curves and the probabilities

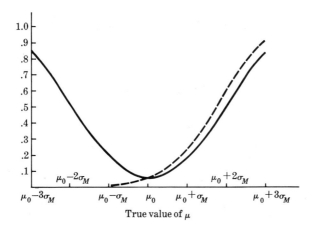

Figure 7.18.1. **Power Curves for a Two-Tailed and a One-Tailed Test of a Mean, $\alpha = .05$**

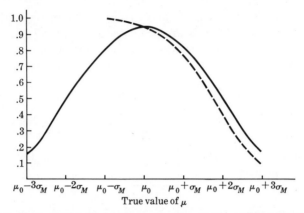

Figure 7.18.2. Operating Characteristic Curves for a Two-Tailed and a One-Tailed Test of a Mean, $\alpha = .05$

of errors, it is clear that while a "high" power curve is desirable, "low" OC and error curves are desirable.

Now let us return to the relative merits of one- and two-tailed tests. In many circumstances calling for one-tailed tests the form of the question or of the sampling distribution makes it clear that the only alternative of logical or practical consequence must lie in a certain direction. For example, when we turn to the problem of testing many means for equality simultaneously it will turn out that the only rejection region making sense lies in one tail of the particular sampling distribution employed. In other circumstances the question involves considerations of "which is better" or "which is more," where the discovery of a difference from H_0 in one direc-

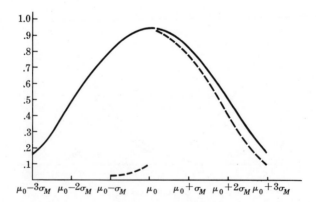

Figure 7.18.3. Error Curves for a Two-Tailed and a One-Tailed Test of a Mean, $\alpha = .05$

tion may have real consequences for practical action, although a mean in another direction from H_0 may indicate nothing. For example, consider a treatment for some disease. The cure rate for the disease is known, and we want to see if the treatment *improves* the cure rate. We have not the slightest practical interest in a possible decrease in cures; this, like no change, leads to nonadoption of the treatment. The thing we want to be sure to detect is whether or not the new treatment is really better than what we have. In many such problems where different practical actions depend on the sign of the deviation from expectation, a one-tailed test is clearly called for. Otherwise, the two-tailed test is often safest for the statistician asking only "What happens?"

7.19 SOME REMARKS ON THE GENERAL THEORY UNDERLYING TESTS OF HYPOTHESES

For purposes of exposition the kinds of hypotheses considered thus far in this chapter have been very elementary ones concerning the mean μ or the population proportion p. The hypotheses about μ have actually been composite as stated, but these have, in effect, been turned into simple hypotheses by our assumption that both σ^2 and the form of the population distribution are known. Working with these elementary examples has permitted us to use the familiar normal and binomial sampling distributions, and to illustrate decision-rules, error probabilities, and power in the simplest possible terms. Finally, and most important, we pretended that we knew that the best way to test a particular hypothesis was in terms either of the statistic M or the statistic P; that is, the question of the relevant and appropriate statistic was not allowed to come up.

As it happens, M and P actually *are* the appropriate statistics for the particular hypotheses tested; however, as we shall see in succeeding sections, in testing other hypotheses (particularly composite hypotheses), various other statistics that are not necessarily estimators for any population parameters may provide us the best ways of carrying out the tests. Such statistics, whose primary role is that of providing a test of some hypothesis, are usually called **test statistics,** and the hypothesis test itself is carried out in terms of the sampling distribution of the appropriate test statistic, rather than that of some estimator such as M, $\hat{s}^2$, or P. We have already encountered the most important test statistics: t, χ^2, and F.

In mathematical statistics a general theory of hypothesis testing exists that extends the elementary ideas discussed here to much more complex situations. Among other things, this theory specifies desirable characteristics that a test statistic should have and indicates methods for finding such test statistics, all within the general framework outlined here. Some particular

H_0 and H_1 are formulated and a value of α is chosen arbitrarily by the experimenter; some appropriate test statistic is selected and the actual decision-rule is formulated in terms of rejection regions in the sampling distribution of this statistic. Much of the theory of hypothesis testing deals with the choice of the *right* statistic for the purpose.

Among the desirable properties of a test based on some statistic is that of **unbiasedness**: a test statistic (and decision-rule) is said to afford an unbiased test of the hypothesis H_0 if the use of this statistic (and rule) makes the probability of rejecting H_0 be at its smallest when H_0 is actually true. An unbiased test will have *minimum* power when H_0 is true.

Another property of a good test is **consistency**: in somewhat imprecise terms, a consistent test is one for which the probability of rejecting a *false* H_0 approaches 1.00 as the sample size approaches ∞. That is, a consistent test always gains power against any true alternative as the sample size is increased.

Still another criterion of a good test is that it be **most powerful for the particular true alternative** to the hypothesis H_0. That is, among all of the different ways that one could devise to test some particular H_0 against some particular true alternative value covered by H_1, the most powerful test for any fixed value of α affords the smallest probability of Type II error, other things such as sample size being equal. **Uniformly most powerful tests** for any fixed value of α are those which give a smaller probability of Type II error than any other test regardless of the value which happens to be true among those covered by H_1, other things being equal.

The general theory of hypothesis testing first took form under the hand of Sir Ronald Fisher in the 1920s, but it was carried to a high state of development in the work of J. Neyman and E. S. Pearson, beginning about 1928. Neyman and Pearson introduced a device for finding "good" tests of hypotheses, a device known as the "likelihood-ratio" test. This is a general procedure for finding the test statistic that will have optimal properties for testing any of a broad class of hypotheses, particularly the statistic that will give a most powerful test for a specific hypothesis. The theory underlying the likelihood ratio is very closely allied to the maximum-likelihood principle discussed in Section 6.7, and involves the maximum likelihood of the particular sample result *given* the hypothesis H_0 relative to the maximum likelihood of the sample result over all possible values of the relevant parameters.

We will consider the use of likelihood ratios in testing and decision procedures in Chapters 8 and 9. For now, suffice it to say that the "classical" tests of hypotheses to be discussed in this and later chapters are all equivalent to likelihood-ratio tests. These are the "best" available tests for the hypotheses considered, meaning that when the assumptions underlying these tests are true, no other available procedure would answer our question about H_0 and H_1 better than the one we will use. After the next section, which involves the reporting of the results of tests of hypotheses, the remainder of the chapter will deal with hypothesis testing in specific situations.

7.20 REPORTING THE RESULTS OF TESTS

The general problem facing the statistician in hypothesis testing should be clear to you by now. He wants to know if the sample results deviate from the results expected under some hypothesis H_0 to such a degree that he should choose to reject H_0. First, he decides upon a sample statistic which will be the basis for the test and determines the sampling distribution of this statistic under the assumption that H_0 is true. Then, he determines α, either by considering the relative seriousness of the losses associated with the two types of errors or by choosing to follow the "classical" statistician's convention. From the sampling distribution and the value of α, he can find a rejection region. Finally, he observes the sample results, calculates the actual value of the sample statistic, and checks to see if it is in the rejection region. If it is, he rejects the hypothesis; if it is not, he accepts the hypothesis (or, of course, he could choose a third option: the option of suspending judgment until more information is obtained).

The end result of this testing procedure is expressed by the decision: accept, reject, or suspend judgment. If the statistician reports this decision and nothing else, he is telling us very little about the sample. Suppose that he rejects H_0 at the .05 level. What does this tell us? It tells us that the sample result fell in a category of results that would be expected to occur by chance only 5 percent of the time, given the truth of the hypothesis. Was the result "just barely" in the rejection region? Was it such an extreme result with relation to H_0 that it would have been in a rejection region of size .01? A region of size .001? We cannot say.

Consider a concrete example, the machine example of Section 7.15. In terms of the standardized normal random variable, the rejection region for $\alpha = .01$ consisted of all z-values greater than or equal to 2.33. The z-value computed from the sample turned out to be 3.30, so that the hypothesis was rejected at the .01 level. Consider this question: "What is the smallest value of α for which the hypothesis would have been rejected?" This is obviously just the probability of a z-value greater than or equal to 3.30; this probability is .0005. **We shall call this probability the p-value associated with this specific hypothesis and sample result.** Figure 7.20.1 illustrates the calculation of this p-value.

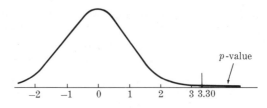

Figure 7.20.1

First, what does the p-value tell us? It tells us how "unusual" the sample result is as compared with the sampling distribution under the assumption that H_0 is true. It answers our question: "How unlikely is this sample result given the situation initially postulated?"

Second, of what importance is the p-value for reporting purposes? It provides more information than the simple reporting of acceptance or rejection at some level of significance. Furthermore, it enables the reader to choose his own level of significance (perhaps on the basis of his losses, which may be quite different from those of the person reporting the results) and to determine whether H_0 would have been rejected at *that* level of significance. For any value of α,

if $\alpha \geq p$-value, the hypothesis should be rejected,

and   if $\alpha < p$-value, the hypothesis should be accepted.

Finding a p-value from a given sample result is just the opposite of the procedure for finding a rejection region from a given value of α. It can be found by asking this question: "If the result obtained fell exactly on the borderline between the region of rejection and the region of acceptance, what would the level of significance be?" The resulting number is equal to the p-value. You should be careful to note that the calculation of a p-value depends on whether the test is one-tailed or two-tailed. In the above example, suppose that the production manager wanted to make a two-tailed test of the hypothesis that the machine would produce at a rate of 100 units per minute. In this case, a rejection region would include portions of both tails of the sampling distribution, and the p-value corresponding to a z-value of 3.30 would be twice as large as for the one-tailed test. The p-value would be $2(.0005) = .001$. For both the one- and two-tailed tests, the sample results obtained would be extremely unlikely if H_0 were true. A low p-value, then, casts doubt upon the hypothesis H_0; conversely, a high p-value tends to support the hypothesis.

Consider one final example, concerning the hypotheses:

$$H_0 : p \geq .5$$

$$H_1 : p < .5.$$

We assume that we are sampling from a stationary and independent Bernoulli process, and that a sample of size 10 results in 4 successes. Since we have a directional test, with the lower tail being the tail of interest,

$$p\text{-value} = P(X \leq 4 \mid N, p) = \sum_{x=0}^{4} \binom{10}{x} (.5)^x (.5)^{10-x}.$$

From tables of the binomial distribution, this cumulative probability is found to be .3770. Now, for any α,

$$\text{if} \quad \alpha \geq .3770, \quad \text{reject } H_0,$$

and $\quad$ if $\quad \alpha < .3770, \quad$ accept H_0.

It is interesting to note that if the test had been directional but in the opposite direction, then the p-value would have been

$$p\text{-value} = P(X \geq 4 \mid N, p) = \sum_{x=4}^{10} \binom{10}{x} (.5)^x (.5)^{10-x} = .8281.$$

Finally, had the test been two-tailed, the p-value would have been

$$p\text{-value} = 2P(X \leq 4 \mid N, p) = .7540.$$

In summary, then, the p-value concept is quite valuable in hypothesis testing. If the statistician does not have to make a decision, but he merely wants some idea of how likely or unlikely the observed sample result is if some hypothesis H_0 is true, then the p-value is just what he needs. It provides him with a measure of how likely or unlikely the observed result is if H_0 is true. He need not even specify a particular alternative, although he does have to specify whether the alternative is directional or non-directional, and if directional, in which direction. The use of p-values in this sense is in the spirit of Section 7.11, in which the option of suspending judgment was discussed.

Suppose that the statistician *does* have to make a decision, and that after evaluating the losses associated with the two types of errors, he determines a value of α. By computing the p-value from the observed sample statistic and comparing this with α, he arrives at the same decision that he would arrive at by determining a rejection region and checking to see if the observed value falls in this region. Furthermore, it is just as easy to find a p-value from the tables as it is to find a rejection region. For the "classical" statistician who always uses the same α, this might not be true, since he could commit the critical values (say, z-values) to memory. The person who determines α from a consideration of losses, however, is not likely always to arrive at the same α. At any rate, in most cases the two approaches are equally time-consuming. Then what is the advantage of the p-value approach? If, in addition to making a decision, the statistician intends to report his findings, then the p-value is very useful as a reporting device. It allows others to make their own decisions, based on their own choices of α. Furthermore, it gives the readers of the report an idea of how "unusual" the observed result was in light of the tested hypothesis.

7.21 SAMPLE SIZE AND p-VALUES

Statistically, a p-value always means the same thing with regard to H_0. For instance, a p-value of .01 indicates that H_0 is unlikely to be true. However, the *practical* interpretation of a p-value depends crucially on the sample size. For a test of

$$H_0 : \mu = 100$$

against
$$H_1 : \mu \neq 100,$$

for example, a p-value of .01 associated with a sample of size 9 would imply quite different things than a p-value of .01 associated with a sample of size 900. A small sample leading to a p-value of .01 might imply not only that $\mu = 100$ was unlikely, but also that other values around 100 (for example, 101) were also unlikely. On the other hand, a p-value of .01 associated with a large sample would imply that μ was not exactly equal to 100 but would not be as likely to rule out values of μ close to but not equal to 100.

Suppose that the population in question is normally distributed with a standard deviation of 10. For a sample of size 9, a sample mean of 108.6 would imply a p-value of .01. Notice that this sample result tends to make $\mu = 101$ seem unlikely, although not quite as unlikely as $\mu = 100$. If 108.6 and 91.4 were taken as the critical values of M, the power of the test against the alternative that $\mu = 101$ would be .0133. Thus, although α is very small (.01), the β corresponding to a true mean of 101 is very large (.9867).

On the other hand, if $N = 900$, a sample mean of 100.86 would imply a p-value of .01. Notice that for $\alpha = .01$, the critical values of M are much closer to 100 in this case than in the case with $N = 9$. Although H_0 is just as unlikely in each case, some alternative hypotheses close to 100 are more likely with the larger sample size. With $N = 900$ and $\alpha = .01$, the power of the test against the alternative that $\mu = 101$ is .6628 and the corresponding value of β is .3372.

By observing that the critical values of M will get closer and closer to 100 as N is increased and that the power against any specified alternative will be greater as N is increased, we note the following fact: if the statistician wants to "disprove" H_0, he just needs to take a large sample. A large sample will be quite sensitive to slight deviations from H_0, whereas a smaller sample will only be sensitive to larger deviations from H_0. If the true mean is 101, for example, it will generally take a larger sample before H_0 is rejected than it would if the true mean were, say, 110.

This result makes intuitive sense: it takes a larger sample to detect small differences than to detect large differences. In a way, this is unfortunate, since it means that we can be almost sure to reject an exact hypothesis such as $\mu = 100$ simply by taking a large enough sample. If this is the case, then it may not be a good idea to take *too* large of a sample, particularly if you are not concerned with small deviations from the hypothesized value. Admittedly, this statement sounds almost like heresy! Statisticians are often trained to think that large samples are good things, and many statistical procedures are valid only for large samples. Furthermore, it is true that *as long as the statistician's primary interest is in precise estimation, then the larger the sample the better*. When he wants to come as close as he possibly can to the true parameter values, he can always do better by increasing sample size.

On the other hand, many statistical studies are exploratory in nature. The statistician wants to investigate associations, to map out the main relationships in some area. He does not want to waste his time and effort by concluding that an association exists when the degree of prediction actually afforded by that association is negligible. In short, the experimenter would like a "significant" result to represent not only a nonzero association (a deviation from the "null" hypothesis), but an association of considerable size. *When the sample size is very large, there is a real danger of detecting trivial associations as significant results.* If the statistician wants significance to be very likely to reflect a sizable association in his data and also wants to be sure that he will not be led by a significant result into some blind alley, then he should pay attention to both aspects of sample size. He would like the sample size to be *large* enough to give confidence that the associations of interest will indeed show up, while being *small* enough so that trivial associations will be excluded from significance.

Stripped of the language of decision theory, all that a "significant" result or a small p-value implies is that one has observed something relatively unlikely given the hypothetical situation. Statistical significance is a statement about conditional probability, nothing else. It does not guarantee that something important, or even meaningful, has been found. Conventions about significant results should not be turned into canons of good scientific practice. It is interesting to speculate how many of the early discoveries in science would have been statistically significant in the experiments in which they were first observed. Even in the crude and poorly controlled experiment, some departures from expectation stand out simply because they are interesting and suggest things that the experimenter might not be able to explain. These are matters that warrant looking into further regardless of what the conventional rule says to decide.

The careful consideration of *both* types of error will help the statistician to decide how large a sample he needs. By using the "classical" convention, it is too easy to decide on α without paying much attention to possible values of β. Of course, if hypothesis testing is viewed as a decision-making procedure and losses are formally brought into the analysis, it is necessary to consider both types of error. Furthermore, the modification of H_0 to make it more realistic reduces the problem of detecting trivial differences; for instance, we might consider $H_0: 99 \leq \mu \leq 101$. We will discuss this problem from a decision-theoretic viewpoint in Chapters 8 and 9.

Now that the general theory and philosophy of hypothesis testing has been presented, the rest of the chapter will deal with specific applications of the theory: tests of particular types of hypotheses, such as hypotheses concerning means.

7.22 TESTS CONCERNING MEANS

Some of the examples presented earlier in the chapter involved means of normal distributions. In these examples, we "fudged" a bit on the usual situation; we assumed that σ^2 is somehow known, so that the standard error of the mean is also known exactly. Now we must face the cold facts of the matter: for inferences about the population mean, σ^2 is seldom known. Instead, we must use the only substitute available for σ^2, which is our unbiased estimate $\hat{s}^2$, calculated from the sample.

Recalling that if the population is normally distributed, then the random variable

$$t = \frac{M - \mu}{\text{est. } \sigma_M} = \frac{M - \mu}{\hat{s}/\sqrt{N}} \qquad (7.22.1^*)$$

has the t distribution with $N - 1$ degrees of freedom, we see that the only change in the testing procedure will be to calculate t instead of z and to look in the tables of the t distribution rather than the normal tables. Otherwise, the rejection regions (or p-values) are determined in exactly the same manner as when the population variance was assumed to be known. Of course, if the sample size is large enough for the t distribution to be well-approximated by the normal distribution, we can calculate t but look in the normal tables.

For example, suppose that we wished to test the following hypotheses:

$$H_0 : \mu = 10$$

$$H_1 : \mu \neq 10.$$

We take a sample of size 16 from the population, which is assumed to be normally distributed, and observe a sample mean μ of 10.24 and an unbiased sample variance $\hat{s}^2$ of .36. The value of t is then

$$t = \frac{10.24 - 10}{.6/4} = 1.6.$$

Looking in the t tables under 9 degrees of freedom, we see that for a two-tailed test, this value of t corresponds to a p-value between .10 and .20. This is as precise a statement as we can make about the p-value, given the accuracy of the tables, unless we try to interpolate. What decision does this p-value imply for any given choice of α? If $\alpha \geq .20$, we will reject H_0; if $\alpha \leq .10$, we will accept H_0; and if $.10 < \alpha < .20$, we cannot say for sure what the decision should be.

When the t distribution is used to test a hypothesis, the assumptions should be kept in mind, primarily the assumption that the population is normally distributed. If large samples are available, this should be no problem, but for small samples the statistician should exercise caution in this regard (see Section 6.15).

7.23 TESTS CONCERNING DIFFERENCES BETWEEN MEANS

As we pointed out in Section 6.19, statisticians are often interested not just in a single mean, but in the difference between means. For example, suppose that the president of a chain of retail stores is interested in the difference in mean sales at two of the stores in the chain. He formulates the following hypotheses:

$$H_0 : \mu_1 - \mu_2 = d_0$$

$$H_1 : \mu_1 - \mu_2 \neq d_0.$$

For instance, he might let $d_0 = 0$, so H_0 corresponds to the hypothesis that there is no difference between the means. Although this is the most common choice of d_0, any other value could be chosen. Furthermore, the president is willing to assume that the distributions of sales at the two stores are both normal, with respective known variances $\sigma_1^2 = 10,000$ and $\sigma_2^2 = 7200$. He then takes independent random samples of size $N_1 = 10$

and $N_2 = 12$ from the sales records of the two stores, observing the following sample results:

$$M_1 = 8000 \qquad M_2 = 8050$$

$$s_1^2 = 9000 \qquad s_2^2 = 8000.$$

Since the variances are known, the test statistic is simply

$$z = \frac{(M_1 - M_2) - d_0}{\sigma_{\text{diff.}}} = \frac{(M_1 - M_2) - d_0}{\sqrt{\sigma_1^2/N_1 + \sigma_2^2/N_2}}. \qquad (7.23.1^*)$$

If $d_0 = 0$,

$$z = \frac{8000 - 8050}{\sqrt{10{,}000/10 + 7200/12}} = \frac{-50}{40} = -1.25.$$

The p-value corresponding to this z-value for a two-tailed test is

$$P(|z| \geq 1.25) = .212.$$

Thus the president will reject H_0 if $\alpha \geq .212$ and will accept H_0 otherwise.

Suppose, however, that the president had not known σ_1^2 and σ_2^2, but that he felt that they could be assumed to be equal. This is much more realistic, and requires the use of a t statistic with $N_1 + N_2 - 2$ degrees of freedom (see Section 6.20):

$$t = \frac{(M_1 - M_2) - d_0}{\text{est. } \sigma_{\text{diff.}}} = \frac{(M_1 - M_2) - d_0}{\sqrt{\left(\dfrac{N_1 s_1^2 + N_2 s_2^2}{N_1 + N_2 - 2}\right)\left(\dfrac{N_1 + N_2}{N_1 N_2}\right)}}. \qquad (7.23.2^*)$$

For this example, $N_1 + N_2 - 2 = 20$, and

$$t = \frac{8000 - 8050}{\sqrt{1705}} = \frac{-50}{41.3} = -1.21.$$

Looking up the corresponding p-value in the t tables, we see that for a two-tailed test, the p-value is between .20 and .50, with the indication being that it is near the lower end-point, .20. If the alternative hypothesis H_1 had been $H_1 : \mu_1 - \mu_2 < 0$, the p-value would be between .10 and .25. On the other hand, if H_1 was of the form $H_1 : \mu_1 - \mu_2 > 0$, then the p-value would be between .75 and .90.

As noted above, it is not necessary for d_0 to be equal to zero. If the manager of store 2 claimed that his sales were at least 200 units higher than the sales at store 1, the president might set up the hypotheses as follows:

$$H_0 : \mu_1 - \mu_2 \leq -200$$

$$H_1 : \mu_1 - \mu_2 > -200.$$

Now, given the sample results, he calculates t:

$$t = \frac{(8000 - 8050) - (-200)}{\text{est. } \sigma_{\text{diff.}}} = \frac{150}{41.3} = 3.63.$$

For this example, the p-value is less than .001, casting much doubt on the claim of the manager of store 2.

Note that the t test in this section required three major assumptions: (a) normality of the population distributions, (b) independence of the two samples as well as independence within each sample, and (c) equality of the population variances. Assumptions (a) and (c) are more important for small samples than for large samples. Incidentally, as for all t statistics, the distribution of t becomes very close to the standard normal distribution for large samples. As for assumption (b), in some situations the statistician intentionally takes samples that are related rather than independent from the two populations. This case is considered in the following section.

7.24 PAIRED OBSERVATIONS

Sometimes it happens that samples from two populations are "paired" samples. By this, we mean that each observation in the first sample is related in some way to exactly one observation in the second sample, so the samples are not independent. For example, suppose that the two samples are two examinations given to the same class, so that we can pair the score of any single member of the class on the first examination with his or her score on the second examination. Or, consider a questionnaire administered to a group of married couples. One sample could be the wives, and one the husbands, in which case a wife could be thought of as "paired" with her husband if husband-wife differences were of interest.

Given two samples matched in this pairwise way, either by the statistician or otherwise, it is still true that the difference between the means is an unbiased estimate of the population difference (in two matched populations):

$$E(M_1 - M_2) = \mu_1 - \mu_2.$$

However, the matching, and the consequent *dependence* within the pairs, changes the standard error of the difference. This can be shown quite simply. By definition, the variance of the difference between two sample means is

$$\sigma_{\text{diff.}}^2 = E(M_1 - M_2 - \mu_1 + \mu_2)^2,$$

which is the same as

$$E[(M_1 - \mu_1) - (M_2 - \mu_2)]^2.$$

Expanding the square, we have

$$E(M_1 - \mu_1)^2 + E(M_2 - \mu_2)^2 - 2E(M_1 - \mu_1)(M_2 - \mu_2).$$

The first of these terms is just $\sigma_{M_1}^2$, and the second is $\sigma_{M_2}^2$. However, what of the third term? This term is cov (M_1, M_2), the covariance of the means (Section 3.24). Then, for matched samples,

$$\sigma_{\text{diff.}}^2 = \sigma_{M_1}^2 + \sigma_{M_2}^2 - 2 \text{ cov } (M_1, M_2).$$

This unknown value of cov (M_1, M_2) could be something of a problem, but actually it is quite easy to bypass this difficulty altogether. Instead of regarding this as two samples, we simply think of the data coming from one sample of *pairs*. Associated with each pair i is a difference

$$D_i = (Y_{i1} - Y_{i2}), \tag{7.24.1}$$

where Y_{i1} is the score of the member of pair i who is in group 1, and Y_{i2} is the score of the member of pair i who is in group 2. Then an ordinary t test for a *single* mean is carried out using the scores D_i. That is,

$$M_D = \frac{\sum_i D_i}{N}$$

and

$$s_D^2 = \frac{\sum_i D_i^2}{N} - M_D^2.$$

Then

$$\text{est. } \sigma_{MD} = \frac{s_D}{\sqrt{N-1}} = \frac{\hat{s}_D}{\sqrt{N}},$$

and t is found from

$$t = \frac{M_D - E(M_D)}{\text{est. } \sigma_{MD}} \tag{7.24.2}$$

with $N - 1$ degrees of freedom. *Be sure to notice that here N stands for the number of differences, which is the number of pairs.*

Naturally, the hypothesis is about the true value of $E(M_D)$, which is always $\mu_1 - \mu_2$. Thus any hypothesis about a difference can be tested in this way, provided that the samples used are matched *pairwise*. Similarly, confidence limits are found just as for a single mean, using M_D and σ_{MD} in place of M and σ_M.

For example, suppose that a company president is concerned about the value of a training program that the company runs for new employees. He decides to administer an examination to the 25 members of a particular class in the training program. Furthermore, he decides to administer the same test twice: once at the beginning of the training program, and once at the end. It is thought that the training program should result in an average score increase of more than 30 points; if it does not, the president may eliminate the program. The hypotheses the president is interested in are as follows:

$$H_0: \mu_2 - \mu_1 \leq 30$$

$$H_1: \mu_2 - \mu_1 > 30.$$

The subscripts 1 and 2 refer to the two times the examination is given. The class of 25 takes the examination twice, the difference in score is computed for each member, and the results are summarized as follows:

$$M_D = 36.16,$$

$$\text{est. } \sigma_{M_D} = 4.24.$$

From these results, t can be calculated:

$$t = \frac{36.16 - 30}{4.24} = 1.45.$$

Looking in the t tables under $N - 1 = 24$ degrees of freedom, we find that the p-value is between .05 and .10. The evidence appears to favor H_1 somewhat. Of course, in order to make a decision one way or the other about the hypotheses, the president would have to consider his losses and determine α. If $\alpha \leq .05$, he should accept H_0 and seriously consider eliminating the training program; if $\alpha \geq .10$, he should reject H_0 and continue the program; if $.05 < \alpha < .10$, we need more precise tables to tell us what he should do.

7.25 THE POWER OF t TESTS

The idea of the power of a statistical test was discussed in detail in preceding sections only in terms of the normal distribution. Nevertheless, the same general considerations apply to the power of tests based on the t distribution. Thus, the power of a t test increases with sample size, increases with the discrepancy between the null hypothesis value and the true value of a mean or a difference, increases with any reduction in the true value of σ, and increases with any increase in the size of α, given a true value covered by H_1.

Unfortunately, the actual determination of the power for a t test against any given true alternative is more complicated than for the normal distribution. The reason is that when the null hypothesis is false, each t ratio computed involves $E(M)$ or $E(M_1 - M_2)$, which is the exact value given by the null (and false) hypothesis. If the true value of the expectation could be calculated into each t ratio, then the distribution would follow the

t function tabled in the appendix. However, when H_0 is false, each t value involves a false expectation; this results in a somewhat different distribution, called the **noncentral** t **distribution.** The probabilities of the various t's cannot be known unless one more parameter, δ, is specified besides ν. This is the so-called noncentrality parameter

$$\delta = \frac{\mu - \mu_0}{\sigma_M}, \qquad (7.25.1)$$

which expresses the difference between the true expectation μ and that given by the null hypothesis, or μ_0, in terms of σ_M. For a hypothesis about a difference and for samples of equal size,

$$\delta = \frac{(\mu_1 - \mu_2) - (\mu_{01} - \mu_{02})}{\sigma_{\text{diff.}}}. \qquad (7.25.2)$$

This makes the determination of the power of a t test considerably more troublesome than for tests using the normal distribution. In particular, information about σ for each population is required. However, for most uses that are likely to be made of the power concept in applications, especially in determining sample size, the normal approximation may still suffice. For more accurate work, most advanced texts in statistics give tables of the power function of t in terms of both ν and δ.

7.26 TESTING HYPOTHESES ABOUT A SINGLE VARIANCE

Just as for the mean, it is possible to test hypotheses about a single population variance (and, of course, a standard deviation). The exact hypothesis tested is

$$H_0 : \sigma^2 = \sigma_0^2,$$

where σ_0^2 is some specific positive number. The alternative hypothesis may be either directional or nondirectional, depending, as always, on the original question.

As usual, some value of α is decided upon, and a region of rejection is adopted depending both on α and the alternative H_1. The test statistic itself is

$$\chi^2_{(N-1)} = \frac{(N-1)\hat{s}^2}{\sigma_0^2}. \qquad (7.26.1)$$

The value of this test statistic, computed from the $\hat{s}^2$ actually obtained and the σ_0^2 dictated by the null hypothesis, is referred to the distribution of χ^2 for $N - 1$ degrees of freedom.

For example, there is some evidence that women tend to be a less variable, more homogenous group than do men. One might ask this question about height: "It is well known that men and women in the United States differ in terms of their mean height; is it true, however, that women show less variability in height than do men?" Now let us assume that from the records of the Selective Service System we actually *know* the mean and standard deviation of height for American men between the ages of twenty and twenty-five years. However, such complete evidence is lacking for women. Assume that the standard deviation of height for the population of men twenty to twenty-five years is 2.5 inches. For this same age range, we want to ask if the population of women shows this same standard deviation, or if women are *less* variable, with their distribution having a smaller σ. The null and alternative hypotheses can be framed as

$$H_0 : \sigma^2 \geq 6.25 \qquad (\text{or } [2.5]^2)$$

$$H_1 : \sigma^2 < 6.25.$$

Imagine that we plan to draw a sample of 30 women at random, each between the ages of twenty and twenty-five years, and measure the height of each. The test statistic will be

$$\chi^2_{(29)} = \frac{(29)\hat{s}^2}{6.25}.$$

What is the region of rejection? Here, *small* values of χ^2 tend to favor H_1, that women actually are less variable than men. Hence, we want to use a region of rejection on the left (or small-value) tail of the chi-square distribution with 29 degrees of freedom. If $\alpha = .01$, then from Table III in the Appendix, the value of χ^2 leading to rejection of H_0 should be less than the value given by the row with $\nu = 29$, and the column for $Q = .99$. This value is 14.257.

Now the actual value of $\hat{s}^2$ obtained turns out to be 4.55, so that

$$\chi^2_{(29)} = \frac{(29)(4.55)}{6.25} = 21.11.$$

This value is larger than the critical value decided upon, and we cannot reject H_0 if α is to be .01. In fact, the *p*-value in this case is found from the tables to be between .10 and .25.

This example illustrates that, as with the t and the normal distribution, either or both tails of the chi-square distribution can be used in testing a hypothesis about a variance. Had the alternative hypothesis in this problem been

$$H_1 : \sigma^2 \neq 6.25,$$

then the rejection region would lie in both tails of the chi-square distribution. The rejection region on the lower tail of the distribution would be bounded by a chi-square value corresponding to $\nu = 29$ and $Q = .995$, which is 13.121; the rejection region on the upper tail would be bounded by the value for $\nu = 29$ and $Q = .005$, which is 52.336. Any obtained χ^2 value falling *below* 13.121 or *above* 52.336 would let one reject H_0 beyond the .01 level. For this two-tailed test, the p-value would be twice as large as for the above one-tailed test; that is, it would be between .20 and .50.

A final note concerning the use of the χ^2 statistic to test hypotheses concerning a single variance is that the population must be normally distributed. If the population is not normally distributed, then the χ^2 statistic might not be applicable unless the sample size is very large (see Section 6.26).

7.27 TESTING HYPOTHESES ABOUT TWO VARIANCES

In many situations the statistician is concerned with hypotheses involving two variances. In the example of the preceding section, if the standard deviation of height for men was not known, then the hypothesis of interest would be the hypothesis that the variance of height for women is greater than or equal to the variance of height for men. To test this hypothesis, we would take independent samples from the population of men and the population of women.

Suppose that independently of the sample results for women presented in Section 7.26, a sample of 21 men was taken, resulting in a value of $\hat{s}^2$ equal to 9.10. If the subscripts 1 and 2 refer to men and women, respectively, then the hypotheses of interest are

$$H_0: \sigma_2{}^2 \geq \sigma_1{}^2$$

$$H_1: \sigma_2{}^2 < \sigma_1{}^2.$$

If it is assumed that the two populations are normally distributed, then the ratio

$$F = \frac{\hat{s}_1{}^2/\sigma_1{}^2}{\hat{s}_2{}^2/\sigma_2{}^2} \tag{7.27.1*}$$

has an F distribution with $N_1 - 1$ and $N_2 - 1$ degrees of freedom (Section 6.28). But the dividing line between the two hypotheses is at $\sigma_1{}^2 = \sigma_2{}^2$, so

$$F = \frac{\hat{s}_1{}^2}{\hat{s}_2{}^2} = \frac{9.10}{4.55} = 2.00,$$

with 20 and 29 degrees of freedom. From the F tables, we find that this is slightly greater than the value required for one-tailed significance at the .05 level. Thus the p-value is slightly less than .05.

Had the alternative hypothesis been one-tailed in the opposite direction, then we would have had to consider the required value of F on the lower tail of the distribution, using the procedure given in Section 6.29. Since the choice of the sample variance to put in the numerator is arbitrary (we could just as easily have chosen $F = \hat{s}_2^2/\hat{s}_1^2$ with $N_2 - 1$ and $N_1 - 1$ degrees of freedom), it is convenient for one-tailed hypotheses to put the larger sample variance in the numerator, so that it is only necessary to consider the upper tail of the F distribution. In doing this, of course, it is imperative that you make sure that the larger $\hat{s}^2$ is the one which would be *expected to be larger in light of the alternative hypothesis*, H_1.

Once again, you should note the caution that the use of the F distribution in inferences about two variances presupposes that the two populations are normally distributed (refer to Section 6.30). In fact, all of the specific testing procedures discussed in Sections 7.22–7.27 require the assumption that the populations of interest are normally distributed. For small samples, the statistician should investigate the applicability of this assumption in any specific situation. If the assumption seems unreasonable, then he may not be justified in using the z, t, χ^2, and F tests. For large samples, the central limit theorem comes to the rescue; by now you should see what we meant when we said that the central limit theorem has important implications for statistical inference.

EXERCISES

1. Explain the difference between a statistical hypothesis and what is generally thought of as a scientific hypothesis.

2. In Section 7.5, Decision-rule 3 was declared to be inadmissible. Suppose that this rejection region or its equivalent were used in a test of $H_0\colon \mu = 0$ against $H_1\colon \mu \neq 0$. What property or properties of a "good" test would be violated, according to Section 7.19?

3. Make up an example of a decision problem in your area of interest which is somewhat similar to the example in Section 7.6. Formulate several decision rules and try to decide among them on the basis of
   (a) the minimax-loss criterion
   (b) the expected loss criterion
   (c) the subjective expected loss criterion.

4. Discuss the propositions "Personal probabilities of the statistician have no place in rigorous scientific investigation", and "All probabilities are interpretable as personal probabilities."

5. Distinguish between the two types of errors which are possible in hypothesis testing. In general, is it reasonable to assume that one type of error is always more serious than the other? If there are two types of error, why does the "classical" convention presented in Section 7.9 only involve one type of error? After the statistician has fixed α, does he have any control over β? Explain.

6. Some statisticians claim that one must *never*, under any circumstances, *accept* the "null" hypothesis as true. One can reject the null hypothesis with sufficient evidence, they agree, but apparently one is in some sort of limbo when a result fails to be significant. Comment on this stand. Does it ever seem reasonable to believe that an hypothesis such as H_0: $\mu = 0$ is *exactly* true? If not, how could H_0 be modified to make it seem reasonable that it *could* be exactly true?

7. Discuss the following proposition: "Much of the controversy surrounding the use of conventions such as that given in Section 7.9 can be explained in terms of whether hypothesis testing is viewed as a "signaling device," as discussed in Section 7.11, or as a formal decision-making procedure."

8. Suppose that a bookbag were filled with 100 poker chips. You know that either 70 of the chips are red and the remainder blue, or that 70 are blue and the remainder red. Ten chips are to be drawn successively with replacement. You must decide whether the composition of the bookbag is 70R — 30B or 70B — 30R. Find a decision rule that will make the probabilities of the possible errors in inference small and equal. [*Hint*: Use the binomial distribution with $N = 10$, first with $p = .70$ and then with $p = .30$.]

9. In Exercise 8, suppose that you stand to lose 5 cents if you decide that the bookbag contains 70R — 30B when it actually contains 70B — 30R, and you stand to lose 10 cents if the opposite error is made. Construct a table showing the two expected loss values, and consider the following decision-rule: decide 70R — 30B if 7 or more out of the sample of 10 turn out to be red, and decide 70B — 30R otherwise. Compare this decision-rule with the one you found in Exercise 8.

10. If everything else is held constant, why does making α smaller lead to a larger β, and vice versa? In what ways can the experimenter make *both* α and β smaller?

11. Suppose that you know that exactly one of the following two hypotheses must be true:

$$H_0 : \mu = 100, \quad \sigma = 15$$

$$H_1 : \mu = 105, \quad \sigma = 15.$$

The population sampled has a normal distribution, and the sample size is to be 25. If the probability of a Type I error is to be .10, what is the critical value of M? If M turns out to be 103, what is the p-value?

12. Suppose that X is the height, in inches, of a population consisting of male basketball players, and you are interested in testing

$$H_0 : \mu \leq 75$$

against $H_1 : \mu > 75.$

Assume that X is normally distributed with a variance of 9. A sample of size 100 is to be drawn from the population. If $\alpha = .05$, what is the critical value of M? What is the power of this test if the true value of μ is equal to 75.5? If the true value of μ is equal to 76? Draw the power curve for this test.

13. In Exercise 12, find the critical value of M and draw the power curve if $\alpha = .20$. Compare this power curve with the one obtained in Exercise 12.

14. In Exercise 12, find the critical value of M and draw the power curve if $\alpha = .05$ and you are testing

$$H_0 : \mu \geq 75$$

against $H_1 : \mu < 75.$

Also, find the region of rejection and draw the power function if $\alpha = .05$ and you are testing

$$H_0 : \mu = 75$$

against $H_1 : \mu \neq 75.$

15. Suppose that you want to test the hypothesis

$$H_0 : \mu = 3$$

against $H_1 : \mu \neq 3$

for a normally distributed population with variance 1.44. Given $\alpha = .10$ and $N = 36$, how powerful would this test be if $\mu = 4$? Given $\alpha = .10$ and $N = 16$, how powerful would this test be if $\mu = 4$? If the statistician wanted his test to have power of .80 if $\mu = 3.5$, what sample size should he use, given that $\alpha = .10$?

16. Suppose that you are interested in X, the gasoline mileage (in miles per gallon) of a particular brand of gasoline. You decide to test

$$H_0 : \mu = 20$$

against $H_1 : \mu \neq 20.$

Assuming that X is normally distributed with variance 16, that $N = 25$, and that $\alpha = .01$, find the region of rejection. What is the power of this test against a true mean of 22? Against what other possible value of μ does the test have exactly the same power?

17. Do Exercise 16, considering

$$H_0 : \mu \leq 20$$

against $H_1 : \mu > 20.$

18. For the test in Exercise 16,
    (a) draw the power curve
    (b) draw the operating characteristic curve
    (c) draw the error curve.
    Do the same for the test in Exercise 17. Explain the relationship among the three types of curves in each case.

19. Draw the "ideal" power curve for the test of

$$H_0 : \mu \leq \mu_0$$

against $H_1 : \mu > \mu_0.$

(That is, consider the best possible situation as far as power is concerned; for example, what would you like the power to be against any $\mu > \mu_0$?)

20. Suppose that X has an exponential distribution with parameter λ:

$$f(x) = \begin{cases} \lambda e^{-\lambda x} & \text{if } x > 0, \\ 0 & \text{elsewhere.} \end{cases}$$

You want to test the hypothesis $H_0 : \lambda = 1$ against the hypothesis $H_1 : \lambda = 2$, based on a sample of size 1. If $\alpha = .10$, find the region of rejection. What is the probability of making a Type II error? [*Hint*: Be careful; as the observed value of X gets larger and larger, what is implied about the value of the parameter λ?]

21. Suppose that X is normally distributed with mean μ and variance 400, and you want to test

$$H_0 : \mu = 50$$

against $H_1 : \mu = 60.$

If $\alpha = .10$ and $N = 4$, find the critical value of M and compute β. Now suppose that you wanted to test H_0 against $H_1 : \mu = 80$. For the same values of α and N, find the critical value of M and compute β. In the light of these results, comment on the "classical" convention and how the test is affected by changes in H_1.

22. An experimenter found that in a random sample of 300 women students in college, 35 percent smoked cigarettes. Estimate the 95 percent confidence interval for the true proportion of women college students who smoke cigarettes, based on this data. Can one reject the hypothesis $H_0 : p = .40$ in favor of the alternative hypothesis $H_1 : p \neq .40$ at the .05 significance level?

23. In a random sample of 400 consumers in a given city, 250 preferred Brand A cars to Brand B cars. Find an 80 percent confidence interval for the true proportion of consumers who prefer Brand A to Brand B. Use this to test the hypothesis

$$H_0 : p = .50$$

against $H_1 : p \neq .50$

with $\alpha = .20.$

24. Explain the relationship between confidence intervals and two-tailed tests of hypotheses. Could you develop "one-sided" confidence intervals to relate to one-tailed tests? Explain.

25. Discuss the relationship between predetermined significance levels (values of α) and the use of p-values in hypothesis testing. What are the advantages of the use of p-values?

26. Discuss the statement: "Ideally, the choice of α should be based on the relative seriousness of the losses associated with the two possible errors. The larger the loss associated with a Type I error is, or the smaller the loss associated with a Type II error is, the smaller α should be, all other things being equal."

27. In Exercise 16, suppose that $M = 21$. What is the p-value associated with this M? If you decided that $\alpha = .01$, would you reject H_0 in favor of H_1? If someone else decided that $\alpha = .35$, would he reject H_0 in favor of H_1 on the basis of the same data?

28. Suppose that you want to test

$$H_0: \mu = 300$$

against $H_1: \mu \neq 300,$

where the population in question is normal with variance 10,000. You plan to take a sample of size 4. Consider the following two rejection regions:

Region 1: $M \leq 200$ or $M \geq 400$.

Region 2: $297 \leq M \leq 303$.

For each of these two rejection regions, draw the power curve. Comment on the difference between the two curves.

29. In Exercise 12, find the p-value if
(a) $M = 75.3$ (b) $M = 76.0$ (c) $M = 74.5$ (d) $M = 75.0$.

30. In Exercise 15, find the p-value if
(a) $M = 2.9$ (b) $M = 3.1$ (c) $M = 3.5$ (d) $M = 2.7$.

31. A psychologist in a public school system is asked by the superintendent to provide evidence to the school board that the students will profit from more intensive training. There are 4000 students in the city school system and 794 in this particular high school. The psychologist decides to test the hypothesis that the mean IQ of the students is 110. He has very good reason to believe, from the standardization of the test used, that whatever the mean IQ value is, the variance of IQ scores for these students is 116. Furthermore, in consultation with the superintendent, the psychologist decides that the probability of Type I error should be .05. Since the only definite course of action to be taken is to recommend special training if there is evidence that the mean IQ is greater than 110, and no particular action will be taken if the mean IQ is less than or equal to 110, the psychlogist decides to do a one-tailed test.

Some 30 of the high school students are selected completely at random, given the test, and found to have a mean IQ of 121. He reports to the superintendent that the students have an IQ significantly greater than 110, and recommends that the board be requested to proceed with the plans for accelerated programs.

Answer the following questions briefly, with particular attention to the italicized words:
(a)   Is his conclusion at all warranted?
(b)   What *sampling distribution* is relevant to the statistical test, and what *assumptions* was it necessary to make?
(c)   What, precisely, is the *null* hypothesis? the alternative hypothesis?
(d)   What is the relevant *statistic*?
(e)   What is the *test statistic*?
(f)   What is the value of the relevant *parameter*?
(g)   What, really, is the *population* under consideration?
(h)   What is the *p*-value of his result? What is the probability of *Type I error* that is being tolerated when the psychologist rejects the null hypothesis?
(i)   What is the *probability of Type II error* in this test if the true IQ mean is 113?
(j)   What is the *99 percent confidence interval for the mean* IQ as calculated from these data?
(k)   Could the psychologist reject the hypothesis that the mean IQ is actually 108? If so, at what level?
(l)   What does it mean, in this situation, to say that the statistical test basically gives a *conditional probability of the form* $P\,(121 \leq M \mid \mu = 100, \sigma^2 = 116)$?
(m)   What is the *region of rejection*, and the *critical value* for this test?

32.   List the assumptions underlying the small sample test of a hypothesis about a single mean and the small sample test of a difference between two means. Evaluate the apparent importance of these assumptions.

33.   Suppose that $\alpha = .05$ and that K separate tests of hypotheses are made, quite independently of each other. Show that when all of the null hypotheses are true, the probability of one or more significant results *by chance alone* (that is, even if all of the hypotheses being tested are exactly true) is equal to

$$1 - (.95)^K.$$

Also, show that in general, for any value of α, the corresponding probability is

$$1 - \alpha^K.$$

34.   Why is it that a t test cannot be applied to test a hypothesis about a proportion?

35.   A new drug manufacturer is concerned over the possibility that a new medication might have the undesirable side-effect of elevating a person's body temperature. He wants to market the product only if he could be quite sure

that the mean temperature of healthy individuals taking the medication would be 98.6 degrees Fahrenheit or less, but withhold the drug otherwise. He administers the drug to a random sample of 60 healthy persons and finds $M = 98.4$ and $s^2 = .38$. Assuming that the distribution of temperatures is normal, what is the p-value? If $\alpha = .10$, what should the drug manufacturer do?

36. How would your answers to Exercise 35 change if $N = 20$ but everything else was the same?

37. In Exercise 35, if $\alpha = .10$, find the region of rejection and determine the power of the test if the true average temperature of healthy persons taking the drug is 99.0. Draw the power curve for the test.

38. Suppose that a statistician wants to test the hypothesis that the mean of a normally distributed population is .536 against the alternative that the mean is not equal to .536. He decides to use $\alpha = .01$, and he wants $\beta = .05$ when the true value of the mean is .540. Approximately how large should his sample size be if he assumes that $\sigma = .01$?

39. A random sample of ten independent observations from a normal distribution produced the following values:

$$51, 50, 49, 43, 56, 46, 45, 30, 55, 52.$$

If $\alpha = .20$, test the hypothesis that the mean is 50 against the alternative that it is not 50. Also, determine a 80 percent confidence interval for the mean. What is the p-value?

40. A statistician is contemplating the use of two independent groups of equal size in a study. He wants his test to have power .90 for detecting a difference of $\pm\sigma/20$ between the two means, and he chooses $\alpha = .01$. Given that the two groups chosen are from populations with the same value of σ, how large should each sample be?

41. Suppose that you are interested in comparing the gasoline mileage of two brands of gasoline. You hypothesize that there is no difference in the mileages, and you assume that the mileage for each brand is normally distributed with unknown variance σ^2. A sample is taken from each brand, with the following results:

Brand A	Brand B
16	13
18	15
15	11
23	17
17	12
14	13
19	
21	
16	

What is the p-value? If $\alpha = .10$, would you decide that the mileages were the same for the two brands or not?

42. If in Exercise 41 you are given the additional information that $\sigma = 3.5$, how would your answers change? Under what conditions might it be reasonable for you to know σ before taking the samples?

43. Suppose that the output of each of two machines is normally distributed. If the new machine has a higher mean output than the old machine, you will purchase the new machine. Thus, you wish to test $H_0:\mu_N \leq \mu_0$ against $H_1:\mu_N > \mu_0$. You are willing to assume that the variability of output is the same for both machines. You observe the following sample results:

   Old machine: 27, 44, 35, 37, 56, 19, 32, 45

   New machine: 50, 33, 43, 61, 25, 51, 38, 41, 35, 27.

What is the p-value? What other facts might you want to know before you make the final decision concerning the purchase of the new machine?

44. In Exercise 43, suppose that you decide that the mean output of the new machine would have to be at least 5 units higher than the mean output of the old machine in order to justify considering its purchase. What are the hypotheses of interest and what is the p-value?

45. A detergent manufacturer (the White Soap Co.) is concerned about claims made by a rival firm (Bright Detergent, Inc.) that Bright detergent results in cleaner clothes than White detergent. Twelve dirty sheets are randomly assigned to the two detergents and washed (in identical washing machines). After washing, the cleanliness of each shirt is measured on a "Bright-O-Meter," with the following results (a cleaner sheet will have a higher rating):

Bright detergent	White detergent
8	7
7	5
9	8
8	10
6	6
10	6

Assuming normality and equality of variances, state the relevant hypotheses and determine the p-value. On the basis of this, what can the White Soap Co. infer? Should they publicly claim that Bright's claim is false?

46. Suppose that you open 16 soft-drink bottles and find that the contents average 11.6 ounces, with a sample variance of 1.0. What can you say about the statement "contents 12 oz." which is on each bottle? State the relevant hypotheses and determine the p-value. What assumptions did you make in computing this p-value?

47. The same test is given to two classes of students in statistics. The first class (41 students) has a mean score of 75 with a sample variance of 100; the second class (21 students) has a mean score of 80 with a sample variance of 121. Did one class perform substantially better than the other? State the relevant hypotheses and the assumptions which you are making, and compute the p-value. Also, compute a 90 percent confidence interval for the difference in the means.

48. An educator has developed a new IQ test and would like to compare it to a certain widely used IQ test. He gives both tests to a randomly chosen group of 5 students, with the following results:

Child	Old test	New test
1	128	123
2	103	113
3	110	115
4	120	130
5	115	135

The new test is much shorter than the old test and it would be easier to use. Of course, this makes no difference unless the results of the two tests are consistent. State the relevant hypotheses and assumptions and determine the p-value.

49. The government is interested in changes in research and development expenditures in the past decade. They randomly select 8 large manufacturing companies and collect the following data on expenditures on research and development (in units of $100,000):

Company	1960	1970
A	12	14
B	8	9
C	14	17
D	9	8
E	17	20
F	20	28
G	7	11
H	11	14

Test the hypothesis that mean R&D expenditures have increased by less than $200,000 against the alternative that the mean has increased by more than that. What assumptions have you made, and what is the p-value? Determine a 90 percent confidence interval for the difference in means.

50. For many years a program of psychotherapy has been conducted at a state hospital. It has been found that among patients treated successfully in the past, the average time under treatment was 220 hours, with a standard deviation s of 40 hours. A new treatment has just been developed, and for a sample of size 25, the mean was approximately the same although the sample

standard deviation s was only 25 hours. Can we say that the standard deviation for the new treatment is significantly less than 40 hours if $\alpha = .05$? What is the p-value?

51. A manufacturer makes parts under a government contract which specifies a mean weight of 100 grams and a tolerance of 1 gram. That is, the government will reject parts unless the weight is between 99 and 101 grams. To meet these tolerances, the manufacturer wants the variance of weight to be no more than 0.5 gram. He samples five parts and observes weights 99.6, 101.2, 100.4, 98.8, and 100.0. He is willing to assume that the weights are normally distributed. Find the p-value for the test of

$$H_0:\sigma^2 \leq .5$$

against $\quad H_1:\sigma^2 > .5.$

52. In Exercise 51, suppose also that the manufacturer decides that it is too costly to reduce the variance below 0.5 gram, so that he wants to test

$$H_0:\sigma^2 = .5$$

against $\quad H_1:\sigma^2 \neq .5.$

Find the p-value. If $\alpha = .20$, would you accept or reject H_0? If $\alpha = .01$?

53. A randomly selected group of 30 boys of junior high-school age and an independent group of 30 senior high-school boys were each given the same test of mechanical ability. The results were as follows:

Junior High	Senior High
$N_1 = 30$	$N_2 = 30$
$M_1 = 168$	$M_2 = 173$
$s_1 = 39$	$s_2 = 33$

At the .01 level of significance, can we say that the two groups are significantly different in variability?

54. National norms for the test mentioned in Exercise 53 exist and show that the adult male population has a standard deviation of 28 on this test. Can we say, with $\alpha = .05$, that junior high-school boys are more variable in their test performance than the population of adult males? How about the senior high-school boys? The combined junior and senior high-school groups?

55. A random sample of 25 U.S. men born in 1935 showed a distribution of height with a standard deviation of 3.1 inches. On the other hand, a random sample of 43 U.S. men born in 1945 showed a standard deviation s of 2.5 inches. Is this evidence (at the .05 level) that these two generations of men have different degrees of variability in height?

56. The mean daily temperatures in a midwestern city was found for a sample of 30 days in the summers of 1930–1945. These temperatures showed $s = 4.8$ degrees Fahrenheit. Another random sample of 20 daily temperatures taken

in the same city but during the summers of 1946–1961 showed $s = 8.6$ degrees Fahrenheit. Can we say that the mean daily temperature in the summers of the second 16-year period was significantly more variable than in the first 16-year period? Use $\alpha = .01$.

57. In Exercise 41, are the variances significantly different at the .05 level?

58. In Exercise 43, are the variances significantly different at the .01 level?

59. Suppose that two independent random samples were to be compared in terms of their variability. If the values in one of these samples were multiplied by the constant k, what would be the effect on the F-ratio based on the variances of the two samples? State whether this suggests a way to test a hypothesis of the form

$$H_0:\sigma_1 = k\sigma_2.$$

How would you go about it?

60. Discuss the importance of the normality assumption in each of the tests discussed in this chapter which involve such an assumption. Is the assumption more crucial for some tests than for others? What other assumptions are important?

8

BAYESIAN
INFERENCE

8.1 INTRODUCTION

The inferential procedures presented in Chapters 6 and 7 are based on
sample information. In general, certain assumptions are made about the
population or process being sampled from, and these assumptions determine sampling distributions for statistics such as the sample mean, sample
variance, or sample proportion. On the basis of these sampling distributions, properties of estimators and tests of hypotheses can be investigated,
and "good" estimators and tests can be found. This approach is empirical,
since it uses only empirical evidence: the evidence contained in samples
from the population or process of interest. It is called the **sampling-theory**
or **classical** approach to statistical inference. The term "sampling theory"
describes the methods better than "classical," which is not very meaningful
in this context. The latter term, however, is used more frequently in the
statistical literature, so we will use it in this book. In its broadest sense,
classical statistics consists of inferential and decision-making procedures
which are (formally) based solely on sample evidence.

A second approach to statistical inference and decision, different from
the classical approach primarily in that information other than sample
information is formally utilized, has gained many new followers in recent
years. In this approach, sample information is combined with other available information, and the resulting combination of information is the
basis for inferential and decision-making procedures. The mechanism used
to combine information is Bayes' theorem, which was encountered in
Chapter 2. As a result, the term **"Bayesian"** is used to describe this general
approach to statistics.

The motivation for Bayesian methods is essentially the desire to base infer-ences and decisions on any and all available information, whether it be sample information or information of some other nature. This motivation is partic-ularly strong in decision theory. Erroneous decisions may be quite costly, and it may not be reasonable to ignore pertinent information just because it is not "objective" sample information. As this last statement implies, the classical approach is associated with the frequency interpretation of probability. Because it is impossible to talk about information in terms of long-run frequency probabilities unless the information arises from a ran-dom sample from a well-defined population or process, classical statisticians formally admit only sample information in inferential and decision-making procedures. Information other than sample information is rejected as "nonobjective."

In order to be able to express all information in probabilistic terms, most (but not all) Bayesians follow the subjective interpretation of probability, as presented in Sections 2.18–2.20. Often, much of the avail-able information consists of the judgments of one or more persons. If probabilities are interpreted as degrees of belief of individuals, this sort of information can be formally included in the analysis. In our discussion of Bayesian methods, we will assume that probabilities do in fact represent "degrees of belief." This assumption does not prevent the utilization of the sampling distributions which have been discussed in previous chapters. If a statistician subjectively accepts the assumptions (such as normality, known variance, and so forth) underlying a particular sampling distribu-tion, then that distribution can be given a subjective interpretation. Since the Bayesian *will* utilize the sampling distributions most commonly en-countered in classical statistics and, in addition, will use other information, the Bayesian approach can be thought of as an *extension* of the classical approach. In this and the following chapter, frequent comparisons of Bayesian and classical methods will be made so that the student can readily understand the similarities and differences between the methods. We will see that under certain conditions, Bayesian and classical methods produce similar or even identical results. Of course, there is a difference in interpretation even though the numbers may be the same. As you may have guessed, the "certain conditions" amount essentially to the following: there is no information available other than sample information, or any such information is of little importance relative to the sample information.

A few brief historical notes might be of interest. Bayes' theorem dates back to the eighteenth century, when it was discovered by the English clergyman Thomas Bayes. After its discovery, it was often misapplied and fell into disrepute. Following the formal development of the theory of subjective probability, due to Bruno de Finetti, Leonard J. Savage, and others, the Bayesian approach to statistics as it now stands has evolved since the late 1950s. Since the motivation behind the Bayesian approach is primarily decision-oriented, Bayesian methods have been of most interest in business and the social sciences, although there is a growing number of adherents in the behavioral and physical sciences. Two business-school statisticians who helped to develop and popularize the Bayesian theory in the late 1950s and early 1960s are Howard Raiffa and Robert Schlaifer; other statisticians who have contributed significantly to Bayesian statistics include Harold Jeffreys, I. J. Good, and Dennis V. Lindley.

A word on the organization of topics in Chapters 8 and 9 is necessary at this point. In this chapter we will be concerned primarily with the combination of sample information with other available information, and with inferential procedures based on the combined information. Insofar as possible, we will attempt to avoid decision-theoretic considerations in discussing these procedures and comparing them with the corresponding classical procedures. In Chapter 9, decision theory will be discussed, and we will re-examine Bavesian procedures from the standpoint of decision theory.

8.2 BAYES' THEOREM FOR DISCRETE RANDOM VARIABLES

In Section 2.25, Bayes' theorem was discussed in terms of events. It is also possible to interpret Bayes' theorem in terms of the conditional distribution of a random variable, discrete or continuous. In this section we present Bayes' theorem in terms of a discrete random variable.

Recall that in Chapter 2 we stated the general form of Bayes' theorem as follows: If A_1, A_2, $\cdots$, A_J represents a set of J mutually exclusive events, and if $B \subset A_1 \cup A_2 \cup \cdots \cup A_J$, so that $P(B) \leq \sum_{j=1}^{J} P(A_j)$, then

$$P(A_j \mid B) = \frac{P(B \mid A_j)P(A_j)}{\sum_{i=1}^{J} P(B \mid A_i)P(A_i)}.$$

In this form, Bayes' theorem provides a convenient way to determine conditional probabilities in some situations. If the conditional probabilities $P(B \mid A_j)$ and the probabilities $P(A_j)$ are known, it is a simple matter to apply the above formula to determine the conditional probabilities $P(A_j \mid B)$.

Suppose, however, that the events A_j, $j = 1$, $\cdots$, J, represent the J

possible values of a random variable X, and B represents a possible value of a second random variable, Y. Then the above equation can be rewritten in the form

$$P(X = A_j \mid Y = B) = \frac{P(Y = B \mid X = A_j)P(X = A_j)}{\sum\limits_{i=1}^{J} P(Y = B \mid X = A_i)P(X = A_i)}. \quad (8.2.1^*)$$

Now, the set of probabilities $P(X = A_j \mid Y = B)$, for $j = 1, \cdots, J$, corresponds to the conditional distribution of the random variable X, given that the random variable Y takes on the value B. Thus, Bayes' theorem can be used to determine the conditional distribution of a discrete random variable. It should be pointed out that it is not necessary for Y to be discrete. If Y is not discrete, we can think of B as an interval of values of Y rather than a single value. Then, of course, it is necessary to replace "$Y = B$" with "Y takes on a value in the interval B" in Equation (8.2.1). In most applications to be considered in this book, B will correspond to a single value, although it is important to note that it *could* represent an entire interval of values.

How, then, can Bayes' theorem be used to revise probabilities (or entire probability distributions) in the light of sample information? At this point we will introduce some new terminology. Suppose that the random variable X is of interest to us, and that we would like to make some inferences about X or to make some decisions related to X. Suppose that our information concerning X is summarized in the set of probabilities $P(A_j)$, which represent the **prior distribution of** X. We then observe a sample, and the outcome of the sample can be summarized by B, the observed value of the sample statistic Y. The probability $P(B \mid A_j)$ is determined from the *sampling distribution* of Y, *evaluated at* $Y = B$ (this is the **likelihood function**). Using Bayes' theorem, the prior information (represented by the prior distribution) and the sample information (represented by the sampling distribution) can be combined to form the **posterior distribution of** X, the set of probabilities $P(X = A_j \mid Y = B)$. The posterior distribution, then, summarizes our information concerning X, given the sample outcome B. The adjectives prior and posterior are relative terms relating to the observed sample. If after observing a particular sample and computing the posterior distribution of X, we decided to take another sample, the distribution just computed would now be considered to be the *prior distribution relative to the new sample.*

As we have seen in Chapters 6 and 7, inferences in statistics generally concern one or more population parameters. Since Greek letters are usually used to represent such population parameters, let us replace X with θ and A_j with θ_j in Equation (8.2.1):

$$P(\theta = \theta_j \mid Y = B) = \frac{P(Y = B \mid \theta = \theta_j)P(\theta = \theta_j)}{\sum\limits_{i=1}^{J} P(Y = B \mid \theta = \theta_i)P(\theta = \theta_i)}$$

$$= \frac{P(B \mid \theta_j)P(\theta_j)}{\sum\limits_{i=1}^{J} P(B \mid \theta_i)P(\theta_i)}. \qquad (8.2.2^*)$$

This equation enables us to revise probabilities concerning a parameter θ on the basis of new sample information, which may be summarized by the sample statistic Y.

8.3 INTERPRETATION OF THE PRIOR AND POSTERIOR DISTRIBUTIONS

In Chapters 6 and 7, inferences about a population value θ were based solely on sample information. The inclusion of the prior distribution enables us to base inferences and decisions on *all* of the information which is available concerning the population value, whether or not the information is in the form of sample information. Often the statistician has some information about a parameter θ prior to taking a sample. This information may be of a subjective nature, in which case the prior distribution will consist of a set of subjective probabilities. In the classical approach to statistics, all probabilities should be based on the long-run frequency interpretation of probability, and hence subjective probabilities are not admissible. Most statisticians advocating the Bayesian viewpoint contend that all available information about a parameter, whether it be of an "objective" or "subjective" nature, should be utilized in making inferences or decisions. As we indicated in Section 8.1, Bayesian methods have been used primarily in business and economics because of their implications for decision theory. In these areas, decisions often must be made in situations where there is little or no sample information but a great deal of information of a more subjective nature. We shall show in later sections that the Bayesian techniques can be thought of as an extension of classical statistics, and as such are applicable to any problem that can be treated by classical statistical procedures.

Since followers of the frequency interpretation of probability do not admit subjective probabilities, they do not admit probability statements about a population parameter θ. Frequentists contend that θ has a certain value, which may be unknown, and that it is senseless to talk of the probability that θ equals some number; either it does or it does not. This was pointed out in Section 6.11 when we discussed the interpretation of confidence interval statements. The subjectivist, on the other hand, *does* think of θ as a random variable and thus allows probability statements concerning θ. He would interpret a confidence interval for θ as a probability statement concerning θ, for example. Because of this, he admits the probability statements which we have called the prior distribution and the posterior distribution; these are both distributions of a random variable θ, which may be a population parameter. It should be noted at this point that the frequency and subjective interpretations are not the only interpretations of probabilities, although they are the most commonly encountered interpretations. Some followers of the frequency school of thought do interpret some probability statements (notably, interval estimation statements) as probability statements about a population parameter, calling such probabilities "fiducial" probabilities. Another interpretation is called the "necessary" view of probability; this approach is based on logic. We will not concern ourselves with interpretations other than the frequency and subjective interpretations. Furthermore, since we saw in Chapter 2 that the subjective approach utilizes all information, including observed frequencies, we shall assume that the probability statements in this and the next chapter are subjective in nature. It should be remembered, however, that the terms "Bayesian" and "subjective" are not synonymous; it is possible to develop the Bayesian approach without using subjective probability.

In classical statistics, all inferences are based on the sampling distribution. In particular, the sampling distribution is used in many instances to determine what we have called the *likelihood function*. In estimation, the likelihood function is used to find maximum likelihood estimators (Section 6.7); in hypothesis testing, the likelihood function is used to develop likelihood ratio tests. In Chapter 7 it was mentioned that all of the specific tests presented were based on the likelihood ratio approach. The input to Bayes' theorem which is of the form $P(Y = B \mid \theta = \theta_j)$ is a likelihood function, and the numerical probability obtained from this form can be thought of as the likelihood of the sample result B, given a particular value of the parameter θ. The inputs to Bayes' theorem for discrete random variables are thus a prior distribution and a likelihood function; the result of the application of Bayes' theorem is a posterior distribution. The likelihood function is determined just as in the application of the principle of maximum likelihood; it is the product of the distributions of the individual sample values. In most situations the likelihood function can be related to the sampling distribution of a particular sample statistic. So far we have not discussed the determination of the prior distribution, the other input to a formal Bayesian analysis. Before doing so, we will present an example of the application of the Bayesian approach.

8.4   AN EXAMPLE OF BAYESIAN INFERENCE AND DECISION

The example to be considered in this section comes from the area of quality control. Suppose that a statistician is concerned about the items produced by a certain manufacturing process. More specifically, he is concerned about the proportion of these items which are defective. From past experience with the process, he feels that the proportion defective, p, takes on four possible values: .01, .05, .10, and .25. These are the four possible values of the population parameter p, or, in decision-theoretic terminology, the four possible states of nature. This is obviously a simplification of the real situation. It is highly unlikely that p would take on only the four given values and no intermediate values. It might be more realistic to think of p as a continuous random variable rather than a discrete random variable, but for the purposes of this example p will be discrete.

Although the statistician has never formally determined the proportion of defectives in any "batch" from the manufacturing process, he has observed the process and he has some information concerning p. Suppose that he feels that this information can be summarized in terms of the following degree-of-belief, or subjective, probabilities for the four possible values of p:

$$P(p = .01) = .60,$$

$$P(p = .05) = .30,$$

$$P(p = .10) = .08,$$

and $\qquad\qquad P(p = .25) = .02.$

These four probabilities constitute the statistician's prior distribution of p.

In addition to this prior information, the statistician decides to obtain some sample information from the process. In doing so, he assumes that the process can be thought of as a Bernoulli process, with the assumptions of stationarity and independence appearing reasonable. That is, the probability that any one item is defective remains constant for all items produced and is independent of the past history of defectives from the process.

A sample of 5 items is taken from the production process, and one of the five is found to be defective. How can this information be combined with the prior information? Under the assumptions that the statistician has made, the sampling distribution of the number of defectives in five trials, given any particular value of p, is a binomial distribution. The likelihoods are thus

$$P(r = 1 \mid n = 5, p = .01) = \binom{5}{1} (.01)(.99)^4 = .0480,$$

$$P(r = 1 \mid n = 5, p = .05) = \binom{5}{1} (.05)(.95)^4 = .2036,$$

$$P(r = 1 \mid n = 5, p = .10) = \binom{5}{1} (.10)(.90)^4 = .3280,$$

and $\quad P(r = 1 \mid n = 5, p = .25) = \binom{5}{1} (.25)(.75)^4 = .3955.$

In this example, Bayes' theorem can be written in the form

$$P(p_i \mid B) = \frac{P(B \mid p_i)P(p_i)}{\sum\limits_i P(B \mid p_i)P(p_i)} ,$$

where B represents the sample result: one defective in five trials. Notice that the numerator is the product of a likelihood and a prior probability, and that the denominator is just the sum of all of the numerators. To determine the posterior distribution, it is convenient to set up the following table:

(1)	(2)	(3)	(4)	(5)
p	Prior probability	Likelihood	(Prior probability) × (likelihood)	Posterior probability
.01	.60	.0480	.02880	.232
.05	.30	.2036	.06108	.492
.10	.08	.3280	.02624	.212
.25	.02	.3955	.00791	.064
	1.00		.12403	1.00

The first column in the table lists the possible values of p (that is, the values of p having a nonzero prior probability). The second and third columns give the prior probabilities and the likelihoods, and the fourth column is the product of these two columns. Finally, the fifth column is obtained from the fourth column by dividing each individual entry by the sum of the four entries. This determines the posterior probabilities, since Bayes' theorem can be written as

Posterior probability = (prior probability)

$$\times \text{ (likelihood)}/\Sigma\text{(prior probability) (likelihood)}.$$

This example illustrates some features of Bayes' theorem. For one thing, the prior probabilities must sum to one (since they form the prior distribution, which must be a proper probability distribution), whereas there is no such restriction on the sum of the likelihoods. Furthermore, the posterior probabilities must also sum to one (since they form the posterior distribution), and this is why we must divide by Σ(prior probability)(likelihood). The sum of the entries in column 4 of the above table is not equal to one, but when we divide each entry by this sum, the resulting values (column 5) do add to one. This process is sometimes referred to as "normalizing" the probabilities—making them sum to one.

Another feature of Bayes' theorem which is illustrated here is that if the prior probability of any value of p is zero, the posterior probability of that value will also be zero, since the posterior probability is a multiple of the prior probability (the multiplier being the likelihood divided by the normalizing sum). Since the prior probability of $p = .20$ is zero, for example, the posterior probability of $p = .20$ is zero. The moral is obvious: when determining prior probabilities in the discrete case, make sure that any value that you consider to be even remotely possible has a nonzero prior probability. We noted above that our example was highly simplified and unrealistic because values such as .02, .09, and .20 were assigned prior probabilities of zero (and thus essentially designated as "impossible" events).

Returning to our statistician, suppose that he decides to take yet another sample in order to obtain more information about the production process. In particular, he takes another sample of size five and obtains two defectives. His prior probabilities are now the posterior probabilities calculated after the first sample, and the likelihoods are once again determined via the binomial distribution, with the following results:

p	Prior probability	Likelihood	(Prior probability) $\times$ (likelihood)	Posterior probability
.01	.232	.0010	.00023	.005
.05	.492	.0214	.01053	.244
.10	.212	.0729	.01545	.359
.25	.064	.2637	.01688	.392
	1.00		.04309	1.00

It is interesting to observe the changes in the probabilities as new sample information is obtained. The original prior distribution indicated that the statistician thought that the low values of p (.01 and .05) were more likely than the higher values. The first sample of size 5, with 1 (that is, 20 percent) defective, shifted the probabilities so that .05 appeared to be the most likely value of p (.01 was the most likely value according to the statistician's original prior distribution). The second sample of size 5, with 2 (that is, 40 percent) defective, shifts the probabilities so that the highest possible value, .25, becomes the most likely, while the lowest value now has a probability of only .005, compared with the original prior probability of .60 assessed by the statistician. In other words, the sample results obtained (a total of 3 defectives in 10 items) changed the probabilities considerably, making the high value of p, .25, much more likely than it had seemed under the original prior distribution. You should be careful not to interpret this as meaning that the sample will always change the probabilities so drastically. If the statistician had observed no defectives or one defective instead of three defectives in the sample of ten items, the posterior probabilities would be much closer to the prior probabilities. You can verify this for yourself by performing the calculations.

In our example, the statistician took two samples and revised his probabilities after each sample. A useful feature of Bayes' theorem is that he could have waited and revised his probabilities just once, using the combined results of the two samples (that is, 3 defectives in 10 trials). When applying Bayes' theorem for independent trials, the same result is obtained whether the calculations are made after each trial or just once at the end of all of the trials. Starting with the original prior probabilities and using all ten trials to determine the likelihoods, we get:

p	Prior probability	Likelihood	(Prior probability) $\times$ (likelihood)	Posterior probability
.01	.60	.0001	.00006	.005
.05	.30	.0105	.00315	.245
.10	.08	.0574	.00459	.359
.25	.02	.2503	.00501	.391
	1.00		.01281	1.00

These are virtually identical to the results obtained from revising twice, once after each sample. The only differences (at the third decimal place) are due to rounding errors in the calculations. This feature of Bayes' theorem is very important, since it reduces the number of applications of the theorem required to revise probabilities on the basis of several samples. Of course, the statistician may want to revise after each sample, primarily because the resulting probabilities may help him to decide whether or not to take another sample, as we shall see in the next chapter.

Now that the posterior probabilities have been calculated, let us consider the following question: why might the statistician have been so interested in p in the first place? There are a number of explanations, but they all relate to the fact that a high value of p could be quite costly to the firm. It could be quite costly to repair defective items, perhaps so costly that they are scrapped entirely. In addition, if any defective items are shipped to the firm's customers, the customers might be displeased and might take their business elsewhere.

Perhaps the quality control statistician (or his boss, the production manager) simply wants an estimate of p. He can determine such an estimate from the posterior distribution by taking some summary measure, such as the mean, median, or mode, of the posterior distribution. The comments in Chapter 3 and Chapter 6 concerning the relative merit of various possible measures (or estimators) apply here. In contrast to the classical statistician (who would base an estimate solely upon the sample information, as reflected by the likelihood function), our Bayesian statistician bases his estimate on all available information, as reflected by the posterior distribution.

It may be that the statistician must make a decision concerning the process rather than just estimating p. Suppose that he has three possible actions: (1) He can leave the process as is, (2) he can have a mechanic make a minor adjustment in the process which will guarantee that p will be no higher than .05 (for example, if p is .10 or .25, the adjustment reduces it to .05; if p is .05 or .01, the adjustment has no effect), or (3) he can have the mechanic make a major adjustment in the process which will guarantee that p will be .01.

The production manager has just received an order for 1000 items, and the statistician must choose one of the above three actions before the production run is started. There are certain costs that are relevant to his decision. First, the profit to the firm is $.50 per item sold, and the cost of replacing defectives is a fixed amount, $2.00 per item. The other relevant costs are the costs of making adjustments in the process; a minor adjustment costs $25, and a major adjustment costs $100. From these figures, it is clear that the profit to be realized from the run of 1000 items is $1000(\$.50) = \$500$. From this amount we must subtract the cost of replacing defectives and the cost of any adjustments. If p is .01, the expected number of defectives is $.01(1000) = 10$, and the cost of replacing these is $2 per item, or a total of $20. Similarly, for the other three possible values of p, the expected costs of replacing defectives are $100, $200, and

$500, respectively. From this, the statistician can determine a "payoff table," which shows the net profit to the firm for each combination of an action and a possible state of nature. Since there are three actions and four states of nature, the payoff table has a total of 12 entries. For example, if $p = .10$ and a minor adjustment is made, the adjustment changes p to .05, so the cost of replacing defectives is $100. The cost of the minor adjustment is $25, so the payoff is $500 − $100 − $25 = $375.

State of Nature = Value of p

	.01	.05	.10	.25
Major Adjustment	$380	$380	$380	$380
Minor Adjustment	$455	$375	$375	$375
No Adjustment	$480	$400	$300	$ 0

In Chapter 9 we will discuss criteria for making decisions. Suffice it to say at this point that a reasonable criterion is to choose the act for which the expected payoff (expected net profit to the firm in this example) is largest. The expectations should be taken with respect to the statistician's posterior distribution of p, since this represents all of the information available to him concerning p, the state of nature. The posterior probabilities of the four states of nature are .005, .245, .359, and .391. The expected payoffs (denoted by EP) are then:

EP (Major Adjustment) = $380,

EP (Minor Adjustment) = .005($455) + .995($375) = $375.50,

and   EP (No Adjustment) = .005($480) + .245($400)

$$+ .359(\$300) = \$208.10.$$

This indicates that the statistician should have the mechanic make a major adjustment. This is reasonable, since the probability of a high proportion of defectives (.10 or .25) is quite high. It is interesting to note that if the statistician had not obtained sample information, he would have based his decision on the prior distribution, in which case no adjustment would have been made.

This quality control example illustrates the use of Bayes' theorem to revise probabilities on the basis of sample information and the use of the resulting posterior probabilities as inputs in a decision-making situation. We emphasize again that the example was purposely simplified, and that we will discuss a more realistic approach to the problem later in the chapter.

8.5 THE ASSESSMENT OF PRIOR PROBABILITIES

We return now to the problem of the determination of the prior distribution, one of the inputs to a formal Bayesian analysis. In the discrete case, the prior distribution consists of a set of probabilities. The only restrictions are that the probabilities must be nonnegative and must sum to one. Under these restrictions, how can the statistician determine a prior distribution?

The prior probabilities should reflect the prior information about the parameter in question. If this information is primarily in the form of sample results, then the prior probabilities should be close to the observed relative frequencies. If there is little or no sample information, then the probabilities should be based on whatever other relevant information is available. The ultimate choice of a prior distribution by the statistician is a subjective choice, no matter what form the prior information is in, so we will assume that the prior probabilities are subjective probabilities. As noted in Chapter 2, a subjective probability reflects the degree of belief of a given person (the statistician) about a given proposition (say, for example, the proposition that $p = .10$). But if probabilities are based on the judgments of a person, what is needed is a way for the person to quantify his judgments, that is, to express them in probabilistic terms.

There are numerous techniques available for the quantification of judgment, some of which are described in Sections 2.18–2.20. A person understanding the concept of probability might simply assign probabilities directly to the various possible values of the random variable of interest. In the example in the previous section, the statistician might have said, "I think the probability is .60 that the proportion of defectives produced by this production process is .01." Alternatively, the statistician might have chosen to think in terms of betting odds rather than probabilities: "I think the odds are 3-to-2 in favor of the proportion defective being .01." In Section 2.19 we discussed the relationship between probabilities and betting odds, stating that if the odds in favor of an event are a-to-b, then the probability of that event is equal to $a/(a + b)$. In the example, $a = 3$ and $b = 2$, so the required probability is $3/(3 + 2)$, or .60. Other techniques of probability assessment involve hypothetical lotteries and betting situations, as suggested in Section 2.19; space is too limited here to provide detailed descriptions of these and other techniques.

It should be pointed out that whatever the techniques used to assess the prior probabilities, the probabilities must be nonnegative and sum to one. If these requirements are not satisfied, we say that the probabilities are *inconsistent*. An assessor who obeys certain reasonable axioms of coherent choice (such as: if he prefers A to B, and B to C, then he should prefer A to C—this is the axiom of transitivity of preferences) will always be consistent in a decision-making sense. Such an assessor (sometimes referred to as the "rational" man) will have no inconsistencies in his probability assessments. Some experiments have indicated that people do not always assess probabilities in a "rational" manner. For example, they frequently assess probabilities which do not sum to one. This may be due to carelessness or to a misunderstanding of the assessment procedure. At any rate, it should not present a major problem, since inconsistencies can be removed if the assessor is made aware of their existence. If someone pointed out an arithmetic error to you, you would no doubt correct the error. Similarly, if someone pointed out that your prior probabilities summed to 1.20, say, you could change the probabilities so that they would sum to one.

In summary, then, anyone who understands the basic concept of probability should be able to assess a prior distribution for a discrete random variable. When we are dealing with continuous random variables, the assessment of a prior distribution may be considerably more difficult, as we will see later in this chapter. The main thing that you should keep in mind is that the prior distribution should reflect all of the relevant information available to the statistician prior to observing a particular sample.

8.6 BAYES' THEOREM FOR CONTINUOUS RANDOM VARIABLES

In many problems involving statistical inference and decision, it is more realistic and more convenient to assume that the random variable of interest is continuous. As we have pointed out, measurement procedures are such that actual observed random variables are never continuous. In many situations, however, the measurement may be precise enough so that for all practical purposes we can assume that a random variable is continuous. Indeed, in many cases it is quite realistic to assume that a random variable is continuous. For example, in the quality control illustration used in Section 8.4, it would be much more realistic to assume that p, the proportion defective, is a continuous random variable than to assume that is just takes on the four values .01, .05, .10, and .25. In Section 8.10, the quality control illustration will be modified so that p will be taken to be continuous.

Suppose that the parameter of interest in some inferential or decision-making procedure is continuous, and call this parameter θ. Furthermore, suppose that sample information involving θ can be summarized by the sample statistic Y. Since θ is continuous, the prior and posterior distributions can be represented by density functions. The posterior distribution, then, is simply the conditional density of θ given the observed value, y, of the sample statistic Y. But, from Equation (3.9.10), this can be written as

$$f(\theta \mid Y = y) = \frac{f(\theta, y)}{f(y)} . \tag{8.6.1}$$

The conditional density of one random variable given a value of a second random variable is simply the joint density of the two random variables divided by the marginal density of the second random variable.

But the joint density $f(\theta, y)$ and the marginal density $f(y)$ are not known; we just know the prior distribution and the likelihood function. Fortunately, the required joint and marginal densities can be written in terms of the prior distribution and the likelihood function (see Section 3.9):

$$f(\theta, y) = f(\theta)f(y \mid \theta), \tag{8.6.2}$$

and

$$f(y) = \int_{-\infty}^{\infty} f(\theta, y) \, d\theta = \int_{-\infty}^{\infty} f(\theta)f(y \mid \theta) \, d\theta. \tag{8.6.3}$$

The posterior density can then be written as follows:

$$f(\theta \mid y) = \frac{f(\theta)f(y \mid \theta)}{\int f(\theta)f(y \mid \theta) \, d\theta} . \tag{8.6.4*}$$

This is Bayes' theorem for continuous random variables. The densities $f(\theta \mid y), f(\theta)$, and $f(y \mid \theta)$ represent the posterior distribution, the prior distribution, and the likelihood function, respectively. These terms have the same interpretation for continuous random variables as they have for discrete random variables. The prior and posterior distributions must be proper density functions. That is, they must be nonnegative and the total area under the curve, determined by integrating the density function over its domain, must be equal to one. The integral in the denominator of Equation (8.6.4) serves the same purpose as the sum in the denominator of Equation (8.2.2): it makes the posterior distribution a proper probability distribution. It can be seen that Bayes' theorem for continuous random variables is analogous to Bayes' theorem for discrete random variables. **In the discrete case, Bayes' theorem can be expressed in words as**

$$\text{Posterior probability} = \frac{(\text{prior probability}) \, (\text{likelihood})}{\Sigma(\text{prior probability}) \, (\text{likelihood})} ,$$

and the corresponding statement for continuous random variables
is

$$\text{Posterior density} = \frac{(\text{prior density})\,(\text{likelihood})}{\int (\text{prior density})\,(\text{likelihood})}.$$

8.7 CONJUGATE PRIOR DISTRIBUTIONS

Conceptually, Equation (8.6.4) provides a convenient way to revise density functions in the light of sample information. In practice, however, it may prove quite difficult to apply this formula. If $f(\theta)$ and $f(y \mid \theta)$ are not fairly simple mathematical functions, it may be a hard task to carry out the integration in the denominator of Equation (8.6.4). In some situations it may be necessary to resort to advanced mathematical techniques to determine the posterior distribution. Because of this, Bayesian statisticians have developed the concept of **conjugate prior distributions**, which essentially are families of distributions that ease the computational burden when they are used as prior distributions. Of course, the posterior distribution depends on the likelihood function as well as on the prior distribution. Usually, once certain assumptions are made about the population or process that is being sampled from (for example, the assumption that the population is normal, or the assumption that the process is a stationary and independent Bernoulli process), the likelihood function is uniquely determined. In other words, once the data-generating model is specified, the likelihood function is known. For any particular "type" of likelihood function, the Bayesian statistician attempts to determine a conjugate family of prior distributions. This is a set of distributions, each of which can be combined with the given likelihood function without too much difficulty.

There are three properties which have been put forth as desirable properties for conjugate families of distributions:

1. Mathematical tractability
2. Richness
3. Ease of interpretation.

The first property, mathematical tractability, is the property which motivated the development of conjugate prior distributions. A prior distribution is mathematically tractable if (1) it is reasonably easy to specify the posterior distribution given the prior distribution and the likelihood, (2) it results in a posterior distribution which is also a member of the same conjugate family, so that successive applications of Bayes' theorem are not difficult, and (3) it is feasible to calculate expectations (such as those required in decision theory) from the prior distribution. The property of mathematical tractability, then, involves the computational burdens involved in Bayesian inference and decision.

The prior distribution should, of course, reflect the statistician's prior information. Since different people may (and no doubt *will*) have different prior information, a conjugate family of distributions should include, insofar as possible, distributions with different location, dispersion, shape, and the like, so as to be able to represent a wide variety of states of prior information. This property is called richness. If a family of distributions is not rich in this sense, there may be no member of the family which accurately reflects a particular person's prior information. If this is the case, then the statistician would not be able to take advantage of the mathematical tractability of the family.

Finally, it should be possible to specify the conjugate family in such a way that it could be readily interpreted by the person whose prior information is of interest. Prior information often consists at least partially of previous sample results. The easiest way to interpret a prior distribution might thus be in terms of previous sample results (whether they be actual results or hypothetical results). In general, conjugate prior distributions can be interpreted in this manner.

It is important to remember that a conjugate prior distribution is "conjugate" only with respect to a given likelihood function. Conjugate families of distributions corresponding to numerous likelihood functions have been developed. We will limit our discussion to two cases. The first case involves sampling from a stationary and independent Bernoulli process, in which case the conjugate family is the family of beta distributions. The second case involves sampling from a normal population with known variance, in which case the conjugate family is the family of normal distributions. These two cases will be discussed in the following sections. The mathematical derivations of these conjugate families of distributions from the given likelihood functions are beyond the level of this book, but an attempt will be made to demonstrate (informally if not formally) how they satisfy the three properties listed in this section.

8.8 BAYESIAN INFERENCE FOR A BERNOULLI PROCESS

In previous chapters we have discussed stationary and independent Bernoulli processes, including estimation and hypothesis testing procedures involving the Bernoulli parameter p, the probability of success on a Bernoulli trial. These procedures utilized only sample information from the Bernoulli process. In Section 8.4 we presented an example of Bayesian techniques involving a Bernoulli process under the assumption that the prior distribution was discrete. It is unrealistic, however, to limit p to a finite number of values. Theoretically, p can assume any real value from zero to one, so that the prior distribution should be continuous rather than discrete. Using Equation (8.6.4), the statistician can determine a continuous prior distribution and then find the posterior distribution following an observed sample. This could be a difficult task unless the prior distribution is a member of the family of distributions which is conjugate relative to the Bernoulli process. This conjugate family is the family of beta distributions.

A word on notation: we will denote prior distributions and parameters of prior distributions (such as the mean of the prior distribution) with single primes. Posterior distributions and parameters of posterior distributions shall be denoted with double primes. For example, $f'(\theta)$ and $f''(\theta \mid y)$ represent the prior and posterior distributions of θ; μ_θ' and μ_θ'' represent the means of the prior and posterior distributions, respectively. Of course, you should remember that the terms "prior" and "posterior" are relative terms, relating to a particular sample result.

Suppose that a statistician is sampling from a stationary and independent Bernoulli process. Then if his prior distribution of p is a beta distribution, his posterior distribution will also be a beta distribution. Recalling the formula for the density function of a beta distribution [Equation (4.26.3)], suppose the prior density is of the form

$$f'(p) = \frac{(N' - 1)!}{(r' - 1)!(N' - r' - 1)!} \, p^{r'-1}(1 - p)^{N'-r'-1} \qquad \text{for } 0 \le p \le 1.$$

$$(8.8.1^*)$$

Furthermore, suppose that the sample results in r "successes" in N trials. Then the posterior density is also a member of the beta family:

$$f''(p \mid y) = \frac{(N'' - 1)!}{(r'' - 1)!(N'' - r'' - 1)!} \, p^{r''-1}(1 - p)^{N''-r''-1}$$

$$\text{for } 0 \le p \le 1, \quad (8.8.2^*)$$

where
$$N'' = N' + N, \qquad (8.8.3^*)$$

$$r'' = r' + r, \qquad (8.8.4^*)$$

and y represents the sample results. To determine the posterior parameters of the beta distribution, N'' and r'', the statistician just adds the prior parameters, N' and r', to the sample statistics, N and r.

For a simple example, suppose that the statistician's prior distribution is a beta distribution with $r' = 5$ and $N' = 10$, and that he observes $r = 15$

successes in $N = 20$ Bernoulli trials. His posterior distribution is then a beta distribution with $r'' = 20$ and $N'' = 30$, as illustrated in Figure 8.8.1. How did the sample information change his distribution? First of all, recalling that the mean of the beta distribution with parameters r_0 and N_0 is just r_0/N_0, we see that the mean of the statistician's prior distribution is $5/10 = 1/2$, and the mean of his posterior distribution is $20/30 = 2/3$. The sample results shifted the mean upward, making higher values of p appear more likely. Note that the sample mean is $r/N = 15/20 = 3/4$. **It will always be true that the posterior mean lies between the prior mean and the sample mean.**

The variance of the beta distribution with parameters r_0 and N_0 is given by the formula

$$\text{var } (p) = \frac{r_0(N_0 - r_0)}{N_0^2(N_0 + 1)}.$$

Applying this for our example, the prior variance is .0227 and the posterior variance is .0072. In most, but not all, cases, the posterior variance will be smaller than the prior variance. This is intuitively reasonable, since the sample provides new information, and increased information should reduce the variance, just as the variance of a sample mean is smaller as the sample size becomes larger. In some situations, notably when the sample information greatly shifts the mean of the distribution toward one-half, the variance of the posterior beta distribution may be larger than the variance of the prior beta distribution.

For sampling from a Bernoulli process, the beta family of distributions clearly satisfies the requirement that a conjugate family of prior distri-

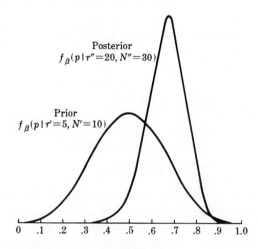

Figure 8.8.1

butions be mathematically tractable. Given the prior distribution and its parameters r' and N', it is very easy to determine the posterior distribution after a sample is observed, using Equations (8.8.2)–(8.8.4). Also, if the prior distribution is a member of the beta family, so is the posterior distribution. Finally, since formulas for the mean and the variance (and other moments as well) of a beta distribution are known, it should not prove too difficult to calculate expectations of value for decision theory. From Figures 4.26.1 and 4.26.2, it appears that the beta family satisfies the requirement of richness. The beta family includes symmetric, skewed, unimodal, U-shaped, J-shaped, and other distributions. By varying the parameters r' and N', the statistician has available a wide variety of shapes of distributions within the beta family.

What about the final requirement of a conjugate family, the requirement that the prior distribution can be interpreted in some manner? As might be suspected from the similarity between r' and N' and r and N, the beta distribution can be interpreted in terms of "equivalent sample information." A beta distribution with parameters r' and N' can be interpreted as follows: the prior information is roughly equivalent to the information contained in a sample of size N' with r' successes from the Bernoulli process of interest. We noted in Section 4.26 that the beta distribution could be thought of as representing a person's knowledge about p, given that his knowledge consisted solely of having observed a sample from the process. We should note that this interpretation should not be taken to imply that the parameters r' and N' must be integers. Since the interpretation is in terms of a "hypothetical" sample, not necessarily an "actual" sample, r' and N' need not be integers. They can take on any real values subject to the restriction that $N' > r' > 0$.

8.9 STOPPING RULES AND THE LIKELIHOOD FUNCTION

Thus far nothing has been said about the observed sample from the process which is used to revise the statistician's prior distribution other than that the sample results are r successes in N trials. In Chapter 4 we discussed two ways of arriving at such sample results. First, the sample size N could be fixed, in which case the distribution of r given p and N is a binomial distribution [Equation (4.3.1)]:

$$P(r \mid N, p) = \binom{N}{r} p^r (1 - p)^{N-r}.$$

A second possibility is that the number of successes r could be fixed (sample until r successes are obtained), in which case the distribution of N given p and r is a Pascal distribution [Equation (4.11.1)]:

$$P(N \mid r, p) = \binom{N - 1}{r - 1} p^r (1 - p)^{N-r}.$$

Notice that these two sampling distributions differ only in the first terms, the combinatorial terms. It turns out that in an application of Bayes' theorem, these terms are irrelevant since they do not involve p, the variable of interest. In Equation (8.6.4), the combinatorial term can be moved outside of the integral sign in the denominator, since it does not involve p. But then the term appears in both the numerator and denominator, so that it can be canceled out. Because of this, the results of Section 8.8 hold whether the sample is taken with N fixed (binomial sampling) or with r fixed (Pascal sampling). The only important part of the sampling distribution for Bayesian purposes is the part involving p. Because of this, the likelihood function for the Bernoulli process is generally taken to be equal to $p^r(1 - p)^{N-r}$, so that it is not even necessary to know whether the sampling is done with N fixed or with r fixed. The procedure used to tell us when to stop sampling is called a **stopping rule,** and if the stopping rule has no effect on the posterior distribution, then it is said to be **noninformative.** The two stopping rules we have discussed (sample until you have N trials, and sample until you have r successes) are both noninformative.

To formally demonstrate this, consider an application of Equation (8.6.4) following binomial sampling (sampling with fixed sample size N). If the prior is a beta distribution with parameters r' and N', then the posterior density is

$$f''(p \mid y) = \frac{f'(p)f(y \mid p)}{\displaystyle\int_0^1 f'(p)f(y \mid p)\ dp} = \frac{f'(p)P(r \mid N, p)}{\displaystyle\int_0^1 f'(p)P(r \mid N, p)\ dp}$$

$$= \frac{\dfrac{(N' - 1)!}{(r' - 1)!(N' - r' - 1)!}\, p^{r'-1}(1 - p)^{N'-r'-1} \dbinom{N}{r} p^r (1 - p)^{N-r}}{\displaystyle\int_0^1 \dfrac{(N' - 1)!}{(r' - 1)!(N' - r' - 1)!}\, p^{r'-1}(1 - p)^{N'-r'-1} \dbinom{N}{r} p^r (1 - p)^{N-r}\ dp}$$

$$\text{(8.9.1*)}$$

The terms not involving p can be moved outside of the integral in the denominator and canceled with the like terms in the numerator, leaving us with

$$f''(p \mid y) = \frac{p^{r'-1}(1 - p)^{N'-r'-1} p^r (1 - p)^{N-r}}{\displaystyle\int_0^1 p^{r'-1}(1 - p)^{N'-r'-1} p^r (1 - p)^{N-r}\ dp}. \qquad \text{(8.9.2)}$$

As in all applications of Bayes' theorem, the denominator serves merely to assure that the posterior distribution will be a proper probability distribution; the denominator of Equation (8.9.2) is a constant with regard to p (since it involves an expression integrated over p), so let us denote it by k:

$$f''(p \mid y) = \frac{1}{k} p^{r'-1}(1-p)^{N'-r'-1}p^{r}(1-p)^{N-r}.$$

Combining terms in this expression, we get

$$f''(p \mid y) = \frac{1}{k} p^{r'+r-1}(1-p)^{N'+N-r'-r-1}. \qquad (8.9.3)$$

Now, if we define $N'' = N' + N$ and $r'' = r' + r$, the posterior density can be written in the form

$$f''(p \mid y) = \frac{1}{k} p^{r''-1}(1-p)^{N''-r''-1}. \qquad (8.9.4^*)$$

But we can see now that if we let

$$k = \frac{(r''-1)!(N''-r''-1)!}{(N''-1)!},$$

then the posterior distribution will simply be a beta distribution with parameters r'' and N''.

Notice that if we had sampled with a fixed value of r (Pascal sampling) rather than a fixed value of N (binomial sampling), the above analysis would not have been affected at all, since the only difference would have been the replacement of $\binom{N}{r}$ in Equation (8.9.1) with $\binom{N-1}{r-1}$. In either case, the term in question cancels out, so the final result is the same.

The above analysis also serves to prove an assertion made in the last section, namely the assertion that if the prior distribution is a beta distribution and the sample is taken from a Bernoulli process, then the posterior distribution is also a beta distribution. Incidentally, it is much more important that you gain an *understanding* of how Bayes' theorem works than that you concern yourself unduly over mathematical details. Now that we have shown that the beta family is conjugate to the Bernoulli process, it will not be necessary to follow steps (8.9.1) through (8.9.4) every time we wish to revise a beta prior distribution on the basis of a sample from a Bernoulli process. Given the prior and the sample, we need simply to note that $r'' = r' + r$ and $N'' = N' + N$ are the parameters of the posterior beta density. This is the beauty of the concept of conjugate prior distributions.

8.10   THE USE OF BETA PRIOR DISTRIBUTIONS: AN EXAMPLE

We are now prepared to modify the example presented in Section 8.4 so that it is somewhat more realistic. Suppose that the statistician feels that his prior information concerning the production process can be well represented by a beta distribution with parameters $r' = 1$ and $N' = 20$. This distribution has a mean of $r'/N' = .05$ and a variance of $r'(N' - r')/N'^2(N' + 1) = .0023$. The graph of the density function is presented in Figure 8.10.1. This prior distribution seems more realistic than the discrete distribution used in Section 8.4, in which only four possible values of p were considered. Notice that the density function reaches its highest value at $p = 0$ and decreases as p increases.

Following the assessment of his prior distribution, the statistician takes a sample of 5 items from the production process, observing one defective item. Under the assumption that the production process behaves as a stationary and independent Bernoulli process, with p representing the probability of a defective item on any trial, the posterior distribution is a beta distribution with parameters

$$r'' = r' + r = 1 + 1 = 2$$

and
$$N'' = N' + N = 20 + 5 = 25.$$

This distribution is shown in Figure 8.10.2. The mean of the distribution is $2/25 = .08$, and the variance is $2(23)/25^2(26) = .0028$. Note that the

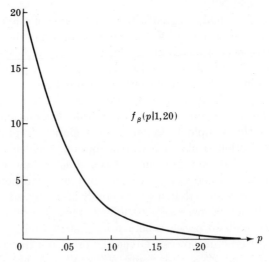

$f_\beta(p|1,20)$

Figure 8.10.1

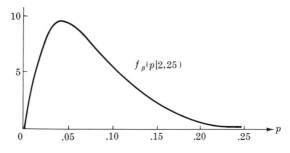

Figure 8.10.2

variance of the posterior distribution is slightly higher than the variance of the prior distribution; this is because of the shift in the mean. The posterior mean is .03 units to the right of the prior mean. Notice also that the mode of the posterior distribution is no longer zero; instead, it is equal to $1/23$, or about .043. In general, if $r_0 > 1$ and $N_0 - r_0 > 1$, the mode of the beta distribution is $r_0 - 1/N_0 - 2$.

Suppose that the statistician decides to take a second sample of size five and observes two defective items. The posterior distribution calculated after the first sample is now the prior distribution with respect to the new sample, so $r' = 2$ and $N' = 25$. After the second sample, the posterior distribution is a beta distribution with parameters

$$r'' = r' + r = 2 + 2 = 4$$

and
$$N'' = N' + N = 25 + 5 = 30.$$

This distribution is shown in Figure 8.10.3. The mean of the distribution is $4/30 = .133$, and the variance is $4(26)/(30)^2(31) = .0037$. Once again, the mean is shifted to the right and the variance increases. The mode is now $3/28 = .107$.

Just as in the discrete case, the statistician could have waited and revised his original prior distribution once after observing both samples. If

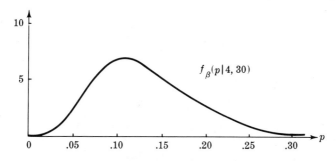

Figure 8.10.3

he did this, he would have $r = 3$ and $N = 10$, and the parameters of the posterior beta distribution would be

$$r'' = r' + r = 1 + 3 = 4$$

and $$N'' = N' + N = 20 + 10 = 30.$$

This is identical to the posterior distribution determined from two successive applications of Bayes' theorem.

If the statistician wants an estimate of p, he should base his estimate on the posterior distribution. For example, he might want to use the mean of the posterior distribution, which in this case is .133. Suppose that instead of estimating p, he must make a decision concerning the process, a decision which depends on the value of p. We will modify the example from Section 8.4 so that the costs are the same but the actions available to the statistician are slightly different. The two actions are: (1) he can leave the process as is, or (2) he can have a mechanic make an adjustment in the process which will halve the proportion defective, p—this adjustment costs $100.

The profit from a run of 1000 items is $1000(\$.50) = \$500$, less the costs of replacing defectives and adjusting the process (if the second action is chosen). The expected number of defectives in a run of 1000 is $1000p$, and the cost of replacing a single defective is $2.00, so the expected cost of replacing defectives is $2000p$ if the process is not adjusted. If the adjustment is made, the proportion defective is cut in half, so the expected cost of replacing defectives is $\$2000p/2 = \$1000p$. Thus, the profit to the firm of each of the two actions is

$$\text{Profit (No Adjustment)} = \$500 - \$2000p$$

and $$\text{Profit (Adjustment)} = \$500 - \$1000p - \$100.$$

The corresponding expected payoff, or expected profit, for each of the two acts is

$$\text{EP (No Adjustment)} = \$500 - \$2000\,E(p)$$

and $$\text{EP (Adjustment)} = \$400 - \$1000\,E(p).$$

From the posterior distribution, $E(p) = .133$, so the respective expected payoffs are $233 and $267. In order to maximize expected profit, the statistician should have the process adjusted. It is interesting to note that if the statistician had not obtained any sample information, the decision would have been based on the prior distribution. The prior mean is $E(p) = .05$, and the expected payoffs are $400 and $350. Solely on the basis of the prior information, then, the statistician should leave the process as is. As in the decision-making example of Section 8.4, the sample information revised the prior distribution in such a way as to make large values of p more likely and small values less likely.

This example illustrates the use of Bayes' theorem to revise a continuous probability distribution and the use of the resulting posterior distribution in a decision-making problem. We will consider decision theory in more detail later, but first we will discuss Bayesian techniques for a normal population or process.

8.11 BAYESIAN INFERENCE FOR A NORMAL PROCESS

The discussion of Bayesian inference for a Bernoulli process, while it may often prove useful, limits us to cases in which the assumptions of a Bernoulli process are satisfied. There must be two possible outcomes at each trial, the probabilities of these two outcomes must remain constant from trial to trial, and the trials must be independent. In many situations the data-generating process is such that there are more than two possible outcomes at each trial; in particular, any real number within some specified interval (which may be the entire real line) may be a potential outcome. An example is sampling from a normal population, or normal process. In previous chapters the wide applicability of the normal distribution has been discussed. In this section we will see how prior information can be easily combined with sample information from a normal process if the prior information can be expressed in the form of a normal distribution.

Suppose that a statistician is sampling from a normal process or population, and that he knows σ^2, the variance of the process. He does not know the mean, μ, however, although he has some prior information concerning μ. **If his prior distribution of μ is a normal distribution, his posterior distribution will also be a normal distribution.** From Equation (4.18.1), suppose that the prior density is of the form

$$f'(\mu) = \frac{1}{\sqrt{2\pi\sigma'^2}}\, e^{-(\mu-M')^2/2\sigma'^2}. \qquad (8.11.1^*)$$

This is a normal density function with mean M' and variance σ'^2. If the statistician takes a sample of size N and observes a sample mean of M, his posterior density is the normal density

$$f''(\mu \mid y) = \frac{1}{\sqrt{2\pi\sigma''^2}}\, e^{-(\mu-M'')^2/2\sigma''^2}, \qquad (8.11.2^*)$$

where y represents the sample results and the posterior parameters M'' and σ''^2 can be determined from the formulas

$$\frac{1}{\sigma''^2} = \frac{1}{\sigma'^2} + \frac{N}{\sigma^2} \qquad (8.11.3^*)$$

and
$$M'' = \frac{\left(\dfrac{1}{\sigma'^2}\right) M' + \left(\dfrac{N}{\sigma^2}\right) M}{\dfrac{1}{\sigma'^2} + \dfrac{N}{\sigma^2}}. \qquad (8.11.4^*)$$

The reciprocal of the posterior variance is equal to the sum of the reciprocal of the prior variance and the reciprocal of the variance of the sampling distribution of M, σ^2/N. The posterior mean is a weighted average of the prior mean and the sample mean, the weights being the reciprocals of the respective variances.

For an example, suppose that the process variance is known to be $\sigma^2 = 400$, but μ is unknown. The prior distribution is a normal distribution with mean 225 and variance 25. A sample of size 4 is taken, and the sample mean is 200. Thus $M' = 225$, $\sigma'^2 = 25$, $M = 200$, and $N = 4$. Using Equations (8.11.3) and (8.11.4),

$$\frac{1}{\sigma''^2} = \frac{1}{25} + \frac{4}{400} = \frac{5}{100} = \frac{1}{20}$$

and
$$M'' = \frac{\left(\dfrac{1}{25}\right) 225 + \left(\dfrac{4}{400}\right) 200}{\dfrac{1}{25} + \dfrac{4}{400}} = \frac{9 + 2}{\dfrac{1}{20}} = 220.$$

The posterior parameters are $M'' = 220$ and $\sigma''^2 = 20$. **As in the Bernoulli case, it will always be true that the posterior mean lies between the prior mean and the sample mean. Unlike the Bernoulli case, the posterior variance will always be smaller than the prior variance.** This can be seen from Equation (8.11.3); as long as $N > 0$, so that the second term on the right-hand side of the equation will be greater than zero, the reciprocal of the posterior variance will be larger than the reciprocal of the prior variance. This implies that the posterior variance will be smaller than the prior variance.

Does the family of normal distributions possess the desirable properties of a conjugate family in this case? Mathematical tractability appears to be satisfied. Although the computations necessary to determine the posterior parameters are more complicated than the simple addition used for the beta distribution and the Bernoulli process, they are still quite simple and should pose no computational problems. The posterior distribution is a member of the same family (the normal family) as the prior distribution. Finally, because of its importance in classical statistics, the normal distribution has been studied extensively. Tables are available showing values of both the cumulative distribution function and the density function, and the moments of the distribution (including higher order moments as well as the mean and variance) are known.

The normal family is reasonably "rich," for the mean and variance can assume any values. However, all members of the family are unimodal and symmetric, so no variation in the basic shape of the density function is possible. As a result, there will be situations in which a person's prior distribution is definitely nonnormal, particularly when the distribution is highly skewed. While it would be hoped that conjugate families of prior distributions would be rich enough to be widely applicable, there will clearly be times when no member of the conjugate family adequately represents the prior information. In such cases, it is necessary to apply Equation (8.6.4) directly rather than use the short cuts provided for by conjugate priors.

8.12   RELATIVE WEIGHTS OF PRIOR AND SAMPLE INFORMATION

In the previous section we avoided the question of the interpretation of a normal prior distribution. A normal prior distribution can be interpreted in terms of "equivalent sample information" from the process of interest, just as the beta distribution is interpreted. To see this in a clear fashion, however, it is necessary to consider a different parametrization of the distribution. The distribution will not be changed, but one of the parameters used to define it will be transformed to a new parameter.

Consider the parameter N' defined by the formula

$$N' = \frac{\sigma^2}{\sigma'^2}. \tag{8.12.1*}$$

Noting that

$$\sigma'^2 = \frac{\sigma^2}{N'}, \tag{8.12.2}$$

we see that the prior variance can be written in terms of N' and the process variance σ^2. The prior distribution is thus a normal distribution with mean M' and variance σ^2/N'. Now, if we let

$$N'' = \frac{\sigma^2}{\sigma''^2}, \tag{8.12.3*}$$

then Equation (8.11.3) becomes

$$\frac{N''}{\sigma^2} = \frac{N'}{\sigma^2} + \frac{N}{\sigma^2},$$

or simply

$$N'' = N' + N. \tag{8.12.4*}$$

Furthermore, Equation (8.11.4) becomes

$$M'' = \frac{\left(\frac{N'}{\sigma^2}\right) M' + \left(\frac{N}{\sigma^2}\right) M}{\frac{N'}{\sigma^2} + \frac{N}{\sigma^2}},$$

or

$$M'' = \frac{N'M' + NM}{N' + N}. \tag{8.12.5*}$$

Comparing Equations (8.12.4) and (8.12.5) with Equations (8.11.3) and (8.11.4), we see that the prior distribution is expressed in terms of M' and N' rather than M' and σ'^2, and the posterior distribution is expressed in terms of M'' and N'' rather than M'' and σ''^2. How can N' (and N'') be interpreted? From Equation (8.12.2), N' appears to be the sample size required to produce a variance of σ'^2 for a sample mean. This is because the variance of the sample mean from a sample of size N' is equal to σ^2/N'. This suggests an interpretation for the prior distribution: the prior distribution is roughly equivalent to the information contained in a sample of size N' with a sample mean of M'. Under this interpretation, Equations (8.12.4) and (8.12.5) can be thought of as formulas for pooling the information from two samples. The pooled (posterior) sample size is equal to the sum of the two individual sample sizes (one from the prior distribution, one from the sample), and the pooled (posterior) sample mean is equal to a weighted average of the two individual sample means (one from the prior, one from the sample). This does not mean that the prior information must consist solely of a prior sample from the process; it merely indicates that the prior information can be thought of as being roughly equivalent to such sample information.

Using the new parameter introduced in Equation (8.12.1), we can talk of the *relative weights* of the prior information and the sample information. If we want to estimate μ, for example, a reasonable estimator is the posterior mean, M''. We know that M'' will always be between the prior mean M' and the sample mean M. Under what conditions will M'' be closer to M' than to M, and under what conditions will M'' be closer to M? From Equation (8.12.5), M'' is a weighted-average of M' and M, with the weights being equal to $N'/(N' + N)$ and $N/(N' + N)$, respectively. Therefore, if $N' > N$, the prior mean will be given more weight, and the posterior mean M'' will be closer to M' than to M. If $N' < N$, the sample mean will be given more weight, and M'' will be closer to M than to M'. If $N' = N$, M'' will be exactly midway between M' and M.

In terms of the above interpretation of the prior distribution, these results appear obvious; in pooling two samples, the one with the larger sample size automatically receives more weight in the determination of a pooled mean. Recalling that the variance of the prior distribution is equal to σ^2/N' and the variance of the sampling distribution of M is equal to σ^2/N, we see that the prior distribution receives more weight than the sampling distribution if the prior variance is less than the variance of the sampling distribution, and vice versa. If we think of this in terms of information, a smaller variance implies more information. In determining the posterior distribution, then, we are pooling information. If there is more prior information than sample information (where information in this context relates to the variance of the prior and sampling distributions), then the posterior distribution will be affected more by the prior distribution than by the sampling distribution, or likelihood function.

Consider the example from the previous section. In this example,

$$N' = \frac{\sigma^2}{\sigma'^2} = \frac{400}{25} = 16,$$

and $N = 4$, so that

$$N'' = N' + N = 16 + 4 = 20$$

and

$$M'' = \frac{N'M' + NM}{N' + N} = \frac{16(225) + 4(200)}{16 + 4} = 220.$$

Notice that the posterior mean, 220, is closer to the prior mean (225) than to the sample mean (200). This is because $N' > N$; the prior information is roughly equivalent to the information contained in a sample of size 16 from the process with sample mean 225. Also, notice that the results are identical to those found by using the original parametrization. The posterior mean is the same in both cases and

$$N'' = \frac{\sigma^2}{\sigma''^2} = \frac{400}{20} = 20.$$

8.13 THE USE OF NORMAL PRIOR DISTRIBUTIONS: AN EXAMPLE

Suppose that a retailer is interested in the distribution of weekly sales at one of his stores. In particular, he is willing to assume that the random variable S, weekly sales (expressed in terms of dollars), is normally distributed with unknown mean μ and known variance $\sigma^2 = 90,000$. He feels that the goods sold at the store in question are such that there is no seasonal effect to worry about. That is, sales generally are not affected by such things as the weather, Christmas, and so on. From informal conversations with the manager of the store and from knowledge about sales at similar stores, the retailer assesses a prior distribution of the parameter μ. The mean of the prior distribution is 1200, and the variance is 2500. Furthermore, the retailer feels that his prior distribution is symmetric about the mean and shaped roughly like a normal density function, so he assumes that the distribution is normal.

The retailer may be interested in the distribution of sales for any one of several reasons. Perhaps he must decide whether to keep the store or to sell it, whether to open more stores of a similar nature or in a similar neighborhood, or whether to bring in a new manager for the store. Such decisions necessitate the consideration of a loss function and of relevant variables other than sales. To keep the example from becoming too complicated, let us assume that the retailer does not have to make a decision at this particular moment. Instead, he just wants some idea about the mean of the distribution of weekly sales, and he decides to look at an interval of values of μ.

Given the prior distribution, an interval can be determined which is centered at M' and which has a probability of .95. This interval is simply

$$(M' - 1.96\sigma', \quad M' + 1.96\sigma'),$$

or $(1200 - 1.96(50)), \quad (1200 + 1.96(50)),$

or $(1102, 1298).$

To differentiate this from a classical "confidence interval," which is based solely on sample information and on the frequency interpretation of probability, it is called a *credible interval*. Based on the prior distribution, then, a 95 percent credible interval for μ is the interval from 1102 to 1298.

The retailer decides that he would like to obtain more information about the sales of the store, so he takes a sample from the past sales records of the store. For a sample of 60 weeks, he finds the average weekly sales to be 1500. The posterior distribution for μ following this sample is a normal distribution with parameters determined from the following equations:

$$\frac{1}{\sigma''^2} = \frac{1}{\sigma'^2} + \frac{N}{\sigma^2} = \frac{1}{2500} + \frac{60}{90,000} = \frac{96}{90,000}.$$

$$M'' = \frac{\left(\frac{1}{\sigma'^2}\right)M' + \left(\frac{N}{\sigma^2}\right)M}{\frac{1}{\sigma'^2} + \frac{N}{\sigma^2}} = \frac{\left(\frac{1}{2500}\right)(1200) + \left(\frac{60}{90,000}\right)(1500)}{\left(\frac{1}{2500}\right) + \left(\frac{60}{90,000}\right)}$$

$$= 1387.5.$$

The mean and variance of the posterior distribution are 1387.5 and
90,000/96 = 937.5. The limits for a 95 percent credible interval based on
the posterior distribution are

$$M'' - 1.96\sigma'' \quad \text{and} \quad M'' + 1.96\sigma'',$$

or $1387.5 - 1.96(30.6)$ and $1387.5 + 1.96(30.6)$,

or 1327.5 and 1447.5.

Notice that the width of the 95 percent credible interval based on the prior
distribution is 196, whereas the width of the interval based on the posterior
distribution is 120. This reflects the fact that the introduction of the sample
information reduced the standard deviation from 50 to 30.6. Incidentally,
the classical 95 percent *confidence* interval, based solely on the sample
information, has limits of approximately 1424 and 1576, which you can
verify using the techniques of Chapter 6.

One aspect of this example seems particularly unrealistic—the assumption that the variance of the process is known. The retailer may have some
information about the variance, but it is highly unlikely that he knows it
for certain. Unfortunately, a formal Bayesian approach to inference for a
normal process is somewhat more complex if the variance is not assumed
known. Since the variance is not known, the retailer would have to assess
a *joint* prior distribution for the variance *and* the mean and modify this
distribution on the basis of the sample information. In this case, the sample
variance will be of interest as well as the sample mean, since it provides
information concerning the process variance. At any rate, assessing and
revising a *joint* prior distribution is more complicated than assessing and
revising a univariate prior distribution, so we will not approach this
problem.

Does this mean that we will be unable to carry out a Bayesian analysis
for the normal process with unknown variance? In some cases, yes. In other
cases, however, the central limit theorem comes to our aid once again.
In Equations (8.11.3) and (8.11.4), σ^2 only appears in connection with
the sample information, not the prior information. If we replace σ^2 by the
unbiased sample variance $\hat{s}^2$, which is an estimator of σ^2, the sampling distribution is a t distribution rather than a normal distribution. If the sample
size N is large enough, though, this t distribution is virtually identical to
the standard normal distribution. If N is large enough, then, we can substitute $\hat{s}^2$ for σ^2 in Equations (8.11.3) and (8.11.4) and proceed as though
σ^2 were known. For example, if the retailer did not know σ^2 but observed
$\hat{s}^2 = 96,000$, you can verify that the posterior mean and variance would be
1383 and 975.6, respectively.

8.14 BAYESIAN INFERENCE FOR OTHER PROCESSES

Because of space limitations, Bayesian inferential procedures have been presented only for two processes, the Bernoulli process and the normal process with known variance. Similar procedures have been developed for numerous other processes, both univariate and multivariate. For each of these processes, the idea is the same as for the Bernoulli and normal processes. The likelihood function is determined, and a family of conjugate distributions is found. If the prior distribution is a member of the conjugate family, the posterior distribution is a member of the same family, and the prior distribution and the likelihood function can be combined through a few reasonably simple formulas such as Equations (8.8.3) and (8.8.4) or (8.11.3) and (8.11.4). The features of Bayes' theorem which have been illustrated in the past few sections will be true for these other processes as well. For instance, the statistician's distribution can be revised trial by trial or just once after all of the trials.

As the process or population being sampled from becomes more complex, so that the mathematical form of the likelihood function becomes more complicated, it may be more difficult to apply Bayesian methods unless approximations (such as the normal distribution) can be used. The difficulties are increased when we are interested in several variables at once. The prior distribution in such situations will be a multivariate distribution. This may well cause pyschological problems to go along with the mathematical complexities. As we will see in the next section, the assessment of prior distributions is often not easy in the univariate case. This is one of the reasons why Bayesian techniques are not more widely applied. Some interest in this problem has been generated in the past few years, and this is reflected in the next section.

8.15 THE ASSESSMENT OF PRIOR DISTRIBUTIONS

In all of the examples, it has been assumed that the prior distribution is given, that is, that the statistician has already determined his prior distribution. Suppose that this is not true. Instead, suppose that the statistician has certain prior information and wants to express this information in terms of a prior distribution. How can he quantify his judgments and express them as a probability distribution? In Section 8.5 we discussed the assessment of individual probabilities. If the random variable in question is discrete, the statistician's distribution will consist of a number of such probabilities. If the random variable is continuous, the task facing the assessor is more difficult. Several techniques have been proposed to aid the assessor in this task. Although we will not be able to go into much detail, we will briefly discuss some of these techniques.

The most obvious way to attempt to assess a continuous probability distribution is simply to specify the density function or the cumulative distribution function. One way to do this is to specify the functional form of the density function. The relationship between subjective judgments and a mathematical function, if such a relationship does exist, is usually not at all obvious. Directly attempting to specify the functional form of the distribution, then, might not be such a good approach. A second approach involves the graph of the density function and/or the cumulative distribution function. The assessor may have some idea of the general shape of the distribution, even if he has no idea about its mathematical form. Perhaps he can draw at least a rough graph of the PDF or CDF. Of these two, the PDF is probably more meaningful to most assessors; that is, they can translate their prior information into a PDF more easily than into a CDF. One way to determine a PDF is to assess probabilities for certain intervals, draw a histogram corresponding to these probabilities, and attempt to fit a smooth density function to the histogram. The relationship between a histogram and a density function was pointed out in Section 5.12. Figure 8.15.1 illustrates the assessment of a PDF by this "grouping and smoothing" technique.

In determining a PDF or CDF, it might be helpful to consider certain summary measures of the distribution. In particular, some measure of location, such as the mean, median, or mode of the prior distribution, would be useful. This would give the assessor some idea of where the "center" of the distribution is. Next, a measure of dispersion might be considered. The variance is probably not a good measure in this context, because it is expressed in terms of squared units and because the relationship between a certain variance and an amount of dispersion is not intuitively obvious. The standard deviation has the advantage of being expressed in the same units as the random variable, but unless the assessor has enough experience with probability and statistics to gain some "feel" for the standard deviation, it might not be too meaningful to him. Measures

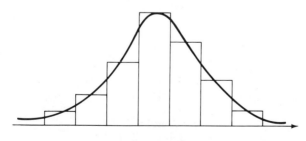

Figure 8.15.1

such as credible intervals are better suited for the assessment of prior distributions. Recall from the previous section that a credible interval is similar to a classical confidence interval except that it can be based on a prior or posterior distribution, whereas the confidence interval is based solely on the sampling distribution. The assessor might ask himself, "Can I find an interval centered at the mean that includes 50 percent, or 75 percent, or 95 percent, and so forth, of the prior probability?"

The use of credible intervals suggests the assessment of fractiles of the prior distribution. Remember that an f fractile of the distribution of the continuous variable X is a point c such that $P(X \leq c) = f$. The .50 fractile, for example, is the median of the distribution, which for a continuous random variable divides the area under the density function into two equal parts, each with area one-half. Intuitively, the median of the assessor's prior distribution is the value that the assessor feels is equally likely to be exceeded or not exceeded. Once the median is assessed, the .25 and .75 fractiles are the values that divide the two halves of the distribution in half again, and so on. This process is illustrated in Figure 8.15.2. The advantage of this technique is that it merely requires the assessor to determine a series of fractiles, and the assessment of any single fractile is similar to the assessment of an individual probability. After a number of fractiles are assessed and plotted on a CDF graph, the assessor can then draw a rough curve through them and thus determine the entire distribution.

For example, suppose that a statistician wants to assess his prior distribution for the proportion of defectives produced by a production process. First of all, he thinks that the proportion, p, is just as likely to be above .05 as below .05. This implies that .05 is the median of his distribution. Then, looking just at the interval below .05 (that is, the interval from 0 to .05), he decides that .035 divides this interval into two equally likely subintervals. Thus, .035 is the .25 fractile of the distribution. Similarly, he assesses the .75, .125, .375, .625, and .875 fractiles. In addition, to get some idea about the extreme tails of the distribution, he assesses the .01 and .99 fractiles. He then plots these assessed points on a graph and draws

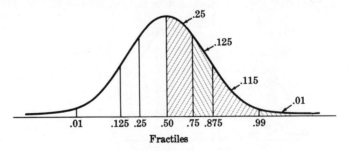

Figure 8.15.2

a rough CDF through them, as illustrated in Figure 8.15.3. To check his assessments, he might look at the curve to see if it reflects his prior information reasonably well.

It should be clear by now that assessing a prior distribution is not an easy task if a careful assessment is desired. To see how difficult it can be, select some variables about which you have some information (some ideas: the proportion of vehicles of a particular make on a given highway; the price, one year from now, of a certain common stock; the total amount of precipitation in the next year in a given city; and so on) and attempt to *carefully* assess a prior distribution in each case. To encourage careful assessment, pretend that your prior distribution in each case will be used as an input for a very important decision-making problem. It might be interesting to select several variables about which you have considerable knowledge and several about which you have very little information, and then to compare the resulting distributions. It is also interesting to assess a prior distribution for a variable which you will be able to observe shortly. You can then evaluate your distribution in the light of the actual event. For example, assess a distribution for the point spread in an upcoming football game.

At this point let us interrupt the discussion of the assessment of prior distributions to ask this question: why do we want a prior distribution in the first place? Presumably, we want to determine an estimate or make a decision and we want to base the estimate or decision on all of the information which happens to be available. We may intend to take a sample and use Bayes' theorem to combine the prior distribution and the likelihood function, forming a posterior distribution. But if that is the case, then it would be nice if the prior distribution were a member of a conjugate family of distributions. If we are sampling from a Bernoulli process, for example, and we are interested in p, the probability of a success on any single trial,

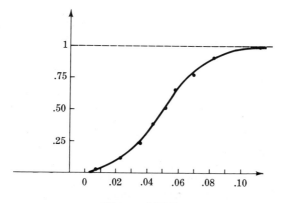

Figure 8.15.3

the analysis would be greatly simplified if the prior distribution happened to be a beta distribution. For normal sampling, the conjugate family is the normal family of distributions.

A subjectively assessed prior distribution will not necessarily be a member of a conjugate family of distributions. More likely than not, it will not follow any mathematical function exactly. However, it may be possible to find a member of the conjugate family which is a "good fit" to the assessed distribution. This is where the property of "richness" of conjugate families is important. Unless the distribution is quite irregular, there is a good chance that a member of the conjugate family will approximate it reasonably well. The assessor could try a number of conjugate distributions to see if any prove satisfactory. We will give a more precise notion of what we mean by "satisfactory" and "good fit" in the next section.

It is possible for the assessor to consider the conjugate family of distributions before assessing his entire distribution. If he feels that his prior information is roughly equivalent to sample information from the process or population of interest, he may be able directly to determine the prior parameters, using the interpretation of conjugate families presented earlier. If he can do this, the assessment procedure is greatly simplified. Suppose that the statistician interested in the proportion of defective items produced by a certain process felt that his prior information was roughly equivalent to a sample of 20 items with 1 defective. His prior distribution would then be a beta distribution with parameters $r' = 1$ and $N' = 20$.

Suppose that the statistician in the above example was not able to translate his prior information into equivalent sample evidence, but that he did think that a beta distribution would adequately represent his information. How could he choose a single beta distribution? If he could determine the mean and variance of the distribution, he could equate these to the formulas for the mean and variance of a beta distribution. Suppose the statistician's mean and standard deviation were .08 and .04, respectively. Then

$$\frac{r'}{N'} = .08$$

and
$$\frac{r'(N' - r')}{N'^2(N' + 1)} = (.04)^2 = .0016.$$

Solving these two equations simultaneously, we find that $r' = 3.6$ and $N' = 45$. Notice that this technique is analogous to the method of moments, which we used to determine estimators.

In the normal case, of course, the mean and variance are themselves the prior parameters, so there are no equations to solve. However, the assessor might wish to assess a pair of fractiles and use the tables of the normal distribution to find M' and σ'. For a very simple example, suppose that the

.50 fractile, or the median, was 86, and the .75 fractile was 92. Since the normal distribution is symmetric, the mean and the median are equal, so that the mean would be 86. From the normal tables, the .75 fractile of the standard normal distribution is approximately .67. This implies that the .75 fractile of any normal distribution is .67 standard deviations, or .67σ', to the right of the mean. But 92 is 6 to the right of 86, so

$$6 = .67\sigma',$$

or
$$\sigma' = 8.96.$$

The prior distribution is thus a normal distribution with mean 86 and standard deviation 8.96.

8.16 SENSITIVITY ANALYSIS

In the preceding section several techniques for assessing prior distributions were discussed. The technique used in any specific situation should depend on the form of the prior information. If the prior information consists of sample information or can be thought of as equivalent to sample information, a conjugate prior distribution can be assessed directly. In other cases, it may be desirable to determine a few summary measures, or perhaps the entire subjective distribution, before attempting to fit a conjugate distribution. In yet other cases we might not be able to find a conjugate distribution that adequately reflects the prior information.

Because of the subjective nature of the assessment process, different techniques of assessment are likely to lead to different distributions. If two such distributions are assessed, we may be interested in how similar they are to each other as well as how similar they are to members of the conjugate family of distributions. In classical statistics, hypothesis testing procedures are available to test the "goodness of fit" of entire probability distributions. We will study some of these procedures in Chapter 12 (Volume II). At this point we are concerned in a decision-theoretic approach to "goodness of fit."

In most applications of Bayes' theorem, the ultimate aim of the statistician is to make a decision on the basis of the posterior distribution. Variations in the prior distribution will cause variations in the posterior distribution. The question the statistician is interested in is this: how will these variations affect the ultimate decision? Or, in other words, how *sensitive* is the decision-making procedure to variations in the prior distribution? An investigation of this question is called a **sensitivity analysis.** If large variations in the prior distribution tend not to affect the decision, then the decision-making procedure is said to be *insensitive* to variations in the prior distribution. If very slight variations in the prior distribution are likely to cause the decision to be changed, the decision-making procedure is said to be highly *sensitive* to such variations.

It should be pointed out that sensitivity depends on the particular decision-making procedure, including the sample which is taken, the actions available to the statistician, and the potential payoffs or losses. Technically, then, we cannot speak of sensitivity except in connection with a particular decision. It is possible, however, to make some generalizations concerning sensitivity. The more sample information we have, the less sensitive the procedure usually is with regard to the prior distribution. As more and more sample information is gathered, the prior distribution is given less and less weight in the computation of the posterior distribution (Section 8.12). As a result, variations in the prior distribution are of less importance. Any time there is a great deal of sample information relative to the prior information, insensitivity can be expected to hold. This fact will be of great importance in the discussion of "informationless" prior distributions in the next section.

Some research concerning sensitivity analysis has indicated that in a wide variety of situations, the decision-making procedure is reasonably insensitive to moderate variations in the prior distribution. This implies that a conjugate prior distribution, while it may not be a perfect fit to a subjectively assessed distribution, will often be a "satisfactory" fit. If there is any doubt in a particular application, it is prudent to investigate before using the conjugate distribution as an approximation. If it is not too burdensome computationally (and it should not be, particularly if a high-speed computer is available), two posterior distributions can be calculated. In this manner, it is possible to see how variations in the prior distribution are reflected in the posterior distribution. If you are still not sure, it is probably best not to use the conjugate distribution.

8.17 REPRESENTING AN "INFORMATIONLESS" PRIOR STATE

Suppose that a statistician wants to assess a prior distribution in a situation where he has very little or no prior information. More specifically, his prior information is such that it is "overwhelmed" by the sample information. Then it is said that the statistician has a **diffuse, or informationless,** state of prior information. The term "diffuse" is preferred to "informationless" because it describes the situation in a better fashion. The situation described is not necessarily an informationless state in the usual meaning of the word; it is informationless in a relative sense. When we say that someone's prior distribution is diffuse, we mean only that it is *diffuse relative to the sample information.*

The classic example (due to L. J. Savage) of diffuseness concerns the determination of the weight of a potato. If you were given a potato and asked to assess a distribution of its weight, you would clearly have some information about the weight. On the other hand, we suspect that your information would be of a rather vague nature. You could probably specify some limits within which you were sure the weight would lie, and your distribution might have a peak, or mode, somewhere within these limits. It is doubtful, however, that your distribution would have a sharp "spike" anywhere.

If the potato were weighed on a balance of known precision (say, a standard error of measurement of $\frac{1}{2}$ gram), what would your posterior distribution look like following a single weighing? Since your distribution is probably quite "spread out," or diffuse relative to the likelihood function, the sample information would receive much more weight than the prior distribution. In fact, the posterior distribution depends almost solely on the sample information. Notice that the sample size here is only one, but the precision of the balance is so high that the prior distribution is much less precise than the sampling distribution. Recall from Chapter 7 that increased precision can result from a larger sample size or a smaller process (or population) variance.

Graphically, the situation is represented in Figure 8.17.1. Let θ represent the weight of the potato, and let y represent the sample result, the weight obtained from one reading of the balance. Relative to the likelihood function, the prior distribution is "flat." It is not strictly a uniform distribution, but it *is* virtually uniform relative to $f(y \mid \theta)$. Recalling that the posterior density function is proportional to the product of the prior density and the likelihood function, where the constant of proportionality is the normalizing integral, we can write

$$f''(\theta \mid y) \propto f'(\theta)f(y \mid \theta), \qquad (8.17.1)$$

where the $\propto$ sign is read "is proportional to." But the prior distribution

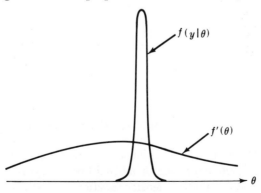

Figure 8.17.1

is essentially "flat," so it can be approximated by a constant function, $f'(\theta) = k$. The posterior density is then proportional to the likelihood function:

$$f''(\theta \mid y) \propto f(y \mid \theta). \tag{8.17.2}$$

This expresses formally what we previously explained in a heuristic manner: **If the prior distribution is diffuse relative to the likelihood function, the posterior distribution depends almost solely on the likelihood function.** The proportionality in Equation (8.17.2) is not a *strict* proportionality because $f'(\theta)$ is not *strictly* uniform. Unless the decision-making problem at hand is quite sensitive to slight variations in the prior distribution, though, we can safely use this approximation.

It is important to remember that diffuseness is a relative term. It is not necessary for a prior distribution to be perfectly "flat" to be diffuse, although it should not have any sharp "spikes," or "peaks." It is also not necessary for the prior distribution to be diffuse over the entire range of values of the random variable (all positive real numbers in the potato example). It *should* be diffuse in the neighborhood of the sample results, that is, in the neighborhood of the likelihood function. Thus, it need only be *locally diffuse*.

The discussion in this section has involved prior distributions of continuous random variables. It is also possible to speak of diffuse *discrete* prior distributions. In this case, the distribution is diffuse (again, relative to the sample information) if the probabilities of the various possible values of the random variable are equal or nearly equal. If θ can take on K possible values $\{\theta_1, \theta_2, \cdots, \theta_K\}$, then one way to represent diffuseness is simply to let

$$P(\theta = \theta_i) = \frac{1}{K} \quad \text{for } i = 1, 2, \cdots, K.$$

In an application of Bayes' theorem for discrete random variables, Equation (8.2.2), the posterior probabilities will then be proportional to their respective likelihoods.

In the next section, members of the conjugate beta and normal families which can be used to represent diffuse prior information will be presented. Then, in Section 8.19, the importance of the concept of diffuseness will be discussed.

8.18 DIFFUSE BETA AND NORMAL PRIOR DISTRIBUTIONS

If the statistician's prior distribution is diffuse relative to the sample information, it should not make too much difference exactly how he specifies the mathematical form of the density function, provided of course that the function chosen is relatively "flat." In order to simplify the process of revising the distribution on the basis of sample information, he would prefer his distribution to be a member of the conjugate family of distributions. We have discussed conjugate distributions for two situations. First, the beta family of distributions is conjugate when we are sampling from a Bernoulli process. Second, the normal family of distributions is conjugate when we are sampling from a normal process. It would therefore be most convenient to have members of these two conjugate families which could be used as diffuse prior distributions. The *exact* form of the prior distribution is not too important because of the diffuseness, so the statistician might as well ease the computational burden by selecting a diffuse conjugate distribution.

An investigation of various beta distributions reveals that when $r' = 1$ and $N' = 2$, the beta distribution is a uniform distribution over the unit interval. This can be demonstrated by substituting $r' = 1$ and $N' = 2$ in the formula for the beta distribution:

$$f'(p) = \frac{(N' - 1)!}{(r' - 1)!(N' - r' - 1)!} \, p^{r'-1}(1 - p)^{N'-r'-1} \quad \text{for} \quad 0 \leq p \leq 1.$$

Since $r' - 1 = 0$ and $N' - r' - 1 = 0$,

$$f'(p) = \frac{1!}{0!0!} \, p^0(1 - p)^0 \quad \text{for} \quad 0 \leq p \leq 1.$$

By convention, $0! = 1$, and anything raised to the zero power is equal to one, so the distribution is simply

$$f'(p) = 1 \quad \text{for} \quad 0 \leq p \leq 1, \tag{8.18.1}$$

which is the uniform distribution defined on the unit interval.

In terms of its density function, then, the beta distribution with $r' = 1$ and $N' = 2$ seems to be a good representation of a diffuse state of prior knowledge. In terms of an interpretation of the distribution, however, some questions arise. According to the "equivalent prior sample information" interpretation, the distribution represents information roughly equivalent to a sample of size 2 with 1 success from the process in question. While this is not a great deal of information, it may be enough information relative to some samples to cast some doubt on the applicability of the distribution as a diffuse distribution. For example, suppose that you have no information whatsoever regarding a Bernoulli process. You decide to let the beta distribution with $r' = 1$ and $N' = 2$ represent this informationless state of affairs. A sample of size three is taken, and one success is observed. Using

Bayes' theorem, your posterior distribution is a beta distribution with parameters

$$r'' = r' + r = 1 + 1 = 2$$

and $$N'' = N' + N = 2 + 3 = 5.$$

The mean of your posterior distribution is $r''/N'' = 2/5 = .40$. But how can this be, if you started out with no information and you then observed one success in three trials? Intuitively, your posterior mean should equal $\frac{1}{3}$, or .33. Here the choice of a prior distribution evidently *did* have some effect on the posterior distribution, in which case it could not be called diffuse.

Difficulties such as this do not arise too often. The sample must be small, and preferably should consist of almost all successes or almost all failures. The fact that it does arise at all is bothersome because it means that the uniform beta distribution cannot be used as a diffuse prior distribution in some situations. It should be noted that if the mode of the posterior distribution is used instead of the mean, this problem does not arise; the mode is $(r'' - 1)/(N'' - 2)$, which is $1/3$ in the above example.

Another approach is to look not at the density function, but at the interpretation of the prior parameters. If the prior distribution is interpreted in terms of equivalent sample information, the obvious choice is $r' = 0$ and $N' = 0$. This should correspond to "no information." Unfortunately, there is a drawback to this choice: the beta distribution is a "proper" distribution only for r' and N' such that $N' > r' > 0$. When $N' = r' = 0$, the total area under the curve is not equal to one; the integral representing this area does not converge. You need not concern yourself with the mathematical details; just remember that the area under the density function is not equal to one.

If we take $r' = 0$ and $N' = 0$ and apply the standard formulas for revising a beta distribution, $r'' = r' + r$ and $N'' = N' + N$, we see that the prior distribution *is* informationless. The posterior parameters are equal to the observed sample results. This result is so intuitively appealing that many Bayesian statisticians take $r' = 0$ and $N' = 0$ as their prior parameters in a diffuse situation, even though the implied prior distribution is "improper." If we only look at the terms involving p in this prior distribution, we see that the density is proportional to $p^{r'-1}(1 - p)^{N'-r'-1}$, or $p^{-1}(1 - p)^{-1}$. A graph of this function looks U-shaped, reaching a minimum at $p = \frac{1}{2}$. At $p = 0$ and $p = 1$, the function is not defined; as p approaches either zero or one, the function increases without bound, becoming infinite. This sort of a function does not exactly agree with a mental picture of what a diffuse prior distribution should look like. But it seems to us that the real test of a diffuse prior distribution in Bayesian inference and decision is whether or not it affects the posterior distribution. This is because, as we noted in the previous section, the ultimate aim of the Bayesian is to use the posterior distribution in an inferential or decision-making situation. As a result, in all examples calling for a diffuse beta distribution, we will use the improper beta distribution with $r' = 0$ and $N' = 0$.

How about the normal process—how can an informationless state of affairs be represented by a conjugate normal prior distribution? The density function of a normal distribution is not nearly as flexible as that of members of the beta family. For any choice of M' and σ'^2, the distribution is symmetric, unimodal, and roughly "bell-shaped." For any given M' the distribution is more "spread out" as the variance, σ'^2, is increased. When σ'^2 is very large, the distribution is almost, but not quite, uniform. For a diffuse normal distribution, then, we need to make σ'^2 large.

How large should σ'^2 be? The answer depends on the variance of the process which is to be sampled. The prior variance should be larger than the variance of the sampling distribution of M, which is σ^2/N. This is still not too satisfactory, for what is needed is a single number which would be applicable in most, if not all, cases. Consider the interpretation of the normal prior distribution in terms of equivalent sample information. To do this, it is necessary to reparametrize and look at N' instead of σ'^2, where

$$N' = \frac{\sigma^2}{\sigma'^2}. \tag{8.18.2}$$

For the prior variance to be large relative to the variance σ^2/N for fixed N, N' should be small. But N' can be interpreted as the sample size of an equivalent prior sample. If no information is available regarding the process, the obvious choice is $N' = 0$.

When $N' = 0$, we run into the same problem as in the case of the beta distribution: the implied distribution is not a proper distribution. The process variance is known and is presumably greater than zero (if it is equal to zero, then a sample of size one will determine μ for certain). From Equation (8.18.2), then, N' can only be zero when the prior variance is infinite. In the mathematical expression for the density of the normal distribution, though, both parameters, the mean and the variance, must be finite numbers.

Since we are primarily interested in the posterior distribution, what happens to the posterior parameters when $N' = 0$? Using Equations (8.12.4) and (8.12.5),

$$N'' = N' + N = 0 + N = N$$

and $$M'' = \frac{N'M' + NM}{N' + N} = \frac{0M' + NM}{0 + N} = \frac{NM}{N} = M.$$

The posterior parameters depend solely on the sample size N and the sample mean M. Notice that nothing has been said about the prior mean M'. When $N' = 0$, it makes no difference what M' is equal to; it receives no weight in the calculation of M'' because it is weighted by N', which is zero. Even though it is an improper density, then, the normal prior density with $N' = 0$ has no effect on the posterior distribution. As a result, we will use this density to represent a diffuse prior state when we are sampling from a normal distribution.

8.19   DIFFUSE PRIOR DISTRIBUTIONS AND CLASSICAL STATISTICS

When a diffuse prior distribution is used in a Bayesian analysis, the posterior distribution is virtually identical to the likelihood function. Thus, any inferences and/or decisions based on the posterior distribution will in reality depend almost solely on the sample information. However, a classical statistician bases inferences and decisions solely on the sample information. Under an "informationless," or diffuse, prior state, then, Bayesian statistical procedures and classical statistical procedures are based on the same set of information. The case in which the prior distribution is diffuse is a special case; if relevant prior information is available, the Bayesian's posterior distribution will reflect both this information *and* that of the sample. In the special case, classical and Bayesian procedures are quite similar, being based on the same information.

As we shall see, under a diffuse prior distribution Bayesian techniques often result in numerical results identical to classical results. However, there is still an important difference in interpretation. The Bayesian thinks of θ as a random variable and is willing to make probability statements concerning θ. The prior and posterior distributions are probability distributions of θ. The classical statistician, on the other hand, claims that θ is a fixed parameter and that it makes no sense to talk of the probability of values of θ occurring. His inferences are based on the sampling distribution, or likelihood function, which is of the form $f(y \mid \theta)$. This is a distribution of possible sample outcomes, y, conditional upon θ. We will explore this difference in interpretation at greater length in succeeding sections. First, the importance of diffuse distributions deserves some discussion. Diffuse distributions are valuable to the Bayesian for several reasons. Their first and most obvious use occurs when the statistician has no prior information or very little information relative to the information contained in the sample. A diffuse conjugate distribution such as those presented in the previous section provides a convenient way to express an "informationless" state. The application of Bayes' theorem is simple, and the posterior distribution is based almost entirely on the sample information.

A second and equally important use of diffuse prior distributions relates to scientific reporting. When reporting the results of a statistical analysis, the statistician is faced with this question: what should be reported? One possible answer is to report the posterior distribution and any resulting inferences or decisions. The entire posterior distribution itself is much more informative than any summarizations in the form of point or interval estimates. Furthermore, the posterior distribution is informative in the sense that it reflects not only the sample results, but also the prior judgments of the statistician. What if someone else wanted to investigate the results of the statistical analysis? He might want to assess his own prior distribution (which could be quite different from our statistician's distribution) and then use Bayes' theorem to revise the distribution. But unless the statistician reports the sample results separately (that is, other than as a part of the posterior), this person will be unable to carry out the analysis using his own prior.

The solution to the problem is for the statistician to carry out a separate analysis using a diffuse prior distribution. This will, by removing the effect of the statistician's own prior judgments, enable anyone to include his own prior judgments and proceed from there. Also, it will allow the reader to see approximately what results would have been obtained using classical techniques. Bear in mind that our statistician should base his *own* inferences and decisions on his own posterior distribution. For the benefit of others, it is useful for him to present the results which follow from the use of a diffuse prior distribution. Such results may be of some interest to the statistician himself, since he can then investigate the influence of his prior distribution on the results.

8.20 THE POSTERIOR DISTRIBUTION AND ESTIMATION

Sometimes the Bayesian statistician does not have to make a decision, but merely wants an estimate of some quantity. His estimate, of course, is based on the posterior distribution. If an estimate of θ is desired, some potential estimators are the posterior mean, median, mode, and so on. For the normal process, the posterior distribution is normal, assuming a conjugate prior distribution. Since the distribution is unimodal and symmetric, the posterior mean, M'', is equal to the posterior median and mode. M'' should thus be a "good" estimate of μ. For the Bernoulli process, the posterior distribution is a beta distribution, assuming a conjugate prior distribution. This distribution is *not* symmetric in general; it is symmetric only when $r''/N'' = \frac{1}{2}$. The mean and mode of the distribution are r''/N'' and, provided that $r'' > 1$ and $N'' - r'' > 1$, $(r'' - 1)/(N'' - 2)$.

If the prior distribution is diffuse, it is interesting to see how estimates based on the posterior distribution relate to classical estimates. One important method for determining "good" classical estimators is the method of maximum likelihood, discussed in Chapter 6. This method finds the value of θ which maximizes the likelihood function. Graphically, in terms of θ, this value is the value corresponding to the highest point on the graph of $f(y \mid \theta)$. But under a diffuse prior distribution, the posterior distribution is essentially identical to the likelihood function, so the maximum-likelihood estimator in this case corresponds to the mode of the posterior distribution. For normal sampling, this is M'', which is equal to M.

Even though the maximum-likelihood estimate and the mode of the posterior distribution are equal under a diffuse prior distribution, their interpretation is not the same. The mode of the posterior distribution is thought of as "the most likely value of θ." The maximum-likelihood estimator is thought of as "the value of θ which makes the observed sample results appear most likely." The difference is subtle, but it is nevertheless a difference. The Bayesian makes a probability statement about θ; the classical statistician makes a probability statement concerning y, the sample results.

How about interval estimation? From the posterior distribution, the probability of any interval of values of θ can be determined. To find an interval containing 90 percent of the probability, one could consider the interval from the .05 fractile to the .95 fractile, or the interval from the .01 fractile to the .91 fractile, and so on. The conventional choice of an interval containing $(1 - \alpha)$ percent of the probability is the interval from the $\alpha/2$ fractile to the $1 - (\alpha/2)$ fractile. Under a diffuse distribution, this corresponds to the classical notion of a $(1 - \alpha)$ percent confidence interval. For a normal process with unknown mean μ and known variance σ^2, the usual classical 95 percent confidence limits for μ are

$$M - 1.96\sigma/\sqrt{N} \quad \text{and} \quad M + 1.96\sigma/\sqrt{N}.$$

The limits of the corresponding Bayesian interval are

$$M'' - 1.96\sigma/\sqrt{N''} \quad \text{and} \quad M'' + 1.96\sigma/\sqrt{N''}.$$

Under a diffuse normal prior distribution with $N' = 0$, $M'' = M$ and $N'' = N$, and thus the Bayesian limits are identical to the classical limits. Once again, the interpretation is different. The Bayesian would say that "the probability is .95 that μ lies within the specified interval," and the classical statistician would say that "in the long run, 95 percent of all such intervals will contain the true value of μ." The classical statement is based on long-run frequency considerations, and the Bayesian statement is a statement which concerns not the long run, but the specific case at hand. To differentiate between the two, we call the classical interval a *"confidence interval"* and the Bayesian interval a *"credible interval."*

We have emphasized that under certain conditions (a diffuse prior distribution), Bayesian and classical procedures may produce similar results, although the interpretations are different. In actual applications, particularly decision-making applications, the assumption of a diffuse prior distribution is seldom reasonable. In most business situations, for example, some prior information is known. For an investment decision, some prior information should be available concerning the past performance of investments similar to those being evaluated. This information may be in the form of recorded statistics, or it may be the subjective information of an investment broker; usually, both types of information will be available. In medicine, a surgeon should have some judgments regarding the probability that a patient will survive a particular operation. Even if this type of operation has never been performed before, the surgeon is familiar with the medical factors involved and with past results of related surgical techniques. In marketing, a marketing manager should have some idea about the proportion of consumers that will purchase a new product. Even in scientific research, some prior information is generally available to the scientist. This prior information may come from some scientific theories, or it may come from previous experimental work. The list of examples is endless, and the point is simply this: diffuse prior distributions are quite useful, primarily for reporting purposes, but they are often quite unrealistic in actual situations.

If the prior distribution is not diffuse, point estimates and interval estimates based on the posterior distribution may differ from similar estimates based on the likelihood function. In the production process example of Section 8.4, the maximum-likelihood estimate of p is $r/N = 3/10 = .30$, and the posterior mean turns out to be .146. In the modification of the same example, presented in Section 8.10, the maximum-likelihood estimate is .30 and the posterior mean is $r''/N'' = 4/30 = .133$. In the example concerning the weekly sales of a store (Section 8.13), the maximum-likelihood estimate, the sample mean, equals 1500. The mean of the posterior distribution is 1387.5. A 95 percent *credible* interval based on the posterior distribution is

$$(M'' - 1.96\sigma'', M'' + 1.96\sigma''),$$

or $(1327.5, 1447.5).$

A 95 percent *confidence* interval based on the sample information alone is

$$\left(M - \frac{1.96\sigma}{\sqrt{N}}, M + \frac{1.96\sigma}{\sqrt{N}}\right),$$

or $(1424, 1576).$

These examples demonstrate that the inclusion of prior information can have quite an effect on point and interval estimates. In the next chapter we will look at estimation in yet another light: estimation as a decision-making procedure. To do this it will be necessary to include not only prior information, but also information concerning the potential "losses" due to errors in estimation. Before discussing decision theory, let us consider other inferential procedures based on the posterior distribution.

8.21 PRIOR AND POSTERIOR ODDS RATIOS

Instead of estimating a certain parameter, a statistician may wish to test a hypothesis concerning that parameter. In Chapter 7 we presented the classical theory of hypothesis testing. If hypothesis testing is thought of as a choice between two actions, accepting or rejecting a given hypothesis, then it is a decision-making procedure. As we pointed out in Chapter 7, a formal decision-theoretic approach to hypothesis testing necessitates the consideration of losses due to erroneous decisions. The decision to accept or reject a hypothesis should depend not only on the sample information, but also on the losses. The formal decision-theoretic approach will be presented in the next chapter. In this and the following sections, we will treat hypothesis testing on a more informal level.

Suppose that a politician is interested in the proportion of votes that he will receive in a forthcoming election. Let us call this proportion p. The politician's advisors think that there are two possible values for p: .40 and .60. Furthermore, the prior probabilities of these two values are

$$P'(.40) = .25$$

and

$$P'(.60) = .75.$$

A sample of 40 voters is taken, and 18 of them claim that they will vote for this politician. A table can be set up for the computation of the posterior probabilities (the likelihoods are based on the assumption that the process is a Bernoulli process):

Value of p	Prior probability	Likelihood	(Prior probability) × (likelihood)	Posterior probability
.40	.25	.10255	.02564	.63
.60	.75	.02026	.01519	.37
	1.00		.04083	1.00

Bayes' theorem can be written in terms of odds when there are only two possible values. Let A correspond to the event that $p = .40$, and let B correspond to the event that $p = .60$. Then the **prior odds ratio of A to B** is defined as the ratio of the prior probabilities,

$$\Omega' = \frac{P'(A)}{P'(B)} = \frac{.25}{.75} = \frac{1}{3}. \tag{8.21.1*}$$

If this ratio is greater than one, then A is more likely than B; if it is less than one, B is more likely than A; and if it is equal to one, A and B are equally likely. The same holds for the **posterior odds ratio of A to B,**

$$\Omega'' = \frac{P''(A \mid y)}{P''(B \mid y)} = \frac{.63}{.57} = 1.70. \tag{8.21.2*}$$

But, from Bayes' theorem, the posterior probability is proportional to the product of the prior probability and the likelihood. Thus a ratio of posterior probabilities should be equal to the product of a ratio of prior probabilities and a ratio of likelihoods. The **likelihood ratio** is of the form

$$LR = \frac{f(y \mid A)}{f(y \mid B)} = \frac{.10255}{.02026} = 5.06. \tag{8.21.3*}$$

We can now write Bayes' theorem in terms of odds:

$$\Omega'' = \Omega' \times (LR). \tag{8.21.4*}$$

In words, **the posterior odds ratio is equal to the product of the prior odds ratio and the likelihood ratio.** You can verify for yourself that this is true for the example.

What implications does Equation (8.21.4) hold for hypothesis testing? We can think of the two possible values of p as two hypotheses:

$$H_0: p = .40$$

$$H_1: p = .60.$$

If we are interested in which hypothesis is more likely, we need only to look at the posterior odds ratio, Ω''. In the example, $\Omega'' = 1.70$, implying that H_0 is more likely than H_1. In fact, it can be said that the odds and in favor of H_0 are 1.7-to-1.

The example is admittedly highly artificial. In this situation, the politician would no doubt assume that p is continuous and determine a continuous posterior distribution for p. If he is primarily interested in whether p is greater than or less than $\frac{1}{2}$, he can determine $P(p > \frac{1}{2})$ and $P(p < \frac{1}{2})$ from the posterior distribution.

It is possible to speak of odds ratios and likelihood ratios when dealing with inexact hypotheses, but their determination may not be nearly as simple as it is for two exact hypotheses. For an inexact hypothesis such as $H_0: p > \frac{1}{2}$, there is no single value of p which can be used to compute the likelihood of any sample result. As a result, the Bayesian approach in such a situation is slightly different from that presented in this section, as we will see in the following sections. The general idea is the same, however.

Despite the artificial nature of the example, the concepts introduced (prior odds ratio, posterior odds ratio, likelihood ratio) are quite important. As we mentioned in Section 7.19, all of the classical tests of hypotheses considered so far are related to the likelihood ratio. We will examine this concept in the next section. The odds ratios are useful in decision-making. If the posterior odds ratio and a ratio of losses are known, it is only necessary to compare the two in order to make a decision, as we shall see in the next chapter.

8.22 THE IMPORTANCE OF THE LIKELIHOOD RATIO IN CLASSICAL TESTS

In the example in the previous section, a classical statistician would have based his inferences solely on the sample information. Given the hypotheses

$$H_0: p = .40$$

and $\qquad\qquad H_1: p = .60,$

the likelihood ratio, LR, summarizes all of the information relevant to these two hypotheses. This ratio is of the form

$$LR = \frac{f(y \mid p = .40)}{f(y \mid p = .60)}. \tag{8.22.1}$$

If $LR > 1$, the observed sample result appears more likely given H_0 than given H_1; if $LR < 1$, the reverse is true. Thus, the larger the likelihood ratio, the more likely it is that H_0 is the true hypothesis; the smaller the likelihood ratio, the more likely it is that H_1 is the true hypothesis.

Because of this behavior of the likelihood ratio, the classical statistician can determine a rejection region in terms of LR which corresponds to some value of α. This rejection region is of the form

$$LR \leq c, \tag{8.22.2}$$

where c is some constant which depends on the choice of α. If c is known and LR is known, the statistician just compares the two to determine whether to accept or reject the hypothesis.

In terms of r, the number of successes in a sample of size N, what would the rejection region look like? Looking at the two hypotheses, it is obvious that the rejection region would be of the form

$$r \geq d, \tag{8.22.3}$$

where d is some constant which depends on the choice of α. Is this equivalent to the rejection region given by Equation (8.22.2)? For sampling from a Bernoulli process with fixed N, the binomial distribution is applicable, and the likelihood ratio is

$$LR = \frac{f(y \mid p = .40)}{f(y \mid p = .60)} = \frac{\binom{N}{r}(.40)^r(.60)^{N-r}}{\binom{N}{r}(.60)^r(.40)^{N-r}}.$$

Canceling and combining terms, we get

$$LR = \frac{(.40)^{2r-N}}{(.60)^{2r-N}} = \left(\frac{.40}{.60}\right)^{2r-N} = \left(\frac{2}{3}\right)^{2r-N}.$$

For a fixed N, what happens to LR as r is varied? As r gets larger, LR gets smaller (because the number being raised to the $2r - N$ power is $\frac{2}{3}$, which is less than one). Conversely, as r gets smaller, LR gets larger. If you are not convinced that this is so, try some specific values of r (for fixed N) and calculate LR. Now, according to Equation (8.22.2), H_0 will be rejected when LR is small. But when LR is small, r must be large, so the rejection regions given by Equations (8.22.2) and (8.22.3) are equivalent. In order for LR to be smaller than some constant c, r must be greater than some other constant d.

In general, the likelihood-ratio approach to hypothesis testing suggests that for two hypothesis H_0 and H_1, H_0 will be rejected if

$$LR = \frac{f(y \mid H_0)}{f(y \mid H_1)} \leq c,$$

where c is some constant which depends on α. As long as H_0 and H_1 are exact hypotheses, LR is uniquely determined. If either H_0 or H_1 is inexact, however, LR is not unique, because the likelihoods are not unique. If the H_0 in the above example were of the form

$$H_0: p \leq .40,$$

how could the likelihood in the numerator of LR be uniquely determined? The solution is to take the largest of the potential likelihoods, that is, to find the value of p which maximizes $f(y \mid p)$, subject to the constraint that p must be no greater than .40. Note the similarity to the principle of maximum likelihood. You need not be concerned about the rationale behind this solution, although it is worthwhile to reiterate at this point that all of the specific tests discussed in Chapter 7 are based on generalizations of this likelihood-ratio approach.

The relation of the Bayesian approach to the likelihood-ratio concept should be clear from Section 8.21. Instead of looking at just LR, the Bayesian considers $\Omega'' = \Omega' \times LR$. This means that the results will be the same only if the prior odds ratio, Ω', is equal to one. This might be thought of as a diffuse prior state of information, since neither hypothesis is favored. If the prior odds ratio differs from one, the posterior odds ratio differs from the likelihood ratio, and hence Bayesian and classical inferences will not necessarily be the same.

8.23   THE POSTERIOR DISTRIBUTION AND HYPOTHESIS TESTING

For exact hypotheses, the Bayesian approach to hypothesis testing can be presented in terms of odds ratios, as we have seen. In virtually all applications, however, at least one of the hypotheses is inexact. Classical procedures for handling such problems were presented in Chapter 7. Can a Bayesian approach be used in such situations? We will first briefly review the classical approach based on p-values and then point out how similar inferences can be made from the posterior distribution.

When we are interested in a particular hypothesis, we ask this question: how likely or unlikely does the hypothesis appear on the basis of the available information? The concept of a p-value, introduced in Section 7.20, enables a classical statistician to answer this question. Essentially, the p-value calculated from a particular sample measures the chance of obtaining a sample result as "unusual" as the one actually observed, given that the hypothesis is true. In this context, "unusual" means "extreme," and depends both on the hypothesis of interest, H_0, and the alternate hypothesis, H_1. For example, suppose that we are sampling from a normal process and the following hypotheses are of interest:

$$H_0 : \mu \le 100$$

$$H_1 : \mu > 100.$$

It is assumed that the variance of the process is known to be equal to 400, so the standard deviation is 20. A sample of size 4 is taken, with $M = 106$. If $\mu = 100$, how "unusual" is this sample result? The corresponding standardized value is

$$z = \frac{M - 100}{20/\sqrt{4}} = \frac{106 - 100}{10} = .60.$$

Since the alternative is one-tailed (to the right), the p-value is equal to $P(z \geq .60)$, which from the tables of the standard normal distribution is .274. It is interesting to note that if the alternative were one-tailed in the other direction, the p-value would be $P(z \leq .60) = .726$. In the case of a two-tailed alternative, the p-value would be $2P(z \geq .60) = 2(.274) = .548$. The p-value clearly depends on the alternative hypothesis as well as on H_0 and on the sample results.

For the moment let us ignore two-tailed tests. Assume that the hypotheses are either of the form

$$H_0: \mu \geq \mu_0$$

and

$$H_1: \mu < \mu_0$$

or of the form

$$H_0: \mu \leq \mu_0$$

and

$$H_1: \mu > \mu_0.$$

Since the Bayesian is willing to make probability statements about μ, he can make direct probability statements concerning the two hypotheses. If he has a probability distribution for μ, he can use this distribution to calculate probabilities such as

$$P(H_0 \text{ is true}) = P(\mu \leq \mu_0)$$

and

$$P(H_1 \text{ is true}) = P(\mu > \mu_0). \tag{8.23.1*}$$

In the example involving weekly sales which was presented in Section 8.13, suppose that the retailer was interested in the hypotheses

$$H_0: \mu \leq 1300$$

and

$$H_1: \mu > 1300,$$

where μ is the mean of the distribution of weekly sales. His posterior distribution is a normal distribution with mean 1387.5 and standard deviation 30.6. From this distribution,

$$P(H_0) = P(\mu \leq 1300) = P\left(\frac{\mu - \mu''}{\sigma''} \leq \frac{1300 - 1387.5}{30.6}\right)$$

$$= P(z \leq -2.86) = .0021$$

and

$$P(H_1) = P(\mu > 1300) = P(z > -2.86) = .9979.$$

Under the retailer's posterior distribution, the hypothesis H_0 appears extremely unlikely.

Let us reconsider the example presented earlier in this section, but from a Bayesian point of view instead of a classical point of view. Under a diffuse prior distribution, will the result be the same as that of the classical p-value approach? If we take as the prior distribution the improper diffuse normal distribution with $N' = 0$, the posterior distribution will be normal with

$$M'' = M = 106 \quad \text{and} \quad \sigma'' = \sigma/\sqrt{N''} = \sigma/\sqrt{N} = 20/2 = 10.$$

Then from the normal tables,

$$P(H_0) = P(\mu \le 100) = P\left(z \le \frac{100 - 106}{10}\right) = P(z \le -.60) = .274,$$

and

$$P(H_1) = P(z > -.60) = .726.$$

The posterior probability that H_0 is true is equal to the classical p-value. This result should not be surprising, since under a diffuse prior distribution the two inferences are based on the same information.

Although the *results* are identical, the *interpretation* is not. In fact, the procedure used to arrive at the results is different in the two instances. The difference is demonstrated in Figures 8.23.1 and 8.23.2. The curve centered about $\mu = 106$ (Figure 8.23.2) is the posterior distribution of the Bayesian, and the shaded area under this curve represent the posterior probability that H_0 is true. The density function centered about $M = 100$ (Figure 8.23.1) is the sampling distribution of M, given that $\mu = 100$. The cross-hatched area represents the p-value, or the "unusualness" of the sample result given the sampling distribution specified by H_0.

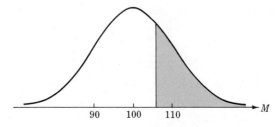

Figure 8.23.1

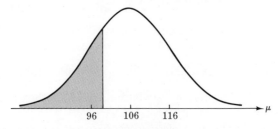

Figure 8.23.2

If the prior distribution is diffuse, then the results obtained from the classical p-value approach in a one-tailed test are identical (or, in some cases, nearly identical) to the results obtained using the posterior distribution. This statement holds for hypotheses concerning parameters other than μ, although the examples given all involve μ, the mean of a normal process with known variance. If the prior distribution is not diffuse, as is usually the case, the two methods may well produce different results. It is important to emphasize that the comparison between Bayesian and classical procedures in this section applies only to one-tailed tests. In the next section we will discuss the problems that arise for two-tailed tests.

8.24 THE POSTERIOR DISTRIBUTION AND TWO-TAILED TESTS

The preceding section involved one-tailed tests, demonstrating that the Bayesian can make probability statements about the truth of the hypotheses H_0 and H_1. Could this apply to two-tailed tests as well? Consider the sales example. What if the retailer had been interested in a two-tailed test of the hypotheses

$$H_0 : \mu = 1300$$

and $$H_1 : \mu \neq 1300?$$

From the posterior distribution,

$$P(H_0) = P(\mu = 1300) = P(z = -2.86) = 0$$

and $$P(H_1) = P(\mu \neq 1300) = P(z \neq -2.86) = 1.$$

At first glance, it appears that H_0 is impossible. The posterior distribution must not agree at all with the hypothesis. But suppose that $M'' = 1300$ (certainly no other value of M'' would make H_0 more likely). Then

$$P(H_0) = P(\mu = 1300) = P(z = 0) = 0$$

and $$P(H_1) = P(\mu \neq 1300) = P(z \neq 0) = 1.$$

H_0 is still an essentially "impossible" event! This is certainly a strange result.

The explanation for the above result is quite simple. The hypothesis H_0 consists of just a single point. The posterior distribution of μ is continuous, and thus the probability of any single value of μ is equal to zero. This does not reflect a weakness in the Bayesian approach; instead, it reflects the absurdity of the hypothesis H_0. As we pointed out in Chapter 7, there is a difference between statistical hypotheses and the corresponding real-world hypotheses, and exact hypotheses such as $\mu = 1300$ are generally unrealistic and should be modified. What the retailer probably means by

H_0 is "the mean is close to the value 1300." He surely cannot literally mean that "the mean is *exactly* equal to the value 1300." Under the classical theory of hypothesis testing, it is convenient to specify the hypothesis as

$$H_0: \mu = 1300$$

even though this statement, taken literally, is quite unrealistic.

If the posterior distribution is strictly continuous, it makes no sense for the Bayesian to consider such a hypothesis. One alternative would be to place a mass of probability at $\mu = 1300$ when assessing the prior distribution. The distribution would then be a "mixture" of distributions, the discrete distribution consisting of the mass at 1300 and the continuous distribution consisting of a normal distribution with mean M' and variance σ'. The resulting posterior distribution would also be a "mixture" of distributions.

This solution is just as unrealistic as H_0 is. Why should the retailer's distribution be altered to fit the hypothesis? It should not, unless such a distribution truly effects his information about μ. A better solution from the Bayesian point of view would be to modify the statistical hypotheses H_0 and H_1 so that they correspond with the real-world hypotheses. The retailer's hypotheses might be expressed as follows:

$$H_0: 1250 \leq \mu \leq 1350$$

$$H_1: \mu < 1250 \text{ or } \mu > 1350.$$

Now, from the retailer's posterior distribution,

$$P(H_0) = P(1250 \leq \mu \leq 1350)$$

$$= P\left(\frac{1250 - 1387.5}{30.6} \leq \frac{\mu - M''}{\sigma''} \leq \frac{1350 - 1387.5}{30.6}\right)$$

$$= P(-4.49 \leq z \leq -1.21) = .1131,$$

and

$$P(H_1) = P(\mu < 1250) + P(\mu > 1350)$$

$$= 1 - P(1250 \leq \mu \leq 1350)$$

$$= 1 - .1131 = .8869.$$

By modifying the hypotheses and making them more realistic, the statistician can make inferences from the posterior distribution when the alternative hypothesis is two-tailed. In general, if he is interested in a parameter θ and specifies the hypotheses

$$H_0: \theta_0 \leq \theta \leq \theta_1$$

and

$$H_1: \theta < \theta_0 \text{ or } \theta > \theta_1, \qquad (8.24.1)$$

where θ_0 and θ_1 are constants, then the posterior distribution $f''(\theta \mid y)$ can be used to determine

$$P(H_0) = P(\theta_0 \leq \theta \leq \theta_1)$$

and
$$P(H_1) = P(\theta < \theta_0) + P(\theta > \theta_1) = 1 - P(H_0).$$

As long as θ_0 and θ_1 are not equal, these probabilities should prove meaningful to the statistician. If they are equal, then $P(H_0) = 0$ regardless of the observed sample outcome if the distribution is continuous.

The hypothesis H_0 need not consist of just a single interval; it could easily consist of a series of intervals. If the entire set of possible values of θ is partitioned into two mutually exclusive and exhaustive sets A and B, the hypotheses can be written in the form

$$H_0:\theta \in A$$

and
$$H_1:\theta \in B. \tag{8.24.2*}$$

Then the posterior probabilities of interest are

$$P(H_0) = P(\theta \in A)$$

and
$$P(H_1) = P(\theta \in B) = 1 - P(H_0). \tag{8.24.3*}$$

No restrictions are placed on A and B other than that they must be mutually exclusive and exhaustive. In the modified two-tailed test,

$$A = \{\theta \mid \theta_0 \leq \theta \leq \theta_1\} \quad \text{and} \quad B = \{\theta \mid \theta < \theta_0 \text{ or } \theta > \theta_1\}.$$

In the example of such a test which is presented above,

$$A = \{\mu \mid 1250 \leq \mu \leq 1350\} \quad \text{and} \quad B = \{\mu \mid \mu < 1250 \text{ or } \mu > 1350\}.$$

The restriction that A and B must be exhausitive is not crucial to the analysis and can be eliminated. Of course, this would mean that the sum of the probabilities $P(H_0)$ and $P(H_1)$ would not necessarily equal one.

It should be pointed out that Equations (8.24.2) and (8.24.3) are applicable whether θ is discrete or continuous. In our examples, it has been convenient to use the continuous case. However, the discrete case can also be handled by this framework. Also, in the discrete case it may be reasonable for one or both of the sets A and B to consist of just a single value of θ. Consider the following hypotheses:

$$H_0:\theta = 10$$

$$H_1:\theta \neq 10.$$

If θ is continuous, then $P(H_0) = P(\theta = 10) = 0$, and it makes no sense to compare these two hypotheses from a Bayesian viewpoint. However, if θ is discrete, then $P(H_0)$ may be greater than zero, in which case it *does* make sense to make inferences about the two hypotheses based on a prior or posterior distribution of θ.

The framework provided by Equations (8.24.2) and (8.24.3) is quite general and is much more flexible than the classical approach. Classical procedures could be developed to handle most tests of this nature, but they would not be as easy to apply or as intuitively appealing as the Bayesian approach. Given the posterior distribution, the Bayesian need only calculate certain posterior probabilities, which is not too difficult, especially if the posterior distribution is a member of a conjugate family (or, for that matter, of *any* family for which tables are available).

8.25 BAYESIAN INFERENCE FOR SEVERAL HYPOTHESES

There is no reason why the statistician should be limited to just two hypotheses. He may have several hypotheses in mind and may want to know which of the hypotheses appears to be the most likely to be true. Extending the results of the previous section, consider the set of hypotheses $\{H_1, H_2, \cdots, H_K\}$, where H_i is of the form

$$H_i : \theta \in A_i, \qquad i = 1, 2, \cdots, K. \tag{8.25.1}$$

The statistician can compute the posterior probabilities of the K hypotheses:

$$P(H_i) = P(\theta \in A_i), \qquad i = 1, 2, \cdots, K. \tag{8.25.2}$$

The K sets $A_1, A_2, \cdots, A_K$ do not have to be mutually exclusive and exhaustive. If they are, of course, the sum of the probabilities,

$$\sum_{i=1}^{K} P(H_i),$$

is equal to one. If the statistician just wants to know which hypothesis is most likely to be true, he can compare the $P(H_i)$ and find the hypothesis with the largest posterior probability. For example, the hypotheses might correspond to competing economic theories, in which case the statistician is interested in knowing which theory seems to be the most likely according to the posterior distribution.

The interest in problems involving several hypotheses is primarily decision-oriented. Recall that a test of a hypothesis H_0 versus an alternative hypothesis H_1 is a decision-making procedure. There are two actions available to the statistician: accept H_0 or reject H_0 in favor of H_1 (assuming that the statistician must make a decision and cannot suspend judgment).

In making a decision, the statistician should consider the probabilities and losses associated with the two possible errors. If there are K hypotheses, we can think of a K-action problem, where the statistician must choose one of the K hypotheses. Again, his decision should be based both on the posterior probabilities associated with the hypotheses, which we have discussed, and on the losses due to erroneous decisions, which we have not discussed.

8.26 THE POSTERIOR DISTRIBUTION AND DECISION THEORY

This chapter has not been decision-oriented, although the potential application of Bayesian techniques in decision-making situations has been suggested in some of the examples. In the next chapter we will discuss the theory of making decisions under conditions of uncertainty. In the example in Section 8.4, a decision has to be made about a production process. The uncertainty involves the proportion of defectives, p, produced by the process. If the statistician knew p for certain, the choice of an action would be easy. He does not know p for certain, but he *does* have some information regarding p. This information is represented by his posterior distribution of p. As a result, the posterior distribution is an important input to the decision problem.

The posterior distribution, then, is one input to a problem of decision-making under uncertainty. A second important input, one which we have not discussed in any detail, is a payoff function or a loss function. These functions tell the decision-maker what the consequences to him will be if he takes a particular action and at the same time a particular value of θ occurs, where θ is the parameter of interest (the posterior distribution will then be a distribution of θ).

Given a set of possible actions, a set of payoffs or losses, and a posterior distribution, the statistician can use decision theory to choose one of the possible actions as the "best" action. It should be mentioned that inferential procedures such as estimation and hypothesis testing can be thought of as problems of decision-making under uncertainty. In estimation, the actions correspond to the available estimators; the statistician must decide, for example, between the mean, median, or mode of his posterior distribution (or, in the case of a classical statistician, his likelihood function). In hypothesis testing, the actions correspond to the hypotheses; the statistician must choose a hypothesis. If there are just two hypotheses, H_0 and H_1, he must decide whether to accept H_0 or to accept H_1. One important input to such decision problems is the posterior distribution, and that is why this chapter has been devoted to the revision of prior probabilities and the determination of posterior distributions. Of course, if no sample information is available, then it is the prior distribution which is an input to the decision-making problem. Prior or posterior, one function of the distribution is to represent the decision maker's state of uncertainty, or state of information, in a situation in which he must make a decision under uncertainty.

EXERCISES

1. Suppose that it is believed that the proportion of consumers in a large population that will purchase a certain product is either .2, .3, .4, or .5. Furthermore, the prior probabilities for these four values are $P(.2) = .2$, $P(.3) = .3$, $P(.4) = .3$, and $P(.5) = .2$. A sample of size 10 is taken from the population, and of the ten consumers, 3 state that they will buy the product. What are the posterior probabilities? In calculating the posterior probabilities, what assumptions did you need to make?

2. You feel that the probability of a head on a toss of a particular coin is either .4, .5, or .6. Your prior probabilities are $P(.4) = .1$, $P(.5) = .7$, $P(.6) = .2$. You toss the coin three times and obtain one head and two tails. What are the posterior probabilities? If you then toss the coin three *more* times and once again obtain one head and two tails, what are the posterior probabilities? Also compute the posterior probabilities by pooling the two samples and revising the original probabilities just once; compare with your previous answer.

3. Explain the statement, "the terms *prior* and *posterior*, when applied to probabilities, are relative terms, relative to a given sample." Is it possible for a set of probabilities to be both prior *and* posterior probabilities? Explain.

4. In Bayes' theorem, why is it necessary to divide by Σ (prior probability) (likelihood) in the discrete case and by $\int$ (prior density) (likelihood) in the continuous case?

5. Discuss the importance of conjugate families of distributions in Bayesian statistics.

6. In Exercise 1, suppose that the prior distribution could have been represented by a beta distribution with $r' = 4$ and $N' = 10$. Find the posterior distribution. Also, find the posterior distribution corresponding to the following beta prior distributions:
   (a) $r' = 2$ $N' = 5$
   (b) $r' = 8$ $N' = 20$
   (c) $r' = 6$ $N' = 15$.
   In each of the four prior distributions considered in this exercise, the mean of the prior distribution is .40. How, then, do you explain the differences in the means of the posterior distributions?

7. In Exercise 2, suppose that you feel that the mean of your prior distribution is 1/2 and the variance of the distribution is 1/20. If your prior distribution is a member of the beta family, find r' and N' and determine the posterior distribution following the sample of size 6.

8. In Exercise 2, suppose that you feel that in comparing $p = .6$ and $p = .4$, the odds are 3-to-2 in favor of $p = .6$. Furthermore, in comparing $p = .4$ and $p = .5$, the odds are 5-to-1 in favor of $p = .5$. On the basis of these odds, find your prior distribution for p. Does this suggest one possible approach to the assessment of a prior distribution in the discrete case?

9. In sampling from a Bernoulli process, the posterior distribution is the same whether we sample with N fixed (binomial sampling) or with r fixed (Pascal sampling). Explain why this is true. Suppose the statistician merely sampled until he was tired and decided to go home—would the posterior distribution still be the same (that is, is the stopping rule noninformative)?

10. Try to assesses a subjective distribution of p, the probability of rain tomorrow. Can you find a beta distribution which expresses your judgments reasonably well?

11. A production manager is interested in the mean weight of items turned out by a particular process. He feels that the weight is normally distributed with standard deviation 2, and his prior distribution for μ is a normal distribution with mean 110 and variance .4. He randomly selects five items from the process and weighs them, with the following results: 108, 109, 107.4, 109.6, 112. What is the production manager's posterior distribution?

12. You are attempting to assess a prior distribution for the mean of a process, and you decide that the .25 fractile if your distribution is 160 and the .60 fractile is 180. If your prior distribution is normal, determine the mean and variance.

13. In reporting the results of a statistical investigation, a statistician reports that his posterior distribution for μ is a normal distribution with mean 52 and variance 10, and that his sample of size 4 with sample mean 55 was taken from a normal population with variance 100. On the basis of this information, can you determine the statistician's prior distribution?

14. Explain the parametrization of a normal prior distribution in terms of N' and M', as given in Section 8.12. How does this parametrization help us to see the relative weights of the prior and the sample information in computing the mean of the posterior distribution? For the prior and posterior distributions in Exercise 11, express the distributions in terms of N' and M'. How could you interpret this prior distribution in terms of an equivalent sample?

15. Suppose that a statistician decides that his prior distribution for a certain parameter θ is an exponential distribution and that the .67 fractile of this distribution is 2. Find the exact form of the distribution.

16. In assessing a distribution for the mean height of a certain population of college students, a person decides that his distribution is normal, the median is 70 inches, the .20 fractile is 67 inches, and the .80 fractile is 72 inches. Can you find a normal distribution with these fractiles? Comment on the ways in which the person could make his assessments more consistent.

17. Comment on the following statement: "We cannot speak of sensitivity except in connection with a particular decision-making situation." Can you think of an example in which the decision-making procedure would be quite insensitive to changes in the prior distribution? An example in which it would be quite sensitive?

18. Comment on the following statement: "An informationless, or diffuse, prior state of information is not informationless in the usual meaning of the word, but rather in a relative sense."

19. Give a few examples of situations in which your prior distribution would effectively be diffuse, and a few examples in which it would definitely *not* be diffuse relative to a given sample.

20. The beta distribution with $r' = N' = 0$ and the beta distribution with $r' = 1$ and $N' = 2$ have both been used by statisticians as "diffuse" beta distributions. Discuss the advantages and disadvantages of each.

21. *Carefully* distinguish between a classical confidence interval and a Bayesian credible interval, and also between the classical and Bayesian approaches to point estimation.

22. In Exercise 11, find
    (a) a point estimate for μ based on the prior distribution alone
    (b) a point estimate for μ based on the sample information alone
    (c) a point estimate for μ based on the posterior distribution
    (d) a 90 percent credible interval for μ based on the prior distribution alone
    (e) a 90 percent confidence interval for μ based on the sample information alone
    (f) a 90 percent credible interval for μ based on the posterior distribution.
    Comment on any differences in your answers to (a)–(c) and to (d)–(f).

23. The random variable X is normally distributed with unknown mean μ and known variance $\sigma^2 = 225$. A sample of size $N = 9$ from X is observed, with the following sample values:

    42, 56, 68, 56, 48, 36, 45, 71, and 64.

    (a) Name three intuitively reasonable point estimators for μ, and calculate their values for this particular sample.
    (b) Find an 86.6 percent confidence interval for μ.
    (c) Suppose that your prior judgments about μ could be well represented by a normal distribution with mean 50 and variance 14. What is your posterior distribution for μ?

(d) Determine a point estimate for μ based on the posterior distribution obtained in (c).

(e) Find a 38.3 percent credible interval for μ.

24. Suppose that a marketing manager is interested in p, the proportion of consumers that will buy a particular new product. He considers the following two hypotheses:

$$H_0: p = .10$$

$$H_1: p = .20.$$

His prior probabilities are $P(p = .10) = .85$ and $P(p = .20) = .15$, and a random sample of 8 consumers results in 3 consumers who state that they will buy the product if it is marketed.

(a) What is the prior odds ratio?

(b) What is the likelihood ratio?

(c) What is the posterior odds ratio?

(d) The manager decides that the posterior probability of H_0 must be no larger than .40 to make it worthwhile to market the product. Should the product be marketed?

(e) On the basis of the sample information alone, should the product be marketed? If the decision is made on this basis, what is implied about the prior distribution?

25. Discuss: "The Bayesian approach to statistics can be thought of as an *extension* of the classical approach."

26. In Exercise 24, it would be more realistic to consider hypotheses such as the following:

$$H_0: p \le .15$$

$$H_1: p > .15.$$

If the prior distribution is a beta distribution with $r' = 1$ and $N' = 9$, find the posterior distribution. How could you use the posterior distribution to find the posterior odds ratio of H_0 to H_1?

27. In Exercise 11, the manager is interested in the hypotheses $H_0: \mu \le 110$ and $H_1: \mu > 110$. From the prior distribution, find the prior odds ratio; also, from the posterior distribution, find the posterior odds ratio. If the losses involved in a decision-making problem involving the production process are such that H_0 should be accepted only if the posterior odds ratio is greater than 3, what decision should be made?

28. A statistician is interested in the mean μ of a normal population, and his prior distribution is normal with mean 800 and variance 12. The variance of the population is known to be 72. How large a sample would be needed to guarantee that the variance of the posterior distribution is no larger than 1? How large a sample would be needed to guarantee that the variance of the posterior distribution is no larger than .1?

29. Prove that the posterior odds ratio of one hypothesis to another is equal to the product of the prior odds ratio and the likelihood ratio.

30. In testing a hypothesis concerning the mean of a normal process with known variance against a one-tailed alternative, discuss the relationship between the classical p-value and the posterior probability $P''(H_0)$, (a) if the prior distribution is diffuse, (b) if the prior distribution is not diffuse.

31. Comment on the following statement: "A hypothesis such as $\mu = 100$ is not realistic and should be modified somewhat; otherwise, the Bayesian approach to hypothesis testing may not be applicable."

32. An automobile manufacturer claims that the average mileage per gallon of gas for a particular model is normally distributed with $M' = 20$ and $\sigma' = 4$, provided that the car is driven on a level road at a constant speed of 30 miles per hour. A rival manufacturer decides to use this as a prior distribution and to obtain additional information by conducting an experiment. In the experiment, it is assumed that the variance of mileage is $\sigma^2 = 96$ and the mileage is normally distributed. The experiment is conducted on 10 randomly chosen cars, with the sample mean (the average mileage in the sample) equaling 18. Find the posterior distribution for μ, the average mileage per gallon of gas. On the basis of this posterior, what is the probability that μ is greater than or equal to 20? What is the probability that μ is between 19 and 21?

33. In Exercise 11, suppose that the manager is interested in the hypotheses $H_0: \mu = 110$ and $H_1: \mu \neq 110$. From the posterior distribution, what is the posterior odds ratio of H_0 to H_1? What is the posterior odds ratio if the hypotheses are $H_0: 109 < \mu < 111$ and $H_1: \mu \leq 109$ or $\mu \geq 111$?

34. Suppose that a statistician is interested in $H_0: \mu = 50$ and $H_1: \mu \neq 50$. His prior distribution consists of a mass of probability of .25 at $\mu = 50$, with the remaining .75 of probability distributed uniformly over the interval from $\mu = 40$ to $\mu = 60$. Find the prior probability that:
    (a) $45 < \mu < 55$
    (b) $47 \leq \mu \leq 50$
    (c) H_0 is exactly true
    (d)   the hypothesis $49 < \mu < 51$ is true.

35. Suppose that X is normally distributed with mean μ and variance $\sigma^2 = 400$. The prior distribution of μ is normally distributed with mean $M' = -60$ and variance $\sigma'^2 = 40$. Furthermore, a sample of size 20 is taken, with sample mean $M = -69$.
    (a)   Find the posterior distribution of μ.
    (b)   Find a 68 percent credible interval for μ.
    (c)   Find $P(H_0)$ from the posterior distribution, where H_0 is the hypothesis that μ is greater than or equal to -70.

36. In Exercise 35, suppose that three hypotheses were under consideration:

$$H_1: \mu \leq -70$$

$$H_2: -70 < \mu < -60$$

$$H_3: \mu \geq -60.$$

   (a) Find the posterior probabilities of these three hypotheses.
   (b) If the decision problem at hand involves three actions (corresponding to the three hypotheses), and the hypothesis to be selected is the one which is most probable, which hypothesis would be selected under the posterior distribution?
   (c) In part (b), which hypothesis would be selected on the basis of the *prior* distribution?

37. Suppose that a statistician is interested in the difference in the means of two normal populations, μ_1 and μ_2. Let $\mu = \mu_1 - \mu_2$. Each population has variance 100, and the two populations are independent. A sample of size 25 is taken from the first population, with sample mean $M_1 = 80$, and a sample of size 25 is taken from the second population, with sample mean $M_2 = 60$.
   (a) Using the classical approach to hypothesis testing presented in Chapter 7, test $H_0: \mu_1 - \mu_2 \leq 0$ against the alternative $H_1: \mu_1 - \mu_2 > 0$, with $\alpha = .10$.
   (b) Let $\mu = \mu_1 - \mu_2$, and suppose that the prior distribution of μ is normally distributed with mean $M' = 10$ and variance $\sigma'^2 = 50$. Find $P'(H_0)$ and $P'(H_1)$.
   (c) Find the posterior distribution of $\mu = \mu_1 - \mu_2$.
   (d) From the posterior distribution, find the posterior odds ratio of H_0 to H_1.

38. Exercise 37 suggests a Bayesian approach to inferences regarding the difference between two means, under the conditions that the two populations of interest are normally distributed and the variances are known. Using the notation of Exercise 37, determine a general formula for finding the posterior distribution of the difference between two means if the prior distribution has mean M' and variance σ'^2.

39. Try to assess a probability distribution for T, the maximum temperature tomorrow in the city where you live. In assessing the distribution, use two or three of the methods proposed in the text and compare the results. Then, save the distribution and look at it again after you find out the true value of T. Do this for three or four consecutive days and comment on any difficulties which you encounter in attempting to express your subjective judgments in probabilistic form.

40. Follow the procedure in Exercise 39 for the variable D, the daily *change* in the price of one share of IBM common stock.

9

DECISION
THEORY

In Chapters 5–8 we discussed sampling distributions, classical estimation and hypothesis testing, and Bayesian methods. At various points in these chapters, we hinted at the decision-theoretic approach to problems of inference and the use of likelihood functions and posterior distributions in making decisions. These references have occurred more and more frequently, and the student may have found himself wondering if he would ever get to the material on decision theory. We feel that a grasp of the concepts of inferential statistics, both classical and Bayesian, is a useful prerequisite to a full understanding of the many statistical implications of decision theory. The student should have that grasp by now, so we are finally ready to take the plunge.

9.1 INTRODUCTION

Why are we interested in decision theory? Life is a constant sequence of decision-making situations. Every action you take, with the exception of a few involuntary physiological actions, such as breathing, can be thought of as a decision. Of course, most of these decisions are quite minor decisions, because the consequences involved are not very important. For example, consider the decision whether to walk up a flight of stairs or take the elevator. Unless you have a heart condition or some other physical reason for taking the elevator, this is not an important decision. The relevant factors entering into the decision include the exertion required to walk up the stairs, the availability of the elevator, whether you are in a hurry, and so on. Minor decisions such as this are usually made intuitively without conscious thought. Other examples are the decision to smoke a cigarette,

510

drive home via a certain route, have a particular beverage with your evening meal, and so on. For some individuals, these decisions may require some thought. A person trying to "break the smoking habit" may find the decision as to whether or not to smoke a cigarette to be a difficult decision. For most people, though, such decisions are usually made out of habit. They do not consciously think about the various possible actions because they have faced the situation (or similar situations) many times in the past.

Other decisions require some thought, but can still be made intuitively. Examples are a choice of brands of a product, a choice of movies on a Saturday night, and so on. Eventually, you will face a decision-making problem which requires some serious thought. Consider a major purchasing decision, such as the decision to buy a new car. In such a situation, you evaluate such factors as the condition of your present car, the cost of a new car, the enjoyment of owning a new car, and so on. You probably would make this decision subjectively, but it is the type of decision for which the formal decision theory to be discussed in this chapter might prove useful. Similar problems are investment decisions (either by individuals or by corporations), medical decisions (whether or not to undergo surgery), and job decisions (for the individual, whether or not to accept a job offer; for the corporation, whether or not to make such an offer). It may not always be easy to formally specify the problem and apply the formal decision-making procedures. Whether to proceed at an informal level or at a formal level is a decision in itself. Whenever the consequences of making a wrong decision could be quite serious, however, decision theory should prove worthwhile.

9.2 CERTAINTY VERSUS UNCERTAINTY

This chapter deals with decision-making under the condition of *uncertainty*. This condition refers to uncertainty about the true value of a variable related to the decision, or, in simpler terms, uncertainty about the actual state of the world. Formally, a consequence of a decision, which may be expressed in terms of a payoff or a loss to the decision maker, is the result of the interaction of two factors: (1) the decision, or the action, selected by the decision maker; and (2) the state of the world which actually occurs. For example, suppose that you have to decide whether or not to carry an umbrella when you leave home in the morning. There are two possible actions: carry the umbrella or leave it at home. Also, there are two relevant states of the world: rain or no rain. If you knew for certain that it would rain, you would probably carry your umbrella. If you knew for certain that it would not rain, you would leave the umbrella at home.

In these situations, you are making a decision under the condition of *certainty*.

For another example, suppose that you are faced with a decision concerning the purchase of common stock. For the sake of simplicity, we shall assume that you intend to invest exactly $1000 in a single common stock and hold the stock for exactly one year, at which time you will sell it at the market price. Furthermore, assume that you are considering only three stocks, A, B, and C, each of which currently sells for $50 a share. Thus, you intend to buy 20 shares of one of the three stocks. Your selection of a single stock from the three constitutes your *decision*, or *action*. The prices of the three stocks one year from now constitute the *state of the world*. The combination of your action and the state of the world determines your payoff. Suppose that at the end of one year, the prices of stocks A, B, and C are $60, $40, and $50. The payoffs for the three possible actions are then +$200, −$200, and $0, respectively. For this state of the world, the best decision is to buy stock A, for this results in the highest payoff. If you knew the state of the world with certainty, the decision could be made in this manner, and this is once again decision-making *under certainty*. In this example, if you knew what the prices of the stocks would be in one year, then you would simply buy the stock which would give you the maximum payoff—that is, the stock which would increase in value the most during the coming year.

In general, the actual state of the world is *not* known with certainty. On many mornings, you may not be at all certain whether or not it will rain that day (technically, you could *never* be *absolutely certain* about the weather, although you may often feel certain for all practical purposes). You listen to the weather report on the morning newscast and you look out the window at the sky, so that you have some information, which may be reflected in probabilistic terms. Subjectively, you can determine your degree-of-belief probabilities for the two events "rain" and "no rain." Or, if your local weather bureau issues "precipitation probabilities" and you have confidence in the weather bureau, you may decide to use these probabilities. At any rate, you are operating under the condition of *uncertainty*, and your uncertainty can be represented by your probabilities for the two possible states of the world.

How about the stock example? Unless you are fortunate enough to receive copies of the *Wall Street Journal* one year in advance or you possess extrasensory perception, you could not know for certain what the prices of the three stocks would be one year hence. Thus, treating the purchase of common stock as a decision under certainty is not at all realistic. You no doubt have some ideas regarding the prices, ideas which may be based on your impressions of the economy in general, particular industries, and specific firms within industries. Your knowledge may stem from years and years of careful study of the stock market, or it may consist of a hot tip from a "friend." At any rate, you have some knowledge, but not perfect knowledge. Once again you are faced with a problem of decision-making *under uncertainty*. Can you express your uncertainty in probabilistic terms, as in the umbrella example? Conceptually, yes; in reality, perhaps. There are three variables of interest here: the prices one year hence of the three stocks. What is needed is a *joint* distribution of the three random variables. If the variables were independent, you could assess three individual marginal distributions and multiply them together to get the joint distribution. For these three variables, the assumption of independence is, to say the least, an unrealistic assumption. There are other more realistic assumptions which provide some shortcuts in the assessment procedure, but this is not the point of this discussion. The point is this: **in general, the problem of making decisions is more complex and more difficult under uncertainty than it is under certainty.** As a result, it is important when formally specifying a decision problem to keep it as simple as possible, only including the most relevant variables. Otherwise it may prove too difficult to apply the techniques of this chapter. This is a drawback which limits the application of decision theory somewhat, although modern high-speed computers permit the solution of complex decision problems that would be impossible to solve manually or on a desk calculator. As in the stock example, the difficulty may lie in determining the inputs to the decision (such as the decision maker's prior or posterior distribution).

9.3 PAYOFFS AND LOSSES

We have noted that the consequences of a decision can be expressed in terms of either payoffs or losses. It is important to define explicitly what is meant by these two terms. A payoff can be interpreted in the usual manner; if it is expressed in monetary units, it represents the net change in your total wealth as a result of your decision and the actual state of the world. This can be either positive or negative. In the umbrella example, we will reproduce the *payoff table* presented in Section 2.20:

		State of the World	
		Rain	*No rain*
Your decision	*Carry umbrella*	−$ 1	−$1
	Do not carry umbrella	−$60	$0

A payoff table consists of the set of payoffs for all possible combinations of actions and states of the world. If there are m actions and n states of the world, the payoff table will have mn entries. In the umbrella example, there are 4 entries, all of which are negative or zero.

How can the above payoff table be converted to an equivalent *loss table*? "Loss" in this context refers to **"opportunity loss."** For any combination of an action and a state of the world, the question is: could you have obtained a higher payoff, given that particular state of the world? If the answer is no, then your loss is zero. If the answer is yes, then your loss is the positive difference between the given payoff and the *highest possible payoff under that state of the world*.

Consider the first entry in the first column of the above payoff table. It is higher than the other entry in that column, so it is the best you could do given that it rains, and the loss is zero. For the second entry in the first column, the payoff is $59 lower than the first entry, so the loss is $59. Similarly, in the second column, the first entry is $1 lower than the second entry, so the corresponding losses are $1 and $0. The **loss table** is thus:

State of the World

		Rain	*No rain*
Your decision	*Carry umbrella*	$ 0	$1
	Do not carry umbrella	$59	$0

Instead of loss table, this is sometimes called a *regret table*, since the entries reflect the decision maker's regret at not having made the decision which turned out to be optimal under the actual state of the world.

For another example, consider the stock-purchasing decision from the previous section. Furthermore, for the sake of simplicity assume that there are only four possible states of the world. In the first (State 1), the prices of the stocks A, B, and C at the end of one year are, respectively, $60, $50, and $40. For State II, the prices are $80, $60, and $40. For State III, the prices are $20, $40, and $10. Finally, for State IV, the prices are $60, $60, and $60. This results in the following payoff table, where the payoffs are expressed in dollars:

State of the World

	I	II	III	IV
Buy A	200	600	−600	200
Actions Buy B	0	200	−200	200
Buy C	−200	−200	−800	200

For example, if Stock A is bought and State II occurs, the price goes up from $50 to $80, or $30 per shape. Since 20 shares will be bought, the payoff is $600.

This payoff table corresponds to the following loss table:

State of the World

	I	II	III	IV
Buy A	0	0	400	0
Action Buy B	200	400	0	0
Buy C	400	800	600	0

The procedure for converting a payoff table to a loss table should be clear by now. First, work with each column of the table separately. In each column, find the highest payoff (which may be negative, zero, or positive). The loss corresponding to this payoff is zero, and the loss corresponding to any other payoff in the same column is the difference between that payoff and the highest payoff.

The loss tables above demonstrate some interesting characteristics of losses. First, losses cannot be negative. This can be seen from the way in which they are defined; each loss is either zero or the (positive) difference between two payoffs. Payoffs can be negative, zero, or positive; losses, on the other hand, can only be zero or positive (there is no such thing as a negative regret!). In addition, if all of the payoffs in a column are equal, the corresponding losses will all be zero. This is because no payoff is higher than another, so that if that particular state of nature occurs, it makes no difference which action was chosen.

One other concept of importance is the concept of **admissible actions,** or **admissible decisions.** An action is said to **dominate** a second action if for *each possible state of the world*, the first action leads to at least as high a payoff (or at least as small a loss) as the second action. If one action dominates the other, then it would never be reasonable to choose the second action (you could always do at least as well with the first action), so the second action is said to be **inadmissible.** Notice that this definition implies that if two actions have the same payoffs or losses for each state of the world, they dominate each other. This is not reasonable, so let us modify the definition of dominance. One action is said to dominate a second action if for each possible state of the world, the first action leads to at least as high a payoff (or at least as small a loss) as the second action, *and for at least one state of the world, the first action leads to a higher payoff* (or a *smaller loss*) *than the second action.*

In the stock purchasing example, the third action (Buy C) is dominated by both of the other two actions. The first action is better than the third action for States I, II, and III, and just as good for State IV. The same is true of the second action as compared with the third action. Hence, the third action is *inadmissible;* there is no need to consider it further. In making a decision, it is necessary to look only at *admissible acts,* acts which are not dominated by any other act.

9.4 NONPROBABILISTIC CRITERIA FOR DECISION UNDER UNCERTAINTY

If a table of payoffs or losses is known in a decision problem, and the actual state of the world is known for certain, the decision is obvious: choose the action giving the highest payoff (smallest loss) for that particular state of the world. In terms of the table, choose the action corresponding to the largest entry in the relevant column of the payoff table or to the smallest entry in the relevant column of the loss table. Because of the way in which we defined losses, the same action is selected in both cases.

Under uncertainty, however, it is not possible to confine our attention to a single column of the table. Because there is uncertainty as to which state of nature will occur, we must consider the entire table. But then it is not obvious that one particular action is a better choice than all of the other actions. It might be the best action given one state of the world, but the worst given another state of the world. The only way to formally introduce the nature of our uncertainty about the state of the world is to assess a probability distribution. Most of the available information is likely to be subjective, so that the probability distribution would be interpreted in terms of degrees of belief. Because followers of the frequency interpretation of probability are unable to use such a distribution, several decision-making criteria which are based solely on the payoff (or loss) table have been developed. These decision rules ignore the probabilistic nature of decision under uncertainty. We will discuss some of these rules in this section and then turn to probabilistic decision criteria in the next section.

Suppose that we consider the stock purchasing example and change the payoff table slightly so that the third action is admissible. Under State III, assume that the price of C is $50 rather than $10. The payoff is then $0 rather than −$800, and C is no longer dominated by A or B. Also, under State II, assume that the price of C is $30 rather than $40, which changes the payoff from −$200 to −$300. The payoff table is as follows:

State of the World

		I	II	III	IV
	Buy A	200	600	−600	200
Action	Buy B	0	200	−200	200
	Buy C	−200	−300	0	200

One decision rule, called the **maximin** rule, says: for each action, find the smallest possible payoff, and then choose the action for which this smallest payoff is largest. The smallest payoffs for the three actions are −600, −200, and −300. According to this rule, we should choose the second action. The maximin rule amounts to maximizing the minimum payoff. In effect, it tells the decision maker to assume that for each action, the worst possible state of the world will occur. This seems to be an extremely conservative approach, since it only looks at the lowest payoff in each row and ignores the magnitudes of the other payoffs. It is easy to construct payoff tables for which the maximin rule gives ridiculous results.

State of the World

100,000	−1
0	0

Action

In this payoff table, the smallest payoffs for the two actions are −1 and 0, so the second action is selected by the maximin criterion. The possible gain of $100,000 is completely ignored. These two actions could be thought of as "do something" and "stay as is." According to the maximin rule, we should never take an action if there is a possibility of a negative payoff, since we could always do better in a maximin sense by doing nothing and obtaining a payoff of zero. This is most unrealistic!

A second decision rule, called the **maximax** rule, assumes that the best will happen, not the worst. According to this rule, the decision maker should find the *largest* possible payoff for each action and then choose the action for which this largest payoff is largest. This maximizes the maximum payoff. In the stock purchasing example, the maximum payoffs are 600, 200, and 200, so that the decision maker should buy stock A. Just as the maximin rule ignores the high payoffs, the maximax rule ignores the low payoffs, assuming that for each action, the best possible state of the world will occur.

State of the World

100,000	99,999
100,001	0

Action

In this payoff table, the highest payoffs for the two actions are 100,000 and 100,001; the maximax criterion chooses the second action. Thus, in order to gain one extra dollar (from $100,000 to $100,001), the decision maker is taking the chance that he could end up with nothing, whereas he is assured of at least $99,999 if he takes the first action. In this example, then, the maximax rule appears to be unreasonable.

The maximin and maximax rules deal with the payoff matrix. A third rule, called the **minimax loss,** or **minimax regret** criterion, says: for each action, find the largest possible loss, and then choose the action for which this largest loss is smallest. For the stock purchasing example, the loss table is:

State of the World

	I	II	III	IV
Buy A	0	0	600	0
Action *Buy B*	200	400	200	0
Buy C	400	900	0	0

The largest losses for the three actions are 600, 400, and 900. The smallest of these three numbers is 400, and thus the second action is chosen by the minimax loss rule. As the name implies, this rule minimizes the maximum possible loss. The minimax loss rule is neither as conservative as the maximin rule nor as risky as the maximax rule. To illustrate this, look at the loss tables corresponding to the payoff tables (other than the stock example) given above. They are:

State of the World   *State of the World*

	0	1			1	0
Action			and	*Action*		
	100,000	0			0	99,999

In each case the minimax loss rule chooses the first action.

In the example presented in Section 8.4 involving a production process, the following payoff table was given:

| | | State of the World (Value of p) | | | |
		.01	.05	.10	.25
	Major adjustment	380	380	380	380
Action	Minor adjustment	455	375	375	375
	No adjustment	480	400	300	0

The minimum payoffs for the three actions are 380, 375, and 0. Thus, the maximin rule would advise the decision maker to make a major adjustment in the production process. This is a conservative decision: "Let us make the major adjustment so we will not have to worry about what could happen if we did not make the adjustment and p happened to be .10 or .25." The maximum payoffs for the three actions are 380, 455, and 480, and the maximax rule chooses action third action, "no adjustment." This is a risky decision: "Let us go for the highest payoff, $480, and hope that p is not .10 or .25." Notice, by the way, that the maximax rule always chooses the action corresponding to the highest payoff in the entire table.

The loss table in the production example is:

| | | State of the World | | | |
		.01	.05	.10	.25
	Major adjustment	100	20	0	0
Action	Minor adjustment	25	25	5	5
	No adjustment	0	0	80	380

The maximum losses for the three actions are 100, 25, and 380. According to the minimax loss rule, a minor adjustment should be made in the production process. In this example, the three decision rules result in three different decisions.

Other nonprobabilistic decision criteria have been proposed, some of which may be better than the above rules. For instance, one rule is a combination of the maximin and maximax rules, using an "optimism-pessimism" index. These rules can all be criticized on the grounds that they ignore the probabilistic nature of uncertainty. As a result, each nonprobabilistic criterion leads to the same decision regardless of how likely the various states of the world are. Looking at the payoff table in the stock purchasing example, the first action looks best if P (State II) $= .99$, and the third action looks best if P (State III) $= .99$. The rules considered in this section ignore such probabilities. The rules in the next section will take into consideration the probabilistic nature of the problem of decision-making under uncertainty.

9.5 PROBABILISTIC CRITERIA FOR DECISION UNDER UNCERTAINTY

If some information is available regarding the states of the world, but not perfect information, then the decision maker is operating under uncertainty. If the information can be expressed in terms of probabilities, then these probabilities can be used as inputs to the decision-making process. In particular, they can be used to calculate expected payoffs or expected losses of the potential actions. The **expected payoff** (EP) criterion says to choose the act with the highest expected payoff. Correspondingly, the **expected loss** (EL) criterion says to choose the act with the smallest expected loss. We have used both of these criteria in various examples in previous chapters.

In the umbrella example in Section 9.3, suppose that you assess

$$P(\text{Rain}) = .20 \quad \text{and} \quad P(\text{No Rain}) = .80$$

on the basis of a weather report and a look at the sky. Then the expected payoffs of the two acts are

$$\text{EP (Carry umbrella)} = .20(-\$1) + .80(-\$1) = -\$1$$

and

$$\text{EP (Do not carry umbrella)} = .20(-\$60) + .80(\$0) = -\$12.$$

Similarly, from the loss table, we have

$$\text{EL (Carry umbrella)} = .20(\$0) + .80(\$1) = \$.80$$

$$\text{EL (Do not carry umbrella)} = .20(\$59) + .80(\$0) = \$11.80.$$

The act "carry umbrella" has the larger expected payoff and the smaller expected loss. Furthermore, its EP is \$11 higher than the EP of the other action, and its EL is \$11 lower than the EL of the other action. This demonstrates a relationship between expected payoffs and expected losses: the action with the highest expected payoff will always have the lowest expected loss, and vice versa. Furthermore, if the expected payoff of one action is A units *higher* than the EP of a second action, then the EL of the first action will be exactly A units *lower* than the EL of the second action. Notice that there is no particular relationship between the *magnitudes* of expected payoffs and the *magnitudes* of expected losses. There is, however, a direct relationship between the *differences* in expected payoffs and *differences* in the corresponding expected losses.

For another example, consider the stock purchasing example, using the payoff table and loss table presented in Section 9.4. Suppose that the probabilities of States I, II, III, and IV are .1, .4, .3, and .2, respectively. The expected payoffs are then

$$\text{EP (Buy A)} = .1(200) + .4(600) + .3(-600) + .2(200) = 120,$$

$$\text{EP (Buy B)} = .1(0) + .4(200) + .3(-200) + .2(200) = 60,$$

and

$$\text{EP (Buy C)} = .1(-200) + .4(-300) + .3(0) + .2(200) = -100.$$

The expected losses are

$$\text{EL (Buy A)} = .1(0) + .4(0) + .3(600) + .2(0) = 180,$$

$$\text{EL (Buy B)} = .1(200) + .4(400) + .3(200) + .2(0) = 240,$$

and

$$\text{EL (Buy C)} = .1(400) + .4(900) + .3(0) + .2(0) = 400.$$

The decision should be to buy A on the basis of EP and EL. Of course, the expected payoffs and losses (and hence the resulting decision) depend on the values of the probabilities. Had the probabilities been .1, .1, .7, and .1, the decision would be to buy C, since the EL's are 420, 200, and 130. If the probabilities were .4, .1, .4, and .1, the decision would be to buy B, since the EL's are 240, 200, and 270. The different sets of probabilities correspond to different information about the state of the world, and different information may lead to a different decision.

How are the probabilities of the states of the world determined? In general, they are what we referred to as *posterior probabilities* in Chapter 8. As such, they represent all relevant information, sample information or otherwise. If no sample information is available, of course, then the prior probabilities are used to calculate expected payoffs and losses.

Consider once again the example involving the proportion of defective items, p, produced by a certain process. In Section 8.4, the statistician's prior probabilities are given as

$$P(p = .01) = .60,$$

$$P(p = .05) = .30,$$

$$P(p = .10) = .08,$$

and

$$P(p = .25) = .02.$$

He then observes a sample of size 10 with 3 defectives. Applying Bayes' theorem, his posterior probabilities are

$$P(p = .01) = .005,$$

$$P(p = .05) = .245,$$

$$P(p = .10) = .359,$$

and

$$P(p = .25) = .391.$$

Using the prior probabilities, the expected payoffs of the three actions are

$$EP \text{ (Major adjustment)} = 380,$$

$$EP \text{ (Minor adjustment)} = 423,$$

and

$$EP \text{ (No adjustment)} = 432.$$

Using the posterior probabilities, the expected payoffs are

$$EP \text{ (Major adjustment)} = 380,$$

$$EP \text{ (Minor adjustment)} = 375.5,$$

and

$$EP \text{ (No adjustment)} = 208.1.$$

The optimal act under the prior distribution is "no adjustment," since there is a high probability that p is small (a probability of .90 that p is either .01 or .05). The particular sample observed makes the higher values of p seem more likely (the posterior probability is only .25 that p is either .01 or .05), and the optimal act under the posterior distribution is "major adjustment."

Because the EP and EL criteria take into consideration the probabilistic nature of uncertainty and because they seem to be intuitively reasonable, we will use them as our criteria for decision-making under uncertainty in the remainder of the chapter. Since they are equivalent, either can be used in any particular example. In some situations it will be convenient to talk in terms of payoffs and in other situations it will be more convenient to talk in terms of losses.

9.6   UTILITY

In all of the examples involving payoff and loss tables, the payoffs and losses have been expressed in monetary terms. This will not always be the case; it is easy to think of examples in which the consequences of a decision are nonmonetary. The consequences may involve quantities of a good or service. If the good or service has a known monetary value, then the payoffs can be expressed in dollars, or pounds, or francs, or whatever. In the production process example, the consequence involving the services of a mechanic was expressed in terms of the cost of these services. In most decisions, however, there are additional factors to consider, such as time, inconvenience, social approval, and so on. In the umbrella example, we said that the consequence of leaving the umbrella home and being caught in the rain is a ruined suit with a value of $60. To simplify the problem we ignored other potential consequences, such as the danger to your health and the discomfort of being caught in the rain with no umbrella or raincoat. If you carry the umbrella and it does not rain, the only cost you suffer is the inconvenience of carrying the umbrella. In the example, this inconvenience was expressed in monetary terms as the amount of money that you would be willing to pay not to have to carry the umbrella. It may not always be easy to express consequences in this manner.

There is another problem concerning the use of money to describe consequences and the related use of EP and EL as decision-making criteria. Suppose you were offered the following bet on one toss of a coin: you win $1 if the coin comes up heads, and you lose $.75 if the coin comes up tails. If you are convinced that the coin is a fair coin, so that

$$P(\text{heads}) = P(\text{tails}) = \tfrac{1}{2},$$

your expected payoff is $\tfrac{1}{2}(\$1) + \tfrac{1}{2}(-\$.75) = \$.125$ if you take the bet and $0 if you do not take the bet. According to the EP criterion, you should take the bet. This is reasonable, since the bet looks intuitively advantageous. Unless you are opposed to gambling in principle, you probably would take the bet. Now suppose that the amounts involved are $10,000 and $7500 rather than $1 and $.75. The expected payoff is now $1250 if you take the bet and $0 if you do not take the bet. According to the EP rule, you should take the bet. Would you do so? We think not, unless you are wealthy enough so that a loss of $7500 would not seriously affect your financial position. The possible gain of $10,000 is tempting, but there is still a 50-50 chance that you will lose $7500. Most persons would be put in deep financial difficulty if faced with a sudden debt of $7500. By the way, if you still would take the bet, change the values to $100,000 and $75,000 and make your choice again.

If you choose not to take the bet in any of these examples, you are violating the EP criterion. Why? The monetary payoffs are unambiguous, so there must be factors involved in this example other than these payoffs. If you lose $7500, the consequence is not just that loss, but the possible accompanying factors, such as the impact of a sharp reduction in your savings account, which is supposed to tide you over in your old age, or the embarrassment when you tell your spouse or your friends that you lost $7500 on the toss of a coin.

Consider a choice between two bets. In the first bet, you win $1 million if a coin comes up heads and you win $1 million if the coin comes up tails. In the second bet, you win $100 million if the coin comes up heads, but you win nothing if the coin comes up tails. The expected payoffs of the two bets are:

$$\text{EP (Bet I)} = \tfrac{1}{2}(\$1 \text{ million}) + \tfrac{1}{2}(\$1 \text{ million}) = \$1 \text{ million}$$

$$\text{EP (Bet II)} = \tfrac{1}{2}(\$100 \text{ million}) + \tfrac{1}{2}(\$0) = \$50 \text{ million}.$$

Bet II has the larger expected payoff. Which bet would you choose? In spite of the great difference in expected payoffs, most persons, if given this opportunity on a one-shot basis, would choose Bet I. This is because $1 million is a very large sum of money, clearly enough to allow a person to lead a quite comfortable life (even at today's prices!). If this amount could be invested at 5 percent interest, you would receive $20,000 per year without touching the principal. It would be even better, of course, to have $100 million, but most people feel that $100 million is not so preferable to $1 million that it is worth a risk of winding up with nothing. In other words, $100 million is not worth 100 times as much as $1 million. If it were, then Bet II would be preferred to Bet I.

A few comments are in order regarding this example. The amounts involved are much larger than they would be in most decision-making situations. Unless the decision at hand is a very important decision involving a large corporation, we would expect the amounts involved to be much smaller. The large amounts were chosen in order to stress the point, but other examples with smaller amounts could be found which give essentially the same results. Also, the bets are to be offered strictly as a one-shot affair, as we pointed out. If the situation were to be repeated several times, Bet II would surely be preferred to Bet I, for the chances of winding up with nothing would be greatly reduced. If the choice of bets is repeated just four times, the probability of winding up with nothing if Bet II is chosen each time is just $(\tfrac{1}{2})^4$, or $\tfrac{1}{16}$. Of course, if the bets are to be repeated, then the decision maker must choose an overall strategy. He might, for instance, choose Bet I on the first trial to assure himself of $1 million, and then choose Bet II thereafter.

At any rate, the examples illustrate the fact that the value of a dollar may differ from person to person and that for any specific person, the value of x dollars is not necessarily x times the value of a single dollar. As a result, the EP rule may not be satisfactory when the payoffs are expressed in terms of money. If we could somehow measure the true relative value to the decision maker of the various possible payoffs, we could take expectations in terms of these "true" values. The theory of **utility** prescribes such a decision-making rule: the **maximization of expected utility,** or the *EU* criterion. In order to understand this criterion, it is first necessary to briefly discuss the concept of utility.

Essentially, the theory of utility makes it possible to measure the *relative* value to a decision maker of the payoffs, or consequences, in a decision problem. Formally, a utility function U can be interpreted in terms of a preference relationship. The two basic axioms of utility are

1. *If payoff P_1 is preferred to payoff P_2, then $U(P_1) > U(P_2)$. If P_2 is preferred to P_1, then $U(P_2) > U(P_1)$; and if neither is preferred to the other, then $U(P_1) = U(P_2)$.*

2. *If you are indifferent between (a) receiving payoff P_1 for certain and (b) taking a bet in which you receive payoff P_2 with probability p and payoff P_3 with probability $(1 - p)$, then $U(P_1) = pU(P_2) + (1 - p)U(P_3)$.*

It is important to note that a utility function is not unique, even for a specific individual. If, for a particular person, a function U satisfies the above axioms, then the function $V = c + dU$ also satisfies the axioms, where c and d are constants with d greater than zero. For Axiom 1, suppose that $U(P_1) > U(P_2)$. Then

$$V(P_1) = c + dU(P_1)$$
and $$V(P_2) = c + dU(P_2).$$

It is obvious that $V(P_1) > V(P_2)$, since $d > 0$. Similarly, if

$$U(P_1) = pU(P_2) + (1 - p)U(P_3),$$

from Axiom 2, then $V(P_1) = pV(P_2) + (1 - p)V(P_3)$. To show this, observe that

$$pV(P_2) + (1 - p)V(P_3)$$
$$= p[c + dU(P_2)] + (1 - p)[c + dU(P_3)]$$
$$= pc + pdU(P_2) + (1 - p)c + (1 - p)dU(P_3)$$
$$= c[p + (1 - p)] + d[pU(P_2) + (1 - p)U(P_3)]$$
$$= c + d[pU(P_2) + (1 - p)U(P_3)].$$

But $pU(P_2) + (1 - p)U(P_3) = U(P_1)$, and therefore the right-hand side of the equation becomes $c + dU(P_1)$, which is equal to $V(P_1)$. This completes the proof that if U satisfies the axioms, then so does $V = c + dU$, where d must be positive. In words, we say that a utility function is only unique up to a positive linear transformation.

It should be emphasized that the development of utility presented in this section is only a brief, rough development of the most important points of the theory of utility, although it will suffice for our purposes. In the next section, we turn to a more practical problem: the assessment of a person's utilities.

9.7 THE ASSESSMENT OF UTILITY FUNCTIONS

Using the axioms of utility, it is possible for an individual to assess a utility function. Suppose that there are a number of possible payoffs in a decision-making problem facing you. First of all, it is necessary to determine what you consider to be the most preferable and the least preferable payoffs. Call the most preferable payoff M and the least preferable payoff L. Since a utility function is unique only up to a positive linear transformation, you can arbitrarily assign any values to $U(M)$ and $U(L)$, provided that $U(M) > U(L)$. Suppose that you let $U(M) = 1$ and $U(L) = 0$. The choice of 1 and 0 is arbitrary; you could just as easily have chosen 3 and -5, or 21.8 and 10.2, or any such pair of values, but 0 and 1 simplify the calculations somewhat.

Now consider any payoff P. From our choice of M and L, it must be true that
$$U(L) \leq U(P) \leq U(M),$$
or
$$0 \leq U(P) \leq 1.$$

To determine the value of $U(P)$ more precisely, consider the following choice of bets:

Bet I: Receive P for certain.

Bet II: Receive M with probability p
and receive L with probability $1 - p$.

How should you choose between these bets? According to the EU criterion, the bet with the higher expected utility should be selected. The expected utilities are
$$EU\ (\text{Bet I}) = U(P),$$
and
$$EU\ (\text{Bet II}) = pU(M) + (1 - p)U(L) = p(1) + (1 - p)(0) = p.$$

Thus, if $U(P) > p$, Bet I should be chosen; if $U(P) < p$, Bet II should be chosen; and if $U(P) = p$, you should be indifferent between the two bets.

This relationship between $U(P)$ and p can be exploited to determine the utility of P. If you can determine a probability p which makes you indifferent between the two bets, then the utility of P must be equal to this value, p. In this manner, you can determine the utility of any consequence, or payoff, once the most and least preferable payoffs, M and L, have been determined.

In the umbrella example, the three possible payoffs are \$0, −\$1, and −\$60. Clearly, M is \$0 and L is −\$60. To find $U(-\$1)$, suppose that $U(M) = 1$ and $U(L) = 0$, and consider these bets:

Bet I: Receive −\$1 for certain.

Bet II: Receive \$0 with probability p
            and receive −\$60 with probability $(1 - p)$.

What value of p makes you indifferent between Bets I and II? Suppose it is $p = .97$. Then $U(-\$1) = .97$, and the payoff table for the umbrella example can be expressed in terms of utilities:

State of the World

	Rain	*No rain*
Carry umbrella	.97	.97
Do not carry umbrella	0	1

Action

If $P(\text{Rain}) = .20$ and $P(\text{No rain}) = .80$, then

$$\text{EU (Carry umbrella)} = .2(.97) + .8(.97) = .97$$

and

$$\text{EU (Do not carry umbrella)} = .2(0) + .8(1) = .80.$$

The EU rule tells you to carry your umbrella.

Note that it is not at all necessary for the consequences, or payoffs, to be stated in terms of money. In the umbrella example, instead of working from the payoff table in terms of dollars, you could have considered the various consequences (getting wet, suit ruined, inconvenience of carrying umbrella, and so on) and determined utilities directly from these consequences. Because of this, it is possible to take into consideration both monetary and nonmonetary factors in determining the utility of a particular consequence. As we have pointed out, there may be nonmonetary factors such as social approval which are relevant to a person in a decision-making situation. In business decisions, such factors can be quite important, and it is useful to be able to consider them formally in decision theory. For instance, a corporation might be concerned about prestige, business ethics, and the satisfaction of employees and customers.

It is important to note that different individuals (or different corporations) may have different utility functions, just as they may have different prior probability distributions. This is demonstrated by the fact that two persons with identical information (and hence identical probabilities) may make two different decisions in the same situation. One person may accept a certain bet while another person rejects it. In any decision-making problem, then, the utility function of interest is the utility function of the person (or firm) that will receive the payoff. In deciding whether or not to take an umbrella with you, you must consider *your own* utilities, not those of anyone else.

9.8   UTILITY AS A FUNCTION OF MONEY

Although nonmonetary factors may be of some importance, it is of some interest to determine the relationship between money and utility. In many situations, the consequences *can* all be expressed in terms of money. In determining the relationship between utility and money over a given interval of monetary values for an individual, the technique introduced in the previous section can be applied. Suppose that you select the interval from $-\$100$ to $+\$100$, and you let $U(-\$100) = 0$ and $U(+\$100) = 1$. Next, you take several values between $-\$100$ and $+\$100$ and determine their utilities. These values can be plotted on a graph, as in Figure 9.8.1, and a rough curve can be drawn through them. This curve is *your utility*

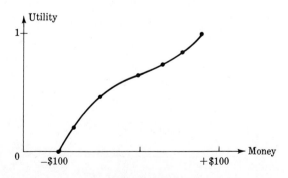

Figure 9.8.1

function for money. The utility of any monetary payoff between $-\$100$ and $+\$100$ can be read directly from the curve.

Several types, or classes, of utility functions can be distinguished, although there are utility functions not falling in any of the classes to be described. In Figures 9.8.2–9.8.4, three utility curves are presented. The curve in Figure 9.8.2 represents the utility function of a *"risk-avoider,"* the curve in Figure 9.8.3 represents the utility function of a *"risk-taker,"* and the curve in Figure 9.8.4 represents the utility function of a person who is neither a "risk-taker" nor a "risk-avoider." To see why these terms aptly describe the curves, consider the following bet: you win $100 with probability $\frac{1}{2}$ and you lose $100 with probability $\frac{1}{2}$. This can be thought of as a bet of $100 on the toss of a fair coin. In terms of expected payoff, a person should be indifferent about the bet, since it has an expected payoff of zero. In terms of expected utility, however, the decision as to whether or not to take the bet depends on the shape of your utility function.

In Figures 9.8.2–9.8.4, the gain in utility if the bet is won is

$$G = U(+\$100) - U(\$0), \tag{9.8.1}$$

and the loss in utility if the bet is lost is

$$L = U(\$0) - U(-\$100). \tag{9.8.2}$$

The expected utility of the bet is

$$\text{EU (Bet)} = \tfrac{1}{2}U(+\$100) + \tfrac{1}{2}U(-\$100).$$

The alternative action is not to bet, and the expected utility of this is just

$$\text{EU (Not bet)} = U(\$0).$$

Under what circumstances would you take the bet? Using the EU rule, you would take the bet if

$$\text{EU (Bet)} > \text{EU (Not bet)},$$

that is, if

$$\text{EU (Bet)} - \text{EU (Not bet)} > 0. \tag{9.8.3}$$

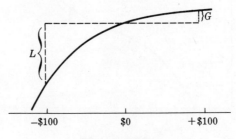

Figure 9.8.2

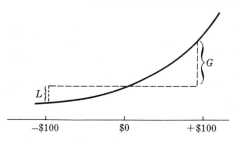

Figure 9.8.3

But from the above equations,

EU (Bet) − EU (Not bet) = $\frac{1}{2}U(+\$100) + \frac{1}{2}U(-\$100) - U(\$0)$.

The right-hand side of this equation can be written as

$$\tfrac{1}{2}U(+\$100) + \tfrac{1}{2}U(-\$100) - \tfrac{1}{2}U(\$0) - \tfrac{1}{2}U(\$0),$$

which is equal to

$$[\tfrac{1}{2}U(+\$100) - \tfrac{1}{2}U(\$0)] - [\tfrac{1}{2}U(\$0) - \tfrac{1}{2}U(-\$100)],$$

or, using Equations (9.8.1) and (9.8.2),

$$\tfrac{1}{2}G - \tfrac{1}{2}L = \tfrac{1}{2}(G - L).$$

From Equation (9.8.3), the decision rule is thus

Take the bet if $\frac{1}{2}(G - L) > 0$,

Do not take the bet if $\frac{1}{2}(G - L) < 0$.

Thus, in order to make a decision, you need only look at the sign of $G - L$.

For the curve in Figure 9.8.2, G is smaller than L, so that $(G - L)$ is negative, and you should not take the bet. Since you will not take a bet with an expected monetary payoff of zero, you are called a **risk-avoider.** In fact, with this curve it is possible to find some bets with expected monetary payoffs *greater than zero* that you would consider unfavorable in terms of expected utility.

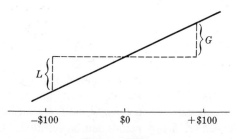

Figure 9.8.4

In Figure 9.8.3, G is greater than L, and as a result you *should* take the bet. Furthermore, we could find some bets with *negative* expected monetary payoffs that you would consider to be favorable bets. As a result, this curve represents the utility function of a **risk-taker.** Finally, $G = L$ in Figure 9.8.4. In this case you are indifferent between taking the bet and not taking it, and thus you are neither a risk-taker nor a risk-avoider. For a person with a linear utility function (that is, the curve is a straight line), maximizing EU is equivalent to maximizing EP. To prove this, note that if the utility curve is linear, it can be written in the form

$$U = a + bP,$$

where P is expressed in terms of money and a and b are constants, with $b > 0$. The requirement that b be greater than zero is necessary because the line must have a positive slope. Since a and b are constants,

$$EU = a + bEP,$$

and EU is at a maximum when EP is at a maximum. This is an important result: **If a person's utility function is linear with respect to money, then the EP criterion and the EU criterion are equivalent.**

The curves discussed above are by no means the only possible forms for utility functions. Several other forms, which we shall not describe, have been proposed, and still other forms, many of which are not mathematically tractable (that is, they cannot be represented by a simple mathematical function), no doubt exist. For instance, some individuals appear to be risk-takers for some decisions (such as gambling) and risk-avoiders for other decisions (such as purchasing insurance). It is an interesting exercise to select a range of values, such as $-\$100$ to $+\$100$, and attempt to determine your own utility curve for money. Just as in the assessment of probabilities, the assessment of utilities is not as easy as it looks. It is often quite difficult to determine how you would act if given a choice between two particular bets.

9.9   UTILITY AND DECISION-MAKING

The criterion which we have adopted for making decisions under uncertainty is the EU rule: select the action for which the expected utility is the highest. If we know the utility of each consequence, this rule is simple to apply. Often, however, we do not know the utility of each consequence. If there are numerous possible consequences, it may be too difficult a task to determine their utilities. If the consequences are all in terms of money, the task is a little easier. A few values can be selected, utilities can be de-determined for these values, and a rough utility curve can be drawn. The approximate utility of any monetary payoff can then be read from the curve. For many payoffs, even this could be laborious. In business decision-making, the difficulties are intensified. Who is to determine the utility function for a firm—the officers, the board of directors, the stockholders?

When the utility function is not known, a convenient assumption is that the function is linear with respect to money. In that case, we have shown that the act with the highest expected utility is also the act with the highest expected monetary payoff. The EP criterion can thus be used to determine the optimal decision. Since the EL rule is equivalent to the EP rule, it too could be used.

How reasonable is the assumption that the utility function is linear with respect to money? Because of the problems mentioned above, this assumption seems as good as any for a firm's utility function. For an individual person's utility function, the applicability of the assumption will vary from person to person. For most people, the assumption should not be too unrealistic unless the amounts of money involved are large. For small amounts, a straight line is probably a good approximation to the utility curve. For large amounts, it is likely to be a very bad approximation. Of course, what is a small amount to one person (say, a millionnaire) may be a very large amount to another person (say, a student). In any decision-making problem, the relevant question concerning the assumption is this: would EU result in a different decision than EP? This is once again the question of the *sensitivity* of the ultimate decision to changes in the inputs. Previously we discussed sensitivity to changes in the prior distribution. Now we are concerned about sensitivity to changes in the utility function.

In the examples involving money in the remainder of this chapter, it will be assumed that utility is linear with respect to money, so that EP and EL can be used as decision-making criteria. Alternatively, when payoffs or losses are discussed, you might think of them as being in terms of utility. In some examples, we will express payoff and loss tables in terms of utilities without worrying about the source of the utilities. These are simplifications which will be useful in allowing us to concentrate on other important points. You should keep in mind that they are simplifications, and that in any actual application of decision theory it is necessary to determine the utilities to be used as inputs to the decision-making problem.

9.10 A FORMAL STATEMENT OF THE DECISION PROBLEM

Now that all of the inputs to a problem of decision under uncertainty have been discussed, we will discuss the use of these inputs and introduce some new notation. In any decision-making problem, we start out with a set of possible actions, or decisions. Call this set A and denote a typical member of this set (that is, an action) by a. In addition, it is presumed that the uncertainty in the problem is uncertainty about the state of the world, θ. The set of possible states of the world, or possible values of θ, will be labeled S. We further assume that the uncertainty about θ can be expressed in probabilistic terms, in the form of a probability distribution of θ. This distribution will be called $P(\theta)$ if θ is discrete and $f(\theta)$ if θ is continuous. Note that this distribution may be a posterior distribution $f''(\theta \mid y)$ following an observed sample result y, or it may be a prior distribution $f'(\theta)$ if there is no sample information. Unless it becomes necessary to do so (and it will when optimal sample size is discussed), we will not be concerned about whether the distribution is prior or posterior. There is also a set of consequences, with one consequence for each combination of an action and a state of the world: (a, θ). Finally, there is a utility function U, giving the utility of each consequence. Since the consequence depends on the combination (a, θ), we will write the utility function in the form $U(a, \theta)$.

Recapitulating, we have the following inputs to the decision problem:

1. A, the set of actions, or acts.

2. S, the set of states of the world.

3. $P(\theta)$ or $f(\theta)$, the probability distribution of θ, the state of the world.

4. $U(a, \theta)$, the utility function which associates a utility with each pair (a, θ), where a is an action and θ is a state of the world.

Given these inputs, we must make a decision, which amounts to selecting an act a from the set A. We will select the act which is optimal under the EU rule. In other words, we will select the act which has the highest expected utility of all acts, or of all members of A. The expected utility of any act a can be calculated by using the definition of expectation, where the expectation is taken with regard to the distribution of θ. We will denote the expected utility of act a by $\mathrm{EU}(a)$. If θ is discrete, the expected utility of a is

$$\mathrm{EU}(a) \;=\; \sum U(a, \theta) P(\theta), \qquad (9.10.1^*)$$

where the summation is taken over all θ in S. If θ is continuous, we have

$$\mathrm{EU}(a) \;=\; \int U(a, \theta) f(\theta)\, d\theta, \qquad (9.10.2^*)$$

where the integration is over the set S. An act a^* is *optimal* with respect to the decision problem if

$$\mathrm{EU}(a^*) \geq \mathrm{EU}(a) \qquad \text{for all acts } a \text{ in } A. \qquad (9.10.3^*)$$

If the utility function U is linear with respect to money, then we may work instead with the payoff function $P(a, \theta)$ or with the loss function

$L(a, \theta)$. If θ is discrete, the expected payoff and expected loss of act a are given by

$$\text{EP}(a) = \sum P(a, \theta) P(\theta) \qquad (9.10.4^*)$$

and
$$\text{EL}(a) = \sum L(a, \theta) P(\theta). \qquad (9.10.5^*)$$

If θ is continuous, we have

$$\text{EP}(a) = \int P(a, \theta) f(\theta) \, d\theta \qquad (9.10.6^*)$$

and
$$\text{EL}(a) = \int L(a, \theta) f(\theta) \, d\theta. \qquad (9.10.7^*)$$

If U is linear with respect to money, the maximization of EU is equivalent to the maximization of EP or to the minimization of EL. In this case, an act a^* is optimal with respect to the decision problem if

$$\text{EP}(a^*) \geq \text{EP}(a) \qquad \text{for all acts } a \text{ in } A, \qquad (9.10.8^*)$$

or, equivalently, if

$$\text{EL}(a^*) \leq \text{EL}(a) \qquad \text{for all acts } a \text{ in } A. \qquad (9.10.9^*)$$

We hope that the student is not bothered by the overly formal nature of this section. The purpose is to take the parts that we have been developing (primarily the prior or posterior distribution and the utility function) and see how they fit together. In the process, the notation is also coordinated. There is nothing in this section that is really new, although we have not yet attacked a decision problem in which the state of the world is a continuous random variable. In the next few sections, we will revisit the inferential procedures of Chapters 5–8, treating them in a decision theory framework. Following that will be a discussion of the value of information and the decision as to whether or not to obtain more information.

9.11 DECISION THEORY AND POINT ESTIMATION

In Chapter 6 we discussed the classical approach to point estimation, which involves certain desirable properties of estimators, such as unbiasedness, consistency, efficiency, and sufficiency. Two methods for determining "good" point estimators are the method of maximum likelihood and the method of moments. In Chapter 8 the Bayesian approach was discussed, and this involves the posterior distribution. Since estimates should reflect all of the available information, according to the Bayesian, they should be based on the posterior distribution. Both of these approaches are informal procedures in the sense that there is no single "best" estimator. In most estimation situations, it is a matter of deciding whether unbiasedness is more important than efficiency, and so on. For example, s^2 is a maximum likelihood estimator of the variance of a normal population (or process), but it is biased. The unbiased estimator $\hat{s}^2$, on the other hand, is not a maximum likelihood estimator of σ^2. Which should be used to estimate σ^2? The choice between the two must be made subjectively. If all that is wanted by the statistician is a "rough and ready" estimate, or a general idea of the value of the parameter of interest, such an informal approach should prove satisfactory. If the estimate is to be used in a situation where errors in estimation may lead to serious consequences, a more formal approach would be desirable.

The choice of an estimate can be thought of as a problem of decision under uncertainty. Here the actions correspond to the possible estimates. If θ is being estimated, and S is the set of values of θ, then the set of actions A is identical to the set S. If a payoff function $P(a, \theta)$ or a loss function $L(a, \theta)$ can be determined which gives the payoff or loss associated with an incorrect estimate, then the problem of point estimation can be treated in accordance with the general framework presented in the previous section.

An example should help to clarify the decision-theoretic approach to point estimation. As usual, the example will be simplified to emphasize the points of interest. Consider an automobile dealer who sells a line of luxury automobiles. He has sold his entire inventory of cars, and therefore he must order some more from the manufacturer. Furthermore, next year's model will be introduced shortly, and this will be his last chance to order the current model. How many automobiles should he order? This is essentially a point estimation problem, for he must estimate the demand for the cars—the number of customers that will come in and want to purchase a car. The economic factors are quite simple: the dealer will make a profit of $500 for every car he sells before the new model comes out. Once the new model comes out, any of the current models that are still in inventory will have to be sold at a large discount, and he will suffer a loss of $300 for each car which must be sold at this discount.

The automobile dealer's loss function can be found from the facts given above. Remember that an action, a, corresponds to the number of cars that he orders from the manufacturer; a state of the world, θ, corresponds to the number of cars that are demanded by his customers. Both a and θ can take on the values 0, 1, 2, $\cdots$ (all nonnegative integers). If $a = \theta$, then the number ordered is exactly equal to the number demanded, and $L(a, \theta) = 0$. This is because the losses are "opportunity losses," or "regrets." Given any value of θ, the best possible action is $a = \theta$. If $a > \theta$, then the number of cars ordered is greater than the number demanded. In this case, there is a loss associated with the "extra" cars. For each car not sold before the new models are introduced, the loss is $300. Thus, if $a > \theta$, the loss function is $L(a, \theta) = 300(a - \theta)$. Finally, suppose that $a < \theta$. In this situation, more cars are demanded than the dealer has available, and for each car demanded but not available he suffers an opportunity loss of \$500, the "lost profit." Therefore, if $a < \theta$, the loss function is $L(a, \theta) = 500(\theta - a)$. The entire loss function is thus

$$L(a, \theta) = \begin{cases} 500(\theta - a) & \text{if } a < \theta, \\ 0 & \text{if } a = \theta, \\ 300(a - \theta) & \text{if } a > \theta. \end{cases} \quad (9.11.1)$$

In order to make a decision, the dealer first must specify the other input to the problem: the probability distribution for demand. Because it is very close to the introduction of next year's model and because the cars are quite expensive, the dealer is certain that demand will be either 0, 1, 2, or 3. His subjective probability distribution is

$$P(\theta = 0) = .1,$$
$$P(\theta = 1) = .2,$$
$$P(\theta = 2) = .3,$$
and
$$P(\theta = 3) = .4.$$

Since there is only a finite number of possible values of θ, there will be only a finite number of possible values of a (the dealer cannot order a negative number of cars, and he would not be foolish enough to order more than 3 as long as he thinks that demand cannot possibly be greater than 3). The loss function in Equation (9.11.1) can be used to find the following loss table:

State of the World
(Number of cars demanded)

		0	1	2	3
	0	0	500	1000	1500
Action (Number of cars ordered)	1	300	0	500	1000
	2	600	300	0	500
	3	900	600	300	0

For any act a, the expected loss is

$$\text{EL}(a) = \sum_{\theta=0}^{3} L(a, \theta) P(\theta). \qquad (9.11.2)$$

The expected losses for the four acts are

$$\text{EL}(a = 0) = 0(.1) + 500(.2) + 1000(.3) + 1500(.4) = 1000,$$

$$\text{EL}(a = 1) = 300(.1) + 0(.2) + 500(.3) + 1000(.4) = 580,$$

$$\text{EL}(a = 2) = 600(.1) + 300(.2) + 0(.3) + 500(.4) = 320,$$

and

$$\text{EL}(a = 3) = 900(.1) + 600(.2) + 300(.3) + 0(.4) = 300.$$

The optimal act is thus $a = 3$; the dealer should order 3 cars from the manufacturer. In a decision-theoretic sense, then, his "point estimate" of demand is 3. This point estimate depends both on the loss function and on the probability distribution. If the last two probabilities were switched around, so that $P(\theta = 2) = .4$ and $P(\theta = 3) = .3$, the expected losses would be 950, 530, 270, and 330. In this case, the optimal point estimate, or number of cars to order, is 2. Incidentally, it can formally be shown that any action other than 0, 1, 2, or 3 is inadmissible in this case. Consider $a = 4$. The values in the loss table for the row $a = 4$ would be 1200, 900, 600, and 300. But in all four columns, these numbers are higher than the corresponding numbers for $a = 3$. Therefore the action $a = 4$ is dominated by $a = 3$, and hence is inadmissible. The same is true for any action $a > 3$.

This example illustrates the decision-theoretic approach to point estimation. Whenever it is possible to determine a loss function $L(a, \theta)$ or a payoff function $P(a, \theta)$ in a point estimation situation, where a is the estimate and θ is the true value of the variable of interest, this procedure is applicable. It is then possible to find an optimal, or "best" (in the sense of minimizing EL or maximizing EP) estimate. In the next section we will present a quick way to determine this optimal estimate under certain assumptions regarding the loss function. Of course, not all point estimation problems involve losses. If a "rough and ready" estimate is all that is wanted, the approaches in Chapters 6 and 8 should prove satisfactory. If losses are involved and the potential losses are important, then it is wise to apply decision theory to the problem.

9.12   POINT ESTIMATION: LINEAR LOSS FUNCTIONS

The example in the last section illustrated the idea of point estimation as decision-making under uncertainty. It is clear, however, that such a problem could be quite difficult to solve if the number of possible values of

θ (and hence the number of possible estimates) were large. Instead of automobiles, suppose that the dealer sells radios, and that the demand for radios could be as low as zero or as high as 100. It would appear to be more trouble than it is worth to calculate the expected losses for all 101 possible actions. Fortunately, if certain assumptions can be made about the loss function, it is not necessary to calculate all of these EL's. If the expected loss can be expressed in a certain form, it is possible to use the calculus to find the point at which EL is smallest. This involves taking the derivative of EL with respect to a, setting it equal to zero, and solving the resulting equation. Because EL is expressed as an integral or as a sum, where the limits of integration or limits of summation are related to a, the differentiation of EL is not as elementary as might be thought. As a result, we will present some general statements about the optimal act without proof.

First of all, suppose that the loss function is of the following form:

$$L(a, \theta) = \begin{cases} k_u(\theta - a) & \text{if } a < \theta, \\ 0 & \text{if } a = \theta, \\ k_o(a - \theta) & \text{if } a > \theta. \end{cases} \qquad (9.12.1^*)$$

Such a loss function is called a **linear loss function** *with respect to a point estimation problem*. This should not be confused with the concept of a utility function which is linear with respect to money. This particular loss function is *linear with respect to the difference between the estimate a and the true state of the world θ.*

Notice that if $a < \theta$, the loss is $k_u(\theta - a)$. In this case, a is smaller than θ, so we say that we have underestimated θ. The *cost per unit of underestimation*, or simply the *cost of underestimation*, is k_u. Similarly, if $a > \theta$, we have overestimated θ, and the *cost per unit of overestimation* is k_o. It can be shown mathematically that **for the loss function given by Equation** (9.12.1), **the optimal point estimate is simply a**

$\dfrac{k_u}{k_u + k_o}$ **fractile of the decision maker's distribution of θ.**

Recalling the definition of a fractile (Section 3.17), this would be a value a such that

$$P(\theta \leq a) \geq \frac{k_u}{k_u + k_o} \quad \text{and} \quad P(\theta \geq a) \geq 1 - \frac{k_u}{k_u + k_o}. \qquad (9.12.2^*)$$

If θ is continuous, this reduces to

$$P(\theta \leq a) = \frac{k_u}{k_u + k_o}.$$

This greatly simplifies the problem from a computational standpoint. Instead of calculating a number of expected losses, it is only necessary, given k_u, k_o, and the distribution of θ, to find a single fractile of the distribution.

In the example presented in the previous section, the loss function is linear with respect to the difference between a and θ. The values of k_u and k_o are 500 and 300, and thus

$$\frac{k_u}{k_u + k_o} = \frac{500}{500 + 300} = \frac{5}{8}.$$

To find the $\frac{5}{8}$ fractile of the dealer's probability distribution $P(\theta)$, consider the corresponding CDF, presented in Figure 9.12.1. The $\frac{5}{8}$ fractile is 3, so this is the optimal value of a, just as we found out by calculating the expected losses for all of the possible actions.

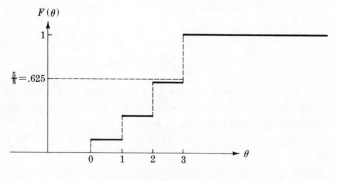

Figure 9.12.1

Equation (9.12.2) is applicable whether the variable θ is discrete or continuous. Suppose that a grocer has to decide how many oranges to order for a particular week. The oranges cost the grocer $.60 per dozen, and he sells them for $.80 per dozen. Any oranges that remain unsold at the end of the week will spoil and therefore will be worthless. If he orders more oranges than are demanded, he suffers a loss of $.60 per dozen, his cost, for all unsold oranges. If he orders fewer than are demanded, he suffers an opportunity loss of $.20 per dozen, reflecting lost profits, for all oranges demanded over and above the amount stocked. His loss function is of the form (9.12.1), with $k_u = .20$ and $k_o = .60$.

Based on past experience, the grocer feels that the weekly demand for oranges is normally distributed with a mean of 80 dozen and a standard deviation of 25 dozen. How many dozen oranges should the grocer order? He needs to find the $k_u/(k_u + k_o) = .2/(.2 + .6) = \frac{1}{4}$ fractile of the distribution $f(\theta)$. From tables of the normal distribution, the $\frac{1}{4}$ fractile of the standard normal distribution is approximately $-.675$. This implies that the $\frac{1}{4}$ fractile of a normal distribution with mean μ and standard deviation σ is simply $.675$ standard deviations to the left of the mean, or $\mu - .675\sigma$. The $\frac{1}{4}$ fractile of the grocer's distribution is then

$$80 - .675(25),$$

or approximately 63 dozen oranges. To minimize expected loss, the grocer should order 63 dozen oranges.

In our two examples, the linear loss function seemed to be realistic. In many cases, especially problems such as these involving inventories, the loss function is linear or almost linear. The result given by Equation (9.12.2) should therefore have wide applicability. Let us see if (9.12.2) is intuitively reasonable. If $k_u = k_o$, the optimal estimate is the .50 fractile, or the median, of the decision maker's distribution. This is reasonable; if the costs of overestimation and underestimation are equal, the optimal act should be somewhere in the "middle" of the distribution. Why specifically the median, rather than the mean? Because if the median is chosen, the probability of overestimating is equal to the probability of underestimating. If $k_u > k_o$, then $k_u/(k_u + k_o)$ is greater than .50. The estimate will be higher than the median. In this case, the cost of underestimation is greater than the cost of overestimation, so we "hedge" our estimate upwards to avoid the greater cost of underestimation. In the automobile dealer example, the cost of ordering too many cars was only $300 per extra car, as compared with the $500 per car opportunity loss if too few cars were ordered. The dealer would rather overestimate slightly than underestimate, so his optimal estimate is 3 cars. In the limiting case, as k_o becomes very small in relation to k_u, the decision maker will move farther and farther toward the right-hand tail of the distribution in an attempt to avoid suffering the high cost of underestimation. If $k_u < k_o$, just the opposite is true. In the grocer example, the cost of not having enough oranges on hand was just the lost profit of $.20 per dozen, while any extra oranges that spoiled would cost him $.60 per dozen. To avoid the higher cost of overestimating he ordered only 63 dozen oranges, although the mean of his distribution was 80 dozen. For highly perishable goods, k_o is often greater than k_u. For nonperishables, k_o is usually lower than k_u, since the only cost of overestimation is the cost of storing the unsold inventory. At any rate, the result (9.12.2) seems to be intuitively reasonable; if $k_o < k_u$, the estimate will be to the right of the median, and if $k_o > k_u$, the estimate will be to the left of the median. In each case, the greater the *relative* difference between k_u and k_o (as represented by the ratio of the larger of the two to the smaller), the farther from the median the estimate will be.

9.13 POINT ESTIMATION: QUADRATIC LOSS FUNCTIONS

A linear loss function of the form (9.12.1) is often quite realistic, as we have seen. There may be situations in which it is not applicable, however. Suppose that the costs of underestimation and overestimation are equal (let $k_o = k_u = k$), and that an error in estimation of 2 units is *more* than twice as costly as an error of just 1 unit. In particular, we will assume that an error of 2 units is four times as costly as an error of just one unit, and that in general, the loss associated with an error is proportional to the *square* of the error. The resulting loss function, which is called a **quadratic loss function** or a **squared-error loss function,** can be written as follows:

$$L(a, \theta) = k(a - \theta)^2. \tag{9.13.1*}$$

For this loss function, the optimal point estimate is the mean of the decision maker's distribution of θ. That is, the optimal estimate is

$$a = E(\theta). \tag{9.13.2*}$$

It is possible to prove the above result without using the calculus. We want to find a so as to minimize

$$\text{EL}(a) = Ek(a - \theta)^2 = kE(a - \theta)^2.$$

Clearly, since k is a constant, if we can find a so as to minimize $E(a - \theta)^2$, this a will also minimize $\text{EL}(a)$. Adding and subtracting $E(\theta)$ within the parentheses, we get

$$
\begin{aligned}
E(a - \theta)^2 &= E[a - E(\theta) + E(\theta) - \theta]^2 \\
&= [(a - E(\theta)) + (E(\theta) - \theta)]^2 \\
&= E[(a - E(\theta))^2 + 2(a - E(\theta))(E(\theta) - \theta) + (E(\theta) - \theta)^2] \\
&= E(a - E(\theta))^2 + 2E(a - E(\theta))(E(\theta) - \theta) + E(E(\theta) - \theta)^2.
\end{aligned}
$$

The expectation is being taken with respect to the distribution of θ, so a and $E(\theta)$ are constants. Thus, using the algebra of expectations,

$$E(a - \theta)^2 = (a - E(\theta))^2 + 2(a - E(\theta))E(E(\theta) - \theta) + E(E(\theta) - \theta)^2.$$

But the middle term is zero, since

$$E(E(\theta) - \theta) = E(\theta) - E(\theta) = 0.$$

The last term does not involve a, so it is irrelevant to our choice of an optimal a. This leaves the first term, which is

$$(a - E(\theta))^2.$$

Since this is a squared term, it must be either positive or zero. But it is zero only when $a = E(\theta)$, so this is the value of a that minimizes EL. Note the similarity between this proof and the proof in Section 3.23 that the expected squared deviation (that is, the variance) is smallest when calculated from the mean. In fact, this is the reason that $E(\theta)$ is the optimal estimate for a quadratic, or squared-error, loss function, and it is also one of the reasons that the mean is considered a "good" estimator in the classical approach to statistics. *In taking the sample mean as an estimator of the population mean, the statistician is essentially acting as though he had a quadratic loss function.* Of course, he is basing his estimate solely on the sample information, rather than the posterior distribution, but the general idea is the same. If the loss function is difficult to define exactly, but appears to be similar to the quadratic loss function in Equation (9.13.1), then the statistician is somewhat justified, in a decision-theoretic sense, in using the mean as his estimator. Similarly, if the loss function appears to be similar to the linear loss function in Equation (9.12.1) with $k_u = k_o$, then there is justification for using the median as an estimator.

In a point estimation problem with a quadratic loss function of the form (9.13.1), the optimal point estimate is the mean of the distribution of θ. Thus, it is not necessary to know the entire distribution; we just need to know the mean in order to make a decision. We can act as though the mean were equal to the true value of θ with certainty; hence the mean is called a **certainty equivalent** in this situation. Similarly, if the loss function is linear of the form (9.12.1), the $k_u/(k_u + k_o)$ fractile of the distribution is a *certainty equivalent*. The use of certainty equivalents simplifies the determination of one of the inputs to the decision problem. In the previous chapter, it was pointed out that the assessment of entire distributions may be a difficult and time-consuming task. If we know that the mean is a certainty equivalent in a given problem, it is just necessary to assess the mean, and the same idea holds for any certainty equivalent. The importance of this is somewhat limited, for the assessed distribution is often combined with sample information to form a posterior distribution. Then the *posterior* mean, or median, or whatever, is a certainty equivalent; but to determine this we generally must assess the entire prior distribution before we can apply Bayes' theorem to get the posterior distribution.

Linear and quadratic loss functions have been presented in algebraic form, and it might be useful to see how they can be represented on a graph. In Figure 9.13.1, a linear loss function with $k_u = k_o = 1$ is presented for three possible acts: a_1, a_2, and a_3. The density function $f(\theta)$ is also included in the graph. When the expected loss of an action is computed, the loss function is "weighted" by the density function. In the graph, a_2 appears to have a lower expected loss than a_1 or a_3. This is because the loss function of a_2 is low when the density function is high, and vice versa. For both a_1 and a_3, the loss function is quite high at the same time the density function is high, so the expected loss would be greater than is the case with a_2. A similar result occurs in Figure 9.13.2 for quadratic loss functions. Once again, a_2 would appear to have the smallest expected loss. Of course, a_1, a_2,

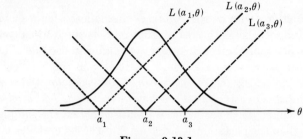

Figure 9.13.1

and a_3 are by no means the only possible estimates. Since θ is continuous, there is an infinite number of potential estimates. In the linear situation, the optimal estimate is the median; in the quadratic situation, the optimal estimate is the mean. Notice that if $f(\theta)$ is a symmetric distribution, the mean and the median will be equal. If $k_o = k_u$, then the difference between

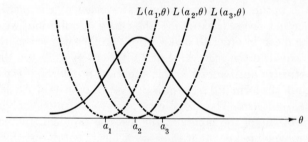

Figure 9.13.2

the linear and quadratic loss functions is only important when the distribution is skewed.

In Figure 9.13.3, loss functions with $k_o = 4$ and $k_u = 1$ are presented for three actions a_1, a_2, and a_3. In this situation, because the cost of overestimation is twice the cost of underestimation, it appears that a_1 is the best of the three estimators. For a_2 and a_3, the line corresponding to the cost of

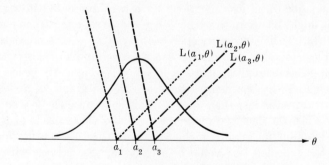

Figure 9.13.3

overestimation (which is the line going to the left, since to the left of a_1, we have $\theta < a_1$) rises so quickly that the loss function is quite high at the same time the density function is reasonably high. We will leave it as an exercise to draw a graph corresponding to the situation where $k_o = 1$ and $k_u = 4$.

9.14 DECISION THEORY AND HYPOTHESIS TESTING

As we suggested in Sections 7.7 and 7.8, hypothesis testing can be thought of as a decision-making problem, where the two decisions are (1) to accept H_0 and (2) to reject H_0 in favor of H_1. Thus the losses associated with the two types of errors should be assessed and the rejection region determined so as to minimize expected loss. Suppose that there are two hypotheses of interest,

$$H_0 : \theta \in A$$

and $$H_1 : \theta \in B. \qquad (9.14.1^*)$$

Furthermore, assume that A and B are mutually exclusive and exhaustive, so that $A \cap B = \varnothing$ and $A \cup B = S$, where S is the set of all possible values of θ. From the decision maker's distribution of θ, it is possible to compute

$$P(H_0) = P(\theta \in A)$$

and $$P(H_1) = P(\theta \in B). \qquad (9.14.2^*)$$

Since A and B are mutually exclusive and exhaustive, $P(H_0) + P(H_1) = 1$. This development is identical to the development of Section 8.24.

The other input to the decision problem is a loss function. In hypothesis testing, there are two types of errors. A Type I error is committed if H_0 is rejected when in fact it is true, and a Type II error is committed if H_0 is accepted when in fact it is false. Denote the loss associated with a Type I error by $L(\mathrm{I})$, and denote the loss associated with a Type II error by $L(\mathrm{II})$. The loss function can be represented by the following loss table, since there are just two actions and two relevant states of the world:

State of the World

		H_0 true $(\theta \in A)$	H_0 false $(\theta \in B)$
	Accept H_0	0	$L(\mathrm{II})$
Action			
	Reject H_0	$L(\mathrm{I})$	0

Calculating expected losses in the usual manner, we find that

$$\text{EL (Accept } H_0) = 0P(H_0) + L(\text{II})P(H_1) = L(\text{II})P(H_1) \qquad (9.14.3^*)$$

and

$$\text{EL (Reject } H_0) = L(\text{I})P(H_0) + 0P(H_1) = L(\text{I})P(H_0). \qquad (9.14.4^*)$$

According to the EL criterion for making decisions, we will reject H_0 if

$$\text{EL (Reject } H_0) < \text{EL (Accept } H_0). \qquad (9.14.5^*)$$

Using Equations (9.14.3) and (9.14.4), this is equivalent to

$$L(\text{I})P(H_0) < L(\text{II})P(H_1).$$

Dividing both sides by $L(\text{I})$ and by $P(H_1)$, we get

$$\frac{P(H_0)}{P(H_1)} < \frac{L(\text{II})}{L(\text{I})} . \qquad (9.14.6^*)$$

If Equation (9.14.6) is satisfied, the optimal decision is to reject H_0. If the inequality is reversed, the optimal decision is to accept H_0, and if the two sides of the equation are equal the two expected losses are equal, so neither action is better than the other.

The left-hand side of Equation (9.14.6) is an odds ratio, and the right-hand side is a ratio of losses, or a *loss ratio*. In order to make a decision in a hypothesis testing situation, it is necessary only to compare an odds ratio with a loss ratio. In fact, suppose that in Equation (9.14.6) we divided both sides of the equation by the loss ratio $L(\text{II})/L(\text{I})$. The result is

$$\frac{P(H_0)}{P(H_1)} \cdot \frac{L(\text{I})}{L(\text{II})} < 1. \qquad (9.14.7^*)$$

Denoting the odds ratio by Ω and the loss ratio by $L_{\text{I/II}}$, the decision rule can be written as follows:

$$\begin{array}{ll} \text{Reject } H_0 & \text{if} \quad \Omega \times L_{\text{I/II}} < 1, \\ \text{Accept } H_0 & \text{if} \quad \Omega \times L_{\text{I/II}} > 1. \end{array} \qquad (9.14.8^*)$$

Let us investigate Equation (9.14.8) to see if it seems reasonable. For a given loss ratio $L_{\text{I/II}}$, we are more likely to reject H_0 when Ω gets smaller, that is, when $P(H_0)$ gets smaller relative to $P(H_1)$. This makes sense; as the probability of H_0 gets smaller, H_0 appears to be less likely to be true, so we should be more likely to reject H_0. For a given odds ratio Ω, we are more likely to reject H_0 when $L(\text{I})$ gets smaller relative to $L(\text{II})$. But $L(\text{I})$ is the loss associated with the error of rejecting H_0 when it is in fact true. As this loss gets smaller relative to $L(\text{II})$, we should be more likely to reject H_0 for the same reason that we are more likely to underestimate in a point estimation problem as the cost of underestimation gets smaller relative to the cost of overestimation.

The framework developed in this section is related to both classical *and* Bayesian hypothesis testing, since we have not specified *exactly* what the odds ratio represents. From Equation (8.21.4), the posterior odds ratio is equal to the product of the prior odds ratio and the likelihood ratio:

$$\Omega'' = \Omega' \times (LR). \tag{9.14.9*}$$

If it is not possible to assess the losses $L(\text{I})$ and $L(\text{II})$, the posterior odds ratio will tell us which hypothesis is more likely but will not enable us to make a decision. If we informally decided to reject H_0 if H_1 was more likely (that is, if $\Omega'' < 1$) and to accept H_0 if H_0 itself was more likely (that is, if $\Omega'' > 1$), then our decision rule would resemble Equation (9.14.8) without the loss ratio. This amounts to assuming that $L(\text{I}) = L(\text{II})$, so that $L_{\text{I/II}} = 1$. The inclusion of the loss ratio, $L_{\text{I/II}}$, represents an *extension* of our informal decision rule to a more formal rule which satisfies the EL criterion.

How does this relate to classical hypothesis testing? We pointed out in Section 8.22 that classical tests of hypotheses are based on the likelihood ratio, LR. In Equation (9.14.9), if the prior odds ratio is equal to one (that is, if the two hypotheses are equal under the prior information), then the posterior odds ratio is equal to the likelihood ratio:

$$\Omega'' = \Omega' \times (LR) = LR.$$

In this case, the decision rule (9.14.8) becomes

$$\text{Reject } H_0 \quad \text{if} \quad LR \times L_{\text{I/II}} < 1,$$
$$\text{Accept } H_0 \quad \text{if} \quad LR \times L_{\text{I/II}} > 1. \tag{9.14.10*}$$

Dividing both sides by $L_{\text{I/II}}$, the rejection region becomes

$$LR < \frac{1}{L_{\text{I/II}}} = \frac{L(\text{II})}{L(\text{I})}. \tag{9.14.11*}$$

Now, from Equation (8.22.2), the rejection region for a classical test is of the form

$$LR \leq c.$$

Thus the classical "likelihood ratio" test is essentially equivalent to the test given by Equation (9.14.11), with $c = L(\text{II})/L(\text{I})$. The difference is that the classical statistician usually does not formally assess $L(\text{I})$ and $L(\text{II})$. Instead, he uses a convention which implies that $L(\text{I})$ is more serious than, and therefore higher than, $L(\text{II})$. The *formal* assessment of losses and the use of the EL criterion is an extension of the informal analysis.

It is not even necessary to assess the two losses $L(\mathrm{I})$ and $L(\mathrm{II})$ separately, since all that is needed is their ratio. Sometimes it is convenient to think in terms of loss ratios. For example, it is difficult to quantify the losses due to (1) convicting an innocent man or (2) setting a guilty man free. We may, however, be able to assess the loss ratio of (1) to (2). Some persons might assess this ratio as 100, others might think it to be 10, and still others might choose 1, saying that the two losses are equal. The point is simply that it is often much easier to assess a loss *ratio* than it is to directly assess the two losses comprising the loss ratio.

9.15 HYPOTHESIS TESTING: AN EXAMPLE

For a simple (and brief) example of hypothesis testing as a decision-making procedure, consider a manufacturer who is concerned about the potential demand for a new product he is thinking of producing. For the sake of simplicity, assume that he is interested in the following two hypotheses, where p is the proportion of consumers that will purchase the new product:

$$H_0 : p = .20$$

$$H_1 : p = .30.$$

If the manufacturer accepts H_0, then he will not produce the new product; if he rejects H_0 in favor of H_1, then the product will be produced.

The manufacturer's prior probabilities are

$$P(H_0) = P(p = .20) = .25$$

and $$P(H_1) = P(p = .30) = .75.$$

He takes a sample of 20 consumers and finds that only 2 will buy the new product. Thus, the likelihood ratio is (from the binomial tables):

$$LR = \frac{P(r = 2 \mid N = 20, p = .20)}{P(r = 2 \mid N = 20, p = .30)} = \frac{.1369}{.0278} = 4.92.$$

Multiplying this by the prior odds ratio of $.25/.75 = \frac{1}{3}$, we get

$$\Omega'' = \Omega'(LR) = \tfrac{1}{3}(4.92) = 1.64.$$

If the manufacturer decided just to accept the most likely hypothesis, then he would accept H_0, since the posterior odds ratio is greater than one. Suppose that he takes potential losses into consideration. If he accepts H_0 (and therefore does not produce the new product), he might suffer a large opportunity loss in the form of lost profits if H_1 is true. This is $L(\mathrm{II})$. On the other hand, if he rejects H_0 (and produces the new product), the new product might result in losses to the firm if the demand is not high enough (that is, if H_0 is true). This is $L(\text{I})$. After some serious thought, the manufacturer decides that $L(\text{II})$ is twice as large as $L(\text{I})$. This means that

$$L_{\text{I/II}} = \frac{L(\text{I})}{L(\text{II})} = \frac{1}{2}.$$

Multiplying the posterior odds ratio by the loss ratio, we get

$$\Omega'' \times L_{\text{I/II}} = (1.64)(\tfrac{1}{2}) = .82.$$

Since this is less than one, the optimal decision is to reject H_0 and to produce the new product. Even though the posterior odds ratio favors H_0, the losses are such that H_1 is the "better" hypothesis from a decision theory standpoint.

This example was purposely made simple in order to clearly demonstrate the idea of hypothesis testing as a decision-making procedure. In particular, both H_0 and H_1 were taken to be exact hypotheses. It should be mentioned that although the general procedure is still valid when the hypotheses are *not* both exact, there are certain subtleties involved in some situations which we will not consider here.

9.16   INFERENCE AND DECISION

In this section we shall attempt to very briefly recapitulate the relationship between what we have labeled classical statistics, Bayesian statistics, and decision theory. First of all, classical statistics prescribes inferential techniques which are based on sample information alone. Inferential methods in Bayesian statistics are based on the posterior distribution, which is a combination of the prior distribution, representing the prior state of information, and the likelihood function, representing the information of a sample. Thus, the prior distribution is an input to Bayesian methods but not to classical methods. In either case, the objective of inferential statistics is to make inferences about some population. These inferences may be in the form of point estimates, interval estimates, p-values, entire probability distributions, and so on. In decision theory, as opposed to inferential statistics, we are interested in taking some *action* rather than just making some inference about a population. As a result, one additional input is required in decision theory: a loss function, or a payoff function.

In the preceding chapter, we mentioned that Bayesian techniques can be thought of as an *extension* of classical techniques, since they utilize both sample information *and* prior information. In the same way, the decision-theoretic approach is yet a further extension, using the sample information, the prior information, and a third input as well: a loss function, or payoff function. The inclusion of the loss function enables the statistician to "optimize" in the sense of minimizing expected loss or maximizing expected payoff. It should be noted at this point that the Bayesian approach has become so closely associated with decision theory that the procedures described in Sections 9.11 through 9.15 are described as Bayesian methods by some authors. In this book we prefer to reserve the term "Bayesian" for the procedures described in Chapter 8, changing to the term "decision theory" when losses or payoffs are introduced into the analysis. This distinction serves a useful pedagogical function, enabling us to consider the introduction of prior information and the introduction of losses or payoffs as separate extensions of classical inferential procedures. We were thus able to compare inferences based on the posterior distribution (Bayesian inferences) with inferences based on the sampling distribution *without* having to be concerned with losses. In this chapter we introduced losses in order to see how this would affect the analysis.

For important decisions, the formal decision-theoretic approach is the most useful of the three approaches because it takes into account all available information which bears on the decision, both information concerning the state of the world and information concerning the consequences of potential decisions. If the decision is not too important, an informal analysis may be preferred, since applying decision theory may be quite time-consuming. It is in this sense that the classical and Bayesian techniques may be useful, since they are somewhat easier to apply, requiring fewer inputs and not specifying a single "optimizing" criterion. They may be useful in other situations as well, even situations in which the decision is important, if the statistician is unable to determine some of the inputs to the problem. In complex decision-making situations, it may not be clear what the relevant losses or payoffs are. If a big decision problem consists of a combination of several smaller problems, it may be difficult to see exactly how the "smaller" decisions relate to the major decision. In a way, this involves model-building: how can we state the problem in decision-theoretic terms and still have it be as realistic as possible? Decision theory provides a useful framework for making decisions under uncertainty. It is up to the statistician to attempt to express real-world problems in this framework without sacrificing realism. For instance, if the statistician makes assumptions which are unrealistic, then the optimal solution to the decision theory problem may not be the same as the best solution to the real-world problem. If this is the case, an informal analysis may prove to be just as useful as a formal decision-theoretic approach. Of course, as we pointed out, the formal approach is particularly valuable for important problems, in which case it might be worthwhile to carefully develop a realistic decision model.

9.17   TERMINAL DECISIONS AND PREPOSTERIOR DECISIONS

It has been assumed in this chapter that a statistician, or a decision maker, as we have often referred to him, is faced with a problem of decision-making under uncertainty, and that he must make a decision on the basis of his current state of information. Such a decision is called a **terminal decision**. If the decision maker's current state of information is represented by a posterior distribution, then his terminal decision will be based on this distribution (and, of course, on a loss function or a payoff function).

Suppose that the decision maker has the option of obtaining more sample information before he makes his terminal decision. Such sample information might be valuable in reducing his uncertainty about the state of the world. There is a cost involved in sampling, so the decision maker must decide if the additional sample information is expected to be useful enough to justify its cost. This type of decision is called a **preposterior decision** because it involves the potential posterior distribution following the proposed sample. Note that this sample has not been observed yet; it is just being contemplated. The value of the sample to the decision maker may depend on the observed result. Before taking the sample, he does not know what this result will be, although he *can* make some probabilistic statements about possible results.

The term "preposterior decision" can refer to such things as sample design, which can get quite complicated. We will only consider decisions concerning sample size. We assume that the decision maker is contemplating a sample of some fixed sample design and wants to know the optimal sample size. Preposterior analysis can determine this optimal sample size. It is possible for the optimal sample size to be zero, in which case the decision maker should not obtain any more sample information; he should make his terminal decision on the basis of his current state of information.

Before we attack the problem of determining optimal sample size, we will investigate a related, but somewhat easier, problem. Suppose the decision maker has the opportunity to purchase "perfect" information (that is, perfect knowledge of the state of the world), so that his problem of decision under uncertainty could be changed into the easier problem of decision under certainty. How much should he be willing to pay for this perfect information?

9.18   THE VALUE OF PERFECT INFORMATION

The term "information" has been used in discussing the state of uncertainty facing the decision maker. Generally, additional information may reduce his uncertainty about the state of the world. In the extreme case, if he were able to get "perfect" information, then the problem of decision-making under uncertainty would become a problem of decision-making under certainty. But decision-making is simpler under certainty, as we pointed out in Section 9.2. In terms of the notation presented in Section 9.10, suppose that we know for certain that the value of the state of the world is $\theta = \theta_v$. Then the optimal act a_v is the act such that

$$U(a_v, \theta_v) \geq U(a, \theta_v) \qquad \text{for all acts } a \text{ in } A. \qquad (9.18.1^*)$$

In terms of payoffs or losses, this becomes

$$P(a_v, \theta_v) \geq P(a, \theta_v) \qquad \text{for all acts } a \text{ in } A, \qquad (9.18.2^*)$$

or $\qquad L(a_v, \theta_v) \leq L(a, \theta_v) \qquad \text{for all acts } a \text{ in } A. \qquad (9.18.3^*)$

It is still possible to think of these as being consistent with the EU (or EP or EL) criterion. Saying that we are certain that $\theta = \theta_v$ implies that

$$P(\theta) = \begin{cases} 1 & \text{if } \theta = \theta_v, \\ 0 & \text{if } \theta \neq \theta_v. \end{cases}$$

Taking expectations with regard to this distribution,

$$\text{EU}(a) = U(a, \theta_v).$$

If the payoffs or losses in a particular problem are expressed in the form of a payoff or loss table, then perfect information enables the decision maker to neglect all of the columns of the table except one. In the umbrella example in Section 9.3, the payoff table was

State of the World

Rain   No rain

		Rain	No rain
	Carry umbrella	− 1	−1
Action			
	Do not carry umbrella	− 60	0

If you knew for certain that it would rain, you would only look at the first column, and your decision would be to carry the umbrella. If you knew for certain that it would not rain, then only the second column would be relevant, in which case you would not carry the umbrella.

For an example using a loss table, consider the stock purchasing example (Section 9.3), with the following loss table:

State of the World

		I	II	III	IV
	Buy A	0	0	400	0
Action	Buy B	200	400	0	0
	Buy C	400	800	600	0

If you knew that State I would occur, then you would just consider the first column. The action "buy A" has the smallest loss in this column, so this is the optimal action. Similarly, the optimal action under perfect knowledge indicating State II (State III) will occur is to buy A (buy B). If you are sure that State IV will occur, then it makes no difference which action you select; the losses are the same.

Note from the above example that because of the way in which we defined "losses," the optimal act under certainty will always have a loss equal to zero. In every column of a loss table, there is at least one zero, and there are never any negative losses. Therefore, if you have perfect information, your loss is zero. The perfect information must have some value to you, then, because under uncertainty (which can be thought of as "imperfect" information), the optimal act has some expected loss which is generally different from zero (although it is smaller than the expected loss of any other action).

It would be nice to find out exactly how much perfect information is worth to the decision maker. Suppose that the action a^* is optimal under the decision maker's current state of information, that is, that

$$\text{EP}(a^*) \geq \text{EP}(a) \qquad \text{for all actions } a \text{ in } A.$$

Furthermore, suppose that the action a_v is optimal under the perfect information that $\theta = \theta_v$. Of what value is this perfect information to the decision maker? If he had taken the action a^* and the state of the world θ_v occurred, his payoff would be $P(a^*, \theta_v)$. With knowledge of the true value of θ_v, he takes the action a_v, so his payoff is $P(a_v, \theta_v)$. The *value of* *the information* is simply the difference between the two payoffs:

$$\text{VPI}(\theta_v) = P(a_v, \theta_v) - P(a^*, \theta_v). \qquad (9.18.4^*)$$

This term, the value of perfect information that $\theta = \theta_v$, is of course dependent upon the distribution $P(\theta)[\text{or } f(\theta)]$ in the sense that it depends upon the optimal act under this distribution, which is a^*.

In the umbrella example, suppose that under your distribution of θ, the optimal act is to carry the umbrella. Then, if you obtain perfect knowledge that it will rain, the optimal act is still to carry the umbrella, so

VPI (Rain) = P(Carry umbrella, Rain) − P(Carry umbrella, Rain) = 0.

If you obtain perfect knowledge that it will *not* rain, your optimal action is to leave the umbrella home, and

VPI (No rain) = P(No umbrella, No rain) − P(Umbrella, No rain)

$$= 0 - (-1) = 1.$$

The value of perfect information clearly depends in this case on whether the perfect information is that there will be rain or whether it is that there will not be rain. How can the decision maker tell how much perfect information is worth to him? He cannot tell until he actually receives the perfect information. But then how can he determine how much he would be willing to pay to receive perfect information? The answer is simply that he should take the expectation of VPI with respect to his distribution of θ. In the umbrella example, suppose that $P(\text{Rain}) = .20$ and $P(\text{No rain}) = .80$. Then

$$E(\text{VPI}) = \text{VPI (Rain)}P(\text{Rain}) + \text{VPI (No rain)}P(\text{No rain})$$

$$= 0(.2) + 1(.8) = .80.$$

You should be willing to pay up to $.80 to find out for sure whether or not it will rain.

In this manner, we can compute the **expected value of perfect information,** EVPI:

$$\text{EVPI} = \sum \text{VPI} (\theta)P(\theta) \qquad (9.18.5^*)$$

or
$$\text{EVPI} = \int \text{VPI} (\theta)f(\theta)\, d\theta, \qquad (9.18.6^*)$$

depending on whether θ is discrete or continuous. Notice that EVPI can never be negative because both $\text{VPI}(\theta)$ and $P(\theta)$ [or $f(\theta)$] can never be negative. By definition, since a_v is the optimal act if $\theta = \theta_v$,

$$P(a_v, \theta_v) \geq P(a^*, \theta_v).$$

Therefore, from Equation (9.18.4), VPI is always greater than or equal to zero.

We have used the payoff function $P(a, \theta)$ to determine EVPI. It is also possible to think of EVPI in terms of the loss function. We noted above that under perfect information, the loss under the optimal action is always equal to zero. If a^* is the optimal act under the decision maker's current state of information, then

$$EL(a^*) \leq EL(a) \qquad \text{for all actions } a \text{ in } A.$$

If the decision maker acts on the basis of his current information, he will choose act a^*, and his expected loss will be equal to $EL(a^*)$. Under perfect information, his loss is zero, so the value of perfect information must be equal to

$$EVPI = EL(a^*) - 0 = EL(a^*). \qquad (9.18.7^*)$$

In words, **the expected value of perfect information to the decision maker is equal to the expected loss of the action which is optimal under his current state of information.**

In the stock-purchasing example, suppose that the probabilities of the four states of the world were .3, .2, .1, and .4. The expected losses of the three actions are then

$$EL \text{ (Buy A)} = 0(.3) + 0(.2) + 400(.1) + 0(.4) = 40,$$

$$EL \text{ (Buy B)} = 200(.3) + 400(.2) + 0(.1) + 0(.4) = 140,$$

and

$$EL \text{ (Buy C)} = 400(.3) + 800(.2) + 600(.1) + 0(.4) = 340.$$

The optimal action, Buy A, has an expected loss of \$40, so this is the EVPI. The decision maker should be willing to pay up to \$40 to obtain perfect information. It is left as an exercise for the reader to show that this is identical to the result obtained from Equation (9.18.5), using the payoff table in Section 9.3.

9.19 EVPI: AN EXAMPLE

To illustrate the calculation of the expected value of perfect information, we return once again to the production process example. There are four states of the world, corresponding to four values of p, the proportion of defectives produced by the process: .01, .05, .10, and .25. The statistician has three actions to choose from: he can have a mechanic make a major adjustment in the production process, he can have a minor adjustment made, or he can leave the process as is. The payoff table is

State of the World (Value of p)

	.01	.05	.10	.25
Major adjustment	380	380	380*	380*
Minor adjustment	455	375	375	375
No adjustment	480*	400*	300	0

The largest entry in each column is starred to show the optimal action for that particular value of p.

The statistician's prior probabilities are

$$P(p = .01) = .60,$$
$$P(p = .05) = .30,$$
$$P(p = .10) = .08,$$
and $$P(p = .25) = .02.$$

Under this prior distribution the expected payoffs of the three actions are

$$\text{EP (Major adjustment)} = 380,$$
$$\text{EP (Minor adjustment)} = 423,$$
and $$\text{EP (No adjustment)} = 432.$$

The optimal act is thus "no adjustment," and the value of perfect information can be calculated:

$$\text{VPI } (p = .01) = 480 - 480 = 0,$$
$$\text{VPI } (p = .05) = 400 - 400 = 0,$$
$$\text{VPI } (p = .10) = 380 - 300 = 80,$$
and $$\text{VPI } (p = .25) = 380 - 0 = 380.$$

The expected value of perfect information is

$$\text{EVPI} = \sum \text{VPI}(p) P(p) = 0(.6) + 0(.3) + 80(.08) + 380(.02) = 14.$$

The loss table corresponding to the above payoff table is

		State of the World (Value of p)			
		.01	.05	.10	.25
	Major adjustment	100	20	0	0
Action	*Minor adjustment*	25	25	5	5
	No adjustment	0	0	80	380

Since the third action, no adjustment, is optimal under the prior distribution,

$$\text{EVPI} = \text{EL (No adjustment)} = 0(.6) + 0(.3) + 80(.08) + 380(.02) = 14,$$

which is equal to the result obtained from the payoff table.

The statistician takes a sample of size five, observing one defective item. Using Bayes' theorem, the following posterior probabilities can be calculated (see Section 8.4 for the calculations):

$$P(p = .01) = .232,$$
$$P(p = .05) = .492,$$
$$P(p = .10) = .212,$$

and
$$P(p = .25) = .064.$$

The expected losses are now

$$\text{EL (Major adjustment)} = 33.04,$$
$$\text{EL (Minor adjustment)} = 19.48,$$

and
$$\text{EL (No adjustment)} = 41.28.$$

The optimal action is now to make a minor adjustment, and the expected value of perfect information is

$$\text{EVPI} = \text{EL (Minor adjustment)} = 19.48.$$

As a result of seeing the sample, the decision maker is now willing to pay up to \$19.48 for perfect information, whereas before the sample he was only willing to pay $14. How can this be? The sample information makes the higher values of p appear more likely and the lower values of p appear less likely. In this example, the posterior distribution has a larger variance than the prior distribution. The variance of the prior distribution is .0011, and the variance of the posterior distribution is .0033. This demonstrates that additional sample information will not always reduce the expected value of perfect information.

Suppose the statistician is concerned about this first sample result, and that he takes another sample of size five, in which he observes two defective items. The new posterior distribution is

$$P(p = .01) = .005,$$
$$P(p = .05) = .245,$$
$$P(p = .10) = .359,$$

and
$$P(p = .25) = .391,$$

and the expected losses under this distribution are

$$EL \text{ (Major Adjustment)} = \quad 5.4,$$
$$EL \text{ (Minor Adjustment)} = \quad 10.0,$$

and
$$EL \text{ (No Adjustment)} = 177.3.$$

The optimal act following the second sample is to make a major adjustment, and the expected value of perfect information is now only $5.40.

Therefore, we see that as new sample information is obtained, the expected value of perfect information may increase or decrease. At any point in time, if the decision maker can obtain perfect information at a cost which is smaller than his EVPI, then he should take advantage of this opportunity. In real-world decision-making situations, however, the decision maker seldom has the opportunity to obtain perfect information. No one knows for sure what the price of a stock will be one year from now, or, for that matter, what it will be one week from now. Similarly, it is doubtful that the statistician in our example would be able to find out *exactly* what the true value of p is. He could obtain more sample information, however, and we now turn to the value of sample information.

9.20 THE VALUE OF SAMPLE INFORMATION

From the expected value of perfect information (EVPI), the decision maker knows how much perfect information is worth to him. In other words, he knows how much he should be willing to pay for perfect information. Unfortunately, perfect information is seldom available, and the decision maker must take a sample if he wants more information. Since sampling involves some cost, it would be helpful if the decision maker could determine the worth (to him) of sample information so that he could decide whether or not to take a sample. In other words, he would like to determine the **expected value of sample information.** Since sample information can never be any better than perfect information, the expected value of sample information (EVSI) will always be less than or equal to the expected value of perfect information. Thus, although it is generally not possible to purchase perfect information, EVPI is still useful as an upper bound for the EVSI. In this section, we will discuss the determination of the exact value of EVSI.

Suppose that the decision maker assesses a prior distribution $P'(\theta)$ or $f'(\theta)$, and that the optimal action under the prior distribution is a':

$$EP'(a') \geq EP'(a) \qquad \text{for all acts } a \text{ in } A, \qquad (9.20.1)$$

or, equivalently,

$$EL'(a') \leq EL'(a) \qquad \text{for all acts } a \text{ in } A,$$

where the prime in EP' and EL' indicates that the expectation is taken with regard to the prior distribution. He then contemplates taking a sample. What would be the value to him of taking a sample and observing the sample result y? He could use this sample result to revise his prior distribution and arrive at a posterior distribution, $P''(\theta \mid y)$ or $f''(\theta \mid y)$. Let us denote the optimal act under this posterior distribution by a'', implying that

$$EP''(a'') \geq EP''(a), \qquad (9.20.2)$$

or $$EL''(a'') \leq EL''(a), \qquad \text{for all } a \text{ in } A. \qquad (9.20.3)$$

The double prime in EP'' and EL'' indicates that the expectation is taken with regard to the posterior distribution.

Under the posterior distribution, the optimal act is a'', whereas under the prior distribution it is a'. The value of this particular sample information to the decision maker is simply

$$VSI = EP''(a'') - EP''(a'), \qquad (9.20.4*)$$

or, equivalently,

$$VSI = EL''(a') - EL''(a''). \qquad (9.20.5*)$$

In words, this is the posterior expected payoff of the now-optimal act a'' minus the posterior expected payoff of the previously optimal act a'.

For example, consider the following experiment. Suppose that there are two decks of cards. Deck A consists of 26 red cards and 26 black cards; Deck B consists of 39 red cards and 26 black cards. A deck is chosen randomly from the two, and you are asked to guess which deck it is, subject to the following payoff table:

State of the World

Deck A   *Deck B*

	Deck A	Deck B
Deck A	4	2
Deck B	1	6

Action   (Your Guess)

Since the only information you have is that the deck has been chosen randomly, your prior probabilities should be

$$P(\text{Deck A}) = P(\text{Deck B}) = \tfrac{1}{2}.$$

Your expected payoffs are thus

$$\text{EP (Guess Deck A)} = 4(\tfrac{1}{2}) + 2(\tfrac{1}{2}) = 3$$

and $$\text{EP (Guess Deck B)} = 1(\tfrac{1}{2}) + 6(\tfrac{1}{2}) = 3.5,$$

so that the optimal act is to guess Deck B. Incidentally, the expected loss of this action is 3/2, so if the payoffs are expressed in terms of dollars, you would be willing to pay up to $1.50 for perfect information.

Suppose that you can purchase sample information in the form of cards drawn randomly (with replacement) from the deck of cards that has been chosen. Each card observed (that is, each trial) will cost you $.20. Should you choose Deck B immediately, or should you pay twenty cents to see a single trial? The two possible sample outcomes are "red" and "black," and for each of these outcomes we can compute the posterior probabilities:

Sample Outcome	State of the World	Prior Probability	Likelihood	Prior Probability × Likelihood	Posterior Probability
Red card	Deck A	.5	.50	.250	.40
	Deck B	.5	.75	.375	.60
				.625	
Black card	Deck A	.5	.50	.250	.67
	Deck B	.5	.25	.125	.33
				.375	

If the sample results in a red card, the posterior probabilities are .40 and .60, and the expected payoffs are

$$\text{EP (Guess Deck A)} = 4(.4) + 2(.6) = 2.8$$

and $\quad\quad\quad\text{EP (Guess Deck B)} = 1(.4) + 6(.6) = 4.0.$

The optimal act is still to guess Deck B, so the value of sample information if we see a red card is

$$\text{VSI (Red card)} = \text{EP}''(\text{Guess B}) - \text{EP}''(\text{Guess B}) = 0.$$

If the sample outcome is a black card, the posterior probabilities are .67 and .33, and the expected payoffs are

$$\text{EP (Guess Deck A)} = 4(.67) + 2(.33) = 3.33$$

and $\quad\quad\quad\text{EP (Guess Deck B)} = 1(.67) + 6(.33) = 2.67.$

The optimal act in this case is to guess Deck A, whereas before the sample guessing B was optimal, so

$$\text{VSI (Black card)} = \text{EP}''(\text{Guess A}) - \text{EP}''(\text{Guess B})$$
$$= 3.33 - 2.67 = .66.$$

The value of sample information (VSI) depends on the specific sample result, just as the value of perfect information (VPI) depends on the specific value θ_v. You must decide whether or not to purchase sample information in advance of the actual sample, so that the values for VSI, taken alone, are of little help. What you need is the *expected value of sample information,* or *EVSI*. In this example, the EVSI is equal to

$$\text{EVSI} = \text{VSI (Red card)} \ P(\text{Red card}) + \text{VSI (Black card)} \ P(\text{Black card}).$$

But what are the probabilities $P(\text{Red card})$ and $P(\text{Black card})$? They can be calculated as follows:

$$P(\text{Red}) = P(\text{Red} \mid \text{Deck A})P(\text{Deck A}) + P(\text{Red} \mid \text{Deck B})P(\text{Deck B})$$
$$= .50(.5) + .75(.5) = .625.$$

$$P(\text{Black}) = P(\text{Black} \mid \text{Deck A})P(\text{Deck A})$$
$$+ P(\text{Black} \mid \text{Deck B})P(\text{Deck B})$$
$$= .50(.5) + .25(.5) = .375.$$

Actually, we had already calculated these probabilities when we determined the posterior probabilities. Each is equal to a sum in the denominator of Bayes' theorem, or the sum of the column headed [(prior probability) $\times$ (likelihood)] in the appropriate part of the above table. Note that these sums are .625 and .375, respectively.

You can now determine the expected value of sample information corresponding to a potential sample of size one:

EVSI = VSI (Red card)P(Red card) + VSI (Black card)P(Black card)

  = 0(.625) + .66(.375)

  = .25.

The cost of taking the sample is \$.20, and thus your expected net gain from the sample is \$.25 − \$.20 = \$.05. Since this is greater than zero, you should take the sample. In general, **we define the expected net gain of sampling** (*ENGS*) **as the difference between the expected value of sample information** (*EVPI*) **and the cost of sampling** (*CS*):

$$\text{ENGS} = \text{EVSI} - \text{CS}. \qquad (9.20.6^*)$$

If the ENGS of a proposed sample is greater than zero, then the sample should be taken.

9.21 SAMPLE SIZE AND ENGS

Equation (9.20.6) gives the expected net gain of sampling for a particular sample. This may differ for different sample designs or different sample sizes. Suppose that the sample design is presumed to be fixed, and that it is necessary to determine how large a sample to take. Recall that in Section 6.13 we discussed the relationship between sample size and accuracy of estimation, noting that if the statistician specified the desired degree of accuracy, he could find out how large a sample would be necessary to attain this accuracy. This is an informal procedure for determining sample size (although not as informal as the more common "oh, an N of about twenty should be sufficient"). Using decision theory, it is possible to determine an optimal sample size.

If the sample design is fixed, and the statistician wishes to determine N, the sample size, he could compute the expected net gain of sampling for various values of N. Equation (9.20.6) could then be written in the form

$$\text{ENGS}(N) = \text{EVSI}(N) - \text{CS}(N). \qquad (9.21.1^*)$$

The optimal sample size is then the value of N for which ENGS is maximized. Denoting this optimal sample size by N^*, we have

$$\text{ENGS}(N^*) \geq \text{ENGS}(N) \qquad \text{for } N = 0, 1, 2, 3, \cdots. \qquad (9.21.2^*)$$

It is entirely possible for N^* to be zero, in which case the statistician should make his terminal decision without further sampling.

A typical ENGS curve is shown in Figure 9.21.1. The ENGS curve represents the difference between the EVSI curve and the CS curve. In this graph, the expected value of sample information rises rather quickly and then levels off. The leveling off reflects the fact that as the sample size gets larger and larger, the additional value of each trial becomes smaller and

smaller. The cost of sampling, on the other hand, is shown as a linear function of N. This would be reasonable if each trial cost a given amount and
there were no "economies of scale" associated with large values of N.
Because the cost of sampling (CS) curve is continually increasing and the
EVPI curve levels off, the ENGS curve rises until $N = N^*$ and then de-
clines, eventually becoming negative. The curve presented in Figure 9.21.1
is by no means the only type of ENGS curve, although it is probably the
most commonly encountered. In some cases the ENGS curve will never
be above the horizontal axis, and thus the statistician should take no
sample. In yet other cases the curve may be below the axis for small N but
may eventually rise above the axis, in which case there is an optimal N^*
different from zero.

A problem with EVSI (and hence ENGS) is that it generally requires
much more computational time and effort than, say, EVPI. In the example
in the previous section, the computations for $N = 1$, while elementary,
were somewhat time-consuming. Furthermore, the difficulty increases as N
increases, for the number of possible sample results increases and each
possible result must be considered. For $N = 10$ in the example, it would
be necessary to calculate posterior probabilities and the value of sample
information for the sample results (10 Red, 0 Black), (9 Red, 1 Black),
(8 Red, 2 Black), and so on, up to (0 Red, 10 Black). If there were more
than 2 States of the World or more than 2 actions, the computational problem would be intensified. In general, it is necessary to program the problem
for a high-speed computer in order to find the optimal sample size.

After the optimal sample size N^* is determined and the sample is taken
(provided that $N^* > 0$), the statistician can use Bayes' theorem to revise
his probability distribution on the basis of the sample results. He then has
two options once again: (1) to make a terminal decision on the basis of his
newly determined posterior distribution, or (2) to take another sample.
The second option involves the calculation of another $ENGS(N)$ curve

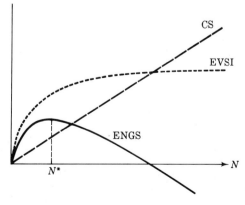

Figure 9.21.1

and the determination of another optimal sample size, N^{**}. Formally, the statistician should continue this process of taking a sample and conducting preposterior analysis until he reaches the point at which the optimal sample size is zero. He should then make a terminal decision. It is important to remember that sample information has no "value" by itself. The "value" in EVSI refers to the expected increase in expected payoff or decrease in expected loss *with respect to the terminal decision*.

9.22   EVSI AND ENGS: AN EXAMPLE

In Section 9.19, the production process example was used to demonstrate the calculation of the expected value of perfect information. Suppose that the statistician is unable to purchase perfect information, but that he *can* purchase sample information from the process. Starting from his original prior distribution (disregard the samples considered in Section 9.19), he contemplates three potential sample sizes: one, three, and five. In an actual situation he might also contemplate other values of N, but for the sake of simplicity, let us assume that he only considers these three values. The payoff table and prior probabilities are given in Section 9.19, and the cost of sampling is $1 per trial. That is, for $1 the statistician can observe an item selected randomly from the items produced by the process, and he can find out if it is defective.

For the sample of size one, the two possible sample results are no defectives and one defective. The posterior probabilities for each of these two situations are calculated as follows:

Sample Outcome (Number of Defectives) in One Trial	State of the World (value of p)	Prior Probability	Likelihood	Prior Probability × Likelihood	Posterior Probability
One defective	.01	.60	.01	.0060	.176
	.05	.30	.05	.0150	.442
	.10	.08	.10	.0080	.235
	.25	.02	.25	.0050	.147
				.0340	
No defectives	.01	.60	.99	.5940	.614
	.05	.30	.95	.2850	.295
	.10	.08	.90	.0720	.075
	.25	.02	.75	.0150	.016
				.9660	

For each value of p, the likelihood is simply p if the sample produces a defective and $(1 - p)$. if the sample does not produce a defective. Note that the probability that the item is defective is .0340 and the probability that it is not defective is .9660. Using the payoff table in Section 9.19, it is possible to calculate the expected payoff for each action and each sample result. If a defective item is observed, the posterior probabilities are .176, .442, .235, and .147, and the expected payoffs are

$$\text{EP (Major adjustment)} = 380.00,$$

$$\text{EP (Minor adjustment)} = 389.08,$$

and $\qquad\qquad$ EP (No adjustment) $= 331.78.$

The optimal decision is to make a minor adjustment, and the expected payoff is 389.08. If a nondefective item is observed, the posterior probabilities are .614, .295, .075, .016, and the expected payoffs are

$$\text{EP (Major adjustment)} = 380.00,$$

$$\text{EP (Minor adjustment)} = 424.12,$$

and $\qquad\qquad$ EP (No adjustment) $= 435.22.$

The optimal decision is to make no adjustment, and the EP is 435.22.

Under the prior distribution, the optimal decision is to make no adjustment. Therefore the value of sample information can be computed [using Equation (9.20.4)]:

VSI (Defective) $=$ EP'' (Minor adjustment) $-$ EP'' (No adjustment)

$$= 389.08 - 331.78 = 57.30.$$

VSI (Nondefective) $=$ EP'' (No adjustment) $-$ EP'' (No adjustment)

$$= 435.22 - 435.22 = 0.$$

The EVSI for a sample of size one is thus

$\text{EVSI}(N = 1)$ $=$ VSI(Defective)P(Defective)

$\qquad\qquad + $ VSI(Nondefective)P(Nondefective)

$\qquad\qquad = 57.30(.0340) + 0(.9660)$

$\qquad\qquad = 1.95.$

The statistician should be willing to pay up to $1.95 for a sample of one item. But the cost of a sample of size one is $1, so it is advantageous to him to take the sample.

For the contemplated sample of size three, there are four possible sample outcomes: 0, 1, 2, or 3 defectives. The likelihoods can be calculated from the binomial distribution with $N = 3$ and p equal to .01, .05, .10, or .25, as the case may be.

Sample Outcome (Number of Defectives in 3 Trials)	State of the World (value of p)	Prior Probability	Likelihood	Prior Probability × Likelihood	Posterior Probability
0 defectives	.01	.60	.9703	.582180	.643
	.05	.30	.8574	.257220	.284
	.10	.08	.7290	.058320	.064
	.25	.02	.4219	.008438	.009
				.906158	
1 defective	.01	.60	.0294	.017640	.205
	.05	.30	.1354	.040620	.472
	.10	.08	.2430	.019440	.225
	.25	.02	.4219	.008438	.098
				.086138	
2 defectives	.01	.60	.0003	.000180	.025
	.05	.30	.0071	.002130	.293
	.10	.08	.0270	.002160	.297
	.25	.02	.1406	.002812	.385
				.007282	
3 defectives	.01	.60	.0000	.000000	.000
	.05	.30	.0001	.000030	.071
	.10	.08	.0010	.000080	.189
	.25	.02	.0156	.000312	.740
				.000422	

For each action and each sample result, we can calculate an expected payoff from the relevant posterior probabilities. These expected payoffs are given in tabular form:

Sample Outcome
Number of Defectives in Sample of Size 3

		0	1	2	3
	Major adjustment	380.00	380.00	380.00*	380.00*
Action	*Minor adjustment*	426.44	391.40*	377.00	375.00
	No adjustment	441.44*	354.70	218.30	85.10

The expected payoff corresponding to the best action for each sample result is starred. To calculate the value of sample information, we note that the act "no adjustment" was optimal under the prior distribution.

VSI (0 defectives) $=$ EP″(No adjustment) $-$ EP″(No adjustment) $= 0$.

VSI (1 defective) $=$ EP″(Minor adjustment) $-$ EP″(No adjustment)

$\qquad = 391.40 - 354.70 = 36.70$.

VSI (2 defectives) $=$ EP″(Major adjustment) $-$ EP″(No adjustment)

$\qquad = 380.00 - 218.30 = 161.70$.

VSI (3 defectives) $=$ EP″(Major adjustment) $-$ EP″(No adjustment)

$\qquad = 380.00 - 85.10 = 294.90$.

The probabilities of the various sample outcomes are

$$P(0 \text{ defectives}) = .906158,$$
$$P(1 \text{ defective}) = .086138,$$
$$P(2 \text{ defectives}) = .007282,$$

and $\qquad\qquad P(3 \text{ defectives}) = .000422,$

so that the expected value of sample information for a sample of size three is

EVSI $(N = 3) = 0(.906158) + 36.70(.086138) + 161.70(.007282)$

$\qquad\qquad + 294.90(.000422) = 4.46$.

Therefore,

$\qquad$ ENGS $(N = 3) = 4.46 - 3(1) = 1.46$.

For the contemplated sample of size five, there are six possible sample outcomes: 0, 1, 2, 3, 4, or 5 defectives. The likelihoods can be calculated from the binomial distribution with $N = 5$ and p equal to .01, .05, .10, or .25, as the case may be.

Sample Outcome (Number of Defectives in 5 Trials)	State of the World (value of p)	Prior Probability	Likelihood	Prior Probability × Likelihood	Posterior Probability
0 defectives	.01	.60	.9510	.570600	.668
	.05	.30	.7738	.232140	.272
	.10	.08	.5905	.047240	.055
	.25	.02	.2373	.004746	.005
				.854726	
1 defective	.01	.60	.0480	.028800	.232
	.05	.30	.2036	.061080	.492
	.10	.08	.3280	.026240	.212
	.25	.02	.3955	.007910	.064
				.124030	
2 defectives	.01	.60	.0010	.000600	.033
	.05	.30	.0214	.006420	.354
	.10	.08	.0729	.005832	.322
	.25	.02	.2637	.005274	.291
				.018126	
3 defectives	.01	.60	.0000	.000000	.000
	.05	.30	.0011	.000033	.014
	.10	.08	.0081	.000648	.266
	.25	.02	.0879	.001758	.720
				.002439	
4 defectives	.01	.60	.0000	.000000	.000
	.05	.30	.0000	.000000	.000
	.10	.08	.0004	.000032	.099
	.25	.02	.0146	.000292	.901
				.000324	
5 defectives	.01	.60	.0000	.000000	.000
	.05	.30	.0000	.000000	.000
	.10	.08	.0000	.000000	.000
	.25	.02	.0010	.000020	1.000
				.000020	

It should be noted that the values of zero occurring in the last three columns are not *exactly* zero. The numbers are very small, and when rounded off to the number of decimal places given in the table, they become zero. For all practical purposes, they can be considered to be equal to zero.

For each action and each sample result, we can calculate an expected payoff from the relevant posterior probabilities. These expected payoffs are presented in the following table:

Sample Outcome

Number of Defectives in Sample of Size 5

		0	1	2	3	4	5
	Major adjustment	380.00	380.00	380.00*	380.00*	380.00*	380.00*
Action	*Minor* adjustment	428.44	393.56*	377.64	375.00	375.00	375.00
	No adjustment	445.94*	371.76	254.04	85.40	29.70	.00

From this table and the probabilities of the various sample outcomes,

$$P(0 \text{ defectives}) = .8547$$
$$P(1 \text{ defective}) = .1240$$
$$P(2 \text{ defectives}) = .0181$$
$$P(3 \text{ defectives}) = .0024$$
$$P(4 \text{ defectives}) = .0003$$
$$P(5 \text{ defectives}) = .0000,$$

the expected value of sample information can be calculated. First,

VSI (0 defectives) = EP''(No adjustment) − EP''(No adjustment) = 0,

VSI (1 defective) = EP''(Minor adjustment) − EP''(No adjustment)
= 393.56 − 371.76 = 21.80,

VSI (2 defectives) = EP''(Major adjustment) − EP''(No adjustment)
= 380.00 − 254.04 = 125.96,

VSI (3 defectives) = EP''(Major adjustment) − EP''(No adjustment)
= 380.00 − 85.40 = 294.60,

VSI (4 defectives) = EP''(Major adjustment) − EP''(No adjustment)
= 380.00 − 29.70 = 350.30,

and

VSI (5 defectives) = EP″(Major adjustment) − EP″(No adjustment)

$$= 380.00 - 0.00 = 380.00.$$

The EVSI is then 5.80. The cost of a sample of size five is \$5, so

ENGS $(N = 5)$ = EVSI $(N = 5)$ − CS $(N = 5)$ = 5.80 − 5.00 = .80.

Recall that

ENGS $(N = 1)$ = EVSI $(N = 1)$ − CS $(N = 1)$ = 1.95 − 1.00 = .95,

and

ENGS $(N = 3)$ = EVSI $(N = 3)$ − CS $(N = 3)$ = 4.46 − 3.00 = 1.46.

Of the three samples considered, then, the sample of size three has the largest ENGS. Of course, it may be that some sample size other than 1, 3, or 5 has a yet larger ENGS. To carry the problem any further would require the use of a high-speed computer, in which case it would be necessary to take into account the cost of using the computer.

The problem of determining optimal sample size can become quite complicated from a computational standpoint, as this example illustrates. As a result, most decisions regarding sample size are made in an informal manner.

9.23 LINEAR PAYOFF FUNCTIONS: THE TWO–ACTION PROBLEM

In examples concerning the value of information, we have only considered the situation in which there are a finite number of states of the world and a finite number of actions. Under these conditions, it is possible to set up a payoff or loss table and use this table in the calculation of EVPI and EVSI. In many decision-making problems, however, the state of the world is represented by a continuous random variable. The production process example would be more realistic if it was assumed that p was continuous and could take on any value between zero and one instead of being limited to .01, .05, .10, and .25. In the stock purchasing example, the four given states of the world are by no means the only possible states; it would be more reasonable to assume that the price of a stock one year from now is a continuous random variable. In numerous such decision problems, the state of the world can be thought of as continuous [although this is an idealization—the state of the world may actually take on a very large (but finite) number of values]. If the state of the world is continuous, it is not possible to represent the payoffs in a table. In general, they should be specified in functional form. In this section, it is assumed that the functional form is *linear in terms of the state of the world*. In other words, if θ is the state of the world, then the payoff function can be expressed in the form

$$P(a, \theta) = r + s\theta, \qquad (9.23.1^*)$$

where r and s are constants.

Payoff functions of the form (9.23.1) greatly simplify the decision-making problem. This is because expected payoffs can be written as

$$EP(a) = E(r + s\theta) = r + sE(\theta). \qquad (9.23.2^*)$$

This means that in order to make a terminal decision, all we need to know about the distribution of θ is the mean, $E(\theta)$. Thus, the mean is a certainty equivalent.

Suppose that there are just two possible actions, a_1 and a_2, and that their payoff functions are

$$P(a_1, \theta) = r_1 + s_1\theta$$
$$\text{and} \qquad P(a_2, \theta) = r_2 + s_2\theta, \qquad \text{where } s_2 > s_1. \qquad (9.23.3)$$

Under what conditions will a_1 be optimal under a prior distribution $f(\theta)$, and under what conditions will a_2 be optimal? The first act, a_1, will be optimal if

$$EP(a_1) > EP(a_2). \qquad (9.23.4)$$

But all we need to know about $f(\theta)$ in order to compute the EP's is the mean, $E(\theta)$, so Equation (9.23.4) is equivalent to

$$r_1 + s_1E(\theta) > r_2 + s_2E(\theta).$$

Subtracting r_2 and $s_1E(\theta)$ from both sides, we get

$$r_1 - r_2 > E(\theta)(s_2 - s_1).$$

Since $s_2 > s_1$, the term $(s_2 - s_1)$ is greater than zero, so dividing both sides by this term will not affect the inequality:

$$\frac{r_1 - r_2}{s_2 - s_1} > E(\theta). \qquad (9.23.5^*)$$

Therefore, if Equation (9.23.5) is satisfied, a_1 is the optimal action; if the inequality is reversed, a_2 is the optimal action. For this decision-making problem, **we define θ_b to be the breakeven value of θ:**

$$\theta_b = \frac{r_1 - r_2}{s_2 - s_1}. \qquad (9.23.6^*)$$

If the expected value of θ is lower than θ_b, then a_1 is optimal; if it is greater than θ_b, then a_2 is optimal.

If the decision maker is basing his terminal decision on a prior distribution with mean $E'(\theta)$, then a_1 is optimal if

$$E'(\theta) < \theta_b$$

and a_2 is optimal if $\qquad E'(\theta) > \theta_b.$

If the decision maker is basing his terminal decision on a posterior distribution with mean $E''(\theta)$, then the above inequalities hold, with $E''(\theta)$ substituted for $E'(\theta)$. Similarly, suppose the decision maker knows for certain that $\theta = \theta_v$. In this case, a_1 is optimal if $\theta_v < \theta_b$ and a_2 is optimal if $\theta_v > \theta_b$. Thus, it is easy to determine which act is optimal under a prior distribution, a posterior distribution, or under certainty.

A graph should serve to illustrate the above points. In Figure 9.23.1, both payoff functions are linear, and hence are represented on the graph by straight lines. The condition $s_2 > s_1$ means that the line corresponding to $P(a_2, \theta)$ has a greater slope than the line corresponding to $P(a_1, \theta)$. This is clearly true, for the former line has a positive slope and the latter line has a negative slope. The breakeven value, θ_b, is the value of θ at which the two lines intersect. To the left of this point, the $P(a_1, \theta)$ line is higher, and to the right of the point the $P(a_2, \theta)$ line is higher.

For a numerical example, suppose that we are interested in the mean μ of a normal process with known variance 100. The prior distribution of μ is a normal distribution with mean 100 and variance 25. There are two possible actions, and their payoff functions are linear:

$$P(a_1, \mu) = -45 + .5\mu$$

$$P(a_2, \mu) = -94 + \mu.$$

The breakeven value of μ is

$$\mu_b = \frac{-45 - (-94)}{1 - .50} = \frac{49}{.5} = 98.$$

Since $E'(\mu) = 100$ and $\mu_b = 98$, the optimal act under the prior distribution is a_2.

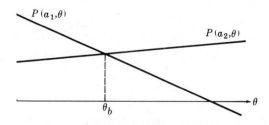

Figure 9.23.1

Consider a sample of size 10 from the above process, with sample mean $M = 95$. Using Bayes' theorem, the posterior distribution is normal with parameters M'' and

$$\frac{1}{\sigma''^2} = \frac{1}{\sigma'^2} + \frac{N}{\sigma^2} = \frac{1}{25} + \frac{10}{100} = .04 + .10 = .14.$$

Thus,

$$\sigma''^2 = \frac{1}{.14} = 7.14$$

and

$$M'' = \frac{\left(\frac{1}{\sigma'^2}\right)M' + \left(\frac{N}{\sigma^2}\right)M}{\frac{1}{\sigma'^2} + \frac{N}{\sigma^2}} = \frac{.04(100) + .10(95)}{.04 + .10} = 96.43.$$

Therefore, the posterior mean is less than the breakeven value, and the optimal act under the posterior distribution is a_1. If we were somehow able to obtain perfect information that $\mu = \mu_v = 97.5$, for instance, then the optimal act under certainty would be a_1. This example is illustrated in Figure 9.23.2.

In summary, then, linear payoff functions are quite useful in two respects. First, payoff functions in real-world decision problems are often linear or very nearly linear. Second, when there are just two actions to choose between, the decision criterion of maximization of expected payoff reduces to the simple process of comparing two numbers: the breakeven value, θ_b, and the mean of the decision maker's probability distribution, $E(\theta)$.

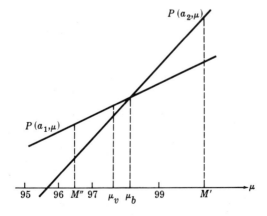

Figure 9.23.2

9.24 LOSS FUNCTIONS AND LINEAR PAYOFF FUNCTIONS

If the payoff functions for both acts in a two-action problem are linear with respect to the state of the world, then the corresponding loss functions are linear also. In the example illustrated in Figure 9.23.1, the optimal action under the prior distribution is a_2. If this action is taken, what does the loss function look like? The loss function, like the payoff function, is a function of μ. If the true value of μ is greater than the breakeven value, 98, then the act chosen under the prior distribution is in fact the better of the two acts, and the loss is zero. On the other hand, suppose that the true value of μ is less than 98. Then a_1 gives the higher payoff, and the loss suffered by choosing a_2 is the difference in payoffs,

$$P(a_1, \mu) - P(a_2, \mu).$$

Using the payoff functions,

$$P(a_1, \mu) = -45 + .5\mu$$

and

$$P(a_2, \mu) = -94 + \mu,$$

the loss function (given that act a_2 is chosen) can be written as

$$L(a_2, \mu) = \begin{cases} (-45 + .5\mu) - (-94 + \mu) = 49 - .5\mu & \text{if } \mu < 98, \\ 0 & \text{if } \mu > 98. \end{cases}$$

This loss function is shown in Figure 9.24.1. If act a_1 is chosen (as it would be under the posterior distribution given in the example), then the loss depends once again on μ. If μ is less than 98, the best act was chosen, and the loss is zero. If μ is greater than the breakeven value, the loss is the difference

$$P(a_2, \mu) - P(a_1, \mu).$$

The loss function for act a_1 can thus be written

$$L(a_1, \mu) = \begin{cases} 0 & \text{if } \mu < 98, \\ (-94 + \mu) - (-45 + .5\mu) = -49 + .5\mu & \text{if } \mu > 98. \end{cases}$$

This loss function is shown in Figure 9.24.2.

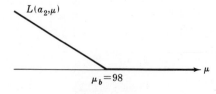

Figure 9.24.1

Figure 9.24.2

Let us look at a specific value or two. If a_2 is chosen and the true value of μ is 95, then the loss incurred is

$$L(a_2, 95) \ = \ 49 \ - \ .5(95) \ = \ 49 \ - \ 47.5 \ = \ 1.5.$$

If a_1 is chosen and the true value of μ is 95, then the loss is zero. If a_1 is chosen and the true value of μ is 110, however, then the loss is

$$L(a_1, 110) \ = \ -49 \ + \ .5(110) \ = \ -49 \ + \ 55 \ = \ 6.$$

In both Figures 9.24.1 and 9.24.2, the loss function is equal to zero on one side of the breakeven value and is a straight line on the other side of the breakeven value. In general, suppose that we have two acts with payoff functions of the form (9.23.3):

$$P(a_1, \theta) \ = \ r_1 \ + \ s_1\theta$$

$$P(a_2, \theta) \ = \ r_2 \ + \ s_2\theta.$$

Let these payoff functions be as shown in Figure 9.24.3. Consider the line which represents the best we can do for any value of θ; this is the V-shaped line with the dots on it in the graph. Now suppose that a_1 is the act which is chosen. The loss suffered at any particular value of θ is just the vertical distance between the $P(a_1, \theta)$ line and the line which represents the best

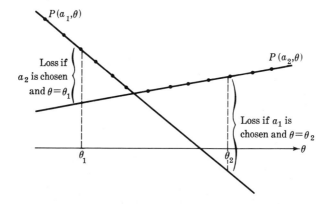

Figure 9.24.3

we can do. But this distance is zero when $\theta < \theta_b$ and is the difference

$$P(a_2, \theta) - P(a_1, \theta)$$

when $\theta > \theta_b$. The loss function is thus of the form

$$L(a_1, \theta) = \begin{cases} 0 \quad \text{if} \quad \theta < \theta_b, \\ (r_2 + s_2\theta) - (r_1 + s_1\theta) = (r_2 - r_1) + (s_2 - s_1)\theta \quad \text{if } \theta > \theta_b. \end{cases}$$

$$(9.24.1^*)$$

The graph of this will look like Figure 9.24.2. Similarly, you can see that

$$L(a_2, \theta) = \begin{cases} (r_1 + s_1\theta) - (r_2 + s_2\theta) = (r_1 - r_2) + (s_1 - s_2)\theta \quad \text{if } \theta < \theta_b, \\ 0 \quad \text{if} \quad \theta > \theta_b. \end{cases}$$

$$(9.24.2^*)$$

The graph of this will look like Figure 9.24.1.

Why are we interested in these loss functions? The decision maker may want to compute EVPI, and we know that this is equal to the expected loss of the act which is optimal under the distribution of θ. For instance, let us assume for the moment that a_1 is optimal, and that the distribution is represented by $f(\theta)$. Then the EVPI is equal to

$$\text{EL}(a_1) = \int_{-\infty}^{\infty} L(a_1, \theta)f(\theta) \, d\theta.$$

Using Equation (9.24.1),

$$\text{EL}(a_1) = \int_{-\infty}^{\theta_b} 0f(\theta) \, d\theta + \int_{\theta_b}^{\infty} [(r_2 - r_1) + (s_2 - s_1)\theta]f(\theta) \, d\theta,$$

or

$$\text{EL}(a_1) = \int_{\theta_b}^{\infty} [(r_2 - r_1) + (s_2 - s_1)\theta]f(\theta) \, d\theta. \qquad (9.24.3^*)$$

Similarly, if a_2 is optimal under $f(\theta)$, then the EVPI is equal to

$$\text{EL}(a_2) = L(a_2, \theta)f(\theta) \, d\theta$$

$$= \int_{-\infty}^{\theta_b} [(r_1 - r_2) + (s_1 - s_2)\theta]f(\theta) \, d\theta + \int_{\theta_b}^{\infty} 0f(\theta) \, d\theta,$$

or

$$\text{EL}(a_2) = \int_{-\infty}^{\theta_b} [(r_1 - r_2) + (s_1 - s_2)\theta]f(\theta) \, d\theta. \qquad (9.24.4^*)$$

The expressions given by Equations (9.24.3) and (9.24.4) are difficult to calculate because the integration in each case is over only part of the real line, representing the part for which the loss function is nonzero. In the next section we present a method for evaluating these integrals under the condition that $f(\theta)$ is a normal distribution.

9.25 LINEAR LOSS FUNCTIONS AND THE NORMAL PROCESS

Under the assumption that $f(\theta)$ is a normal distribution, it is possible to compute the values of Equations (9.24.3) and (9.24.4) with the aid of a table of the "Unit Normal Linear Loss Integral," which is defined as

$$L_N(D) = \int_D^\infty (\theta - D) f_N(\theta) \, d\theta,$$

where $f_N(\theta)$ is the standard normal density function. In addition, this table permits the computation of the expected value of sample information when the loss function is linear and the distribution is normal. The equations by which EVPI and EVSI can be computed from the table will be stated without proof to avoid mathematical details.

Assume that a decision maker is interested in the mean, μ, of a normal process with known variance σ^2. He has a prior distribution $f'(\mu)$ and a two-action problem with linear loss functions (linear in terms of μ) of the forms (9.24.1) and (9.24.2). The prior distribution is a normal distribution with mean M' and variance σ'^2. The EVPI is then

$$\text{EVPI} = |\, s_1 - s_2 \,| \, \sigma' L_N(D), \qquad (9.25.1^*)$$

where
$$D = \left|\, \frac{\mu_b - M'}{\sigma'} \,\right| \qquad (9.25.2^*)$$

and the value $L_N(D)$ can be read from the table of the unit normal linear loss integral, presented in the appendix. First, note that in Equation (9.24.1), the slope of the nonzero portion of the loss function is $(s_2 - s_1)$; similarly, in Equation (9.24.2), the slope of the nonzero portion of the loss function is $(s_1 - s_2)$. In either case, the absolute value of the slope is $|\, s_1 - s_2 \,|$, and that is represented by the first term of Equation (9.25.1). As this slope increases (in absolute value), the EVPI increases, which is entirely reasonable, since a greater slope implies greater losses for wrong decisions. But if the losses due to wrong decisions are greater, then it is more important that the right decision be made, so that perfect information is worth more to the decision maker. The second term in the EVPI equation is the prior standard deviation. The smaller the prior standard deviation, the more current information we have about μ. But the more information we have, the less perfect information is worth to us. Thus, the second term makes sense intuitively. Finally, consider the $L_N(D)$ term. As D increases (note that it will always be positive, since it is an absolute value), $L_N(D)$ decreases, as can be seen from the table of $L_N(D)$. What does D represent? D is the number of prior standard deviations σ' that separate the prior mean, M', and the breakeven value, μ_b. The larger D is, the more certain we are that the optimal decision under the prior distribution is the correct decision. This is because the probability that μ is on the opposite side of the breakeven value from M' decreases. In this case, the EVPI should decrease, and we see from the $L_N(D)$ table that it does decrease as D increases.

Consider the example from the previous section, in which

$$L(a_1, \mu) = \begin{cases} 0 & \text{if } \mu < 98, \\ -49 + .5\mu & \text{if } \mu > 98, \end{cases}$$

and

$$L(a_2, \mu) = \begin{cases} 49 - .5\mu & \text{if } \mu < 98, \\ 0 & \text{if } \mu > 98. \end{cases}$$

The breakeven value is $\mu_b = 98$. The decision maker's prior distribution is a normal distribution with mean 100 and variance 25, and the known variance of the process (or population) is 100.

The slope of $L(a_1, \mu)$ is $+.5$ and the slope of $L(a_2, \mu)$ is -5 (in terms of the payoff functions, these slopes are $s_2 - s_1$ and $s_1 - s_2$). In both cases, the absolute value of the slope, $|s_1 - s_2|$, is equal to .5. The prior standard deviation is $\sqrt{25}$, or 5, and

$$D = \left| \frac{\mu_b - M'}{\sigma'} \right| = \left| \frac{98 - 100}{5} \right| = |-.4| = .4.$$

The expected value of perfect information, from Equation (9.25.1), is

$$\text{EVPI} = .5(5)L_N(.4).$$

From the table of $L_N(D)$, $L_N(.4) = .2304$, so that

$$\text{EVPI} = .5(5)(.2304) = .5760.$$

If the losses are in terms of dollars, then the decision maker should be willing to pay up to 57 cents for perfect information. If the losses are in terms of thousands of dollars, he should be willing to pay up to $576 for perfect information.

Perhaps the decision maker is contemplating a sample of size N and would like to know the expected value of such a sample. The EVSI formula, which is very similar to Equation (9.25.1), is

$$\text{EVSI} = |\, s_1 - s_2 \,| \, \sigma^* L_N(D^*), \qquad (9.25.3^*)$$

where

$$\sigma^{*2} = \frac{N\sigma'^4}{\sigma^2 + N\sigma'^2} \qquad (9.25.4^*)$$

and

$$D^* = \left| \frac{\mu_b - M'}{\sigma^*} \right|. \qquad (9.25.5^*)$$

The term denoted by σ^{*2} is essentially the reduction in variance due to the sample, or the prior variance minus the posterior variance. We know that

$$\frac{1}{\sigma''^2} = \frac{1}{\sigma'^2} + \frac{N}{\sigma^2} = \frac{\sigma^2 + N\sigma'^2}{\sigma'^2 \sigma^2}.$$

Hence,

$$\sigma''^2 = \frac{\sigma'^2 \sigma^2}{\sigma^2 + N\sigma'^2},$$

and the reduction in variance is then

$$\sigma'^2 - \sigma''^2 = \sigma'^2 - \frac{\sigma'^2 \sigma^2}{\sigma^2 + N\sigma'^2}$$

$$= \frac{\sigma'^2 \sigma^2 + N\sigma'^4 - \sigma'^2 \sigma^2}{\sigma^2 + N\sigma'^2}$$

$$= \frac{N\sigma'^4}{\sigma^2 + N\sigma'^2},$$

which is identical with the expression for σ^{*2} given in Equation (9.25.4). The term σ^* is therefore the square root of the reduction in variance due to the sample of size N.

Returning to our example, suppose that the decision maker contemplates a sample of size 10. The reduction in variance would be

$$\sigma^{*2} = \frac{N\sigma'^4}{\sigma^2 + N\sigma'^2} = \frac{10(5)^4}{100 + 10(25)} = \frac{6250}{350} = 17.9,$$

and

$$\sigma^* = \sqrt{17.9} = 4.23.$$

The value of D^* is

$$D^* = \left| \frac{\mu_b - M'}{\sigma^*} \right| = \left| \frac{98 - 100}{4.23} \right| = |-.47| = .47,$$

and from Equation (9.25.3),

$$\text{EVSI} = .5(4.23)L_N(.47) = .5(4.23)(.2072) = .438.$$

This is the value to the decision maker of a sample of size 10. In a similar manner, he could compute the value of a sample of any given sample size. Some values are computed in the following table:

Sample Size (N)	σ^*	D^*	EVSI(N)
1	2.24	.89	.115
2	2.89	.69	.210
3	3.28	.61	.272
4	3.54	.56	.319
5	3.75	.53	.354
6	3.88	.51	.378
7	3.99	.50	.396
8	4.09	.49	.411
9	4.16	.48	.424
10	4.23	.47	.438
11	4.29	.47	.444

Notice that as N gets larger, EVSI(N) also gets larger, but the rate of increase of EVSI(N) becomes smaller. In other words, the EVSI curve is beginning to "level off." It is worthwhile to note that as N gets larger and larger, the sample information is closer and closer to what we have termed "perfect information," and the EVSI gets closer and closer to the EVPI, which in this case has been shown to be .576. For instance, if $N = 100$, EVSI = .556. Obviously EVSI can never exceed EVPI, for sample information can never be more valuable than perfect information.

In general, as N increases, the posterior variance becomes smaller, implying that the reduction in variance, σ^{*2}, becomes larger. As σ^{*2} becomes larger, D^* becomes smaller, since σ^* is in the denominator of D^*. Finally, as D^* becomes smaller, $L_N(D^*)$ becomes larger. The EVSI consists of three factors: $|s_1 - s_2|$, σ^*, and $L_N(D^*)$. As N gets larger, the first term remains constant and the other two terms increase, as the above discussion indicates. Thus, as N gets larger, the EVSI gets larger. By comparing Equations (9.25.1) and (9.25.3) as the sample size increases, you can see that the EVSI approaches the EVPI.

In determining optimal sample size, it is necessary to consider the cost of sampling, CS(N), as well as EVSI(N). Suppose that

$$CS(N) = .03N.$$

That is, the cost of sampling is a constant .03 per trial. Then, for each value of N, we can determine ENGS(N) = EVSI(N) − CS(N): the optimal sample size is $N = 5$. As N increases, the ENGS increases until the EVSI starts to level off while the cost of sampling continues to increase at the same rate. In Figure 9.25.1, the values in the preceding table are graphed, and three curves are drawn through the three sets of points, representing the EVSI, CS, and ENGS curves.

Sample Size (N)	EVSI(N)	CS(N)	ENGS(N)
1	.115	.030	.085
2	.210	.060	.150
3	.272	.090	.182
4	.319	.120	.199
5	.354	.150	.204
6	.378	.180	.198
7	.396	.210	.186
8	.411	.240	.171
9	.424	.270	.154
10	.438	.300	.138
11	.444	.330	.114

In this section we have discussed the determination of the value of perfect information and sample information under the assumptions that the loss functions were linear and the distribution of θ was normal. If $f(\theta)$ is not a normal distribution, the general idea is the same, but different linear loss integrals must be considered. Recall that L_N is the unit *normal* linear loss integral. If $f(\theta)$ is a normal distribution or if it can be closely approximated by a normal distribution, then the table of values of L_N can be used in determining EVPI and EVSI. Otherwise, it is necessary to look at different linear loss integrals (such as the loss integrals associated with a beta distribution). Since the general approach is similar to the approach presented in this section, we will not discuss loss integrals other than L_N, the linear loss integral associated with the normal distribution.

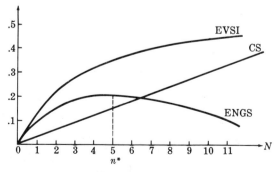

Figure 9.25.1

9.26   THE GENERAL FINITE-ACTION PROBLEM WITH LINEAR PAYOFFS AND LOSSES

In the past three sections we have been concerned with *two-action problems* in which the payoff functions (and hence the loss functions) of both of the actions are linear. In some situations there may be more than two possible actions; as long as the number of possible actions is finite, we refer to these situations as *finite-action problems*. Under linear payoff functions, the *finite-action problem* is similar to its special case, the two-action problem. The formulas for the loss functions, EVPI, and EVSI are fairly complex, however, so we will just attempt to present a brief nonmathematical discussion, illustrating the important points with graphs.

Formally, suppose that there are k possible actions, so that the set of actions A can be represented by $\{a_1, a_2, \cdots, a_k\}$. The payoff function of the action a_i is given by

$$P(a_i, \theta) = r_i + s_i\theta \qquad \text{for } i = 1, 2, \cdots, k. \qquad (9.26.1)$$

It is clear that if $k = 2$, this reduces to a two-action problem. Figure 9.26.1 illustrates the payoff functions in a four-action problem. Notice that there are three "breakeven values," which are labeled θ_1, θ_2, and θ_3. Under perfect information about θ, the following decision rule maximizes EP:

$$\begin{array}{lll}
\text{Choose } a_1 & \text{if} & \theta \leq \theta_1, \\
\text{Choose } a_2 & \text{if} & \theta_1 \leq \theta \leq \theta_2, \\
\text{Choose } a_3 & \text{if} & \theta_2 \leq \theta \leq \theta_3, \\
\text{Choose } a_4 & \text{if} & \theta \geq \theta_3. & (9.26.2)
\end{array}$$

Under uncertainty, the decision rule is as above with $E(\theta)$ substituted for θ. The reason for this is that the payoff functions are linear, and there-

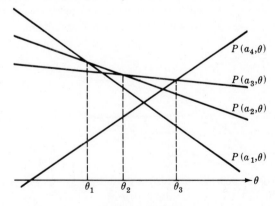

Figure 9.26.1

fore the mean of the decision maker's distribution is a certainty equivalent. In general, for a k-action problem, there will be $(k-1)$ breakeven values. These values will divide the real line into k intervals, and each of the k actions will be optimal over exactly one of the intervals.

The loss function in a k-action problem is a piecewise linear function. That is, it is made up of a series of k line segments with different slopes. In the four-action problem, suppose that a_2 is the optimal act under a prior distribution. From Figure 9.26.1, we see that

$$L(a_2, \theta) = \begin{cases} P(a_1, \theta) - P(a_2, \theta) & \text{if } \theta \leq \theta_1, \\ \\ 0 & \text{if } \theta_1 \leq \theta \leq \theta_2, \\ \\ P(a_3, \theta) - P(a_2, \theta) & \text{if } \theta_2 \leq \theta \leq \theta_3, \\ \\ P(a_4, \theta) - P(a_2, \theta) & \text{if } \theta \geq \theta_3. \end{cases} \tag{9.26.3}$$

At any point, the loss is equal to the difference between the highest payoff for that value of θ and the payoff obtained with action a_2. Figure 9.26.2 demonstrates the shape of $L(a_2, \theta)$. To evaluate the expected loss (and hence the EVPI), it is necessary to consider a series of terms involving linear loss integrals. Since the general idea is similar to the discussion in the previous section, little would be gained by presenting these reasonably complicated formulas. The point of this section is simply to note that the two-action problem with linear losses can be generalized to a finite-action problem with linear losses and to point out the similarity between the two types of problems. It might also be mentioned that we have already considered *infinite-action problems* with linear loss functions in Section 9.12 under the discussion of point estimation as a decision-making problem.

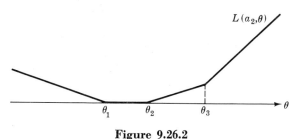

Figure 9.26.2

9.27   DECISION UNDER CERTAINTY

We have indicated how the problem of making decisions under uncertainty is simplified if the uncertainty is removed. If he knows for sure what state of the world will occur, the decision maker does not have to work with expected payoffs or expected losses; he simply chooses the action with the highest payoff (or, equivalently, the smallest loss) for that state of the world. Given a payoff or a loss *table*, it is necessary to determine the action associated with the highest (or lowest, for losses) value in the appropriate column. Given a payoff or loss *function*, it is necessary to determine the action associated with the highest (or lowest) possible value of the function. This requires the use of a class of methods which are called mathematical optimization techniques. You are probably familiar with some of these techniques. The basic optimization technique using the calculus involves taking the derivative of a function, setting it equal to zero, and solving the resulting equation. To see whether the solution is a maximum, a minimum, or neither, it is necessary to also look at the second derivative. We will not discuss this optimization technique in further detail, since anyone familiar with the calculus should have encountered it at some time or other.

For many payoff and loss functions, the above technique will determine the optimal decision, thereby solving the problem of decision under certainty. Sometimes, however, other optimization techniques are needed. The study of certain types of optimization problems arising in decision-making problems is called *mathematical programming*, the simplest example of which is *linear programming*. Such techniques are tangential to the study of statistics, so we will not attempt to describe them. Statistics deals with uncertainty, and these mathematical procedures deal primarily with decision under *certainty*. The point of this section is merely to note that some problems of decision under certainty are difficult problems, and that advanced mathematical optimization techniques are required to solve them.

9.28   GAME THEORY

In the discussion of decision under uncertainty, we have implicitly assumed that the "state of the world" is determined by "nature" or by some disinterested party. Suppose that a manufacturer must decide whether or not to produce a new product, and that the "state of the world" represents the demand for the new product. If the manufacturer can express his state of information about the potential demand in the form of a probability distribution, then he faces a problem of decision under uncertainty. However, there may be other factors which he should consider. If he produces the new product, perhaps a rival firm will decide to follow suit, and the total demand for the product would then be divided (not necessarily equally) between the two firms. The payoff to the manufacturer thus depends not only on the demand for the product, but also on the decision of another firm. The problem is now one of decision-making under competitive conditions. Since this type of problem has traditionally been investigated in the context of "games" between individuals, the study of such problems is called *game theory*.

In game theory the decision maker faces not a disinterested opponent, such as "nature," but an opponent who is interested in maximizing his own payoff. The "state of the world" is thus the action selected by the opponent. Some examples of realistic game-theoretic situations are (1) a bidding situation, in which competing firms bid for a contract and the firm with the lowest bid is awarded the contract; (2) a bargaining situation, in which two parties (for example, union and management) attempt to reach an agreement; and (3) a military situation, in which each side must consider the possible actions that could be taken by the opponents. Much of the study of game theory has not involved the use of probabilistic decision rules. Such rules as the maximin rule have been used with the justification that the decision maker's opponent will try to act so as to maximize his own payoff and hence minimize the decision maker's payoff.

Through the use of subjective probabilities, the game theory problem can be thought of as a problem of decision-making under uncertainty, which can be treated by the methods discussed in this chapter. The decision maker must assess a probability distribution on the set of actions available to his opponent. If the opponent has two actions, and we call these actions θ_1 and θ_2, then the decision maker must subjectively assess $P(\theta_1)$ and $P(\theta_2)$. In order to make such assessments, he must take into consideration his information concerning the opponent and concerning the decision problem facing the opponent. Suppose that a manufacturer's decision as to whether or not to introduce a new product depends on the manufacturer's assessment of the chances that a rival firm will follow suit. By considering past actions of the rival firm in similar situations, their general willingness or lack of willingness to introduce new products, the capabilities of their key personnel, and so on, the manufacturer should be able to represent his information by a set of probabilities. It may be even more difficult to assess probabilities in this case than it is when the state of the world is determined by a "disinterested opponent." If such probabilities can be assessed, though, the game theory problem can be treated as a problem of decision under uncertainty.

9.29 A COMMENT ON THE DIFFERENT APPROACHES TO STATISTICAL PROBLEMS

Chapters 5–9 have covered the basic theory of statistical inference and decision. Briefly reviewing, in the classical approach inferences are based on the information of a sample; Bayesian inference goes one step further by including any other available information; and in decision theory, a third input is included, representing the payoffs or losses associated with various decisions. We have seen how Bayesian inference and decision theory can be thought of as extensions of classical inference. In Volume II (Chapters 10–12), a number of important statistical techniques will be presented. Because many of the Bayesian and decision-theoretic extensions of these techniques are fairly complex, involving such things as the assessment of subjective multivariate distributions, most of what follows will be presented in the classical vein. However, we will attempt to indicate briefly how prior information and losses or payoffs could be brought into the analysis. If you prefer the Bayesian interpretation of classical inferential statements (essentially, this means that you allow probability statements about the unknown parameter θ), then the classical results can be viewed as being roughly (sometimes *very* roughly, especially with small samples) equivalent to Bayesian results under a diffuse prior distribution.

EXERCISES

1. Explain the difference between decision-making under certainty and decision-making under uncertainty.

2. You are given the following payoff table:

State of the World

		A	B	C	D	E
Action	1	−50	80	20	100	0
Action	2	30	40	70	20	50
Action	3	10	30	−30	10	40
Action	4	−10	−50	−70	−20	200

(a) Are any of the actions inadmissible? If so, eliminate them from further consideration.

(b) Find the loss table corresponding to the above payoff table.

3. Consider the following *loss* table:

State of the World

	I	II	III
Action 1	0	3	6
Action 2	1	1	0
Action 3	4	0	1

Given this loss table, complete the corresponding *payoff* table:

State of the World

	I	II	III
Action 1	12	7	9
Action 2			
Action 3			

Are any of the actions inadmissible?

4. For the payoff and loss tables in Exercise 2, find the action which is optimal under each of the following rules:
    (a) maximin
    (b) maximax
    (c) minimax loss.

5. For the payoff and loss tables in Exercise 3, find the action which is optimal under the maximin, maximax, and minimax loss rules.

6. What is the primary disadvantage of decision-making rules such as maximin, maximax, and minimax loss?

7. In Exercise 2, find the expected payoff and the expected loss of each action, and find the action which has the largest expected payoff and smallest expected loss, given that the probabilities of the various states of the world are $P(A) = .10$, $P(B) = .20$, $P(C) = .25$, $P(D) = .10$, and $P(E) = .35$.

8. In Exercise 3, if $P(I) = .25$, $P(B) = .45$, and $P(C) = .30$, find the expected payoff and expected loss of each of the three actions. Using EP (or EL) as a decision-making criterion, which action is optimal?

9. Why is the maximization of expected monetary value (that is, EP with the payoffs expressed in terms of money) not always a reasonable criterion for decision-making? What problems does this create for the decision maker?

10. You must choose between two acts, where the payoff matrix is as follows (in terms of dollars):

State of the World

		A	B	C
	1	100	70	20
Action	2	10	50	120
	3	50	80	30

Find the optimal act according to the expected utility criterion if $P(A) = .3$, $P(B) = .3$, $P(C) = .4$, and the utility function is (assuming M represents dollars):
(a) $U(M) = 50 + 2M$
(b) $U(M) = 50 + 2M^2$
(c) $U(M) = M$
(d) $U(M) = M^2 + 5M + 6$.

11. Suppose that you are offered a choice between Bets A and B:

Bet A: You win $1,000,000 with certainty (that is, with probability 1).

Bet B: You win $5,000,000 with probability .1
You win $1,000,000 with probability .89
You win $0 with probability .01.

Which bet would you choose? Similarly, choose between Bets C and D:

Bet C: You win $1,000,000 with probability .11
        You win $0 with probability .89.

Bet D: You win $5,000,000 with probability .10
        You win $0 with probability .90.

Prove that if you chose Bet A, then you should have chosen Bet C, and if you chose Bet B, then you should have chosen Bet D. If you selected A and D or B and C, explain your choices. In light of the proof, would you change your choices?

12. Suppose that you are contemplating drilling an oil well, with the following payoff table (in terms of thousands of dollars):

State of the World

	Oil	No oil
Drill	100	−40
Do not drill	0	0

If after consulting a geologist you decide that $P(\text{Oil}) = .30$, would you drill or not drill according to the
(a) maximin criterion
(b) maximax criterion
(c) minimax loss criterion
(d) EP criterion
(e) EL criterion
(f) EU criterion, where $U(0) = .40$, $U(100) = +1$, and $U(-40) = 0$?
Explain the differences in the results in parts (a)–(f).

13. Comment on the following statement from the text: "Some individuals appear to be risk-takers for some decisions (such as gambling) and risk-avoiders for other decisions (such as purchasing insurance)." Can you explain why this phenomenon occurs? Can you draw a utility function which would explain it?

14. Attempt to determine your own utility function for money in the range from −$500 to +$500. If you were given actual decision-making situations, would you act in accordance with this utility function?

15. Suppose that a contractor must decide whether or not to build any speculative houses (that is, houses that he would have to find a buyer for), and if so,

how many. The houses which this contractor builds are sold for a price of $30,000, and they cost him $26,000 to build. Since the contractor cannot afford to have too much cash tied up at once, any houses that remain unsold three months after they are completed will have to be sold to a realtor for $25,000. The contractor's prior distribution for the number of houses that will be sold within three months of completion is

x	$P(X = x)$
0	.05
1	.10
2	.10
3	.20
4	.25
5	.20
6	.10

If the contractor's utility function is linear with respect to money, how many houses should he build?

16. A hot dog vendor at a football game must decide in advance how many hot dogs to order. He makes a profit of $.10 on each hot dog which is sold, and he suffers a $.20 loss on hot dogs which are unsold. If his distribution of the number of hot dogs that will be demanded at the football game is a normal distribution with mean 10,000 and standard deviation 2000, how many hot dogs should he order?

17. A sales manager is asked to forecast the total sales of his division for a forth-coming period of time. He feels that his loss function is linear as a function of the difference between his estimate and the true value, but he also feels that an error of overestimation is three times as serious as an error of under-estimation (given that the magnitude of the errors is equal). This is because his superiors will criticize him if the division does worse than he predicts, but they will be happy if the division does better than predicted and will be less likely to be concerned about an error in predicting. If his actual judg-ments can be represented by a normal distribution with mean 50,000 and standard deviation 10,000, what value should he report as his forecast of sales?

18. Prove that if a decision maker's loss function in a point estimation problem is given by Equation (9.12.1), then the $k_u/(k_u + k_o)$ fractile of the decision maker's distribution of θ is the optimal point estimate.

19. An economist is asked to predict a future value of a particular economic indicator, and he thinks that his loss function is linear with $k_u = 4k_o$. If his distribution is a uniform distribution on the interval from 650 to 680, what should his estimate be?

20. In Exercises 15, 16, 17, and 19, what would the optimal estimates be if the loss function were quadratic instead of linear [that is, if the loss function were of the form (9.13.1)]?

21. Comment on the following statement: "In taking the sample mean as an estimator of the population mean, the statistician is acting essentially as though he had a quadratic loss function." Explain the difference between the approach to point estimation taken in Chapter 6 and the approach taken in this chapter.

22. If a person faces a point estimation problem with a linear loss function with $k_u = 4$ and $k_o = 3$, does he need to assess an entire probability distribution or can he determine a certainty equivalent? Explain.

23. If a statistician wishes to estimate a parameter θ subject to a loss function which is linear with $k_o = 2k_u$, and his distribution of θ is an exponential distribution with $\lambda = 4$, what is the optimal estimate of θ?

24. If the loss function in a point estimation problem is of the form

$$L(a, \theta) = \begin{cases} 0 & \text{if } |a - \theta| < k, \\ 1 & \text{otherwise,} \end{cases}$$

where k is some very small positive number, what would the optimal estimate be? Can you think of any realistic situations in which the loss function might be of this form?

25. In Exercise 7, Chapter 8, determine an estimate of p from the posterior distribution if the loss function is

$$L(a, p) = k(a - p)^2.$$

How does this differ from the estimate obtained from the prior distribution using the same loss function? From the estimate obtained from the sample alone?

26. In Exercise 11, Chapter 8, the production manager must make an estimate of the mean weight of items turned out by the process in question. His loss function is linear with $k_o = k_u$. What should his estimate be?

27. In Exercise 24, Chapter 8, the marketing manager decides that the loss which will be suffered if the company markets the product and p is in fact only .10 is three times as great as the loss which will be suffered if the company fails to market the product and p is actually .20. Should the product be marketed?

28. A statistician is interested in testing the hypotheses

$$H_0 : \mu \geq 120$$

and

$$H_1 : \mu < 120,$$

where μ is the mean of a normally distributed population with variance 144. The prior distribution for μ is a normal distribution with mean $M' = 115$ and variance $\sigma'^2 = 36$. A sample of size 8 is taken, with sample mean $M = 121$.
(a)   Find the prior odds ratio of H_0 to H_1.
(b)   Find the posterior odds ratio of H_0 to H_1.
(c)   If $L_I = 4$ and $L_{II} = 6$, which hypothesis should be accepted according to the posterior distribution?

29.   For a sample of size one from a normal population with known variance 25, show that the classical test of

$$H_0 : \mu = 50$$

versus $$H_1 : \mu = 60$$

with $\alpha = .05$ is a likelihood ratio test. That is, show that the rejection region can be expressed in the form $LR \leq c$, where LR is the likelihood ratio and c is some constant. [*Hint:* Consider the equation $LR \leq c$ and attempt to manipulate it algebraically to get the result $X \geq k$, where k is some constant, since you know that the rejection region must be of this form.]

30.   Consider the hypotheses $H_0 : \mu = 10$ and $H_1 : \mu = 12$, where μ is the mean of a normally distributed population with variance 1. The prior distribution is diffuse (take a normal distribution with $N' = 0$), and a sample of size 10 is to be taken. If $L_I = 100$ and $L_{II} = 50$,
(a)   find the region of rejection (using the decision-theoretic approach)
(b)   from this region of rejection, find α and β
(c)   if $L_I = 50$ and $L_{II} = 50$, what would the region of rejection be?

31.   Compare and contrast the determination of α and β in hypothesis testing problems according to (a) the classical "convention" presented in Chapter 7 and (b) the decision-theoretic approach, as illustrated in Exercise 30.

32.   Carefully distinguish between "classical" inferential statistics, Bayesian inferential statistics, and decision theory.

33.   What is the difference between a terminal decision and a preposterior decision? Are they at all related?

34.   In Exercise 12, what is the value of perfect information that there is oil? What is the value of perfect information that there is no oil? From these results, calculate the expected value of perfect information in this example.

35.   In Exercise 8, find the expected value of perfect information. If you could purchase perfect information for $5 (assuming the values in the table are given in terms of dollars), would you do so?

36. A store must decide whether or not to stock a new item. The decision depends
on the reaction of consumers to the item, and the payoff table is as follows:

Proportion of Consumers Purchasing the Item

		.10	.20	.30	.40	.50
	Stock 100	−10	−2	12	22	40
Decision	*Stock* 50	− 4	6	12	16	16
	Do not stock	0	0	0	0	0

If $P(.10) = .2$, $P(.20) = .3$, $P(.30) = .3$, $P(.40) = .1$, and $P(.50) = .1$,
what decision should be made? If perfect information is available, find VPI
for each of the five possible states of the world and compute EVPI.

37. In Exercise 36, suppose that sample information is available in the form of
a random sample of consumers. For a sample of size *one,*
(a) find the posterior distribution if the one person sampled will purchase
the item, and find the value of this sample information
(b) find the posterior distribution if the one person sampled will *not* purchase the item, and find the value of this sample information
(c) find the expected value of sample information.

38. In Exercise 37, suppose that you also wanted to consider samples of other
sizes.
(a) Find EVSI for a sample of size 2.
(b) Find EVSI for a sample of size 5.
(c) Find EVSI for a sample of size 10.
(d) If the cost of sampling is .50 per unit sampled, find the expected net
gain of sampling (ENGS) for samples of size 1, 2, 5, and 10.

39. Consider a bookbag filled with 100 poker chips. You know that either 70 of the
chips are red and the remainder blue, or that 70 are blue and the remainder
red. You must guess whether the bookbag has 70R–30B or 70B–30R. If you
guess correctly, you win $5. If you guess incorrectly, you lose $3. Your prior
probability that the bookbag contains 70R–30B is .40.
(a) If you had to make your guess on the basis of the prior information,
what would you guess?

(b)   If you could purchase perfect information, what is the most that you should be willing to pay for it?

(c)   If you could purchase sample information in the form of one draw from the bookbag, how much should you be willing to pay for it?

(d)   If you could purchase sample information in the form of five draws (with replacement) from the bookbag, how much should you be willing to pay for it?

40.   Do Exercise 39 with the following payoff table (in dollars):

State of the World

		70R–30B	70B–30R
Your Guess	70R–30B	6	−2
	70R–30B	−6	10

41.   In Exercise 36, show that the EVPI is equal to the expected loss of the optimal action under the prior distribution.

42.   Suppose that the payoff functions of two actions are

$$P(a_1, \mu) = 50 + .5\mu$$

and

$$P(a_2, \mu) = 70 - .5\mu,$$

where μ is the mean of a normal process with variance 1200.

(a)   Find the breakeven value, μ_b.

(b)   If the prior distribution is a normal distribution with mean $M' = 25$ and variance $\sigma'^2 = 400$, which action should you choose?

(c)   What is the value of perfect information that $\mu = 115$?

(d)   What is the value of perfect information that $\mu = 121$?

(e)   What is the expected value of perfect information?

(f)   Graph the payoff functions and the associated loss functions.

(g)   What would the expected value of perfect information be if the prior mean $M' = 30$? Explain the difference between this answer and the answer to (e).

43.   The payoff in a certain decision-making problem depends on p, the parameter of a Bernoulli process. The payoff functions are

$$P(a_1, p) = -10 + 100p$$

and

$$P(a_2, p) = 50p.$$

(a) What is the breakeven value of p?
(b) Graph the payoff functions and the associated loss functions.
(c) If the prior distribution is a beta distribution with $r' = 4$ and $N' = 24$, which action should be chosen?
(d) If a sample of size 15 is taken, with 4 successes, find the posterior distribution and the optimal action under this distribution.

44. In Exercise 42, find the EVSI and ENGS for samples of size 1, 2, 3, 4, 5, 6, 7, 8, 9, and 10, assuming that the cost of sampling is 2. On a graph, draw curves representing EVSI, ENGS, and CS.

45. Suppose that an investor has a choice of four investments (assume that he must choose one, and only one, of the four—he cannot divide his money among them):
A: a savings account in a bank.
B: a stock which moves counter to the general stock market.
C: a growth stock which moves at a faster pace than the market.
D: a second growth stock which is similar to C but not quite as speculative.
The investor has decided that his payoff depends on his choice of investment and on the *next year's change* in the Dow-Jones Industrials (DJI), which we shall call θ. (Ignore the fact that he should be interested in the change in price of the investments themselves—for example, by assuming perfect correlations between the DJI and the investments). His payoff functions are as follows:

$$P(A, \theta) = .05$$
$$P(B, \theta) = .05 - \theta$$
$$P(C, \theta) = -.15 + 2\theta$$
$$P(D, \theta) = -.05 + \theta.$$

(a) Draw the graphs of these payoff functions and determine a decision rule for choice of an investment under certainty about θ.
(b) Suppose that the investor's prior distribution is normal with mean .08 and standard deviation .05. Which investment should he choose?
(c) Show graphically and algebraically the CVPI curve. If the investor obtains perfect information and finds that $\theta = .02$, what is the CVPI?
(d) Determine and graph the loss functions for the four actions A, B, C, and D.

46. Explain how game theory problems (that is, decision-making problems in which there is some "opponent") can be analyzed using the same techniques which are used in decision-making under uncertainty. Instead of "states of the world," we have "actions of the opponent." Might this make it more difficult to determine the probabilities necessary to determine expected payoffs and losses? Explain.

47. Consider a situation in which you and a friend both must choose a number
from the two numbers 1 and 2. The relevant payoffs are as follows:

If you choose 1 and he chooses 1,   you both win $10
If you choose 1 and he chooses 2,   he wins $15 and you win $0
If you choose 2 and he chooses 1,   you win $15 and he wins $0
If you choose 2 and he chooses 2,   you both win $5.

(a) How would you go about assigning probabilities to his two actions?
(b) On the basis of the probabilities assigned in (a), what is your optimal
     action?
(c) Could you have determined that this was your optimal action without
     using any probabilities? Why?
(d) Would you expect your action to be different if you could get together
     with your friend and make a bargain with him before the game is
     played? Explain.

APPENDIX: SOME COMMON DIFFERENTIATION AND INTEGRATION FORMULAS

The derivative of y with respect to x is simply the rate of change of y as x changes. In other words, if y can be shown on a graph as a function of x, $y = f(x)$, then the derivative of y with respect to x at the point x_0 is simply the slope of the line which is tangent to the curve $y = f(x)$ at the point x_0. The derivative of y with respect to x is denoted by dy/dx, or $df(x)/dx$, or $f'(x)$, and it is formally defined as follows:

$$\frac{dy}{dx} = \lim_{h \to 0} \frac{f(x+h) - f(x)}{h}.$$

We are not concerned with deriving differentiation formulas here; for derivations the student can refer to a textbook on calculus. For the student's convenience, several common differentiation formulas are presented:

y	dy/dx
c	0
x^k	kx^{k-1}
cx	c
$cf(x)$	$c\,\dfrac{df(x)}{dx}$
$[f(x)]^k$	$k[f(x)]^{k-1}\dfrac{df(x)}{dx}$
$f(x) + g(x)$	$\dfrac{df(x)}{dx} + \dfrac{dg(x)}{dx}$
$f(x) - g(x)$	$\dfrac{df(x)}{dx} - \dfrac{dg(x)}{dx}$

y	dy/dx
$f(x)g(x)$	$f(x)\dfrac{dg(x)}{dx} + g(x)\dfrac{df(x)}{dx}$
$f(x)/g(x)$	$\left[g(x)\dfrac{df(x)}{dx} - f(x)\dfrac{dg(x)}{dx} \right] \Big/ [g(x)]^2$
e^x	e^x
$e^{f(x)}$	$e^{f(x)}\dfrac{df(x)}{dx}$
a^x	$a^x \log_e a$
$\log_e x$	$1/x$
$\log_e f(x)$	$\dfrac{1}{f(x)}\dfrac{df(x)}{dx}$

An integral is simply an antiderivative; that is, integration is essentially the inverse of differentiation. If the derivative of $f(x)$ is $f'(x)$, then the integral of $f'(x)$ with respect to x, written $\int f'(x)\,dx$, is equal to $f(x) + C$, where C is a constant with respect to x. In general, the integral of a function is not unique, for the constant C could be just about anything, provided that it is not a function of x. However, if we specify limits of integration, then the integral is a definite integral, and it is unique. The definite integral of a function over a certain interval can be interpreted graphically as the area between the curve and the x-axis in this interval, where the area is taken to be positive if the curve is above the axis and negative if it is below the axis. This area can be approximated by the sum of the areas of a finite set of rectangles, and this is the basis for the formal definition of a definite integral. Several common integration formulas are presented below (once again the student is referred to a calculus textbook for the derivation of these formulas):

$$\int a\,dx = ax + C$$

$$\int af(x)\,dx = a\int f(x)\,dx$$

$$\int x^k\,dx = \frac{x^{k+1}}{k+1} + C$$

$$\int [f(x) + g(x)] \, dx = \int f(x) \, dx + \int g(x) \, dx$$

$$\int [f(x) - g(x)] \, dx = \int f(x) \, dx - \int g(x) \, dx$$

$$\int e^x \, dx = e^x + C$$

$$\int e^{f(x)} \frac{df(x)}{dx} \, dx = e^{f(x)} + C$$

$$\int \frac{1}{x} \, dx = \log_e |x| + C$$

$$\int \frac{1}{f(x)} \frac{df(x)}{dx} \, dx = \log_e |f(x)| + C$$

$$\int f(x) \frac{dg(x)}{dx} \, dx = f(x) \, g(x) - \int g(x) \frac{df(x)}{dx} \, dx$$

If

$$\int f(x) = F(x) + C, \quad \text{then} \quad \int_a^b f(x) = F(b) - F(a).$$

TABLES

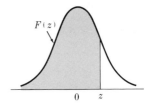

$F(z)$

Table I. Cumulative Normal Probabilities

z	$F(z)$	z	$F(z)$	z	$F(z)$	z	$F(z)$
.00	.5000000	.36	.6405764	.72	.7642375	1.08	.8599289
.01	.5039894	.37	.6443088	.73	.7673049	1.09	.8621434
.02	.5079783	.38	.6480273	.74	.7703500	1.10	.8643339
.03	.5119665	.39	.6517317	.75	.7733726	1.11	.8665005
.04	.5159534	.40	.6554217	.76	.7763727	1.12	.8686431
.05	.5199388	.41	.6590970	.77	.7793501	1.13	.8707619
.06	.5239222	.42	.6627573	.78	.7823046	1.14	.8728568
.07	.5279032	.43	.6664022	.79	.7852361	1.15	.8749281
.08	.5318814	.44	.6700314	.80	.7881446	1.16	.8769756
.09	.5358564	.45	.6736448	.81	.7910299	1.17	.8789995
.10	.5398278	.46	.6772419	.82	.7938919	1.18	.8809999
.11	.5437953	.47	.6808225	.83	.7967306	1.19	.8829768
.12	.5477584	.48	.6843863	.84	.7995458	1.20	.8849303
.13	.5517168	.49	.6879331	.85	.8023375	1.21	.8868606
.14	.5556700	.50	.6914625	.86	.8051055	1.22	.8887676
.15	.5596177	.51	.6949743	.87	.8078498	1.23	.8906514
.16	.5635595	.52	.6984682	.88	.8105703	1.24	.8925123
.17	.5674949	.53	.7019440	.89	.8132671	1.25	.8943502
.18	.5714237	.54	.7054015	.90	.8159399	1.26	.8961653
.19	.5753454	.55	.7088403	.91	.8185887	1.27	.8979577
.20	.5792597	.56	.7122603	.92	.8212136	1.28	.8997274
.21	.5831662	.57	.7156612	.93	.8238145	1.29	.9014747
.22	.5870604	.58	.7190427	.94	.8263912	1.30	.9031995
.23	.5909541	.59	.7224047	.95	.8289439	1.31	.9049021
.24	.9948349	.60	.7257469	.96	.8314724	1.32	.9065825
.25	.5987063	.61	.7290691	.97	.8339768	1.33	.9082409
.26	.6025681	.62	.7323711	.98	.8364569	1.34	.9098773
.27	.6064199	.63	.7356527	.99	.8389129	1.35	.9114920
.28	.6102612	.64	.7389137	1.00	.8413447	1.36	.9130850
.29	.6140919	.65	.7421539	1.01	.8437524	1.37	.9146565
.30	.6179114	.66	.7453731	1.02	.8461358	1.38	.9162067
.31	.6217195	.67	.7485711	1.03	.8484950	1.39	.9177356
.32	.6255158	.68	.7517478	1.04	.8508300	1.40	.9192433
.33	.6293000	.69	.7549029	1.05	.8531409	1.41	.9207302
.34	.6330717	.70	.7580363	1.06	.8554277	1.42	.9221962
.35	.6368307	.71	.7611479	1.07	.8576903	1.43	.9236415

Table I   (continued)

z	$F(z)$	z	$F(z)$	z	$F(z)$	z	$F(z)$
1.44	.9250663	1.77	.9616364	2.10	.9821356	2.43	.9924506
1.45	.9264707	1.78	.9624620	2.11	.9825708	2.44	.9926564
1.47	.9278550	1.79	.9632730	2.12	.9829970	2.45	.9928572
1.47	.9292191	1.80	.9640697	2.13	.9834142	2.46	.9930531
1.48	.9305634	1.81	.9648521	2.14	.9838226	2.47	.9932443
1.49	.9318879	1.82	.9656205	2.15	.9842224	2.48	.9934309
1.50	.9331928	1.83	.9663750	2.16	.9846137	2.49	.9936128
1.51	.9344783	1.84	.9671159	2.17	.9849966	2.50	.9937903
1.52	.9357445	1.85	.9678432	2.18	.9853713	2.51	.9939634
1.53	.9369916	1.86	.9685572	2.19	.9857379	2.52	.9941323
1.54	.9382198	1.87	.9692581	2.20	.9860966	2.53	.9942969
1.55	.9394292	1.88	.9699460	2.21	.9864474	2.54	.9944574
1.56	.9406201	1.89	.9706210	2.22	.9867906	2.55	.9946139
1.57	.9417924	1.90	.9712834	2.23	.9871263	2.56	.9947664
1.58	.9429466	1.91	.9719334	2.24	.9874545	2.57	.9949151
1.59	.9440826	1.92	.9725711	2.25	.9877755	2.58	.9950600
1.60	.9452007	1.93	.9731966	2.26	.9880894	2.59	.9952012
1.61	.9463011	1.94	.9738102	2.27	.9883962	2.60	.9953388
1.62	.9473839	1.95	.9744119	2.28	.9886962	2.70	.9965330
1.63	.9484493	1.96	.9750021	2.29	.9889893	2.80	.9974449
1.64	.9494974	1.97	.9755808	2.30	.9892759	2.90	.9981342
1.65	.9505285	1.98	.9761482	2.31	.9895559	3.00	.9986501
1.66	.9515428	1.99	.9767045	2.32	.9898296	3.20	.9993129
1.67	.9525403	2.00	.9772499	2.33	.9900969	3.40	.9996631
1.68	.9535213	2.01	.9777844	2.34	.9903581	3.60	.9998409
1.69	.9544860	2.02	.9783083	2.35	.9906133	3.80	.9999277
1.70	.9554345	2.03	.9788217	2.36	.9908625	4.00	.9999683
1.71	.9563671	2.04	.9793248	2.37	.9911060	4.50	.9999966
1.72	.9572838	2.05	.9798178	2.38	.9913437	5.00	.9999997
1.73	.9581849	2.06	.9803007	2.39	.9915758	5.50	.9999999
1.74	.9590705	2.07	.9807738	2.40	.9918025		
1.75	.9599408	2.08	.9812372	2.41	.9920237		
1.76	.9607961	2.09	.9816911	2.42	.9922397		

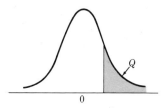

Table II.   Upper Percentage Points of the t Distribution

ν	$Q = 0.4$ $2Q = 0.8$	0.25 0.5	0.1 0.2	0.05 0.1	0.025 0.05	0.01 0.02	0.005 0.01	0.001 0.002
1	0.325	1.000	3.078	6.314	12.706	31.821	63.657	318.31
2	.289	0.816	1.886	2.920	4.303	6.965	9.925	22.326
3	.277	.765	1.638	2.353	3.182	4.541	5.841	10.213
4	.271	.741	1.533	2.132	2.776	3.747	4.604	7.173
5	0.267	0.727	1.476	2.015	2.571	3.365	4.032	5.893
6	.265	.718	1.440	1.943	2.447	3.143	3.707	5.208
7	.263	.711	1.415	1.895	2.365	2.998	3.499	4.785
8	.262	.706	1.397	1.860	2.306	2.896	3.355	4.501
9	.261	.703	1.383	1.833	2.262	2.821	3.250	4.297
10	0.260	0.700	1.372	1.812	2.228	2.764	3.169	4.144
11	.260	.697	1.363	1.796	2.201	2.718	3.106	4.025
12	.259	.695	1.356	1.782	2.179	2.681	3.055	3.930
13	.259	.694	1.350	1.771	2.160	2.650	3.012	3.852
14	.258	.692	1.345	1.761	2.145	2.624	2.977	3.787
15	0.258	0.691	1.341	1.753	2.131	2.602	2.947	3.733
16	.258	.690	1.337	1.746	2.120	2.583	2.921	3.686
17	.257	.689	1.333	1.740	2.110	2.567	2.898	3.646
18	.257	.688	1.330	1.734	2.101	2.552	2.878	3.610
19	.257	.688	1.328	1.729	2.093	2.539	2.861	3.579
20	0.257	0.687	1.325	1.725	2.086	2.528	2.845	3.552
21	.257	.686	1.323	1.721	2.080	2.518	2.831	3.527
22	.256	.686	1.321	1.717	2.074	2.508	2.819	3.505
23	.256	.685	1.319	1.714	2.069	2.500	2.807	3.485
24	.256	.685	1.318	1.711	2.064	2.492	2.797	3.467
25	0.256	0.684	1.316	1.708	2.060	2.485	2.787	3.450
26	.256	.684	1.315	1.706	2.056	2.479	2.779	3.435
27	.256	.684	1.314	1.703	2.052	2.473	2.771	3.421
28	.256	.683	1.313	1.701	2.048	2.467	2.763	3.408
29	.256	.683	1.311	1.699	2.045	2.462	2.756	3.396
30	0.256	0.683	1.310	1.697	2.042	2.457	2.750	3.385
40	.255	.681	1.303	1.684	2.021	2.423	2.704	3.307
60	.254	.679	1.296	1.671	2.000	2.390	2.660	3.232
120	.254	.677	1.289	1.658	1.980	2.358	2.617	3.160
∞	.253	.674	1.282	1.645	1.960	2.326	2.576	3.090

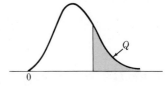

Table III.   Upper Percentage Points
of the χ^2 Distribution

ν \ Q	0.995	0.990	0.975	0.950	0.900	0.750	0.500
1	$392704 \cdot 10^{-10}$	$157088 \cdot 10^{-9}$	$982069 \cdot 10^{-9}$	$393214 \cdot 10^{-8}$	0.0157908	0.1015308	0.454937
2	0.0100251	0.0201007	0.0506356	0.102587	0.210720	0.575364	1.38629
3	0.0717212	0.114832	0.215795	0.351846	0.584375	1.212534	2.36597
4	0.206990	0.297110	0.484419	0.710721	1.063623	1.92255	3.35670
5	0.411740	0.554300	0.831211	1.145476	1.61031	2.67460	4.35146
6	0.675727	0.872085	1.237347	1.63539	2.20413	3.45460	5.34812
7	0.989265	1.239043	1.68987	2.16735	2.83311	4.25485	6.34581
8	1.344419	1.646482	2.17973	2.73264	3.48954	5.07064	7.34412
9	1.734926	2.087912	2.70039	3.32511	4.16816	5.89883	8.34283
10	2.15585	2.55821	3.24697	3.94030	4.86518	6.73720	9.34182
11	2.60321	3.05347	3.81575	4.57481	5.57779	7.58412	10.3410
12	3.07382	3.57056	4.40379	5.22603	6.30380	8.43842	11.3403
13	3.56503	4.10691	5.00874	5.89186	7.04150	9.29906	12.3398
14	4.07468	4.66043	5.62872	6.57063	7.78953	10.1653	13.3393
15	4.60094	5.22935	6.26214	7.26094	8.54675	11.0365	14.3389
16	5.14224	5.81221	6.90766	7.96164	9.31223	11.9122	15.3385
17	5.69724	6.40776	7.56418	8.67176	10.0852	12.7919	16.3381
18	6.26481	7.01491	8.23075	9.39046	10.8649	13.6753	17.3379
19	6.84398	7.63273	8.90655	10.1170	11.6509	14.5620	18.3376
20	7.43386	8.26040	9.59083	10.8508	12.4426	15.4518	19.3374
21	8.03366	8.89720	10.28293	11.5913	13.2396	16.3444	20.3372
22	8.64272	9.54249	10.9823	12.3380	14.0415	17.2396	21.3370
23	9.26042	10.19567	11.6885	13.0905	14.8479	18.1373	22.3369
24	9.88623	10.8564	12.4011	13.8484	15.6587	19.0372	23.3367
25	10.5197	11.5240	13.1197	14.6114	16.4734	19.9393	24.3366
26	11.1603	12.1981	13.8439	15.3791	17.2919	20.8434	25.3364
27	11.8076	12.8786	14.5733	16.1513	18.1138	21.7494	26.3363
28	12.4613	13.5648	15.3079	16.9279	18.9392	22.6572	27.3363
29	13.1211	14.2565	16.0471	17.7083	19.7677	23.5666	28.3362
30	13.7867	14.9535	16.7908	18.4926	20.5992	24.4776	29.3360
40	20.7065	22.1643	24.4331	26.5093	29.0505	33.6603	39.3354
50	27.9907	29.7067	32.3574	34.7642	37.6886	42.9421	49.3349
60	35.5346	37.4848	40.4817	43.1879	46.4589	52.2938	59.3347
70	43.2752	45.4418	48.7576	51.7393	55.3290	61.6983	69.3344
80	51.1720	53.5400	57.1532	60.3915	64.2778	71.1445	79.3343
90	59.1963	61.7541	65.6466	69.1260	73.2912	80.6247	89.3342
100	67.3276	70.0648	74.2219	77.9295	82.3581	90.1332	99.3341
z_Q	-2.5758	-2.3263	-1.9600	-1.6449	-1.2816	-0.6745	0.0000

Table III   (continued)

Q v	0.250	0.100	0.050	0.025	0.010	0.005	0.001
1	1.32330	2.70554	3.84146	5.02389	6.63490	7.87944	10.828
2	2.77259	4.60517	5.99147	7.37776	9.21034	10.5966	13.816
3	4.10835	6.25139	7.81473	9.34840	11.3449	12.8381	16.266
4	5.38527	7.77944	9.48773	11.1433	13.2767	14.8602	18.467
5	6.62568	9.23635	11.0705	12.8325	15.0863	16.7496	20.515
6	7.84080	10.6446	12.5916	14.4494	16.8119	18.5476	22.458
7	9.03715	12.0170	14.0671	16.0128	18.4753	20.2777	24.322
8	10.2188	13.3616	15.5073	17.5346	20.0902	21.9550	26.125
9	11.3887	14.6837	16.9190	19.0228	21.6660	23.5893	27.877
10	12.5489	15.9871	18.3070	20.4831	23.2093	25.1882	29.588
11	13.7007	17.2750	19.6751	21.9200	24.7250	26.7569	31.264
12	14.8454	18.5494	21.0261	23.3367	26.2170	28.2995	32.909
13	15.9839	19.8119	22.3621	24.7356	27.6883	29.8194	34.528
14	17.1170	21.0642	23.6848	26.1190	29.1413	31.3193	36.123
15	18.2451	22.3072	24.9958	27.4884	30.5779	32.8013	37.697
16	19.3688	23.5418	26.2962	28.8454	31.9999	34.2672	39.252
17	20.4887	24.7690	27.5871	30.1910	33.4087	35.7185	40.790
18	21.6049	25.9894	28.8693	31.5264	34.8053	37.1564	42.312
19	22.7178	27.2036	30.1435	32.8523	36.1908	38.5822	43.820
20	23.8277	28.4120	31.4104	34.1696	37.5662	39.9968	45.315
21	24.9348	29.6151	32.6705	35.4789	38.9321	41.4010	46.797
22	26.0393	30.8133	33.9244	36.7807	40.2894	42.7956	48.268
23	27.1413	32.0069	35.1725	38.0757	41.6384	44.1813	49.728
24	28.2412	33.1963	36.4151	39.3641	42.9798	45.5585	51.179
25	29.3389	34.3816	37.6525	40.6465	44.3141	46.9278	52.620
26	30.4345	35.5631	38.8852	41.9232	45.6417	48.2899	54.052
27	31.5284	36.7412	40.1133	43.1944	46.9630	49.6449	55.476
28	32.6205	37.9159	41.3372	44.4607	48.2782	50.9933	56.892
29	33.7109	39.0875	42.5569	45.7222	49.5879	52.3356	58.302
30	34.7998	40.2560	43.7729	46.9792	50.8922	53.6720	59.703
40	45.6160	51.8050	55.7585	59.3417	63.6907	66.7659	73.402
50	56.3336	63.1671	67.5048	71.4202	76.1539	79.4900	86.661
60	66.9814	74.3970	79.0819	83.2976	88.3794	91.9517	99.607
70	77.5766	85.5271	90.5312	95.0231	100.425	104.215	112.317
80	88.1303	96.5782	101.879	106.629	112.329	116.321	124.839
90	98.6499	107.565	113.145	118.136	124.116	128.299	137.208
100	109.141	118.498	124.342	129.561	135.807	140.169	149.449
z_Q	+0.6745	+1.2816	+1.6449	+1.9600	+2.3263	+2.5758	+3.0902

This table is taken from Table 8 of the *Biometrika Tables for Statisticians*, Vol. 1 (1st ed.), edited by E. S. Pearson and H. O. Hartley. Reproduced with the kind permission of E. S. Pearson and the trustees of *Biometrika*.

Table IV. Percentage Points of the F Distribution: Upper 5% Points

v_2 \ v_1	1	2	3	4	5	6	7	8	9	10	12	15	20	24	30	40	60	120	∞
1	161.4	199.5	215.7	224.6	230.2	234.0	236.8	238.9	240.5	241.9	243.9	245.9	248.0	249.1	250.1	251.1	252.2	253.3	254.3
2	18.51	19.00	19.16	19.25	19.30	19.33	19.35	19.37	19.38	19.40	19.41	19.43	19.45	19.45	19.46	19.47	19.48	19.49	19.50
3	10.13	9.55	9.28	9.12	9.01	8.94	8.89	8.85	8.81	8.79	8.74	8.70	8.66	8.64	8.62	8.59	8.57	8.55	8.53
4	7.71	6.94	6.59	6.39	6.26	6.16	6.09	6.04	6.00	5.96	5.91	5.86	5.80	5.77	5.75	5.72	5.69	5.66	5.63
5	6.61	5.79	5.41	5.19	5.05	4.95	4.88	4.82	4.77	4.74	4.68	4.62	4.56	4.53	4.50	4.46	4.43	4.40	4.36
6	5.99	5.14	4.76	4.53	4.39	4.28	4.21	4.15	4.10	4.06	4.00	3.94	3.87	3.84	3.81	3.77	3.74	3.70	3.67
7	5.59	4.74	4.35	4.12	3.97	3.87	3.79	3.73	3.68	3.64	3.57	3.51	3.44	3.41	3.38	3.34	3.30	3.27	3.23
8	5.32	4.46	4.07	3.84	3.69	3.58	3.50	3.44	3.39	3.35	3.28	3.22	3.15	3.12	3.08	3.04	3.01	2.97	2.93
9	5.12	4.26	3.86	3.63	3.48	3.37	3.29	3.23	3.18	3.14	3.07	3.01	2.94	2.90	2.86	2.83	2.79	2.75	2.71
10	4.96	4.10	3.71	3.48	3.33	3.22	3.14	3.07	3.02	2.98	2.91	2.85	2.77	2.74	2.70	2.66	2.62	2.58	2.54
11	4.84	3.98	3.59	3.36	3.20	3.09	3.01	2.95	2.90	2.85	2.79	2.72	2.65	2.61	2.57	2.53	2.49	2.45	2.40
12	4.75	3.89	3.49	3.26	3.11	3.00	2.91	2.85	2.80	2.75	2.69	2.62	2.54	2.51	2.47	2.43	2.38	2.34	2.30
13	4.67	3.81	3.41	3.18	3.03	2.92	2.83	2.77	2.71	2.67	2.60	2.53	2.46	2.42	2.38	2.34	2.30	2.25	2.21
14	4.60	3.74	3.34	3.11	2.96	2.85	2.76	2.70	2.65	2.60	2.53	2.46	2.39	2.35	2.31	2.27	2.22	2.18	2.13
15	4.54	3.68	3.29	3.06	2.90	2.79	2.71	2.64	2.59	2.54	2.48	2.40	2.33	2.29	2.25	2.20	2.16	2.11	2.07
16	4.49	3.63	3.24	3.01	2.85	2.74	2.66	2.59	2.54	2.49	2.42	2.35	2.28	2.24	2.19	2.15	2.11	2.06	2.01
17	4.45	3.59	3.20	2.96	2.81	2.70	2.61	2.55	2.49	2.45	2.38	2.31	2.23	2.19	2.15	2.10	2.06	2.01	1.96
18	4.41	3.55	3.16	2.93	2.77	2.66	2.58	2.51	2.46	2.41	2.34	2.27	2.19	2.15	2.11	2.06	2.02	1.97	1.92
19	4.38	3.52	3.13	2.90	2.74	2.63	2.54	2.48	2.42	2.38	2.31	2.23	2.16	2.11	2.07	2.03	1.98	1.93	1.88
20	4.35	3.49	3.10	2.87	2.71	2.60	2.51	2.45	2.39	2.35	2.28	2.20	2.12	2.08	2.04	1.99	1.95	1.90	1.84
21	4.32	3.47	3.07	2.84	2.68	2.57	2.49	2.42	2.37	2.32	2.25	2.18	2.10	2.05	2.01	1.96	1.92	1.87	1.81
22	4.30	3.44	3.05	2.82	2.66	2.55	2.46	2.40	2.34	2.30	2.23	2.15	2.07	2.03	1.98	1.94	1.89	1.84	1.78
23	4.28	3.42	3.03	2.80	2.64	2.53	2.44	2.37	2.32	2.27	2.20	2.13	2.05	2.01	1.96	1.91	1.86	1.81	1.76
24	4.26	3.40	3.01	2.78	2.62	2.51	2.42	2.36	2.30	2.25	2.18	2.11	2.03	1.98	1.94	1.89	1.84	1.79	1.73
25	4.24	3.39	2.99	2.76	2.60	2.49	2.40	2.34	2.28	2.24	2.16	2.09	2.01	1.96	1.92	1.87	1.82	1.77	1.71
26	4.23	3.37	2.98	2.74	2.59	2.47	2.39	2.32	2.27	2.22	2.15	2.07	1.99	1.95	1.90	1.85	1.80	1.75	1.69
27	4.21	3.35	2.96	2.73	2.57	2.46	2.37	2.31	2.25	2.20	2.13	2.06	1.97	1.93	1.88	1.84	1.79	1.73	1.67
28	4.20	3.34	2.95	2.71	2.56	2.45	2.36	2.29	2.24	2.19	2.12	2.04	1.96	1.91	1.87	1.82	1.77	1.71	1.65
29	4.18	3.33	2.93	2.70	2.55	2.43	2.35	2.28	2.22	2.18	2.10	2.03	1.94	1.90	1.85	1.81	1.75	1.70	1.64
30	4.17	3.32	2.92	2.69	2.53	2.42	2.33	2.27	2.21	2.16	2.09	2.01	1.93	1.89	1.84	1.79	1.74	1.68	1.62
40	4.08	3.23	2.84	2.61	2.45	2.34	2.25	2.18	2.12	2.08	2.00	1.92	1.84	1.79	1.74	1.69	1.64	1.58	1.51
60	4.00	3.15	2.76	2.53	2.37	2.25	2.17	2.10	2.04	1.99	1.92	1.84	1.75	1.70	1.65	1.59	1.53	1.47	1.39
120	3.92	3.07	2.68	2.45	2.29	2.17	2.09	2.02	1.96	1.91	1.83	1.75	1.66	1.61	1.55	1.50	1.43	1.35	1.25
∞	3.84	3.00	2.60	2.37	2.21	2.10	2.01	1.94	1.88	1.83	1.75	1.67	1.57	1.52	1.46	1.39	1.32	1.22	1.00

Table IV (continued) Upper 2.5% Points

ν_2 \ ν_1	1	2	3	4	5	6	7	8	9	10	12	15	20	24	30	40	60	120	∞
1	647.8	799.5	864.2	899.6	921.8	937.1	948.2	956.7	963.3	968.6	976.7	984.9	993.1	997.2	1001	1006	1010	1014	1018
2	38.51	39.00	39.17	39.25	39.30	39.33	39.36	39.37	39.39	39.40	39.41	39.43	39.45	39.46	39.46	39.47	39.48	39.49	39.50
3	17.44	16.04	15.44	15.10	14.88	14.73	14.62	14.54	14.47	14.42	14.34	14.25	14.17	14.12	14.08	14.04	13.99	13.95	13.90
4	12.22	10.65	9.98	9.60	9.36	9.20	9.07	8.98	8.90	8.84	8.75	8.66	8.56	8.51	8.46	8.41	8.36	8.31	8.26
5	10.01	8.43	7.76	7.39	7.15	6.98	6.85	6.76	6.68	6.62	6.52	6.43	6.33	6.28	6.23	6.18	6.12	6.07	6.02
6	8.81	7.26	6.60	6.23	5.99	5.82	5.70	5.60	5.52	5.46	5.37	5.27	5.17	5.12	5.07	5.01	4.96	4.90	4.85
7	8.07	6.54	5.89	5.52	5.29	5.12	4.99	4.90	4.82	4.76	4.67	4.57	4.47	4.42	4.36	4.31	4.25	4.20	4.14
8	7.57	6.06	5.42	5.05	4.82	4.65	4.53	4.43	4.36	4.30	4.20	4.10	4.00	3.95	3.89	3.84	3.78	3.73	3.67
9	7.21	5.71	5.08	4.72	4.48	4.32	4.20	4.10	4.03	3.96	3.87	3.77	3.67	3.61	3.56	3.51	3.45	3.39	3.33
10	6.94	5.46	4.83	4.47	4.24	4.07	3.95	3.85	3.78	3.72	3.62	3.52	3.42	3.37	3.31	3.26	3.20	3.14	3.08
11	6.72	5.26	4.63	4.28	4.04	3.88	3.76	3.66	3.59	3.53	3.43	3.33	3.23	3.17	3.12	3.06	3.00	2.94	2.88
12	6.55	5.10	4.47	4.12	3.89	3.73	3.61	3.51	3.44	3.37	3.28	3.18	3.07	3.02	2.96	2.91	2.85	2.79	2.72
13	6.41	4.97	4.35	4.00	3.77	3.60	3.48	3.39	3.31	3.25	3.15	3.05	2.95	2.89	2.84	2.78	2.72	2.66	2.60
14	6.30	4.86	4.24	3.89	3.66	3.50	3.38	3.29	3.21	3.15	3.05	2.95	2.84	2.79	2.73	2.67	2.61	2.55	2.49
15	6.20	4.77	4.15	3.80	3.58	3.41	3.29	3.20	3.12	3.06	2.96	2.86	2.76	2.70	2.64	2.59	2.52	2.46	2.40
16	6.12	4.69	4.08	3.73	3.50	3.34	3.22	3.12	3.05	2.99	2.89	2.79	2.68	2.63	2.57	2.51	2.45	2.38	2.32
17	6.04	4.62	4.01	3.66	3.44	3.28	3.16	3.06	2.98	2.92	2.82	2.72	2.62	2.56	2.50	2.44	2.38	2.32	2.25
18	5.98	4.56	3.95	3.61	3.38	3.22	3.10	3.01	2.93	2.87	2.77	2.67	2.56	2.50	2.44	2.38	2.32	2.26	2.19
19	5.92	4.51	3.90	3.56	3.33	3.17	3.05	2.96	2.88	2.82	2.72	2.62	2.51	2.45	2.39	2.33	2.27	2.20	2.13
20	5.87	4.46	3.86	3.51	3.29	3.13	3.01	2.91	2.84	2.77	2.68	2.57	2.46	2.41	2.35	2.29	2.22	2.16	2.09
21	5.83	4.42	3.82	3.48	3.25	3.09	2.97	2.87	2.80	2.73	2.64	2.53	2.42	2.37	2.31	2.25	2.18	2.11	2.04
22	5.79	4.38	3.78	3.44	3.22	3.05	2.93	2.84	2.76	2.70	2.60	2.50	2.39	2.33	2.27	2.21	2.14	2.08	2.00
23	5.75	4.35	3.75	3.41	3.18	3.02	2.90	2.81	2.73	2.67	2.57	2.47	2.36	2.30	2.24	2.18	2.11	2.04	1.97
24	5.72	4.32	3.72	3.38	3.15	2.99	2.87	2.78	2.70	2.64	2.54	2.44	2.33	2.27	2.21	2.15	2.08	2.01	1.94
25	5.69	4.29	3.69	3.35	3.13	2.97	2.85	2.75	2.68	2.61	2.51	2.41	2.30	2.24	2.18	2.12	2.05	1.98	1.91
26	5.66	4.27	3.67	3.33	3.10	2.94	2.82	2.73	2.65	2.59	2.49	2.39	2.28	2.22	2.16	2.09	2.03	1.95	1.88
27	5.63	4.24	3.65	3.31	3.08	2.92	2.80	2.71	2.63	2.57	2.47	2.36	2.25	2.19	2.13	2.07	2.00	1.93	1.85
28	5.61	4.22	3.63	3.29	3.06	2.90	2.78	2.69	2.61	2.55	2.45	2.34	2.23	2.17	2.11	2.05	1.98	1.91	1.83
29	5.59	4.20	3.61	3.27	3.04	2.88	2.76	2.67	2.59	2.53	2.43	2.32	2.21	2.15	2.09	2.03	1.96	1.89	1.81
30	5.57	4.18	3.59	3.25	3.03	2.87	2.75	2.65	2.57	2.51	2.41	2.31	2.20	2.14	2.07	2.01	1.94	1.87	1.79
40	5.42	4.05	3.46	3.13	2.90	2.74	2.62	2.53	2.45	2.39	2.29	2.18	2.07	2.01	1.94	1.88	1.80	1.72	1.64
60	5.29	3.93	3.34	3.01	2.79	2.63	2.51	2.41	2.33	2.27	2.17	2.06	1.94	1.88	1.82	1.74	1.67	1.58	1.48
120	5.15	3.80	3.23	2.89	2.67	2.52	2.39	2.30	2.22	2.16	2.05	1.94	1.82	1.76	1.69	1.61	1.53	1.43	1.31
∞	5.02	3.69	3.12	2.79	2.57	2.41	2.29	2.19	2.11	2.05	1.94	1.83	1.71	1.64	1.57	1.48	1.39	1.27	1.00

Table IV (continued) Upper 1% Points

v_2 \ v_1	1	2	3	4	5	6	7	8	9	10	12	15	20	24	30	40	60	120	∞
1	4052	4999.5	5403	5625	5764	5859	5928	5982	6022	6056	6106	6157	6209	6235	6261	6287	6313	6339	6366
2	98.50	99.00	99.17	99.25	99.30	99.33	99.36	99.37	99.39	99.40	99.42	99.43	99.45	99.46	99.47	99.47	99.48	99.49	99.50
3	34.12	30.82	29.46	28.71	28.24	27.91	27.67	27.49	27.35	27.23	27.05	26.87	26.69	26.60	26.50	26.41	26.32	26.22	26.13
4	21.20	18.00	16.69	15.98	15.52	15.21	14.98	14.80	14.66	14.55	14.37	14.20	14.02	13.93	13.84	13.75	13.65	13.56	13.46
5	16.26	13.27	12.06	11.39	10.97	10.67	10.46	10.29	10.16	10.05	9.89	9.72	9.55	9.47	9.38	9.29	9.20	9.11	9.06
6	13.75	10.92	9.78	9.15	8.75	8.47	8.26	8.10	7.98	7.87	7.72	7.56	7.40	7.31	7.23	7.14	7.06	6.97	6.88
7	12.25	9.55	8.45	7.85	7.46	7.19	6.99	6.84	6.72	6.62	6.47	6.31	6.16	6.07	5.99	5.91	5.82	5.74	5.65
8	11.26	8.65	7.59	7.01	6.63	6.37	6.18	6.03	5.91	5.81	5.67	5.52	5.36	5.28	5.20	5.12	5.03	4.95	4.86
9	10.56	8.02	6.99	6.42	6.06	5.80	5.61	5.47	5.35	5.26	5.11	4.96	4.81	4.73	4.65	4.57	4.48	4.40	4.31
10	10.04	7.56	6.55	5.99	5.64	5.39	5.20	5.06	4.94	4.85	4.71	4.56	4.41	4.33	4.25	4.17	4.08	4.00	3.91
11	9.65	7.21	6.22	5.67	5.32	5.07	4.89	4.74	4.63	4.54	4.40	4.25	4.10	4.02	3.94	3.86	3.78	3.69	3.60
12	9.33	6.93	5.95	5.41	5.06	4.82	4.64	4.50	4.39	4.30	4.16	4.01	3.86	3.78	3.70	3.62	3.54	3.45	3.36
13	9.07	6.70	5.74	5.21	4.86	4.62	4.44	4.30	4.19	4.10	3.96	3.82	3.66	3.59	3.51	3.43	3.34	3.25	3.17
14	8.86	6.51	5.56	5.04	4.69	4.46	4.28	4.14	4.03	3.94	3.80	3.66	3.51	3.43	3.35	3.27	3.18	3.09	3.00
15	8.68	6.36	5.42	4.89	4.56	4.32	4.14	4.00	3.89	3.80	3.67	3.52	3.37	3.29	3.21	3.13	3.05	2.96	2.87
16	8.53	6.23	5.29	4.77	4.44	4.20	4.03	3.89	3.78	3.69	3.55	3.41	3.26	3.18	3.10	3.02	2.93	2.84	2.75
17	8.40	6.11	5.18	4.67	4.34	4.10	3.93	3.79	3.68	3.59	3.46	3.31	3.16	3.08	3.00	2.92	2.83	2.75	2.65
18	8.29	6.01	5.09	4.58	4.25	4.01	3.84	3.71	3.60	3.51	3.37	3.23	3.08	3.00	2.92	2.84	2.75	2.66	2.57
19	8.18	5.93	5.01	4.50	4.17	3.94	3.77	3.63	3.52	3.43	3.30	3.15	3.00	2.92	2.84	2.76	2.67	2.58	2.49
20	8.10	5.85	4.94	4.43	4.10	3.87	3.70	3.56	3.46	3.37	3.23	3.09	2.94	2.86	2.78	2.69	2.61	2.52	2.42
21	8.02	5.78	4.87	4.37	4.04	3.81	3.64	3.51	3.40	3.31	3.17	3.03	2.88	2.80	2.72	2.64	2.55	2.46	2.36
22	7.95	5.72	4.82	4.31	3.99	3.76	3.59	3.45	3.35	3.26	3.12	2.98	2.83	2.75	2.67	2.58	2.50	2.40	2.31
23	7.88	5.66	4.76	4.26	3.94	3.71	3.54	3.41	3.30	3.21	3.07	2.93	2.78	2.70	2.62	2.54	2.45	2.35	2.26
24	7.82	5.61	4.72	4.22	3.90	3.67	3.50	3.36	3.26	3.17	3.03	2.89	2.74	2.66	2.58	2.49	2.40	2.31	2.21
25	7.77	5.57	4.68	4.18	3.85	3.63	3.46	3.32	3.22	3.13	2.99	2.85	2.70	2.62	2.54	2.45	2.36	2.27	2.17
26	7.72	5.53	4.64	4.14	3.82	3.59	3.42	3.29	3.18	3.09	2.96	2.81	2.66	2.58	2.50	2.42	2.33	2.23	2.13
27	7.68	5.49	4.60	4.11	3.78	3.56	3.39	3.26	3.15	3.06	2.93	2.78	2.63	2.55	2.47	2.38	2.29	2.20	2.10
28	7.64	5.45	4.57	4.07	3.75	3.53	3.36	3.23	3.12	3.03	2.90	2.75	2.60	2.52	2.44	2.35	2.26	2.17	2.06
29	7.60	5.42	4.54	4.04	3.73	3.50	3.33	3.20	3.09	3.00	2.87	2.73	2.57	2.49	2.41	2.33	2.23	2.14	2.03
30	7.56	5.39	4.51	4.02	3.70	3.47	3.30	3.17	3.07	2.98	2.84	2.70	2.55	2.47	2.39	2.30	2.21	2.11	2.01
40	7.31	5.18	4.31	3.83	3.51	3.29	3.12	2.99	2.89	2.80	2.66	2.52	2.37	2.29	2.20	2.11	2.02	1.92	1.80
60	7.08	4.98	4.13	3.65	3.34	3.12	2.95	2.82	2.72	2.63	2.50	2.35	2.20	2.12	2.03	1.94	1.84	1.73	1.60
120	6.85	4.79	3.95	3.48	3.17	2.96	2.79	2.66	2.56	2.47	2.34	2.19	2.03	1.95	1.86	1.76	1.66	1.53	1.38
∞	6.63	4.61	3.78	3.32	3.02	2.80	2.64	2.51	2.41	2.32	2.18	2.04	1.88	1.79	1.70	1.59	1.47	1.32	1.00

n=r

Table V.   Binomial Probabilities $\binom{N}{r}p^r q^{N-r}$

N	r	.05	.10	.15	.20	.25	.30	.35	.40	.45	.50
1	0	.9500	.9000	.8500	.8000	.7500	.7000	.6500	.6000	.5500	.5000
	1	.0500	.1000	.1500	.2000	.2500	.3000	.3500	.4000	.4500	.5000
2	0	.9025	.8100	.7225	.6400	.5625	.4900	.4225	.3600	.3025	.2500
	1	.0950	.1800	.2550	.3200	.3750	.4200	.4550	.4800	.4950	.5000
	2	.0025	.0100	.0225	.0400	.0625	.0900	.1225	.1600	.2025	.2500
3	0	.8574	.7290	.6141	.5120	.4219	.3430	.2746	.2160	.1664	.1250
	1	.1354	.2430	.3251	.3840	.4219	.4410	.4436	.4320	.4084	.3750
	2	.0071	.0270	.0574	.0960	.1406	.1890	.2389	.2880	.3341	.3750
	3	.0001	.0010	.0034	.0080	.0156	.0270	.0429	.0640	.0911	.1250
4	0	.8145	.6561	.5220	.4096	.3164	.2401	.1785	.1296	.0915	.0625
	1	.1715	.2916	.3685	.4096	.4219	.4116	.3845	.3456	.2995	.2500
	2	.0135	.0486	.0975	.1536	.2109	.2646	.3105	.3456	.3675	.3750
	3	.0005	.0036	.0115	.0256	.0469	.0756	.1115	.1536	.2005	.2500
	4	.0000	.0001	.0005	.0016	.0039	.0081	.0150	.0256	.0410	.0625
5	0	.7738	.5905	.4437	.3277	.2373	.1681	.1160	.0778	.0503	.0312
	1	.2036	.3280	.3915	.4096	.3955	.3602	.3124	.2592	.2059	.1562
	2	.0214	.0729	.1382	.2048	.2637	.3087	.3364	.3456	.3369	.3125
	3	.0011	.0081	.0244	.0512	.0879	.1323	.1811	.2304	.2757	.3125
	4	.0000	.0004	.0022	.0064	.0146	.0284	.0488	.0768	.1128	.1562
	5	.0000	.0000	.0001	.0003	.0010	.0024	.0053	.0102	.0185	.0312
6	0	.7351	.5314	.3771	.2621	.1780	.1176	.0754	.0467	.0277	.0156
	1	.2321	.3543	.3993	.3932	.3560	.3025	.2437	.1866	.1359	.0938
	2	.0305	.0984	.1762	.2458	.2966	.3241	.3280	.3110	.2780	.2344
	3	.0021	.0146	.0415	.0819	.1318	.1852	.2355	.2765	.3032	.3125
	4	.0001	.0012	.0055	.0154	.0330	.0595	.0951	.1382	.1861	.2344
	5	.0000	.0001	.0004	.0015	.0044	.0102	.0205	.0369	.0609	.0938
	6	.0000	.0000	.0000	.0001	.0002	.0007	.0018	.0041	.0083	.0156
7	0	.6983	.4783	.3206	.2097	.1335	.0824	.0490	.0280	.0152	.0078
	1	.2573	.3720	.3960	.3670	.3115	.2471	.1848	.1306	.0872	.0547
	2	.0406	.1240	.2097	.2753	.3115	.3177	.2985	.2613	.2140	.1641
	3	.0036	.0230	.0617	.1147	.1730	.2269	.2679	.2903	.2918	.2734
	4	.0002	.0026	.0109	.0287	.0577	.0972	.1442	.1935	.2388	.2734
	5	.0000	.0002	.0012	.0043	.0115	.0250	.0466	.0774	.1172	.1641
	6	.0000	.0000	.0001	.0004	.0013	.0036	.0084	.0172	.0320	.0547
	7	.0000	.0000	.0000	.0000	.0001	.0002	.0006	.0016	.0037	.0078
8	0	.6634	.4305	.2725	.1678	.1001	.0576	.0319	.0168	.0084	.0039
	1	.2793	.3826	.3847	.3355	.2760	.1977	.1373	.0896	.0548	.0312
	2	.0515	.1488	.2376	.2936	.3115	.2965	.2587	.2090	.1569	.1094
	3	.0054	.0331	.0839	.1468	.2076	.2541	.2786	.2787	.2568	.2188
	4	.0004	.0046	.0185	.0459	.0865	.1361	.1875	.2322	.2627	.2734
	5	.0000	.0004	.0026	.0092	.0231	.0467	.0808	.1239	.1719	.2188
	6	.0000	.0000	.0002	.0011	.0038	.0100	.0217	.0413	.0703	.1094
	7	.0000	.0000	.0000	.0001	.0004	.0012	.0033	.0079	.0164	.0312
	8	.0000	.0000	.0000	.0000	.0000	.0001	.0002	.0007	.0017	.0039

p

Table V   (continued)

N	r	.05	.10	.15	.20	.25	.30	.35	.40	.45	.50
						p					
9	0	.6302	.3874	.2316	.1342	.0751	.0404	.0277	.0101	.0046	.0020
	1	.2985	.3874	.3679	.3020	.2253	.1556	.1004	.0605	.0339	.0176
	2	.0629	.1722	.2597	.3020	.3003	.2668	.2162	.1612	.1110	.0703
	3	.0077	.0446	.1069	.1762	.2336	.2668	.2716	.2508	.2119	.1641
	4	.0006	.0074	.0283	.0661	.1168	.1715	.2194	.2508	.2600	.2461
	5	.0000	.0008	.0050	.0165	.0389	.0735	.1181	.1672	.2128	.2461
	6	.0000	.0001	.0006	.0028	.0087	.0210	.0424	.0743	.1160	.1641
	7	.0000	.0000	.0000	.0003	.0012	.0039	.0098	.0212	.0407	.0703
	8	.0000	.0000	.0000	.0000	.0001	.0004	.0013	.0035	.0083	.0176
	9	.0000	.0000	.0000	.0000	.0000	.0000	.0001	.0003	.0008	.0020
10	0	.5987	.3487	.1969	.1074	.0563	.0282	.0135	.0060	.0025	.0010
	1	.3151	.3874	.3474	.2684	.1877	.1211	.0725	.0403	.0207	.0098
	2	.0746	.1937	.2759	.3020	.2816	.2335	.1757	.1209	.0763	.0439
	3	.0105	.0574	.1298	.2013	.2503	.2668	.2522	.2150	.1665	.1172
	4	.0010	.0112	.0401	.0881	.1460	.2001	.2377	.2508	.2384	.2051
	5	.0001	.0015	.0085	.0264	.0584	.1029	.1536	.2007	.2340	.2461
	6	.0000	.0001	.0012	.0055	.0162	.0368	.0689	.1115	.1596	.2051
	7	.0000	.0000	.0001	.0008	.0031	.0090	.0212	.0425	.0746	.1172
	8	.0000	.0000	.0000	.0001	.0004	.0014	.0043	.0106	.0229	.0439
	9	.0000	.0000	.0000	.0000	.0000	.0001	.0005	.0016	.0042	.0098
	10	.0000	.0000	.0000	.0000	.0000	.0000	.0000	.0001	.0003	.0016
11	0	.5688	.3138	.1673	.0859	.0422	.0198	.0088	.0036	.0014	.0005
	1	.3293	.3835	.3248	.2362	.1549	.0932	.0518	.0266	.0125	.0054
	2	.0867	.2131	.2866	.2953	.2581	.1998	.1395	.0887	.0513	.0269
	3	.0137	.0710	.1517	.2215	.2581	.2568	.2254	.1774	.1259	.0806
	4	.0014	.0158	.0536	.1107	.1721	.2201	.2428	.2365	.2060	.1611
	5	.0001	.0025	.0132	.0388	.0803	.1231	.1830	.2207	.2360	.2256
	6	.0000	.0003	.0023	.0097	.0268	.0566	.0985	.1471	.1931	.2256
	7	.0000	.0000	.0003	.0017	.0064	.0173	.0379	.0701	.1128	.1611
	8	.0000	.0000	.0000	.0002	.0011	.0037	.0102	.0234	.0462	.0806
	9	.0000	.0000	.0000	.0000	.0001	.0005	.0018	.0052	.0126	.0269
	10	.0000	.0000	.0000	.0000	.0000	.0000	.0002	.0007	.0021	.0054
	11	.0000	.0000	.0000	.0000	.0000	.0000	.0000	.0000	.0002	.0005
12	0	.5404	.2824	.1422	.0687	.0317	.0138	.0057	.0022	.0008	.0002
	1	.3413	.3766	.3012	.2062	.1267	.0712	.0368	.0174	.0075	.0029
	2	.0988	.2301	.2924	.2835	.2323	.1678	.1088	.0639	.0339	.0161
	3	.0173	.0852	.1720	.2362	.2581	.2397	.1954	.1419	.0923	.0537
	4	.0021	.0213	.0683	.1329	.1936	.2311	.2367	.2128	.1700	.1208
	5	.0002	.0038	.0193	.0532	.1032	.1585	.2039	.2270	.2225	.1934
	6	.0000	.0005	.0040	.0155	.0401	.0792	.1281	.1766	.2124	.2256
	7	.0000	.0000	.0006	.0033	.0115	.0291	.0591	.1009	.1489	.1934
	8	.0000	.0000	.0001	.0005	.0024	.0078	.0199	.0420	.0762	.1208
	9	.0000	.0000	.0000	.0001	.0004	.0015	.0048	.0125	.0277	.0537
	10	.0000	.0000	.0000	.0000	.0000	.0002	.0008	.0025	.0068	.0161
	11	.0000	.0000	.0000	.0000	.0000	.0000	.0001	.0003	.0010	.0029
	12	.0000	.0000	.0000	.0000	.0000	.0000	.0000	.0000	.0001	.0002

Table V (continued)

N	r	.05	.10	.15	.20	.25	.30	.35	.40	.45	.50
13	0	.5133	.2542	.1209	.0550	.0238	.0097	.0037	.0013	.0004	.0001
	1	.3512	.3672	.2774	.1787	.1029	.0540	.0259	.0113	.0045	.0016
	2	.1109	.2448	.2937	.2680	.2059	.1388	.0836	.0453	.0220	.0095
	3	.0214	.0997	.1900	.2457	.2517	.2181	.1651	.1107	.0660	.0349
	4	.0028	.0277	.0838	.1535	.2097	.2337	.2222	.1845	.1350	.0873
	5	.0003	.0055	.0266	.0691	.1258	.1803	.2154	.2214	.1989	.1571
	6	.0000	.0008	.0063	.0230	.0559	.1030	.1546	.1968	.2169	.2095
	7	.0000	.0001	.0011	.0058	.0186	.0442	.0833	.1312	.1775	.2095
	8	.0000	.0000	.0001	.0011	.0047	.0142	.0336	.0656	.1089	.1571
	9	.0000	.0000	.0000	.0001	.0009	.0034	.0101	.0243	.0495	.0873
	10	.0000	.0000	.0000	.0000	.0001	.0006	.0022	.0065	.0162	.0349
	11	.0000	.0000	.9000	.0000	.0000	.0001	.0003	.0012	.0036	.0095
	12	.0000	.0000	.0000	.0000	.0000	.0000	.0000	.0001	.0005	.0016
	13	.0000	.0000	.0000	.0000	.0000	.0000	.0000	.0000	.0000	.0001
14	0	.4877	.2288	.1028	.0440	.0178	.0068	.0024	.0008	.0002	.0001
	1	.3593	.3559	.2539	.1539	.0832	.0407	.0181	.0073	.0027	.0009
	2	.1229	.2570	.2912	.2501	.1802	.1134	.0634	.0317	.0141	.0056
	3	.0259	.1142	.2056	.2501	.2402	.1943	.1366	.0845	.0462	.0222
	4	.0037	.0349	.0998	.1720	.2202	.2290	.2022	.1549	.1040	.0611
	5	.0004	.0078	.0352	.0860	.1468	.1963	.2178	.2066	.1701	.1222
	6	.0000	.0013	.0093	.0322	.0734	.1262	.1759	.2066	.2088	.1833
	7	.0000	.0002	.0019	.0092	.0280	.0618	.1082	.1574	.1952	.2095
	8	.0000	.0000	.0003	.0020	.0082	.0232	.0510	.0918	.1398	.1833
	9	.0000	.0000	.0000	.0003	.0018	.0066	.0183	.0408	.0762	.1222
	10	.0000	.0000	.0000	.0000	.0003	.0014	.0049	.0136	.0312	.0611
	11	.0000	.0000	.0000	.0000	.0000	.0002	.0010	.0033	.0093	.0222
	12	.0000	.0000	.0000	.0000	.0000	.0000	.0001	.0005	.0019	.0056
	13	.0000	.0000	.0000	.0000	.0000	.0000	.0000	.0001	.0002	.0009
	14	.0000	.0000	.0000	.0000	.0000	.0000	.0000	.0000	.0000	.0001
15	0	.4633	.2059	.0874	.0352	.0134	.0047	.0016	.0005	.0001	.0000
	1	.3658	.3432	.2312	.1319	.0668	.0305	.0126	.0047	.0016	.0005
	2	.1348	.2669	.2856	.2309	.1559	.0916	.0476	.0219	.0090	.0032
	3	.0307	.1285	.2184	.2501	.2252	.1700	.1110	.0634	.0318	.0139
	4	.0049	.0428	.1156	.1876	.2252	.2186	.1792	.1268	.0780	.0417
	5	.0006	.0105	.0449	.1032	.1651	.2061	.2123	.1859	.1404	.0916
	6	.0000	.0019	.0132	.0430	.0917	.1472	.1906	.2066	.1914	.1527
	7	.0000	.0003	.0030	.0138	.0393	.0811	.1319	.1771	.2013	.1964
	8	.0000	.0000	.0005	.0035	.0131	.0348	.0710	.1181	.1647	.1964
	9	.0000	.0000	.0001	.0007	.0034	.0116	.0298	.0612	.1048	.1527
	10	.0000	.0000	.0000	.0001	.0007	.0030	.0096	.0245	.0515	.0916
	11	.0000	.0000	.0000	.0000	.0001	.0006	.0024	.0074	.0191	.0417
	12	.0000	.0000	.0000	.0000	.0000	.0001	.0004	.0016	.0052	.0139
	13	.0000	.0000	.0000	.0000	.0000	.0000	.0001	.0003	.0010	.0032
	14	.0000	.0000	.0000	.0000	.0000	.0000	.0000	.0000	.0001	.0005
	15	.0000	.0000	.0000	.0000	.0000	.0000	.0000	.0000	.0000	.0000
16	0	.4401	.1853	.0743	.0281	.0100	.0033	.0010	.0003	.0001	.0000
	1	.3706	.3294	.2097	.1126	.0535	.0228	.0087	.0030	.0009	.0002
	2	.1463	.2745	.2775	.2111	.1336	.0732	.0353	.0150	.0056	.0018

Table V   (continued)

N	r	.05	.10	.15	.20	.25	.30	.35	.40	.45	.50
16	3	.0359	.1423	.2285	.2463	.2079	.1465	.0888	.0468	.0215	.0085
	4	.0061	.0514	.1311	.2001	.2252	.2040	.1553	.1014	.0572	.0278
	5	.0008	.0137	.0555	.1201	.1802	.2099	.2008	.1623	.1123	.0667
	6	.0001	.0028	.0180	.0550	.1101	.1649	.1982	.1983	.1684	.1222
	7	.0000	.0004	.0045	.0197	.0524	.1010	.1524	.1889	.1969	.1746
	8	.0000	.0001	.0009	.0055	.0197	.0487	.0923	.1417	.1812	.1964
	9	.0000	.0000	.0001	.0012	.0058	.0185	.0442	.0840	.1318	.1746
	10	.0000	.0000	.0000	.0002	.0014	.0056	.0167	.0392	.0755	.1222
	11	.0000	.0000	.0000	.0000	.0002	.0013	.0049	.0142	.0337	.0667
	12	.0000	.0000	.0000	.0000	.0000	.0002	.0011	.0040	.0115	.0278
	13	.0000	.0000	.0000	.0000	.0000	.0000	.0002	.0008	.0029	.0085
	14	.0000	.0000	.0000	.0000	.0000	.0000	.0000	.0001	.0005	.0018
	15	.0000	.0000	.0000	.0000	.0000	.0000	.0000	.0000	.0001	.0002
	16	.0000	.0000	.0000	.0000	.0000	.0000	.0000	.0000	.0000	.0000
17	0	.4181	.1668	.0631	.0225	.0075	.0023	.0007	.0002	.0000	.0000
	1	.3741	.3150	.1893	.0957	.0426	.0169	.0060	.0019	.0005	.0001
	2	.1575	.2800	.2673	.1914	.1136	.0581	.0260	.0102	.0035	.0010
	3	.0415	.1556	.2359	.2393	.1893	.1245	.0701	.0341	.0144	.0052
	4	.0076	.0605	.1457	.2093	.2209	.1868	.1320	.0796	.0411	.0182
	5	.0010	.0175	.0668	.1361	.1914	.2081	.1849	.1379	.0875	.0472
	6	.0001	.0039	.0236	.0680	.1276	.1784	.1991	.1839	.1432	.0944
	7	.0000	.0007	.0065	.0267	.0668	.1201	.1685	.1927	.1841	.1484
	8	.0000	.0001	.0014	.0084	.0279	.0644	.1143	.1606	.1883	.1855
	9	.0000	.0000	.0003	.0021	.0093	.0276	.0611	.1070	.1540	.1855
	10	.0000	.0000	.0000	.0004	.0025	.0095	.0263	.0571	.1008	.1484
	11	.0000	.0000	.0000	.0001	.0005	.0026	.0090	.0242	.0525	.0944
	12	.0000	.0000	.0000	.0000	.0001	.0006	.0024	.0081	.0215	.0472
	13	.0000	.0000	.0000	.0000	.0000	.0001	.0005	.0021	.0068	.0182
	14	.0000	.0000	.0000	.0000	.0000	.0000	.0001	.0004	.0016	.0052
	15	.0000	.0000	.0000	.0000	.0000	.0000	.0000	.0001	.0003	.0010
	16	.0000	.0000	.0000	.0000	.0000	.0000	.0000	.0000	.0000	.0001
	17	.0000	.0000	.0000	.0000	.0000	.0000	.0000	.0000	.0000	.0000
18	0	.3972	.1501	.0536	.0180	.0056	.0016	.0004	.0001	.0000	.0000
	1	.3763	.3002	.1704	.0811	.0338	.0126	.0042	.0012	.0003	.0001
	2	.1683	.2835	.2556	.1723	.0958	.0458	.0190	.0069	.0022	.0006
	3	.0473	.1680	.2406	.2297	.1704	.1046	.0547	.0246	.0095	.0031
	4	.0093	.0700	.1592	.2153	.2130	.1681	.1104	.0614	.0291	.0117
	5	.0014	.0218	.0787	.1507	.1988	.2017	.1664	.1146	.0666	.0327
	6	.0002	.0052	.0310	.0816	.1436	.1873	.1941	.1655	.1181	.0708
	7	.0000	.0010	.0091	.0350	.0820	.1376	.1792	.1892	.1657	.1214
	8	.0000	.0002	.0022	.0120	.0376	.0811	.1327	.1734	.1864	.1669
	9	.0000	.0000	.0004	.0033	.0139	.0386	.0794	.1284	.1694	.1855
18	10	.0000	.0000	.0001	.0008	.0042	.0149	.0385	.0771	.1248	.1669
	11	.0000	.0000	.0000	.0001	.0010	.0046	.0151	.0374	.0742	.1214
	12	.0000	.0000	.0000	.0000	.0002	.0012	.0047	.0145	.0354	.0708
	13	.0000	.0000	.0000	.0000	.0000	.0002	.0012	.0045	.0134	.0327
	14	.0000	.0000	.0000	.0000	.0000	.0000	.0002	.0011	.0039	.0117

Table V   (continued)

N	r	.05	.10	.15	.20	.25	.30	.35	.40	.45	.50
	15	.0000	.0000	.0000	.0000	.0000	.0000	.0000	.0002	.0009	.0031
	16	.0000	.0000	.0000	.0000	.0000	.0000	.0000	.0000	.0001	.0006
	17	.0000	.0000	.0000	.0000	.0000	.0000	.0000	.0000	.0000	.0001
	18	.0000	.0000	.0000	.0000	.0000	.0000	.0000	.0000	.0000	.0000
19	0	.3774	.1351	.0456	.0144	.0042	.0011	.0003	.0001	.0000	.0000
	1	.3774	.2852	.1529	.0685	.0268	.0093	.0029	.0008	.0002	.0000
	2	.1787	.2852	.2428	.1540	.0803	.0358	.0138	.0046	.0013	.0003
	3	.0533	.1796	.2428	.2182	.1517	.0869	.0422	.0175	.0062	.0018
	4	.0112	.0798	.1714	.2182	.2023	.1491	.0909	.0467	.0203	.0074
	5	.0018	.0266	.0907	.1636	.2023	.1916	.1468	.0933	.0497	.0222
	6	.0002	.0069	.0374	.0955	.1574	.1916	.1844	.1451	.0949	.0518
	7	.0000	.0014	.0122	.0443	.0974	.1525	.1844	.1797	.1443	.0961
	8	.0000	.0002	.0032	.0166	.0487	.0981	.1489	.1797	.1771	.1442
	9	.0000	.0000	.0007	.0051	.0198	.0514	.0980	.1464	.1771	.1762
	10	.0000	.0000	.0001	.0013	.0066	.0220	.0528	.0976	.1449	.1762
	11	.0000	.0000	.0000	.0003	.0018	.0077	.0233	.0532	.0970	.1442
	12	.0000	.0000	.0000	.0000	.0004	.0022	.0083	.0237	.0529	.0961
	13	.0000	.0000	.0000	.0000	.0001	.0005	.0024	.0085	.0233	.0518
	14	.0000	.0000	.0000	.0000	.0000	.0001	.0006	.0024	.0082	.0222
	15	.0000	.0000	.0000	.0000	.0000	.0000	.0001	.0005	.0022	.0074
	16	.0000	.0000	.0000	.0000	.0000	.0000	.0000	.0001	.0005	.0018
	17	.0000	.0000	.0000	.0000	.0000	.0000	.0000	.0000	.0001	.0003
	18	.0000	.0000	.0000	.0000	.0000	.0000	.0000	.0000	.0000	.0000
	19	.0000	.0000	.0000	.0000	.0000	.0000	.0000	.0000	.0000	.0000
20	0	.3585	.1216	.0388	.0115	.0032	.0008	.0002	.0000	.0000	.0000
	1	.3774	.2702	.1368	.0576	.0211	.0068	.0020	.0005	.0001	.0000
	2	.1887	.2852	.2293	.1369	.0669	.0278	.0100	.0031	.0008	.0002
	3	.0596	.1901	.2428	.2054	.1339	.0716	.0323	.0123	.0040	.0011
	4	.0133	.0898	.1821	.2182	.1897	.1304	.0738	.0350	.0139	.0046
	5	.0022	.0319	.1028	.1746	.2023	.1789	.1272	.0746	.0365	.0148
	6	.0003	.0089	.0454	.1091	.1686	.1916	.1712	.1244	.0746	.0370
	7	.0000	.0020	.0160	.0545	.1124	.1643	.1844	.1659	.1221	.0739
	8	.0000	.0004	.0046	.0222	.0609	.1144	.1614	.1797	.1623	.1201
	9	.0000	.0001	.0011	.0074	.0271	.0654	.1158	.1597	.1771	.1602
	10	.0000	.0000	.0002	.0020	.0099	.0308	.0686	.1171	.1593	.1762
	11	.0000	.0000	.0000	.0005	.0030	.0120	.0336	.0710	.1185	.1602
	12	.0000	.0000	.0000	.0001	.0008	.0039	.0136	.0355	.0727	.1201
	13	.0000	.0000	.0000	.0000	.0002	.0010	.0045	.0146	.0366	.0739
	14	.0000	.0000	.0000	.0000	.0000	.0002	.0012	.0049	.0150	.0370
	15	.0000	.0000	.0000	.0000	.0000	.0000	.0003	.0013	.0049	.0148
	16	.0000	.0000	.0000	.0000	.0000	.0000	.0000	.0003	.0013	.0046
	17	.0000	.0000	.0000	.0000	.0000	.0000	.0000	.0000	.0002	.0011
	18	.0000	.0000	.0000	.0000	.0000	.0000	.0000	.0000	.0000	.0002
	19	.0000	.0000	.0000	.0000	.0000	.0000	.0000	.0000	.0000	.0000
	20	.0000	.0000	.0000	.0000	.0000	.0000	.0000	.0000	.0000	.0000

This table is reproduced by permission from R. S. Burington and D. C. May, *Handbook of Probability and Statistics with Tables.* McGraw-Hill Book Company, 1953.

Table VI.   Poisson Probabilities $e^{-\lambda}\lambda^r/r!$

λ

r	0.1	0.2	0.3	0.4	0.5	0.6	0.7	0.8	0.9	1.0
0	.9048	.8187	.7408	.6703	.6065	.5488	.4966	.4493	.4066	.3679
1	.0905	.1637	.2222	.2681	.3033	.3293	.3476	.3595	.3659	.3679
2	.0045	.0164	.0333	.0536	.0758	.0988	.1217	.1438	.1647	.1839
3	.0002	.0011	.0033	.0072	.0126	.0198	.0284	.0383	.0494	.0613
4	.0000	.0001	.0002	.0007	.0016	.0030	.0050	.0077	.0111	.0153
5	.0000	.0000	.0000	.0001	.0002	.0004	.0007	.0012	.0020	.0031
6	.0000	.0000	.0000	.0000	.0000	.0000	.0001	.0002	.0003	.0005
7	.0000	.0000	.0000	.0000	.0000	.0000	.0000	.0000	.0000	.0001

λ

r	1.1	1.2	1.3	1.4	1.5	1.6	1.7	1.8	1.9	2.0
0	.3329	.3012	.2725	.2466	.2231	.2019	.1827	.1653	.1496	.1353
1	.3662	.3614	.3543	.3452	.3347	.3230	.3106	.2975	.2842	.2707
2	.2014	.2169	.2303	.2417	.2510	.2584	.2640	.2678	.2700	.2707
3	.0738	.0867	.0998	.1128	.1255	.1378	.1496	.1607	.1710	.1804
4	.0203	.0260	.0324	.0395	.0471	.0551	.0636	.0723	.0812	.0902
5	.0045	.0062	.0084	.0111	.0141	.0176	.0216	.0260	.0309	.0361
6	.0008	.0012	.0018	.0026	.0035	.0047	.0061	.0078	.0098	.0120
7	.0001	.0002	.0003	.0005	.0008	.0011	.0015	.0020	.0027	.0034
8	.0000	.0000	.0001	.0001	.0001	.0002	.0003	.0005	.0006	.0009
9	.0000	.0000	.0000	.0000	.0000	.0000	.0001	.0001	.0001	.0002

λ

r	2.1	2.2	2.3	2.4	2.5	2.6	2.7	2.8	2.9	3.0
0	.1225	.1108	.1003	.0907	.0821	.0743	.0672	.0608	.0550	.0498
1	.2572	.2438	.2306	.2177	.2052	.1931	.1815	.1703	.1596	.1494
2	.2700	.2681	.2652	.2613	.2565	.2510	.2450	.2384	.2314	.2240
3	.1890	.1966	.2033	.2090	.2138	.2176	.2205	.2225	.2237	.2240
4	.0992	.1082	.1169	.1254	.1336	.1414	.1488	.1557	.1622	.1680
5	.0417	.0476	.0538	.0602	.0668	.0735	.0804	.0872	.0940	.1008
6	.0146	.0174	.0206	.0241	.0278	.0319	.0362	.0407	.0455	.0540
7	.0044	.0055	.0068	.0083	.0099	.0118	.0139	.0163	.0188	.0216
8	.0011	.0015	.0019	.0025	.0031	.0038	.0047	.0057	.0068	.0081
9	.0003	.0004	.0005	.0007	.0009	.0011	.0014	.0018	.0022	.0027
10	.0001	.0001	.0001	.0002	.0002	.0003	.0004	.0005	.0006	.0008
11	.0000	.0000	.0000	.0000	.0000	.0001	.0001	.0001	.0002	.0002
12	.0000	.0000	.0000	.0000	.0000	.0000	.0000	.0000	.0000	.0001

λ

r	3.1	3.2	3.3	3.4	3.5	3.6	3.7	3.8	3.9	4.0
0	.0450	.0408	.0369	.0344	.0302	.0273	.0247	.0224	.0202	.0183
1	.1397	.1304	.1217	.1135	.1057	.0984	.0915	.0850	.0789	.0733
2	.2165	.2087	.2008	.1929	.1850	.1771	.1692	.1615	.1539	.1465
3	.2237	.2226	.2209	.2186	.2158	.2125	.2087	.2046	.2001	.1954
4	.1734	.1781	.1823	.1858	.1888	.1912	.1931	.1944	.1951	.1954

Table VI   (continued)

r	3.1	3.2	3.3	3.4	3.5	3.6	3.7	3.8	3.9	4.0
5	.1075	.1140	.1203	.1264	.1322	.1377	.1429	.1477	.1522	.1563
6	.0555	.0608	.0662	.0716	.0771	.0826	.0881	.0936	.0989	.1042
7	.0246	.0278	.0312	.0348	.0385	.0425	.0466	.0508	.0551	.0595
8	.0095	.0111	.0129	.0148	.0169	.0191	.0215	.0241	.0269	.0298
9	.0033	.0040	.0047	.0056	.0066	.0076	.0089	.0102	.0116	.0132
10	.0010	.0013	.0016	.0019	.0023	.0028	.0033	.0039	.0045	.0053
11	.0003	.0004	.0005	.0006	.0007	.0009	.0011	.0013	.0016	.0019
12	.0001	.0001	.0001	.0002	.0002	.0003	.0003	.0004	.0005	.0006
13	.0000	.0000	.0000	.0000	.0001	.0001	.0001	.0001	.0002	.0002
14	.0000	.0000	.0000	.0000	.0000	.0000	.0000	.0000	.0000	.0001

λ

r	4.1	4.2	4.3	4.4	4.5	4.6	4.7	4.8	4.9	5.0
0	.0166	.0150	.0136	.0123	.0111	.0101	.0091	.0082	.0074	.0067
1	.0679	.0630	.0583	.0540	.0500	.0462	.0427	.0395	.0365	.0337
2	.1393	.1323	.1254	.1188	.1125	.1063	.1005	.0948	.0894	.0842
3	.1904	.1852	.1798	.1743	.1687	.1631	.1574	.1517	.1460	.1404
4	.1951	.1944	.1933	.1917	.1898	.1875	.1849	.1820	.1789	.1755
5	.1600	.1633	.1662	.1687	.1708	.1725	.1738	.1747	.1753	.1755
6	.1093	.1143	.1191	.1237	.1281	.1323	.1362	.1398	.1432	.1462
7	.0640	.0686	.0732	.0778	.0824	.0869	.0914	.0959	.1002	.1044
8	.0328	.0360	.0393	.0428	.0463	.0500	.0537	.0575	.0614	.0653
9	.0150	.0168	.0188	.0209	.0232	.0255	.0280	.0307	.0334	.0363
10	.0061	.0071	.0081	.0092	.0104	.0118	.0132	.0147	.0164	.0181
11	.0023	.0027	.0032	.0037	.0043	.0049	.0056	.0064	.0073	.0082
12	.0008	.0009	.0011	.0014	.0016	.0019	.0022	.0026	.0030	.0034
13	.0002	.0003	.0004	.0005	.0006	.0007	.0008	.0009	.0011	.0013
14	.0001	.0001	.0001	.0001	.0002	.0002	.0003	.0003	.0004	.0005
15	.0000	.0000	.0000	.0000	.0001	.0001	.0001	.0001	.0001	.0002

λ

r	5.1	5.2	5.3	5.4	5.5	5.6	5.7	5.8	5.9	6.0
0	.0061	.0055	.0050	.0045	.0041	.0037	.0033	.0030	.0027	.0025
1	.0311	.0287	.0265	.0244	.0225	.0207	.0191	.0176	.0162	.0149
2	.0793	.0746	.0701	.0659	.0618	.0580	.0544	.0509	.0477	.0446
3	.1348	.1293	.1239	.1185	.1133	.1082	.1033	.0985	.0938	.0892
4	.1719	.1681	.1641	.1600	.1558	.1515	.1472	.1428	.1383	.1339
5	.1753	.1748	.1740	.1728	.1714	.1697	.1678	.1656	.1632	.1606
6	.1490	.1515	.1537	.1555	.1571	.1584	.1594	.1601	.1605	.1606
7	.1086	.1125	.1163	.1200	.1234	.1267	.1298	.1326	.1353	.1377
8	.0692	.0731	.0771	.0810	.0849	.0887	.0925	.0962	.0998	.1033
9	.0392	.0423	.0454	.0486	.0519	.0552	.0586	.0620	.0654	.0688

Table VI (continued)

λ

r	5.1	5.2	5.3	5.4	5.5	5.6	5.7	5.8	5.9	6.0
10	.0200	.0220	.0241	.0262	.0285	.0309	.0334	.0359	.0386	.0413
11	.0093	.0104	.0116	.0129	.0143	.0157	.0173	.0190	.0207	.0225
12	.0039	.0045	.0051	.0058	.0065	.0073	.0082	.0092	.0102	.0113
13	.0015	.0018	.0021	.0024	.0028	.0032	.0036	.0041	.0046	.0052
14	.0006	.0007	.0008	.0009	.0011	.0013	.0015	.0017	.0019	.0022
15	.0002	.0002	.0003	.0003	.0004	.0005	.0006	.0007	.0008	.0009
16	.0001	.0001	.0001	.0001	.0001	.0002	.0002	.0002	.0003	.0003
17	.0000	.0000	.0000	.0000	.0000	.0001	.0001	.0001	.0001	.0001

λ

r	6.1	6.2	6.3	6.4	6.5	6.6	6.7	6.8	6.9	7.0
0	.0022	.0020	.0018	.0017	.0015	.0014	.0012	.0011	.0010	.0009
1	.0137	.0126	.0116	.0106	.0098	.0090	.0082	.0076	.0070	.0064
2	.0417	.0390	.0364	.0340	.0318	.0296	.0276	.0258	.0240	.0223
3	.0848	.0806	.0765	.0726	.0688	.0652	.0617	.0584	.0552	.0521
4	.1294	.1249	.1205	.1162	.1118	.1076	.1034	.0992	.0952	.0912
5	.1579	.1549	.1519	.1487	.1454	.1420	.1385	.1349	.1314	.1277
6	.1605	.1601	.1595	.1586	.1575	.1562	.1546	.1529	.1511	.1490
7	.1399	.1418	.1435	.1450	.1462	.1472	.1480	.1486	.1489	.1490
8	.1066	.1099	.1130	.1160	.1188	.1215	.1240	.1263	.1284	.1304
9	.0723	.0757	.0791	.0825	.0858	.0891	.0923	.0954	.0985	.1014
10	.0441	.0469	.0498	.0528	.0558	.0588	.0618	.0649	.0679	.0710
11	.0245	.0265	.0285	.0307	.0330	.0353	.0377	.0401	.0426	.0452
12	.0124	.0137	.0150	.0164	.0179	.0194	.0210	.0227	.0245	.0264
13	.0058	.0065	.0073	.0081	.0089	.0098	.0108	.0119	.0130	.0142
14	.0025	.0029	.0033	.0037	.0041	.0046	.0052	.0058	.0064	.0071
15	.0010	.0012	.0014	.0016	.0018	.0020	.0023	.0026	.0029	.0033
16	.0004	.0005	.0005	.0006	.0007	.0008	.0010	.0011	.0013	.0014
17	.0001	.0002	.0002	.0002	.0003	.0003	.0004	.0004	.0005	.0006
18	.0000	.0001	.0001	.0001	.0001	.0001	.0001	.0002	.0002	.0002
19	.0000	.0000	.0000	.0000	.0000	.0000	.0000	.0001	.0001	.0001

λ

r	7.1	7.2	7.3	7.4	7.5	7.6	7.7	7.8	7.9	8.0
0	.0008	.0007	.0007	.0006	.0006	.0005	.0005	.0004	.0004	.0003
1	.0059	.0054	.0049	.0045	.0041	.0038	.0035	.0032	.0029	.0027
2	.0208	.0194	.0180	.0167	.0156	.0145	.0134	.0125	.0116	.0107
3	.0492	.0464	.0438	.0413	.0389	.0366	.0345	.0324	.0305	.0286
4	.0874	.0836	.0799	.0764	.0729	.0696	.0663	.0632	.0602	.0573
5	.1241	.1204	.1167	.1130	.1094	.1057	.1021	.0986	.0951	.0916
6	.1468	.1445	.1420	.1394	.1367	.1339	.1311	.1282	.1252	.1221
7	.1489	.1486	.1481	.1474	.1465	.1454	.1442	.1428	.1413	.1396
8	.1321	.1337	.1351	.1363	.1373	.1382	.1388	.1392	.1395	.1396
9	.1042	.1070	.1096	.1121	.1144	.1167	.1187	.1207	.1224	.1241
10	.0740	.0770	.0800	.0829	.0858	.0887	.0914	.0941	.0967	.0993
11	.0478	.0504	.0531	.0558	.0585	.0613	.0640	.0667	.0695	.0722

Table VI   (continued)

λ

r	7.1	7.2	7.3	7.4	7.5	7.6	7.7	7.8	7.9	8.0
12	.0283	.0303	.0323	.0344	.0366	.0388	.0411	.0434	.0457	.0481
13	.0154	.0168	.0181	.0196	.0211	.0227	.0243	.0260	.0278	.0296
14	.0078	.0086	.0095	.0104	.0113	.0123	.0134	.0145	.0157	.0169
15	.0037	.0041	.0046	.0051	.0057	.0062	.0069	.0075	.0083	.0090
16	.0016	.0019	.0021	.0024	.0026	.0030	.0033	.0037	.0041	.0045
17	.0007	.0008	.0009	.0010	.0012	.0013	.0015	.0017	.0019	.0021
18	.0003	.0003	.0004	.0004	.0005	.0006	.0006	.0007	.0008	.0009
19	.0001	.0001	.0001	.0002	.0002	.0002	.0003	.0003	.0003	.0004
20	.0000	.0000	.0001	.0001	.0001	.0000	.0001	.0001	.0001	.0002
21	.0000	.0000	.0000	.0000	.0000	.0000	.0000	.0000	.0001	.0001

λ

r	8.1	8.2	8.3	8.4	8.5	8.6	8.7	8.8	8.9	9.0
0	.0003	.0003	.0002	.0002	.0002	.0002	.0002	.0002	.0001	.0001
1	.0025	.0023	.0021	.0019	.0017	.0016	.0014	.0013	.0012	.0011
2	.0100	.0092	.0086	.0079	.0074	.0068	.0063	.0058	.0054	.0050
3	.0269	.0252	.0237	.0222	.0208	.0195	.0183	.0171	.0160	.0150
4	.0544	.0517	.0491	.0466	.0443	.0420	.0398	.0377	.0357	.0337
5	.0882	.0849	.0816	.0784	.0752	.0722	.0692	.0663	.0635	.0607
6	.1191	.1160	.1128	.1097	.1066	.1034	.1003	.0972	.0941	.0911
7	.1378	.1358	.1338	.1317	.1294	.1271	.1247	.1222	.1197	.1171
8	.1395	.1392	.1388	.1382	.1375	.1366	.1356	.1344	.1332	.1318
9	.1256	.1269	.1280	.1290	.1299	.1306	.1311	.1315	.1317	.1318
10	.1017	.1040	.1063	.1084	.1104	.1123	.1140	.1157	.1172	.1186
11	.0749	.0776	.0802	.0828	.0853	.0878	.0902	.0925	.0948	.0970
12	.0505	.0530	.0555	.0579	.0604	.0629	.0654	.0679	.0703	.0728
13	.0315	.0334	.0354	.0374	.0395	.0416	.0438	.0459	.0481	.0504
14	.0182	.0196	.0210	.0225	.0240	.0256	.0272	.0289	.0306	.0324
15	.0098	.0107	.0116	.0126	.0136	.0147	.0158	.0169	.0182	.0194
16	.0050	.0055	.0060	.0066	.0072	.0079	.0086	.0093	.0101	.0109
17	.0024	.0026	.0029	.0033	.0036	.0040	.0044	.0048	.0053	.0058
18	.0011	.0012	.0014	.0015	.0017	.0019	.0021	.0024	.0026	.0029
19	.0005	.0005	.0006	.0007	.0008	.0009	.0010	.0011	.0012	.0014
20	.0002	.0002	.0002	.0003	.0003	.0004	.0004	.0005	.0005	.0006
21	.0001	.0001	.0001	.0001	.0001	.0002	.0002	.0002	.0002	.0003
22	.0000	.0000	.0000	.0000	.0001	.0001	.0001	.0001	.0001	.0001

λ

r	9.1	9.2	9.3	9.4	9.5	9.6	9.7	9.8	9.9	10
0	.0001	.0001	.0001	.0001	.0001	.0001	.0001	.0001	.0001	.0000
1	.0010	.0009	.0009	.0008	.0007	.0007	.0006	.0005	.0005	.0005
2	.0046	.0043	.0040	.0037	.0034	.0031	.0029	.0027	.0025	.0023
3	.0140	.0131	.0123	.0115	.0107	.0100	.0093	.0087	.0081	.0076
4	.0319	.0302	.0285	.0269	.0254	.0240	.0226	.0213	.0201	.0189

Table VI   (continued)

					λ					
r	9.1	9.2	9.3	9.4	9.5	9.6	9.7	9.8	9.9	10
5	.0581	.0555	.0530	.0506	.0483	.0460	.0439	.0418	.0398	.0378
6	.0881	.0851	.0822	.0793	.0764	.0736	.0709	.0682	.0656	.0631
7	.1145	.1118	.1091	.1064	.1037	.1010	.0982	.0955	.0928	.0901
8	.1302	.1286	.1269	.1251	.1232	.1212	.1191	.1170	.1148	.1126
9	.1317	.1315	.1311	.1306	.1300	.1293	.1284	.1274	.1263	.1251
10	.1198	.1210	.1219	.1228	.1235	.1241	.1245	.1249	.1250	.1251
11	.0991	.1012	.1031	.1049	.1067	.1083	.1098	.1112	.1125	.1137
12	.0752	.0776	.0799	.0822	.0844	.0866	.0888	.0908	.0928	.0948
13	.0526	.0549	.0572	.0594	.0617	.0640	.0662	.0685	.0707	.0729
14	.0342	.0361	.0380	.0399	.0419	.0439	.0459	.0479	.0500	.0521
15	.0208	.0221	.0235	.0250	.0265	.0281	.0297	.0313	.0330	.0347
16	.0118	.0127	.0137	.0147	.0157	.0168	.0180	.0192	.0204	.0217
17	.0063	.0069	.0075	.0081	.0088	.0095	.0103	.0111	.0119	.0128
18	.0032	.0035	.0039	.0042	.0046	.0051	.0055	.0060	.0065	.0071
19	.0015	.0017	.0019	.0021	.0023	.0026	.0028	.0031	.0034	.0037
20	.0007	.0008	.0009	.0010	.0011	.0012	.0014	.0015	.0017	.0019
21	.0003	.0003	.0004	.0004	.0005	.0006	.0006	.0007	.0008	.0009
22	.0001	.0001	.0002	.0002	.0002	.0002	.0003	.0003	.0004	.0004
23	.0000	.0001	.0001	.0001	.0001	.0001	.0001	.0001	.0002	.0002
24	.0000	.0000	.0000	.0000	.0000	.0000	.0000	.0001	.0001	.0001

					λ					
r	11	12	13	14	15	16	17	18	19	20
0	.0000	.0000	.0000	.0000	.0000	.0000	.0000	.0000	.0000	.0000
1	.0002	.0001	.0000	.0000	.0000	.0000	.0000	.0000	.0000	.0000
2	.0010	.0004	.0002	.0001	.0000	.0000	.0000	.0000	.0000	.0000
3	.0037	.0018	.0008	.0004	.0002	.0001	.0000	.0000	.0000	.0000
4	.0102	.0053	.0027	.0013	.0006	.0003	.0001	.0001	.0000	.0000
5	.0224	.0127	.0070	.0037	.0019	.0010	.0005	.0002	.0001	.0001
6	.0411	.0255	.0152	.0087	.0048	.0026	.0014	.0007	.0004	.0002
7	.0646	.0437	.0281	.0174	.0104	.0060	.0034	.0018	.0010	.0005
8	.0888	.0655	.0457	.0304	.0194	.0120	.0072	.0042	.0024	.0013
9	.1085	.0874	.0661	.0473	.0324	.0213	.0135	.0083	.0050	.0029
10	.1194	.1048	.0859	.0663	.0486	.0341	.0230	.0150	.0095	.0058
11	.1194	.1144	.1015	.0844	.0663	.0496	.0355	.0245	.0164	.0106
12	.1094	.1144	.1099	.0984	.0829	.0661	.0504	.0368	.0259	.0176
13	.0926	.1056	.1099	.1060	.0956	.0814	.0658	.0509	.0378	.0271
14	.0728	.0905	.1021	.1060	.1024	.0930	.0800	.0655	.0541	.0387
15	.0534	.0724	.0885	.0989	.1024	.0992	.0906	.0786	.0650	.0516
16	.0367	.0543	.0719	.0866	.0960	.0992	.0963	.0884	.0772	.0646
17	.0237	.0383	.0550	.0713	.0847	.0934	.0963	.0936	.0863	.0760
18	.0145	.0256	.0397	.0554	.0706	.0830	.0909	.0936	.0911	.0844
19	.0084	.0161	.0272	.0409	.0557	.0699	.0814	.0887	.0911	.0888
20	.0046	.0097	.0177	.0286	.0418	.0559	.0692	.0798	.0866	.0888
21	.0024	.0055	.0109	.0191	.0299	.0426	.0560	.0684	.0783	.0846
22	.0012	.0030	.0065	.0121	.0204	.0310	.0433	.0560	.0676	.0769
23	.0006	.0016	.0037	.0074	.0133	.0216	.0320	.0438	.0559	.0669
24	.0003	.0008	.0020	.0043	.0083	.0144	.0226	.0328	.0442	.0557

Table VI   (continued)

r	11	12	13	14	15	16	17	18	19	20
					λ					
25	.0001	.0004	.0010	.0024	.0050	.0092	.0154	.0237	.0336	.0446
26	.0000	.0002	.0005	.0013	.0029	.0057	.0101	.0164	.0246	.0343
27	.0000	.0001	.0002	.0007	.0016	.0034	.0063	.0109	.0173	.0254
28	.0000	.0000	.0001	.0003	.0009	.0019	.0038	.0070	.0117	.0181
29	.0000	.0000	.0001	.0002	.0004	.0011	.0023	.0044	.0077	.0125
30	.0000	.0000	.0000	.0001	.0002	.0006	.0013	.0026	.0049	.0083
31	.0000	.0000	.0000	.0000	.0001	.0003	.0007	.0015	.0030	.0054
32	.0000	.0000	.0000	.0000	.0001	.0001	.0004	.0009	.0018	.0034
33	.0000	.0000	.0000	.0000	.0000	.0001	.0002	.0005	.0010	.0020
34	.0000	.0000	.0000	.0000	.0000	.0000	.0001	.0002	.0006	.0012
35	.0000	.0000	.0000	.0000	.0000	.0000	.0000	.0001	.0003	.0007
36	.0000	.0000	.0000	.0000	.0000	.0000	.0000	.0001	.0002	.0004
37	.0000	.0000	.0000	.0000	.0000	.0000	.0000	.0000	.0001	.0002
38	.0000	.0000	.0000	.0000	.0000	.0000	.0000	.0000	.0000	.0001
39	.0000	.0000	.0000	.0000	.0000	.0000	.0000	.0000	.0000	.0001

This table is reproduced by permission from R. S. Burington and D. C. May, *Handbook of Probability and Statistics with Tables.* McGraw-Hill Book Company, Inc., 1953.

Table VII. Unit Normal Linear Loss Integral $L_N(D)$

D	.00	.01	.02	.03	.04	.05	.06	.07	.08	.09
.0	.3989	.3940	.3890	.3841	.3793	.3744	.3697	.3649	.3602	.3556
.1	.3509	.3464	.3418	.3373	.3328	.3284	.3240	.3197	.3154	.3111
.2	.3069	.3027	.2986	.2944	.2904	.2863	.2824	.2784	.2745	.2706
.3	.2668	.2630	.2592	.2555	.2518	.2481	.2445	.2409	.2374	.2339
.4	.2304	.2270	.2236	.2203	.2169	.2137	.2104	.2072	.2040	.2009
.5	.1978	.1947	.1917	.1887	.1857	.1828	.1799	.1771	.1742	.1714
.6	.1687	.1659	.1633	.1606	.1580	.1554	.1528	.1503	.1478	.1453
.7	.1429	.1405	.1381	.1358	.1334	.1312	.1289	.1267	.1245	.1223
.8	.1202	.1181	.1160	.1140	.1120	.1100	.1080	.1061	.1042	.1023
.9	.1004	.09860	.09680	.09503	.09328	.09156	.08986	.08819	.08654	.08491
1.0	.08332	.08174	.08019	.07866	.07716	.07568	.07422	.07279	.07138	.06999
1.1	.06862	.06727	.06595	.06465	.06336	.06210	.06086	.05964	.05844	.05726
1.2	.05610	.05496	.05384	.05274	.05165	.05059	.04954	.04851	.04750	.04650
1.3	.04553	.04457	.04363	.04270	.04179	.04090	.04002	.03916	.03831	.03748
1.4	.03667	.03587	.03508	.03431	.03356	.03281	.03208	.03137	.03067	.02998
1.5	.02931	.02865	.02800	.02736	.02674	.02612	.02552	.02494	.02436	.02380
1.6	.02324	.02270	.02217	.02165	.02114	.02064	.02015	.01967	.01920	.01874
1.7	.01829	.01785	.01742	.01699	.01658	.01617	.01578	.01539	.01501	.01464
1.8	.01428	.01392	.01357	.01323	.01290	.01257	.01226	.01195	.01164	.01134
1.9	.01105	.01077	.01049	.01022	$.0^2 9957$	$.0^2 9698$	$.0^2 9445$	$.0^2 9198$	$.0^2 8957$	$.0^2 8721$
2.0	$.0^2 8491$	$.0^2 8266$	$.0^2 8046$	$.0^2 7832$	$.0^2 7623$	$.0^2 7418$	$.0^2 7219$	$.0^2 7024$	$.0^2 6835$	$.0^2 6649$
2.1	$.0^2 6468$	$.0^2 6292$	$.0^2 6120$	$.0^2 5952$	$.0^2 5788$	$.0^2 5628$	$.0^2 5472$	$.0^2 5320$	$.0^2 5172$	$.0^2 5028$
2.2	$.0^2 4887$	$.0^2 4750$	$.0^2 4616$	$.0^2 4486$	$.0^2 4358$	$.0^2 4235$	$.0^2 4114$	$.0^2 3996$	$.0^2 3882$	$.0^2 3770$
2.3	$.0^2 3662$	$.0^2 3556$	$.0^2 3453$	$.0^2 3352$	$.0^2 3255$	$.0^2 3159$	$.0^2 3067$	$.0^2 2977$	$.0^2 2889$	$.0^2 2804$
2.4	$.0^2 2720$	$.0^2 2640$	$.0^2 2561$	$.0^2 2484$	$.0^2 2410$	$.0^2 2337$	$.0^2 2267$	$.0^2 2199$	$.0^2 2132$	$.0^2 2067$
2.5	$.0^2 2004$	$.0^2 1943$	$.0^2 1883$	$.0^2 1826$	$.0^2 1769$	$.0^2 1715$	$.0^2 1662$	$.0^2 1610$	$.0^2 1560$	$.0^2 1511$
2.6	$.0^2 1464$	$.0^2 1418$	$.0^2 1373$	$.0^2 1330$	$.0^2 1288$	$.0^2 1247$	$.0^2 1207$	$.0^2 1169$	$.0^2 1132$	$.0^2 1095$

D										
2.7	$.0^2 1060$	$.0^2 1026$	$.0^3 9928$	$.0^3 9607$	$.0^3 9295$	$.0^3 8992$	$.0^3 8699$	$.0^3 8414$	$.0^3 8138$	$.0^3 7870$
2.8	$.0^3 7611$	$.0^3 7359$	$.0^3 7115$	$.0^3 6879$	$.0^3 6650$	$.0^3 6428$	$.0^3 6213$	$.0^3 6004$	$.0^3 5802$	$.0^3 5606$
2.9	$.0^3 5417$	$.0^3 5233$	$.0^3 5055$	$.0^3 4883$	$.0^3 4716$	$.0^3 4555$	$.0^3 4398$	$.0^3 4247$	$.0^3 4101$	$.0^3 3959$
3.0	$.0^3 3822$	$.0^3 3689$	$.0^3 3560$	$.0^3 3436$	$.0^3 3316$	$.0^3 3199$	$.0^3 3087$	$.0^3 2978$	$.0^3 2873$	$.0^3 2771$
3.1	$.0^3 2673$	$.0^3 2577$	$.0^3 2485$	$.0^3 2396$	$.0^3 2311$	$.0^3 2227$	$.0^3 2147$	$.0^3 2070$	$.0^3 1995$	$.0^3 1922$
3.2	$.0^3 1852$	$.0^3 1785$	$.0^3 1720$	$.0^3 1657$	$.0^3 1596$	$.0^3 1537$	$.0^3 1480$	$.0^3 1426$	$.0^3 1373$	$.0^3 1322$
3.3	$.0^3 1273$	$.0^3 1225$	$.0^3 1179$	$.0^3 1135$	$.0^3 1093$	$.0^3 1051$	$.0^3 1012$	$.0^4 9734$	$.0^4 9365$	$.0^4 9009$
3.4	$.0^4 8666$	$.0^4 8335$	$.0^4 8016$	$.0^4 7709$	$.0^4 7413$	$.0^4 7127$	$.0^4 6852$	$.0^4 6587$	$.0^4 6331$	$.0^4 6085$
3.5	$.0^4 5848$	$.0^4 5620$	$.0^4 5400$	$.0^4 5188$	$.0^4 4984$	$.0^4 4788$	$.0^4 4599$	$.0^4 4417$	$.0^4 4242$	$.0^4 4073$
3.6	$.0^4 3911$	$.0^4 3755$	$.0^4 3605$	$.0^4 3460$	$.0^4 3321$	$.0^4 3188$	$.0^4 3059$	$.0^4 2935$	$.0^4 2816$	$.0^4 2702$
3.7	$.0^4 2592$	$.0^4 2486$	$.0^4 2385$	$.0^4 2287$	$.0^4 2193$	$.0^4 2103$	$.0^4 2016$	$.0^4 1933$	$.0^4 1853$	$.0^4 1776$
3.8	$.0^4 1702$	$.0^4 1632$	$.0^4 1563$	$.0^4 1498$	$.0^4 1435$	$.0^4 1375$	$.0^4 1317$	$.0^4 1262$	$.0^4 1208$	$.0^4 1157$
3.9	$.0^4 1108$	$.0^4 1061$	$.0^4 1016$	$.0^5 9723$	$.0^5 9307$	$.0^5 8908$	$.0^5 8525$	$.0^5 8158$	$.0^5 7806$	$.0^5 7469$
4.0	$.0^5 7145$	$.0^5 6835$	$.0^5 6538$	$.0^5 6253$	$.0^5 5980$	$.0^5 5718$	$.0^5 5468$	$.0^5 5227$	$.0^5 4997$	$.0^5 4777$
4.1	$.0^5 4566$	$.0^5 4364$	$.0^5 4170$	$.0^5 3985$	$.0^5 3807$	$.0^5 3637$	$.0^5 3475$	$.0^5 3319$	$.0^5 3170$	$.0^5 3027$
4.2	$.0^5 2891$	$.0^5 2760$	$.0^5 2635$	$.0^5 2516$	$.0^5 2402$	$.0^5 2292$	$.0^5 2188$	$.0^5 2088$	$.0^5 1992$	$.0^5 1901$
4.3	$.0^5 1814$	$.0^5 1730$	$.0^5 1650$	$.0^5 1574$	$.0^5 1501$	$.0^5 1431$	$.0^5 1365$	$.0^5 1301$	$.0^5 1241$	$.0^5 1183$
4.4	$.0^5 1127$	$.0^5 1074$	$.0^5 1024$	$.0^6 9756$	$.0^6 9296$	$.0^6 8857$	$.0^6 8437$	$.0^6 8037$	$.0^6 7655$	$.0^6 7290$
4.5	$.0^6 6942$	$.0^6 6610$	$.0^6 6294$	$.0^6 5992$	$.0^6 5704$	$.0^6 5429$	$.0^6 5167$	$.0^6 4917$	$.0^6 4679$	$.0^6 4452$
4.6	$.0^6 4236$	$.0^6 4029$	$.0^6 3833$	$.0^6 3645$	$.0^6 3467$	$.0^6 3297$	$.0^6 3135$	$.0^6 2981$	$.0^6 2834$	$.0^6 2694$
4.7	$.0^6 2560$	$.0^6 2433$	$.0^6 2313$	$.0^6 2197$	$.0^6 2088$	$.0^6 1984$	$.0^6 1884$	$.0^6 1790$	$.0^6 1700$	$.0^6 1615$
4.8	$.0^6 1533$	$.0^6 1456$	$.0^6 1382$	$.0^6 1312$	$.0^6 1246$	$.0^6 1182$	$.0^6 1122$	$.0^6 1065$	$.0^6 1011$	$.0^7 9588$
4.9	$.0^7 9096$	$.0^7 8629$	$.0^7 8185$	$.0^7 7763$	$.0^7 7362$	$.0^7 6982$	$.0^7 6620$	$.0^7 6276$	$.0^7 5950$	$.0^7 5640$

EXAMPLES: $L_N(-D) = D + L_N(D)$
$L_N(3.57) = .0^4 4417 = .00004417$
$L_N(-3.57) = 3.57004417$

This table is reproduced by permission from Robert Schlaifer, *Probability and Statistics for Business Decisions*. McGraw-Hill Book Company, Inc., 1959.

Table VIII.   Random Digits

10	09	73	25	33	76	52	01	35	86	34	67	35	48	76	80	95	90	91	17	39	29	27	49	45
37	54	20	48	05	64	89	47	42	96	24	80	52	40	37	20	63	61	04	02	00	82	29	16	65
08	42	26	89	53	19	64	50	93	03	23	20	90	25	60	15	95	33	47	64	35	08	03	36	06
99	01	90	25	29	09	37	67	07	15	38	31	13	11	65	88	67	67	43	97	04	43	62	76	59
12	80	79	99	70	80	15	73	61	47	64	03	23	66	53	98	95	11	68	77	12	17	17	68	33
66	06	57	47	17	34	07	27	68	50	36	69	73	61	70	65	81	33	98	85	11	19	92	91	70
31	06	01	08	05	45	57	18	24	06	35	30	34	26	14	86	79	90	74	39	23	40	30	97	32
85	26	97	76	02	02	05	16	56	92	68	66	57	48	18	73	05	38	52	47	18	62	38	85	79
63	57	33	21	35	05	32	54	70	48	90	55	35	75	48	28	46	82	87	09	83	49	12	56	24
73	79	64	57	53	03	52	96	47	78	35	80	83	42	82	60	93	52	03	44	35	27	38	84	35
98	52	01	77	67	14	90	56	86	07	22	10	94	05	58	60	97	09	34	33	50	50	07	39	98
11	80	50	54	31	39	80	82	77	32	50	72	56	82	48	29	40	52	42	01	52	77	56	78	51
83	45	29	96	34	06	28	89	80	83	13	74	67	00	78	18	47	54	06	10	68	71	17	78	17
88	68	54	02	00	86	50	75	84	01	36	76	66	79	51	90	36	47	64	93	29	60	91	10	62
99	59	46	73	48	87	51	76	49	69	91	82	60	89	28	93	78	56	13	68	23	47	83	41	13
65	48	11	76	74	17	46	85	09	50	58	04	77	69	74	73	03	95	71	86	40	21	81	65	44
80	12	43	56	35	17	72	70	80	15	45	31	82	23	74	21	11	57	82	53	14	38	55	37	63
74	35	09	98	17	77	40	27	72	14	43	23	60	02	10	45	52	16	42	37	96	28	60	26	55
69	91	62	68	03	66	25	22	91	48	36	93	68	72	03	76	62	11	39	90	94	40	05	64	18
09	89	32	05	05	14	22	56	85	14	46	42	75	67	88	96	29	77	88	22	54	38	21	45	98
91	49	91	45	23	68	47	92	76	86	46	16	28	35	54	94	75	08	99	23	37	08	92	00	48
80	33	69	45	98	26	94	03	68	58	70	29	73	41	35	53	14	03	33	40	42	05	08	23	41
44	10	48	19	49	85	15	74	79	54	32	97	92	65	75	57	60	04	08	81	22	22	20	64	13
12	55	07	37	42	11	10	00	20	40	12	86	07	46	97	96	64	48	94	39	28	70	72	58	15
63	60	64	93	29	16	50	53	44	84	40	21	95	25	63	43	65	17	70	82	07	20	73	17	90
61	19	69	04	46	26	45	74	77	74	51	92	43	37	29	65	39	45	95	93	42	58	26	05	27
15	47	44	52	66	95	27	07	99	53	59	36	78	38	48	82	39	61	01	18	33	21	15	94	66
94	55	72	85	73	67	89	75	43	87	54	62	24	44	31	91	19	04	25	92	92	92	74	59	73
42	48	11	62	13	97	34	40	87	21	16	86	84	87	67	03	07	11	20	29	25	70	14	66	70
23	52	37	83	17	73	20	88	98	37	68	93	59	14	16	26	25	22	96	63	05	52	28	25	62
04	49	35	24	94	75	24	63	38	24	45	86	25	10	25	61	96	27	93	35	65	33	71	24	72
00	54	99	76	54	64	05	18	81	59	96	11	96	38	96	54	69	28	23	91	23	28	72	95	29
35	96	31	53	07	26	89	80	93	54	33	35	13	54	62	77	97	45	00	24	90	10	33	93	33
59	80	80	83	91	45	42	72	68	42	83	60	94	97	00	13	02	12	48	92	78	56	52	01	06
46	05	88	52	36	01	39	09	22	86	77	28	14	40	77	93	91	08	36	47	70	61	74	29	41
32	17	90	05	97	87	37	92	52	41	05	56	70	70	07	86	74	31	71	57	85	39	41	18	38
69	23	46	14	06	20	11	74	52	04	15	95	66	00	00	18	74	39	24	23	97	11	89	63	38
19	56	54	14	30	01	75	87	53	79	40	41	92	15	85	66	67	43	68	06	84	96	28	52	07
45	15	51	49	38	19	47	60	72	46	43	66	79	45	43	59	04	79	00	33	20	82	66	95	41
94	86	43	19	94	36	16	81	08	51	34	88	88	15	53	01	54	03	54	56	05	01	45	11	76
98	08	62	48	26	45	24	02	84	04	44	99	90	88	96	39	09	47	34	07	35	44	13	18	80
33	18	51	62	32	41	94	15	09	49	89	43	54	85	81	88	69	54	19	94	37	54	87	30	43
80	95	10	04	06	96	38	27	07	74	20	15	12	33	87	25	01	62	52	98	94	62	46	11	71
79	75	24	91	40	71	96	12	82	96	69	86	10	25	91	74	85	22	05	39	00	38	75	95	79
18	63	33	25	37	98	14	50	65	71	31	01	02	46	74	05	45	56	14	27	77	93	89	19	36
74	02	94	39	02	77	55	73	22	70	97	79	01	71	19	52	52	75	80	21	80	81	45	17	48
54	17	84	56	11	80	99	33	71	43	05	33	51	29	69	56	12	71	92	55	36	04	09	03	24
11	66	44	98	83	52	07	98	48	27	59	38	17	15	39	09	97	33	34	40	88	46	12	33	56
48	32	47	79	28	31	24	96	47	10	02	29	53	68	70	32	30	75	75	46	15	02	00	99	94
69	07	49	41	38	87	63	79	19	76	35	58	40	44	01	10	51	82	16	15	01	84	87	69	38

This table is reproduced here by permission from The RAND Corporation, *A Million Random Digits*. The Free Press, New York. 1955.

Table IX. The Transformation of r to $Z = \frac{1}{2} \log \left[(1 + r)/(1 - r) \right]$

r	r (3rd decimal)					r	r (3rd decimal)				
	.000	.002	.004	.006	.008		.000	.002	.004	.006	.008
.00	.0000	.0020	.0040	.0060	.0080	.35	.3654	.3677	.3700	.3723	.3746
1	.0100	.0120	.0140	.0160	.0180	6	.3769	.3792	.3815	.3838	.3861
2	.0200	.0220	.0240	.0260	.0280	7	.3884	.3907	.3931	.3954	.3977
3	.0300	.0320	.0340	.0360	.0380	8	.4001	.4024	.4047	.4071	.4094
4	.0400	.0420	.0440	.0460	.0480	9	.4118	.4142	.4165	.4189	.4213
.05	.0500	.0520	.0541	.0561	.0581	.40	.4236	.4260	.4284	.4308	.4332
6	.0601	.0621	.0641	.0661	.0681	1	.4356	.4380	.4404	.4428	.4453
7	.0701	.0721	.0741	.0761	.0782	2	.4477	.4501	.4526	.4550	.4574
8	.0802	.0822	.0842	.0862	.0882	3	.4599	.4624	.4648	.4673	.4698
9	.0902	.0923	.0943	.0963	.0983	4	.4722	.4747	.4772	.4797	.4822
.10	.1003	.1024	.1044	.1064	.1084	.45	.4847	.4872	.4897	.4922	.4948
1	.1104	.1125	.1145	.1165	.1186	6	.4973	.4999	.5024	.5049	.4075
2	.1206	.1226	.1246	.1267	.1287	7	.5101	.5126	.5152	.5178	.5204
3	.1307	.1328	.1348	.1368	.1389	8	.5230	.5256	.5282	.5308	.5334
4	.1409	.1430	.1450	.1471	.1491	9	.5361	.5387	.5413	.5440	.5466
.15	.1511	.1532	.1552	.1573	.1593	.50	.5493	.5520	.5547	.5573	.5600
6	.1614	.1634	.1655	.1676	.1696	1	.5627	.5654	.5682	.5709	.5736
7	.1717	.1737	.1758	.1779	.1799	2	.5763	.5791	.5818	.5846	.5874
8	.1820	.1841	.1861	.1882	.1903	3	.5901	.5929	.5957	.5985	.6013
9	.1923	.1944	.1965	.1986	.2007	4	.6042	.6070	.6098	.6127	.6155
.20	.2027	.2048	.2069	.2090	.2111	.55	.6194	.6213	.6241	.6270	.6299
1	.2132	.2153	.2174	.2195	.2216	6	.6328	.6358	.6387	.6416	.6446
2	.2237	.2258	.2279	.2300	.2321	7	.6475	.6505	.6535	.6565	.6595
3	.2342	.2363	.2384	.2405	.2427	8	.6625	.6655	.6685	.6716	.6746
4	.2448	.2469	.2490	.2512	.2533	9	.6777	.6807	.6838	.6869	.6900
.25	.2554	.2575	.2597	.2618	.2640	.60	.6931	.6963	.6994	.7026	.7057
6	.2661	.2683	.2704	.2726	.2747	1	.7089	.7121	.7153	.7185	.7218
7	.2769	.2790	.2812	.2833	.2855	2	.7250	.7283	.7315	.7348	.7381
8	.2877	.2899	.2920	.2942	.2964	3	.7414	.7447	.7481	.7514	.7548
9	.2986	.3008	.3029	.3051	.3073	4	.7582	.7616	.7650	.7684	.7718
.30	.3095	.3117	.3139	.3161	.3183	.65	.7753	.7788	.7823	.7858	.7893
1	.3205	.3228	.3250	.3272	.3294	6	.7928	.7964	.7999	.8035	.8071
2	.3316	.3339	.3361	.3383	.3406	7	.8107	.8144	.8180	.8217	.8254
3	.3428	.3451	.3473	.3496	.3518	8	.8291	.8328	.8366	.8404	.8441
4	.3541	.3564	.3586	.3609	.3632	9	.8480	.8518	.8556	.8595	.8634

Table IX   (continued)

r	r (3rd decimal)					r	r (3rd decimal)				
	.000	**.002**	**.004**	**.006**	**.008**		**.000**	**.002**	**.004**	**.006**	**.008**
.70	.8673	.8712	.8752	.8792	.8832	.85	1.256	1.263	1.271	1.278	1.286
1	.8872	.8912	.8953	.8994	.9035	6	1.293	1.301	1.309	1.317	1.325
2	.9076	.9118	.9160	.9202	.9245	7	1.333	1.341	1.350	1.358	1.367
3	.9287	.9330	.9373	.9417	.9461	8	1.376	1.385	1.394	1.403	1.412
4	.9505	.9549	.9549	.9639	.9684	9	1.422	1.432	1.442	1.452	1.462
.75	0.973	0.978	0.982	0.987	0.991	.90	1.472	1.483	1.494	1.505	1.516
6	0.996	1.001	1.006	1.011	1.015	1	1.528	1.539	1.551	1.564	1.576
7	1.020	1.025	1.030	1.035	1.040	2	1.589	1.602	1.616	1.630	1.644
8	1.045	1.050	1.056	1.061	1.066	3	1.658	1.673	1.689	1.705	1.721
9	1.071	1.077	1.082	1.088	1.093	4	1.738	1.756	1.774	1.792	1.812
.80	1.099	1.104	1.110	1.116	1.121	.95	1.832	1.853	1.874	1.897	1.921
1	1.127	1.133	1.139	1.145	1.151	6	1.946	1.972	2.000	2.029	2.060
2	1.157	1.163	1.169	1.175	1.182	7	2.092	2.127	2.165	2.205	2.249
3	1.188	1.195	1.201	1.208	1.214	8	2.298	2.351	2.410	2.477	2.555
4	1.221	1.228	1.235	1.242	1.249	9	2.647	2.759	2.903	3.106	3.453

Table X.   Upper Percentage Points of D in the Kolmogorov-Smirnov One-Sample Test

Q / N	.20	.15	.10	.05	.01
1	.900	.925	.950	.975	.995
2	.684	.726	.776	.842	.929
3	.565	.597	.642	.708	.828
4	.494	.525	.564	.624	.733
5	.446	.474	.510	.565	.669
6	.410	.436	.470	.521	.618
7	.381	.405	.438	.486	.577
8	.358	.381	.411	.457	.543
9	.339	.360	.388	.432	.514
10	.322	.342	.368	.410	.490
11	.307	.326	.352	.391	.468
12	.295	.313	.338	.375	.450
13	.284	.302	.325	.361	.433
14	.274	.292	.314	.349	.418
15	.266	.283	.304	.338	.404
16	.258	.274	.295	.328	.392
17	.250	.266	.280	.318	.381
18	.244	.259	.278	.309	.371
19	.237	.252	.272	.301	.363
20	.231	.246	.264	.294	.356
25	.21	.22	.24	.27	.32
30	.19	.20	.22	.24	.29
35	.18	.19	.21	.23	.27
Over 35	$1.07/\sqrt{N}$	$1.14/\sqrt{N}$	$1.22/\sqrt{N}$	$1.36/\sqrt{N}$	$1.63/\sqrt{N}$

This table is reproduced by permission, adapted from F. J. Massey, "The Kolmogorov-Smirnov Test for Goodness of Fit," *Journal of the American Statistical Association*, **46,** 70 (1951).

Table XI.   Binomial Coefficients, $\binom{N}{r}$

N \ r	0	1	2	3	4	5	6	7	8	9	10
1	1	1									
2	1	2	1								
3	1	3	3	1							
4	1	4	6	4	1						
5	1	5	10	10	5	1					
6	1	6	15	20	15	6	1				
7	1	7	21	35	35	21	7	1			
8	1	8	28	56	70	56	28	8	1		
9	1	9	36	84	126	126	84	36	9	1	
10	1	10	45	120	210	252	210	120	45	10	1
11	1	11	55	165	330	462	462	330	165	55	11
12	1	12	66	220	495	792	924	792	495	220	66
13	1	13	78	286	715	1287	1716	1716	1287	715	286
14	1	14	91	364	1001	2002	3003	3432	3003	2002	1001
15	1	15	105	455	1365	3003	5005	6435	6435	5005	3003
16	1	16	120	560	1820	4368	8008	11440	12870	11440	8008
17	1	17	136	680	2380	6188	12376	19448	24310	24310	19448
18	1	18	153	816	3060	8568	18564	31824	43758	48620	43758
19	1	19	171	969	3876	11628	27132	50388	75582	92378	92378
20	1	20	190	1140	4845	15504	38760	77520	125970	167960	184756

Table XII.   Factorials of Integers

n	$n!$	n	$n!$
1	1	26	4.03291×10^{26}
2	2	27	1.08889×10^{28}
3	6	28	3.04888×10^{29}
4	24	29	8.84176×10^{30}
5	120	30	2.65253×10^{32}
6	720	31	8.22284×10^{33}
7	5040	32	2.63131×10^{35}
8	40320	33	8.68332×10^{36}
9	362880	34	2.95233×10^{38}
10	3.62880×10^{6}	35	1.03331×10^{40}
11	3.99168×10^{7}	36	3.71993×10^{41}
12	4.79002×10^{8}	37	1.37638×10^{43}
13	6.22702×10^{9}	38	5.23023×10^{44}
14	8.71783×10^{10}	39	2.03979×10^{46}
15	1.30767×10^{12}	40	8.15915×10^{47}
16	2.09228×10^{13}	41	3.34525×10^{49}
17	3.55687×10^{14}	42	1.40501×10^{51}
18	6.40327×10^{15}	43	6.04153×10^{52}
19	1.21645×10^{17}	44	2.65827×10^{54}
20	2.43290×10^{18}	45	1.19622×10^{56}
21	5.10909×10^{19}	46	5.50262×10^{57}
22	1.12400×10^{21}	47	2.58623×10^{59}
23	2.58520×10^{22}	48	1.24139×10^{61}
24	6.20448×10^{23}	49	6.08282×10^{62}
25	1.55112×10^{25}	50	3.04141×10^{64}

Table XIII. Powers and Roots

N	N^2	$\sqrt{N}$	$\sqrt{10N}$
1	1	1.00 000	3.16 228
2	4	1.41 421	4.47 214
3	9	1.73 205	5.47 723
4	16	2.00 000	6.32 456
5	25	2.23 607	7.07 107
6	36	2.44 949	7.74 597
7	49	2.64 575	8.36 660
8	64	2.82 843	8.94 427
9	81	3.00 000	9.48 683
10	100	3.16 228	10.00 00
11	121	3.31 662	10.48 81
12	144	3.46 410	10.95 45
13	169	3.60 555	11.40 18
14	196	3.74 166	11.83 22
15	225	3.87 298	12.24 74
16	256	4.00 000	12.64 91
17	289	4.12 311	13.03 84
18	324	4.24 264	13.41 64
19	361	4.35 890	13.78 40
20	400	4.47 214	14.14 21
21	441	4.58 258	14.49 14
22	484	4.69 042	14.83 24
23	529	4.79 583	15.16 58
24	576	4.89 898	15.49 19
25	625	5.00 000	15.81 14
26	676	5.09 902	16.12 45
27	729	5.19 615	16.43 17
28	784	5.29 150	16.73 32
29	841	5.38 516	17.02 94
30	900	5.47 723	17.32 05
31	961	5.56 776	17.60 68
32	1 024	5.65 685	17.88 85
33	1 089	5.74 456	18.16 59
34	1 156	5.83 095	18.43 91
35	1 225	5.91 608	18.70 83
36	1 296	6.00 000	18.97 37
37	1 369	6.08 276	19.23 54
38	1 444	6.16 441	19.49 36
39	1 521	6.24 500	19.74 84
40	1 600	6.32 456	20.00 00
41	1 681	6.40 312	20.24 85
42	1 764	6.48 074	20.49 39
43	1 849	6.55 744	20.73 64
44	1 936	6.63 325	20.97 62
45	2 025	6.70 820	21.21 32
46	2 116	6.78 233	21.44 76
47	2 209	6.85 565	21.67 95
48	2 304	6.92 820	21.90 89
49	2 401	7.00 000	22.13 59
50	2 500	7.07 107	22.36 07
N	N^2	$\sqrt{N}$	$\sqrt{10N}$

N	N^2	$\sqrt{N}$	$\sqrt{10N}$
50	2 500	7.07 107	22.36 07
51	2 601	7.14 143	22.58 32
52	2 704	7.21 110	22.80 35
53	2 809	7.28 011	23.02 17
54	2 916	7.34 847	23.23 79
55	3 025	7.41 620	23.45 21
56	3 136	7.48 331	23.66 43
57	3 249	7.54 983	23.87 47
58	3 364	7.61 577	24.08 32
59	3 481	7.68 115	24.28 99
60	3 600	7.74 597	24.49 49
61	3 721	7.81 025	24.69 82
62	3 844	7.87 401	24.89 98
63	3 969	7.93 725	25.09 98
64	4 096	8.00 000	25.29 82
65	4 225	8.06 226	25.49 51
66	4 356	8.12 404	25.69 05
67	4 489	8.18 535	25.88 44
68	4 624	8.24 621	26.07 68
69	4 761	8.30 662	26.26 79
70	4 900	8.36 660	26.45 75
71	5 041	8.42 615	26.64 58
72	5 184	8.48 528	26.83 28
73	5 329	8.54 400	27.01 85
74	5 476	8.60 233	27.20 29
75	5 625	8.66 025	27.38 61
76	5 776	8.71 780	27.56 81
77	5 929	8.77 496	27.74 89
78	6 084	8.83 176	27.92 85
79	6 241	8.88 819	28.10 69
80	6 400	8.94 427	28.28 43
81	6 561	9.00 000	28.46 05
82	6 724	9.05 539	28.63 56
83	6 889	9.11 043	28.80 97
84	7 056	9.16 515	28.98 28
85	7 225	9.21 954	29.15 48
86	7 396	9.27 362	29.32 58
87	7 569	9.32 738	29.49 58
88	7 744	9.38 083	29.66 48
89	7 921	9.43 398	29.83 29
90	8 100	9.48 683	30.00 00
91	8 281	9.53 939	30.16 62
92	8 464	9.59 166	30.33 15
93	8 649	9.64 365	30.49 59
94	8 836	9.69 536	30.65 94
95	9 025	9.74 679	30.82 21
96	9 216	9.79 796	30.98 39
97	9 409	9.84 886	31.14 48
98	9 604	9.89 949	31.30 50
99	9 801	9.94 987	31.46 43
100	10 000	10.00 000	31.62 28
N	N^2	$\sqrt{N}$	$\sqrt{10N}$

Table XIII   (continued)

N	N²	√N	√10N		N	N²	√N	√10N
100	10 000	10.00 00	31.62 28		**150**	22 500	12.24 74	38.72 98
101	10 201	10.04 99	31.78 05		151	22 801	12.28 82	38.85 87
102	10 404	10.09 95	31.93 74		152	23 104	12.32 88	38.98 72
103	10 609	10.14 89	32.09 36		153	23 409	12.36 93	39.11 52
104	10 816	10.19 80	32.24 90		154	23 716	12.40 97	39.24 28
105	11 025	10.24 70	32.40 37		155	24 025	12.44 99	39.37 00
106	11 236	10.29 56	32.55 76		**156**	24 336	12.49 00	39.49 68
107	11 449	10.34 41	32.71 09		157	24 649	12.53 00	39.62 32
108	11 664	10.39 23	32.86 34		158	24 964	12.56 98	39.74 92
109	11 881	10.44 03	33.01 51		159	25 281	12.60 95	39.87 48
110	12 100	10.48 81	33.16 62		160	25 600	12.64 91	40.00 00
111	12 321	10.53 57	33.31 67		**161**	25 921	12.68 86	40.12 48
112	12 544	10.58 30	33.46 64		162	26 244	12.72 79	40.24 92
113	12 769	10.63 01	33.61 55		163	26 569	12.76 71	40.37 33
114	12 996	10.67 71	33.76 39		164	26 896	12.80 62	40.49 69
115	13 225	10.72 38	33.91 16		165	27 225	12.84 52	40.62 02
116	13 456	10.77 03	34.05 88		**166**	27 556	12.88 41	40.74 31
117	13 689	10.81 67	34.20 53		167	27 889	12.92 28	40.86 56
118	13 924	10.86 28	34.35 11		168	28 224	12.96 15	40.98 78
119	14 161	10.90 87	34.49 64		169	28 561	13.00 00	41.10 96
120	14 400	10.95 45	34.64 10		170	28 900	13.03 84	41.23 11
121	14 641	11.00 00	34.78 51		**171**	29 241	13.07 67	41.35 21
122	14 884	11.04 54	34.92 85		172	29 584	13.11 49	41.47 29
123	15 129	11.09 05	35.07 14		173	29 929	13.15 29	41.59 33
124	15 376	11.13 55	35.21 36		174	30 276	13.19 09	41.71 33
125	15 625	11.18 03	35.35 53		175	30 625	13.22 88	41.83 30
126	15 876	11.22 50	35.49 65		**176**	30 976	13.26 65	41.95 24
127	16 129	11.26 94	35.63 71		177	31 329	13.30 41	42.07 14
128	16 384	11.31 37	35.77 71		178	31 684	13.34 17	42.19 00
129	16 641	11.35 78	35.91 66		179	32 041	13.37 91	42.30 84
130	16 900	11.40 18	36.05 55		180	32 400	13.41 64	42.42 64
131	17 161	11.44 55	36.19 39		**181**	32 761	13.45 36	42.54 41
132	17 424	11.48 91	36.33 18		182	33 124	13.49 07	42.66 15
133	17 689	11.53 26	36.46 92		183	33 489	13.52 77	42.77 85
134	17 956	11.57 58	36.60 60		184	33 856	13.56 47	42.89 52
135	18 225	11.61 90	36.74 23		185	34 225	13.60 15	43.01 16
136	18 496	11.66 19	36.87 82		**186**	34 596	13.63 82	43.12 77
137	18 769	11.70 47	37.01 35		187	34 969	13.67 48	43.24 35
138	19 044	11.74 73	37.14 84		188	35 344	13.71 13	43.35 90
139	19 321	11.78 98	37.28 27		189	35 721	13.74 77	43.47 41
140	19 600	11.83 22	37.41 66		190	36 100	13.78 40	43.58 90
141	19 881	11.87 43	37.55 00		**191**	36 481	13.82 03	43.70 35
142	20 164	11.91 64	37.68 29		192	36 864	13.85 64	43.81 78
143	20 449	11.95 83	37.81 53		193	37 249	13.89 24	43.93 18
144	20 736	12.00 00	37.94 73		194	37 636	13.92 84	44.04 54
145	21 025	12.04 16	38.07 89		195	38 025	13.96 42	44.15 88
146	21 316	12.08 30	38.20 99		**196**	38 416	14.00 00	44.27 19
147	21 609	12.12 44	38.34 06		197	38 809	14.03 57	44.38 47
148	21 904	12.16 55	38.47 08		198	39 204	14.07 12	44.49 72
149	22 201	12.20 66	38.60 05		199	39 601	14.10 67	44.60 94
150	22 500	12.24 74	38.72 98		200	40 000	14.14 21	44.72 14
N	N²	√N	√10N		N	N²	√N	√10N

Table XIII   (continued)

N	N^2	$\sqrt{N}$	$\sqrt{10N}$
200	40 000	14.14 21	44.72 14
201	40 401	14.17 74	44.83 30
202	40 804	14.21 27	44.94 44
203	41 209	14.24 78	45.05 55
204	41 616	14.28 29	45.16 64
205	42 025	14.31 78	45.27 69
206	42 436	14.35 27	45.38 72
207	42 849	14.38 75	45.49 73
208	43 264	14.42 22	45.60 70
209	43 681	14.45 68	45.71 65
210	44 100	14.49 14	45.82 58
211	44 521	14.52 58	45.93 47
212	44 944	14.56 02	46.04 35
213	45 369	14.59 45	46.15 19
214	45 796	14.62 87	46.26 01
215	46 225	14.66 29	46.36 81
216	46 656	14.69 69	46.47 58
217	47 089	14.73 09	46.58 33
218	47 524	14.76 48	46.69 05
219	47 961	14.79 86	46.79 74
220	48 400	14.83 24	46.90 42
221	48 841	14.86 61	47.01 06
222	49 284	14.89 97	47.11 69
223	49 729	14.93 32	47.22 29
224	50 176	14.96 66	47.32 86
225	50 625	15.00 00	47.43 42
226	51 076	15.03 33	47.53 95
227	51 529	15.06 65	47.64 45
228	51 984	15.09 97	47.74 93
229	52 441	15.13 27	47.85 39
230	52 900	15.16 58	47.95 83
231	53 361	15.19 87	48.06 25
232	53 824	15.23 15	48.16 64
233	54 289	15.26 43	48.27 01
234	54 756	15.29 71	48.37 35
235	55 225	15.32 97	48.47 68
236	55 696	15.36 23	48.57 98
237	56 169	15.39 48	48.68 26
238	56 644	15.42 72	48.78 52
239	57 121	15.45 96	48.88 76
240	57 600	15.49 19	48.98 98
241	58 081	15.52 42	49.09 18
242	58 564	15.55 63	49.19 35
243	59 049	15.58 85	49.29 50
244	59 536	15.62 05	49.39 64
245	60 025	15.65 25	49.49 75
246	60 516	15.68 44	49.59 84
247	61 009	15.71 62	49.69 91
248	61 504	15.74 80	49.79 96
249	62 001	15.77 97	49.89 99
250	62 500	15.81 14	50.00 00
N	N^2	$\sqrt{N}$	$\sqrt{10N}$

N	N^2	$\sqrt{N}$	$\sqrt{10N}$
250	62 500	15.81 14	50.00 00
251	63 001	15.84 30	50.09 99
252	63 504	15.87 45	50.19 96
253	64 009	15.90 60	50.29 91
254	64 516	15.93 74	50.39 84
255	65 025	15.96 87	50.49 75
256	65 536	16.00 00	50.59 64
257	66 049	16.03 12	50.69 52
258	66 564	16.06 24	50.79 37
259	67 081	16.09 35	50.89 20
260	67 600	16.12 45	50.99 02
261	68 121	16.15 55	51.08 82
262	68 644	16.18 64	51.18 59
263	69 169	16.21 73	51.28 35
264	69 696	16.24 81	51.38 09
265	70 225	16.27 88	51.47 82
266	70 756	16.30 95	51.57 52
267	71 289	16.34 01	51.67 20
268	71 824	16.37 07	51.76 87
269	72 361	16.40 12	51.86 52
270	72 900	16.43 17	51.96 15
271	73 441	16.46 21	52.05 77
272	73 984	16.49 24	52.15 36
273	74 529	16.52 27	52.24 94
274	75 076	16.55 29	52.34 50
275	75 625	16.58 31	52.44 04
276	76 176	16.61 32	52.53 57
277	76 729	16.64 33	52.63 08
278	77 284	16.67 33	52.72 57
279	77 841	16.70 33	52.82 05
280	78 400	16.73 32	52.91 50
281	78 961	16.76 31	53.00 94
282	79 524	16.79 29	53.10 37
283	80 089	16.82 26	53.19 77
284	80 656	16.85 23	53.29 17
285	81 225	16.88 19	53.38 54
286	81 796	16.91 15	53.47 90
287	82 369	16.94 11	53.57 24
288	82 944	16.97 06	53.66 56
289	83 521	17.00 00	53.75 87
290	84 100	17.02 94	53.85 16
291	84 681	17.05 87	53.94 44
292	85 264	17.08 80	54.03 70
293	85 849	17.11 72	54.12 95
294	86 436	17.14 64	54.22 18
295	87 025	17.17 56	54.31 39
296	87 616	17.20 47	54.40 59
297	88 209	17.23 37	54.49 77
298	88 804	17.26 27	54.58 94
299	89 401	17.29 16	54.68 09
300	90 000	17.32 05	54.77 23
N	N^2	$\sqrt{N}$	$\sqrt{10N}$

Table XIII   (continued)

N	N^2	$\sqrt{N}$	$\sqrt{10N}$	N	N^2	$\sqrt{N}$	$\sqrt{10N}$
300	90 000	17.32 05	54.77 23	**350**	122 500	18.70 83	59.16 08
301	90 601	17.34 94	54.86 35	351	123 201	18.73 50	59.24 53
302	91 204	17.37 81	54.95 45	352	123 904	18.76 17	59.32 96
303	91 809	17.40 69	55.04 54	353	124 609	18.78 83	59.41 38
304	92 416	17.43 56	55.13 62	354	125 316	18.81 49	59.49 79
305	93 025	17.46 42	55.22 68	355	126 025	18.84 14	59.58 19
306	93 636	17.49 29	55.31 73	**356**	126 736	18.86 80	59.66 57
307	94 249	17.52 14	55.40 76	357	127 449	18.89 44	59.74 95
308	94 864	17.54 99	55.49 77	358	128 164	18.92 09	59.83 31
309	95 481	17.57 84	55.58 78	359	128 881	18.94 73	59.91 66
310	96 100	17.60 68	55.67 76	360	129 600	18.97 37	60.00 00
311	96 721	17.63 52	55.76 74	**361**	130 321	19.00 00	60.08 33
312	97 344	17.66 35	55.85 70	362	131 044	19.02 63	60.16 64
313	97 969	17.69 18	55.94 64	363	131 769	19.05 26	60.24 95
314	98 596	17.72 00	56.03 57	364	132 496	19.07 88	60.33 24
315	99 225	17.74 82	56.12 49	365	133 225	19.10 50	60.41 52
316	99 856	17.77 64	56.21 39	**366**	133 956	19.13 11	60.49 79
317	100 489	17.80 45	56.30 28	367	134 689	19.15 72	60.58 05
318	101 124	17.83 26	56.39 15	368	135 424	19.18 33	60.66 30
319	101 761	17.86 06	56.48 01	369	136 161	19.20 94	60.74 54
320	102 400	17.88 85	56.56 85	370	136 900	19.23 54	60.82 76
321	103 041	17.91 65	56.65 69	**371**	137 641	19.26 14	60.90 98
322	103 684	17.94 44	56.74 50	372	138 384	19.28 73	60.99 18
323	104 329	17.97 22	56.83 31	373	139 129	19.31 32	61.07 37
324	104 976	18.00 00	56.92 10	374	139 876	19.33 91	61.15 55
325	105 625	18.02 78	57.00 88	375	140 625	19.36 49	61.23 72
326	106 276	18.05 55	57.09 64	**376**	141 376	19.39 07	61.31 88
327	106 929	18.08 31	57.18 39	377	142 129	19.41 65	61.40 03
328	107 584	18.11 08	57.27 13	378	142 884	19.44 22	61.48 17
329	108 241	18.13 84	57.35 85	379	143 641	19.46 79	61.56 30
330	108 900	18.16 59	57.44 56	380	144 400	19.49 36	61.64 41
331	109 561	18.19 34	57.53 26	**381**	145 161	19.51 92	61.72 52
332	110 224	18.22 09	57.61 94	382	145 924	19.54 48	61.80 61
333	110 889	18.24 83	57.70 62	383	146 689	19.57 04	61.88 70
334	111 556	18.27 57	57.79 27	384	147 456	19.59 59	61.96 77
335	112 225	18.30 30	57.87 92	385	148 225	19.62 14	62.04 84
336	112 896	18.33 03	57.96 55	**386**	148 996	19.64 69	62.12 89
337	113 569	18.35 76	58.05 17	387	149 769	19.67 23	62.20 93
338	114 244	18.38 48	58.13 78	388	150 544	19.69 77	62.28 96
339	114 921	18.41 20	58.22 37	389	151 321	19.72 31	62.36 99
340	115 600	18 43 91	58.30 95	390	152 100	19.74 84	62.45 00
341	116 281	18.46 62	58.39 52	**391**	152 881	19.77 37	62.53 00
342	116 964	18.49 32	58.48 08	392	153 664	19.79 90	62.60 99
343	117 649	18.52 03	58.56 62	393	154 449	19.82 42	62.68 97
344	118 336	18.54 72	58.65 15	394	155 236	19.84 94	62.76 94
345	119 025	18.57 42	58.73 67	395	156 025	19.87 46	62.84 90
346	119 716	18.60 11	58.82 18	**396**	156 816	19.89 97	62.92 85
347	120 409	18.62 79	58.90 67	397	157 609	19.92 49	63.00 79
348	121 104	18.65 48	58.99 15	398	158 404	19.94 99	63.08 72
349	121 801	18.68 15	59.07 62	399	159 201	19.97 50	63.16 64
350	122 500	18.70 83	59.16 08	400	160 000	20.00 00	63.24 56
N	N^2	$\sqrt{N}$	$\sqrt{10N}$	N	N^2	$\sqrt{N}$	$\sqrt{10N}$

Table XIII   (continued)

N	N^2	$\sqrt{N}$	$\sqrt{10N}$
400	160 000	20.00 00	63.24 56
401	160 801	20.02 50	63.32 46
402	161 604	20.04 99	63.40 35
403	162 409	20.07 49	63.48 23
404	163 216	20.09 98	63.56 10
405	164 025	20.12 46	63.63 96
406	164 836	20.14 94	63.71 81
407	165 649	20.17 42	63.79 66
408	166 464	20.19 90	63.87 49
409	167 281	20.22 37	63.95 31
410	168 100	20.24 85	64.03 12
411	168 921	20.27 31	64.10 93
412	169 744	20.29 78	64.18 72
413	170 569	20.32 24	64.26 51
414	171 396	20.34 70	64.34 28
415	172 225	20.37 15	64.42 05
416	173 056	20.39 61	64.49 81
417	173 889	20.42 06	64.57 55
418	174 724	20.44 50	64.65 29
419	175 561	20.46 95	64.73 02
420	176 400	20.49 39	64.80 74
421	177 241	20.51 83	64.88 45
422	178 084	20.54 26	64.96 15
423	178 929	20.56 70	65.03 85
424	179 776	20.59 13	65.11 53
425	180 625	20.61 55	65.19 20
426	181 476	20.63 98	65.26 87
427	182 329	20.66 40	65.34 52
428	183 184	20.68 82	65.42 17
429	184 041	20.71 23	65.49 81
430	184 900	20.73 64	65.57 44
431	185 761	20.76 05	65.65 06
432	186 624	20.78 46	65.72 67
433	187 489	20.80 87	65.80 27
434	188 356	20.83 27	65.87 87
435	189 225	20.85 67	65.95 45
436	190 096	20.88 06	66.03 03
437	190 969	20.90 45	66.10 60
438	191 844	20.92 84	66.18 16
439	192 721	20.95 23	66.25 71
440	193 600	20.97 62	66.33 25
441	194 481	21.00 00	66.40 78
442	195 364	21.02 38	66.48 31
443	196 249	21.04 76	66.55 82
444	197 136	21.07 13	66.63 33
445	198 025	21.09 50	66.70 83
446	198 916	21.11 87	66.78 32
447	199 809	21.14 24	66.85 81
448	200 704	21.16 60	66.93 28
449	201 601	21.18 96	67.00 75
450	202 500	21.21 32	67.08 20
N	N^2	$\sqrt{N}$	$\sqrt{10N}$

N	N^2	$\sqrt{N}$	$\sqrt{10N}$
450	202 500	21.21 32	67.08 20
451	203 401	21.23 68	67.15 65
452	204 304	21.26 03	67.23 09
453	205 209	21.28 38	67.30 53
454	206 116	21.30 73	67.37 95
455	207 025	21.33 07	67.45 37
456	207 936	21.35 42	67.52 78
457	208 849	21.37 76	67.60 18
458	209 764	21.40 09	67.67 57
459	210 681	21.42 43	67.74 95
460	211 600	21.44 76	67.82 33
461	212 521	21.47 09	67.89 70
462	213 444	21.49 42	67.97 06
463	214 369	21.51 74	68.04 41
464	215 296	21.54 07	68.11 75
465	216 225	21.56 39	68.19 09
466	217 156	21.58 70	68.26 42
467	218 089	21.61 02	68.33 74
468	219 024	21.63 33	68.41 05
469	219 961	21.65 64	68.48 36
470	220 900	21.67 95	68.55 65
471	221 841	21.70 25	68.62 94
472	222 784	21.72 56	68.70 23
473	223 729	21.74 86	68.77 50
474	224 676	21.77 15	68.84 77
475	225 625	21.79 45	68.92 02
476	226 576	21.81 74	68.99 28
477	227 529	21.84 03	69.06 52
478	228 484	21.86 32	69.13 75
479	229 441	21.88 61	69.20 98
480	230 400	21.90 89	69.28 20
481	231 361	21.93 17	69.35 42
482	232 324	21.95 45	69.42 62
483	233 289	21.97 73	69.49 82
484	234 256	22.00 00	69.57 01
485	235 225	22.02 27	69.64 19
486	236 196	22.04 54	69.71 37
487	237 169	22.06 81	69.78 54
488	238 144	22.09 07	69.85 70
489	239 121	22.11 33	69.92 85
490	240 100	22.13 59	70.00 00
491	241 081	22.15 85	70.07 14
492	242 064	22.18 11	70.14 27
493	243 049	22.20 36	70.21 40
494	244 036	22.22 61	70.28 51
495	245 025	22.24 86	70.35 62
496	246 016	22.27 11	70.42 73
497	247 009	22.29 35	70.49 82
498	248 004	22.31 59	70.56 91
499	249 001	22.33 83	70.63 99
500	250 000	22.36 07	70.71 07
N	N^2	$\sqrt{N}$	$\sqrt{10N}$

Table XIII   (continued)

N	N²	√N	√10N		N	N²	√N	√10N
500	250 000	22.36 07	70.71 07		**550**	302 500	23.45 21	74.16 20
501	251 001	22.38 30	70.78 14		551	303 601	23.47 34	74.22 94
502	252 004	22.40 54	70.85 20		552	304 704	23.49 47	74.29 67
503	253 009	22.42 77	70.92 25		553	305 809	23.51 60	74.36 40
504	254 016	22.44 99	70.99 30		554	306 916	23.53 72	74.43 12
505	255 025	22.47 22	71.06 34		555	308 025	23.55 84	74.49 83
506	256 036	22.49 44	71.13 37		**556**	309 136	23.57 97	74.56 54
507	257 049	22.51 67	71.20 39		557	310 249	23.60 08	74.63 24
508	258 064	22.53 89	71.27 41		558	311 364	23.62 20	74.69 94
509	259 081	22.56 10	71.34 42		559	312 481	23.64 32	74.76 63
510	260 100	22.58 32	71.41 43		560	313 600	23.66 43	74.83 31
511	261 121	22.60 53	71.48 43		**561**	314 721	23.68 54	74.89 99
512	262 144	22.62 74	71.55 42		562	315 844	23.70 65	74.96 67
513	263 169	22.64 95	71.62 40		563	316 969	23.72 76	75.03 33
514	264 196	22.67 16	71.69 38		564	318 096	23.74 87	75.09 99
515	265 225	22.69 36	71.76 35		565	319 225	23.76 97	75.16 65
516	266 256	22.71 56	71.83 31		**566**	320 356	23.79 08	75.23 30
517	267 289	22.73 76	71.90 27		567	321 489	23.81 18	75.29 94
518	268 324	22.75 96	71.97 22		568	322 624	23.83 28	75.36 58
519	269 361	22.78 16	72.04 17		569	323 761	23.85 37	75.43 21
520	270 400	22.80 35	72.11 10		570	324 900	23.87 47	75.49 83
521	271 441	22.82 54	72.18 03		**571**	326 041	23.89 56	75.56 45
522	272 484	22.84 73	72.24 96		572	327 184	23.91 65	75.63 07
523	273 529	22.86 92	72.31 87		573	328 329	23.93 74	75.69 68
524	274 576	22.89 10	72.38 78		574	329 476	23.95 83	75.76 28
525	275 625	22.91 29	72.45 69		575	330 625	23.97 92	75.82 88
526	276 676	22.93 47	72.52 59		**576**	331 776	24.00 00	75.89 47
527	277 729	22.95 65	72.59 48		577	332 929	24.02 08	75.96 05
528	278 784	22.97 83	72.66 36		578	334 084	24.04 16	76.02 63
529	279 841	23.00 00	72.73 24		579	335 241	24.06 24	76.09 20
530	280 900	23.02 17	72.80 11		580	336 400	24.08 32	76.15 77
531	281 961	23.04 34	72.86 97		**581**	337 561	24.10 39	76.22 34
532	283 024	23.06 51	72.93 83		582	338 724	24.12 47	76.28 89
533	284 089	23.08 68	73.00 68		583	339 889	24.14 54	76.35 44
534	285 156	23.10 84	73.07 53		584	341 056	24.16 61	76.41 99
535	286 225	23.13 01	73.14 37		585	342 225	24.18 68	76.48 53
536	287 296	23.15 17	73.21 20		**586**	343 396	24.20 74	76.55 06
537	288 369	23.17 33	73.28 03		587	344 569	24.22 81	76.61 59
538	289 444	23.19 48	73.34 85		588	345 744	24.24 87	76.68 12
539	290 521	23.21 64	73.41 66		589	346 921	24.26 93	76.74 63
540	291 600	23.23 79	73.48 47		590	348 100	24.28 99	76.81 15
541	292 681	23.25 94	73.55 27		**591**	349 281	24.31 05	76.87 65
542	293 764	23.28 09	73.62 06		592	350 464	24.33 11	76.94 15
543	294 849	23.30 24	73.68 85		593	351 649	24.35 16	77.00 65
544	295 936	23.32 38	73.75 64		594	352 836	24.37 21	77.07 14
545	297 025	23.34 52	73.82 41		595	354 025	24.39 26	77.13 62
546	298 116	23.36 66	73.89 18		**596**	355 216	24.41 31	77.20 10
547	299 209	23.38 80	73.95 94		597	356 409	24.43 36	77.26 58
548	300 304	23.40 94	74.02 70		598	357 604	24.45 40	77.33 05
549	301 401	23.43 07	74.09 45		599	358 801	24.47 45	77.39 51
550	302 500	23.45 21	74.16 20		600	360 000	24.49 49	77.45 97
N	N²	√N	√10N		N	N²	√N	√10N

Table XIII   (continued)

N	N^2	$\sqrt{N}$	$\sqrt{10N}$
600	360 000	24.49 49	77.45 97
601	361 201	24.51 53	77.52 42
602	362 404	24.53 57	77.58 87
603	363 609	24.55 61	77.65 31
604	364 816	24.57 64	77.71 74
605	366 025	24.59 67	77.78 17
606	367 236	24.61 71	77.84 60
607	368 449	24.63 74	77.91 02
608	369 664	24.65 77	77.97 44
609	370 881	24.67 79	78.03 85
610	372 100	24.69 82	78.10 25
611	373 321	24.71 84	78.16 65
612	374 544	24.73 86	78.23 04
613	375 769	24.75 88	78.29 43
614	376 996	24.77 90	78.35 82
615	378 225	24.79 92	78.42 19
616	379 456	24.81 93	78.48 57
617	380 689	24.83 95	78.54 93
618	381 924	24.85 96	78.61 30
619	383 161	24.87 97	78.67 66
620	384 400	24.89 98	78.74 01
621	385 641	24.91 99	78.80 36
622	386 884	24.93 99	78.86 70
623	388 129	24.96 00	78.93 03
624	389 376	24.98 00	78.99 37
625	390 625	25.00 00	79.05 69
626	391 876	25.02 00	79.12 02
627	393 129	25.04 00	79.18 33
628	394 384	25.05 99	79.24 65
629	395 641	25.07 99	79.30 95
630	396 900	25.09 98	79.37 25
631	398 161	25.11 97	79.43 55
632	399 424	25.13 96	79.49 84
633	400 689	25.15 95	79.56 13
634	401 956	25.17 94	79.62 41
635	403 225	25.19 92	79.68 69
636	404 496	25.21 90	79.74 96
637	405 769	25.23 89	79.81 23
638	407 044	25.25 87	79.87 49
639	408 321	25.27 84	79.93 75
640	409 600	25.29 82	80.00 00
641	410 881	25.31 80	80.06 25
642	412 164	25.33 77	80.12 49
643	413 449	25.35 74	80.18 73
644	414 736	25.37 72	80.24 96
645	416 025	25.39 69	80.31 19
646	417 316	25.41 65	80.37 41
647	418 609	25.43 62	80.43 63
648	419 904	25.45 58	80.49 84
649	421 201	25.47 55	80.56 05
650	422 500	25.49 51	80.62 26
N	N^2	$\sqrt{N}$	$\sqrt{10N}$

N	N^2	$\sqrt{N}$	$\sqrt{10N}$
650	422 500	25.49 51	80.62 26
651	423 801	25.51 47	80.68 46
652	425 104	25.53 43	80.74 65
653	426 409	25.55 39	80.80 84
654	427 716	25.57 34	80.87 03
655	429 025	25.59 30	80.93 21
656	430 336	25.61 25	80.99 38
657	431 649	25.63 20	81.05 55
658	432 964	25.65 15	81.11 72
659	434 281	25.67 10	81.17 88
660	435 600	25.69 05	81.24 04
661	436 921	25.70 99	81.30 19
662	438 244	25.72 94	81.36 34
663	439 569	25.74 88	81.42 48
664	440 896	25.76 82	81.48 62
665	442 225	25.78 76	81.54 75
666	443 556	25.80 70	81.60 88
667	444 889	25.82 63	81.67 01
668	446 224	25.84 57	81.73 13
669	447 561	25.86 50	81.79 24
670	448 900	25.88 44	81.85 35
671	450 241	25.90 37	81.91 46
672	451 584	25.92 30	81.97 56
673	452 929	25.94 22	82.03 66
674	454 276	25.96 15	82.09 75
675	455 625	25.98 08	82.15 84
676	456 976	26.00 00	82.21 92
677	458 329	26.01 92	82.28 00
678	459 684	26.03 84	82.34 08
679	461 041	26.05 76	82.40 15
680	462 400	26.07 68	82.46 21
681	463 761	26.09 60	82.52 27
682	465 124	26.11 51	82.58 33
683	466 489	26.13 43	82.64 38
684	467 856	26.15 34	82.70 43
685	469 225	26.17 25	82.76 47
686	470 596	26.19 16	82.82 51
687	471 969	26.21 07	82.88 55
688	473 344	26.22 98	82.94 58
689	474 721	26.24 88	83.00 60
690	476 100	26.26 79	83.06 62
691	477 481	26.28 69	83.12 64
692	478 864	26.30 59	83.18 65
693	480 249	26.32 49	83.24 66
694	481 636	26.34 39	83.30 67
695	483 025	26.36 29	83.36 67
696	484 416	26.38 18	83.42 66
697	485 809	26.40 08	83.48 65
698	487 204	26.41 97	83.54 64
699	488 601	26.43 86	83.60 62
700	490 000	26.45 75	83.66 60
N	N^2	$\sqrt{N}$	$\sqrt{10N}$

Table XIII   (continued)

N	N^2	$\sqrt{N}$	$\sqrt{10N}$	N	N^2	$\sqrt{N}$	$\sqrt{10N}$
700	490 000	26.45 75	83.66 60	**750**	562 500	27.38 61	86.60 25
701	491 401	26.47 64	83.72 57	751	564 001	27.40 44	86.66 03
702	492 804	26.49 53	83.78 54	752	565 504	27.42 26	86.71 79
703	494 209	26.51 41	83.84 51	753	567 009	27.44 08	86.77 56
704	495 616	26.53 30	83.90 47	754	568 516	27.45 91	86.83 32
705	497 025	26.55 18	83.96 43	755	570 025	27.47 73	86.89 07
706	498 436	26.57 07	84.02 38	**756**	571 536	27.49 55	86.94 83
707	499 849	26.58 95	84.08 33	757	573 049	27.51 36	87.00 57
708	501 264	26.60 83	84.14 27	758	574 564	27.53 18	87.06 32
709	502 681	26.62 71	84.20 21	759	576 081	27.55 00	87.12 06
710	504 100	26.64 58	84.26 15	760	577 600	27.56 81	87.17 80
711	505 521	26.66 46	84.32 08	**761**	579 121	27.58 62	87.23 53
712	506 944	26.68 33	84.38 01	762	580 644	27.60 43	87.29 26
713	508 369	26.70 21	84.43 93	763	582 169	27.62 25	87.34 99
714	509 796	26.72 08	84.49 85	764	583 696	27.64 05	87.40 71
715	511 225	26.73 95	84.55 77	765	585 225	27.65 86	87.46 43
716	512 656	26.75 82	84.61 68	**766**	586 756	27.67 67	87.52 14
717	514 089	26.77 69	84.67 59	767	588 289	27.69 48	87.57 85
718	515 524	26.79 55	84.73 49	768	589 824	27.71 28	87.63 56
719	516 961	26.81 42	84.79 39	769	591 361	27.73 08	87.69 26
720	518 400	26.83 28	84.85 28	770	592 900	27.74 89	87.74 96
721	519 841	26.85 14	84.91 17	**771**	594 441	27.76 69	87.80 66
722	521 284	26.87 01	84.97 06	772	595 984	27.78 49	87.86 35
723	522 729	26.88 87	85.02 94	773	597 529	27.80 29	87.92 04
724	524 176	26.90 72	85.08 82	774	599 076	27.82 09	87.97 73
725	525 625	26.92 58	85.14 69	775	600 625	27.83 88	88.03 41
726	527 076	26.94 44	85.20 56	**776**	602 176	27.85 68	88.09 09
727	528 529	26.96 29	85.26 43	777	603 729	27.87 47	88.14 76
728	529 984	26.98 15	85.32 29	778	605 284	27.89 27	88.20 43
729	531 441	27.00 00	85.38 15	779	606 841	27.91 06	88.26 10
730	532 900	27.01 85	85.44 00	780	608 400	27.92 85	88.31 76
731	534 361	27.03 70	85.49 85	**781**	609 961	27.94 64	88.37 42
732	535 824	27.05 55	85.55 70	782	611 524	27.96 43	88.43 08
733	537 289	27.07 40	85.61 54	783	613 089	27.98 21	88.48 73
734	538 756	27.09 24	85.67 38	784	614 656	28.00 00	88.54 38
735	540 225	27.11 09	85.73 21	785	616 225	28.01 79	88.60 02
736	541 696	27.12 93	85.79 04	**786**	617 796	28.03 57	88.65 66
737	543 169	27.14 77	85.84 87	787	619 369	28.05 35	88.71 30
738	544 644	27.16 62	85.90 69	788	620 944	28.07 13	88.76 94
739	546 121	27.18 46	85.96 51	789	622 521	28.08 91	88.82 57
740	547 600	27.20 29	86.02 33	790	624 100	28.10 69	88.88 19
741	549 081	27.22 13	86.08 14	**791**	625 681	28.12 47	88.93 82
742	550 564	27.23 97	86.13 94	792	627 264	28.14 25	88.99 44
743	552 049	27.25 80	86.19 74	793	628 849	28.16 03	89.05 05
744	553 536	27.27 64	86.25 54	794	630 436	28.17 80	89.10 67
745	555 025	27.29 47	86.31 34	795	632 025	28.19 57	89.16 28
746	556 516	27.31 30	86.37 13	**796**	633 616	28.21 35	89.21 88
747	558 009	27.33 13	86.42 92	797	635 209	28.23 12	89.27 49
748	559 504	27.34 96	86.48 70	798	636 804	28.24 89	89.33 08
749	561 001	27.36 79	86.54 48	799	638 401	28.26 66	89.38 68
750	562 500	27.38 61	86.60 25	800	640 000	28.28 43	89.44 27
N	N^2	$\sqrt{N}$	$\sqrt{10N}$	N	N^2	$\sqrt{N}$	$\sqrt{10N}$

Table XIII   (continued)

N	N^2	$\sqrt{N}$	$\sqrt{10N}$
800	640 000	28.28 43	89.44 27
801	641 601	28.30 19	89.49 86
802	643 204	28.31 96	89.55 45
803	644 809	28.33 73	89.61 03
804	646 416	28.35 49	89.66 60
805	648 025	28.37 25	89.72 18
806	649 636	28.39 01	89.77 75
807	651 249	28.40 77	89.83 32
808	652 864	28.42 53	89.88 88
809	654 481	28.44 29	89.94 44
810	656 100	28.46 05	90.00 00
811	657 721	28.47 81	90.05 55
812	659 344	28.49 56	90.11 10
813	660 969	28.51 32	90.16 65
814	662 596	28.53 07	90.22 19
815	664 225	28.54 82	90.27 74
816	665 856	28.56 57	90.33 27
817	667 489	28.58 32	90.38 81
818	669 124	28.60 07	90.44 34
819	670 761	28.61 82	90.49 86
820	672 400	28.63 56	90.55 39
821	674 041	28.65 31	90.60 91
822	675 684	28.67 05	90.66 42
823	677 329	28.68 80	90.71 93
824	678 976	28.70 54	90.77 44
825	680 625	28.72 28	90.82 95
826	682 276	28.74 02	90.88 45
827	683 929	28.75 76	90.93 95
828	685 584	28.77 50	90.99 45
829	687 241	28.79 24	91.04 94
830	688 900	28.80 97	91.10 43
831	690 561	28.82 71	91.15 92
832	692 224	28.84 44	91.21 40
833	693 889	28.86 17	91.26 88
834	695 556	28.87 91	91.32 36
835	697 225	28.89 64	91.37 83
836	698 896	28.91 37	91.43 30
837	700 569	28.93 10	91.48 77
838	702 244	28.94 82	91.54 23
839	703 921	28.96 55	91.59 69
840	705 600	28.98 28	91.65 15
841	707 281	29.00 00	91.70 61
842	708 964	29.01 72	91.76 06
843	710 649	29.03 45	91.81 50
844	712 336	29.05 17	91.86 95
845	714 025	29.06 89	91.92 39
846	715 716	29.08 61	91.97 83
847	717 409	29.10 33	92.03 26
848	719 104	29.12 04	92.08 69
849	720 801	29.13 76	92.14 12
850	722 500	29.15 48	92.19 54
N	N^2	$\sqrt{N}$	$\sqrt{10N}$

N	N^2	$\sqrt{N}$	$\sqrt{10N}$
850	722 500	29.15 48	92.19 54
851	724 201	29.17 19	92.24 97
852	725 904	29.18 90	92.30 38
853	727 609	29.20 62	92.35 80
854	729 316	29.22 33	92.41 21
855	731 025	29.24 04	92.46 62
856	732 736	29.25 75	92.52 03
857	734 449	29.27 46	92.57 43
858	736 164	29.29 16	92.62 83
859	737 881	29.30 87	92.68 23
860	739 600	29.32 58	92.73 62
861	741 321	29.34 28	92.79 01
862	743 044	29.35 98	92.84 40
863	744 769	29.37 69	92.89 78
864	746 496	29.39 39	92.95 16
865	748 225	29.41 09	93.00 54
866	749 956	29.42 79	93.05 91
867	751 689	29.44 49	93.11 28
868	753 424	29.46 18	93.16 65
869	755 161	29.47 88	93.22 02
870	756 900	29.49 58	93.27 38
871	758 641	29.51 27	93.32 74
872	760 384	29.52 96	93.38 09
873	762 129	29.54 66	93.43 45
874	763 876	29.56 35	93.48 80
875	765 625	29.58 04	93.54 14
876	767 376	29.59 73	93.59 49
877	769 129	29.61 42	93.64 83
878	770 884	29.63 11	93.70 17
879	772 641	29.64 79	93.75 50
880	774 400	29.66 48	93.80 83
881	776 161	29.68 16	93.86 16
882	777 924	29.69 85	93.91 49
883	779 689	29.71 53	93.96 81
884	781 456	29.73 21	94.02 13
885	783 225	29.74 89	94.07 44
886	784 996	29.76 58	94.12 76
887	786 769	29.78 25	94.18 07
888	788 544	29.79 93	94.23 38
889	790 321	29.81 61	94.28 68
890	792 100	29.83 29	94.33 98
891	793 881	29.84 96	94.39 28
892	795 664	29.86 64	94.44 58
893	797 449	29.88 31	94.49 87
894	799 236	29.89 98	94.55 16
895	801 025	29.91 66	94.60 44
896	802 816	29.93 33	94.65 73
897	804 609	29.95 00	94.71 01
898	806 404	29.96 66	94.76 29
899	808 201	29.98 33	94.81 56
900	810 000	30.00 00	94.86 83
N	N^2	$\sqrt{N}$	$\sqrt{10N}$

Table XIII   (continued)

N	N²	√N	√10N
900	810 000	30.00 00	94.86 83
901	811 801	30.01 67	94.92 10
902	813 604	30.03 33	94.97 37
903	815 409	30.05 00	95.02 63
904	817 216	30.06 66	95.07 89
905	819 025	30.08 32	95.13 15
906	820 836	30.09 98	95.18 40
907	822 649	30.11 64	95.23 65
908	824 464	30.13 30	95.28 90
909	826 281	30.14 96	95.34 15
910	828 100	30.16 62	95.39 39
911	829 921	30.18 28	95.44 63
912	831 744	30.19 93	95.49 87
913	833 569	30.21 59	95.55 10
914	835 396	30.23 24	95.60 33
915	837 225	30.24 90	95.65 56
916	839 056	30.26 55	95.70 79
917	840 889	30.28 20	95.76 01
918	842 724	30.29 85	95.81 23
919	844 561	30.31 50	95.86 45
920	846 400	30.33 15	95.91 66
921	848 241	30.34 80	95.96 87
922	850 084	30.36 45	96.02 08
923	851 929	30.38 09	96.07 29
924	853 776	30.39 74	96.12 49
925	855 625	30.41 38	96.17 69
926	857 476	30.43 02	96.22 89
927	859 329	30.44 67	96.28 08
928	861 184	30.46 31	96.33 28
929	863 041	30.47 95	96.38 46
930	864 900	30.49 59	96.43 65
931	866 761	30.51 23	96.48 83
932	868 624	30.52 87	96.54 01
933	870 489	30.54 50	96.59 19
934	872 356	30.56 14	96.64 37
935	874 225	30.57 78	96.69 54
936	876 096	30.59 41	96.74 71
937	877 969	30.61 05	96.79 88
938	879 844	30.62 68	96.85 04
939	881 721	30.64 31	96.90 20
940	883 600	30.65 94	96.95 36
941	885 481	30.67 57	97.00 52
942	887 364	30.69 20	97.05 67
943	889 249	30.70 83	97.10 82
944	891 136	30.72 46	97.15 97
945	893 025	30.74 09	97.21 11
946	894 916	30.75 71	97.26 25
947	896 809	30.77 34	97.31 39
948	898 704	30.78 96	97.36 53
949	900 601	30.80 58	97.41 66
950	902 500	30.82 21	97.46 79
N	N²	√N	√10N

N	N²	√N	√10N
950	902 500	30.82 21	97.46 79
951	904 401	30.83 83	97.51 92
952	906 304	30.85 45	97.57 05
953	908 209	30.87 07	97.62 17
954	910 116	30.88 69	97.67 29
955	912 025	30.90 31	97.72 41
956	913 936	30.91 92	97.77 53
957	915 849	30.93 54	97.82 64
958	917 764	30.95 16	97.87 75
959	919 681	30.96 77	97.92 85
960	921 600	30.98 39	97.97 96
961	923 521	31.00 00	98.03 06
962	925 444	31.01 61	98.08 16
963	927 369	31.03 22	98.13 26
964	929 296	31.04 83	98.18 35
965	931 225	31.06 44	98.23 44
966	933 156	31.08 05	98.28 53
967	935 089	31.09 66	98.33 62
968	937 024	31.11 27	98.38 70
969	938 961	31.12 88	98.43 78
970	940 900	31.14 48	98.48 86
971	942 841	31.16 09	98.53 93
972	944 784	31.17 69	98.59 01
973	946 729	31.19 29	98.64 08
974	948 676	31.20 90	98.69 14
975	950 625	31.22 50	98.74 21
976	952 576	31.24 10	98.79 27
977	954 529	31.25 70	98.84 33
978	956 484	31.27 30	98.89 39
979	958 441	31.28 90	98.94 44
980	960 400	31.30 50	98.99 49
981	962 361	31.32 09	99.04 54
982	964 324	31.33 69	99.09 59
983	966 289	31.35 28	99.14 64
984	968 256	31.36 88	99.19 68
985	970 225	31.38 47	99.24 72
986	972 196	31.40 06	99.29 75
987	974 169	31.41 66	99.34 79
988	976 144	31.43 25	99.39 82
989	978 121	31.44 84	99.44 85
990	980 100	31.46 43	99.49 87
991	982 081	31.48 02	99.54 90
992	984 064	31.49 60	99.59 92
993	986 049	31.51 19	99.64 94
994	988 036	31.52 78	99.69 95
995	990 025	31.54 36	99.74 97
996	992 016	31.55 95	99.79 98
997	994 009	31.57 53	99.84 99
998	996 004	31.59 11	99.89 99
999	998 001	31.60 70	99.95 00
1000	1000 000	31.62 28	100.00 00
N	N²	√N	√10N

REFERENCES AND SUGGESTIONS FOR FURTHER READING

Probability Theory

Feller, W., *An Introduction to Probability Theory and Its Applications,* Vol. I, 3rd Ed. New York: John Wiley & Sons, Inc., 1968.

Fellner, W., *Probability and Profit.* Homewood, Ill.: Richard D. Irwin, Inc., 1965.

Jeffreys, H., *Theory of Probability.* Oxford: Clarendon Press, 1961.

Kyburg, H. E., and Smokler, H. E., *Studies in Subjective Probability.* New York: John Wiley & Sons, Inc., 1964.

Mosteller, F., Rourke, R. E. K., and Thomas, G. B., *Probability with Statistical Applications.* Reading, Mass.: Addison-Wesley Publishing Company, Inc., 1961.

Parzen, E., *Modern Probability Theory and Its Applications.* New York: John Wiley & Sons, Inc., 1960.

Statistical Inference and Decision

1. Primarily Classical Inference

Brunk, H. D., *An Introduction to Mathematical Statistics.* Boston: Ginn & Company, 1960.

Cramér, H., *Mathematical Methods of Statistics.* Princeton, N. J.: Princeton University Press, 1946.

Freund, J. E., *Mathematical Statistics.* Englewood Cliffs, N. J.: Prentice-Hall, Inc., 1962.

Hodges, J. L., and Lehmann, E. L., *Basic Concepts of Probability and Statistics.* San Francisco: Holden-Day, Inc., 1964.

Hoel, P. G., *Introduction to Mathematical Statistics*, 3rd Ed. New York: John Wiley & Sons, Inc., 1962.

Hogg, R. V., and Craig, A. T., *Introduction to Mathematical Statistics*, 2d Ed. New York: The Macmillan Company, 1965.

Kendall, M. G., and Stuart, A., *The Advanced Theory of Statistics*, Vols. I and II. London: Charles Griffin & Co., Ltd., 1958 and 1961.

Lehmann, E. L., *Testing Statistical Hypotheses*. New York: John Wiley & Sons, Inc., 1959.

Lindgren, B. W., *Statistical Theory*. New York: The Macmillan Company, 1960.

Mood, A. M., and Graybill, F. A., *Introduction to the Theory of Statistics*, 2d Ed. New York: McGraw-Hill, Inc., 1963.

Wallis, W. A., and Roberts, H. V., *Statistics: A New Approach*. Glencoe, Ill: Free Press, 1956.

Wilks, S. S., *Mathematical Statistics*. New York: John Wiley & Sons, Inc., 1962.

2. Primarily Bayesian Inference and/or Decision Theory

Blackwell, D., and Girshick, M. A., *Theory of Games and Statistical Decisions*. New York: John Wiley & Sons, Inc., 1954.

Chernoff, H., and Moses, L. E., *Elementary Decision Theory*. New York: John Wiley & Sons, Inc., 1959.

Edwards, W., Lindman, H., and Savage, L. J., Bayesian statistical inference for psychological research. *Psychological Review*, 1963, **70**, 193–242.

Ferguson, T. S., *Mathematical Statistics: A Decision-Theoretic Approach*. New York: Academic Press, Inc., 1967.

Fishburn, P. C., *Decision and Value Theory*. New York: John Wiley & Sons, Inc., 1964.

Good, I. J., *The Estimation of Probabilities*. Cambridge, Mass.: M.I.T. Press, 1965.

Lindley, D. V., *Introduction to Probability and Statistics from a Bayesian Viewpoint* (2 Vols.). Cambridge: Cambridge University Press, 1965.

Luce, R. D., and Raiffa, H., *Games and Decisions*. New York: John Wiley & Sons, Inc., 1957.

Pratt, J. W., Raiffa, H., and Schlaifer, R., *Introduction to Statistical Decision Theory*. New York: McGraw-Hill, Inc., 1965.

Raiffa, H., *Decision Analysis*. Reading, Mass.: Addison-Wesley Publishing Company, Inc., 1968.

Raiffa, H., and Schlaifer, R., *Applied Statistical Decision Theory*. Boston: Graduate School of Business, Harvard University, 1961.

Savage, L. J., *The Foundations of Statistics*. New York: John Wiley & Sons, Inc., 1954.

Schlaifer, R., *Probability and Statistics for Business Decisions*. New York: McGraw-Hill, Inc., 1959.

Schlaifer, R., *Analysis of Decisions Under Uncertainty*. New York: McGraw-Hill, Inc., 1969.

Schmitt, S. A., *Measuring Uncertainty: An Elementary Introduction to Bayesian Statistics*. Reading, Mass.: Addison-Wesley Publishing Company, Inc., 1969.

Von Neumann, J., and Morgenstern, O., *Theory of Games and Economic Behavior*. Princeton, N. J.: Princeton University Press, 1944.

Wald, A., *Statistical Decision Functions*. New York: John Wiley & Sons, Inc., 1950.

3. General

Clelland, R. C., deCani, J. S., Brown, F. E., Bursk, J. P., and Murray, D. S., *Basic Statistics With Business Applications*. New York: John Wiley & Sons, Inc., 1966.

Dyckman, T. R., Smidt, S., and McAdams, A. K., *Management Decision Making Under Uncertainty*. New York: The Macmillan Company, 1969.

Hadley, G., *Introduction to Probability and Statistical Decision Theory*. San Francisco: Holden-Day, Inc., 1967.

Sasaki, K., *Statistics for Modern Business Decision Making*. Belmont, Calif.: Wadsworth, 1968.

Spurr, W. A., and Bonini, C. P., *Statistical Analysis for Business Decisions*. Homewood, Ill.: Richard D. Irwin, Inc., 1967.

Tables

Burington, R. S., and May, D. C., *Handbook of Probability and Statistics with Tables*. New York: McGraw-Hill, Inc., 1953.

Fisher, R. A., and Yates, F., *Statistical Tables for Biological, Agricultural and Medical Research*, 6th Ed. Edinburgh: Oliver and Boyd, 1963

Owen, D. B., *Handbook of Statistical Tables*. Reading, Mass.: Addison-Wesley Publishing Company, Inc., 1962.

Pearson, E. S., and Hartley, H. O., *Biometrika Tables for Statisticians*, Vol. I, 3rd Ed. Cambridge: Cambridge University Press, 1967.

INDEX